Electron Microscopy of Subcellular Dynamics

Editor

Helmut Plattner, Ph.D.
Professor
Faculty of Biology
University of Konstanz
Konstanz, West Germany

CRC Press, Inc.
Boca Raton, Florida

Library of Congress Cataloging-in-Publication Data

Electron microscopy of subcellular dynamics / editor, Helmut Plattner.
p. cm.
Bibliography: p.
ISBN 0-8493-6079-X
1. Electron microscopy--Technique. 2. Ultrastructure (Biology)
3. Cell membranes. I. Plattner, Helmut, 1939-
QH212.E4E398 1989
574.87′028--dc19

89-492
CIP

This book represents information obtained from authentic and highly regarded sources. Reprinted material is quoted with permission, and sources are indicated. A wide variety of references are listed. Every reasonable effort has been made to give reliable data and information, but the author and the publisher cannot assume responsibility for the validity of all materials or for the consequences of their use.

Direct all inquiries to CRC Press, Inc., 2000 Corporate Blvd., N.W., Boca Raton, Florida, 33431.

International Standard Book Number 0-8493-6079-X

Library of Congress Card Number 89-492
Printed in the United States

PREFACE

Dynamic phenomena in cells that can be analyzed on the electron microscope (EM) level comprise quite different aspects, ranging from ion shifts and conformational changes of macromolecules to intracellular transport phenomena and restructuring of cytoskeletal elements, etc. Therefore, the methodical approaches to be used have to be widely different depending on the actual problem to be solved. Methods range from fast freezing, specific labeling, low temperature processing and/or analysis, X-ray microanalysis, high voltage EM, low voltage scanning EM, etc., which all are covered in this volume by leading experts. An integrative approach of different EM preparation and analysis techniques now allows the analysis of dynamic cellular processes with high temporal and spatial resolution on the EM level. This volume intends to give a survey of the correlated use of such EM techniques, their rationale, and the latest developments. Not only currently available methods, but also the goals and perspectives for future developments are discussed. This volume, which contains numerous figures and tables, should thus be of interest to all scientists and students dealing with the analysis of problems in the field of molecular and cell biology by EM methods (in the broadest sense). It addresses biochemists and biophysicists, and cell and molecular biologists, wherever they intend to exploit the high capacity of EM techniques. The editor also wants to express his gratitude to the CRC Press staff for their cooperation.

Helmut Plattner
Konstanz, Federal Republic of Germany

THE EDITOR

Helmut Plattner, Ph.D., heads the Chair of Ultrastructure Research (as a professor of cell biology) at the Faculty of Biology of the University of Konstanz, Konstanz, Federal Republic of Germany.

Dr. Plattner obtained his Ph.D. in 1965 from the University of Innsbruck, Austria, for a thesis dealing with the radiation dose distribution as a consequence of the natural radioactivity from thorium decay products. He then worked in different laboratories devoted to studies in cell biology, where the emphasis was the development and application of electron microscopic techniques, first at the University of Innsbruck (1965—1968), then as a Research Associate at the Department of Applied Physics, Cornell University (Ithaca, New York), in the years 1969 and 1970. After another year at Innsbruck, he worked from 1971—1975 at the Institute of Cell Biology, University of Munich (Federal Republic of Germany), and then returned for three years to Innsbruck, during which time period he also was a guest research fellow at the CNRS in Gif-sur-Yvette, France. In 1978 he was offered a chair at the University of Konstanz, which he still holds.

In the last decade, the topics of Dr. Plattner's work have been mainly exocytosis and aspects pertinent to the regulation of membrane fusion during exocytosis, combined with fast freezing techniques for the analysis of different dynamic subcellular processes on the electron microscope level. His group attempts to correlate biochemical and biophysical data on the cellular level by combining quite different methodical approaches.

Publications have appeared in leading journals of cell biology, cytochemistry, and electron microscopy. Dr. Plattner is also the author of numerous invited review articles and of a book on *Electron Microscopic Techniques in Cell and Molecular Biology*. In 1981/82 he was a dean of his faculty and in 1984/85 the president of the German Society for Electron Microscopy.

CONTRIBUTORS

Wolfgang Bohn, Ph.D.
Doctor
Department of Cytology and Virology
Heinrich-Pette Institute
Hamburg, West Germany

Dennis Brown, Ph.D.
Associate Professor of Pathology
Harvard University Medical School
and Massachusetts General Hospital
Boston, Massachusetts

Chantal De Chastellier Ph.D.
Doctor
Department of Molecular Biology
Institut Pasteur
Paris, France

Etienne de Harven, M.D.
Director, Electron Microscope Laboratory
Department of Pathology
University of Toronto
Toronto, Ontario, Canada

Ludwig Edelmann, Ph.D.
Research Scientist
Department of Medical Biology
University of Saarland
Homburg, West Germany

Adolf Ellinger, Ph.D.
Institute of Micromorphology
and Electron Microscopy
University of Vienna
Vienna, Austria

Joachim Greule, Dipl. Biol.
Department of Biophysics
University of Stuttgart
Stuttgart, West Germany

Jean Hubert, Dr. ès Sci. Natur.
Chief of Research
Department of Biochemistry
of Glycoconjugates
CBM — CNRS
Orléans, France

Sek Wen Hui, Ph.D.
Director
Membrane Biophsyics Laboratory
Roswell Park Memorial Institute
and Research Professor
State University of New York at Buffalo
Buffalo, New York

Dieter F. Hülser, Ph.D.
Professor
Department of Biophysics
University of Stuttgart
Stuttgart, West Germany

Gerrit Isenberg, M.D.
Professor
Department of Applied Physiology
University of Cologne
Cologne, West Germany

Gerd Knoll, Ph.D.
Doctor
Faculty of Biology
University of Konstanz
Konstanz, West Germany

Paul Messner, Dipl. Ing. Dr.
Associate Professor
Center for Ultrastructure Research
University of Agriculture
Vienna, Austria

Michel Monsigny, Dr. ès Sciences
Professor
Head of Department of Biochemistry
of Glycoconjugates
CBN — CNRS University of Orléans,
Orléans, France

Eberhard Morgenstern, M.D.
Professor
Department of Medical Biology
University of Saarland
Homburg/Saar, West Germany

Angèle Obrénovitch, Dr. ès Sciences
Chief of Research
Department of Biochemistry
of Glycoconjugates
CBM — CNRS
Orléans, France

Dietmar Paschke
Assessor
Department of Biophysics
University of Stuttgart
Stuttgart, West Germany

Margit Pavelka, M.D.
Univ. — Doz.
Institute of Micromorphology
and Electron Microscopy
University of Vienna
Vienna, Austria

James B. Pawley, Ph.D.
Associate Professor
Department of Zoology
University of Wisconsin
Madison, Wisconsin

Helmut Plattner, Ph.D.
Professor
Faculty of Biology
University of Konstanz
Konstanz, West Germany

Hans Ris, Ph.D.
Professor Emeritus
Department of Zoology
University of Wisconsin
Madison, Wisconsin

Horst Robenek, Dr. rer. nat.
Priv. — Doz.
Cell Biology Group
University of Münster
Münster, West Germany

Jürgen Roth, M.D., Ph.D.
Professor
Interdepartmental Electron Microscopy
Biocenter
University of Basel
Basel, Switzerland

Manfred Schmitz, Ph.D.
Federal Minister for Youth, Family Affairs,
Women and Health
Bonn, West Germany

Joseph Schrével, Dr. ès Sciences Naturelles
Professor
Laboratory of Cellular Biology
University of Poitiers
Poitiers, France

Uwe Sleytr, Dipl. Ing. Dr.
Professor
Center for Ultrastructure Research
University of Agriculture
Vienna, Austria

J. Victor Small, Ph.D.
Department Head
Institute of Molecular Biology
Austrian Academy of Sciences
Salzburg, Austria

Douglas J. Taatjes, Ph.D.
Director of Electron Microscopy Facility
Department of Pathology
University of Vermont
Burlington, Vermont

Lutz Thilo, D. Ret. Nat.
Associate Professor
Department of Medical Biochemistry
University of Cape Town Medical School
Cape Town, South Africa

Arie J. Verkleij, Ph.D.
Professor
Department of Molecular Cell Biology
University of Utrecht
Utrecht, Netherlands

Maria Wendt-Gallitelli, Ph.D.
Univ. — Doz.
Department of Physiology II
University of Tübingen
Tübingen, West Germany

Karl Zierold, Ph.D.
Research Scientist
Central Laboratory for Functional Morphology
and System Physiology
Max-Planck Institut
Dortmund, West Germany

Dedicated to my children

TABLE OF CONTENTS

Chapter 1

CURRENT TRENDS IN THE ELECTRON MICROSCOPIC ANALYSIS OF DYNAMIC PROCESSES IN THE FIELD OF CELL AND MOLECULAR BIOLOGY

Helmut Plattner

TABLE OF CONTENTS*

* Numbers mentioned in this article are rough key values only, since detailed experimental specifications would be required.

I. INTRODUCTION

Despite its high spatial resolution, the electron microscope (EM) per se is evidently not able to provide appropriate time resolution required for directly resolving dynamic processes as they occur in cell and molecular biology. Table 1 presents a selection of actual values of dynamic events; in each case the time resolution required would have to be, of course, at least of the same order of magnitude (or deliberately better). In all cases dynamic processes can be analyzed only by static snapshots after the application of appropriate labeling and/or other preparation procedures. Due to the inherent pitfalls and limitations of each preparation method available, EM data thus obtained have to be controlled by independent alternative methods (Table 2); these, in turn, should, as far as possible, operate by a widely different principle. A more extensive recent review on this topic is available from Plattner and Knoll.[1]

In the subsequent sections II and III, dynamic aspects pertinent to cell or molecular biology, respectively, are considered. In Tables 1 to 6 aspects relevant for both these fields are summarized; Table 7 concentrates on molecular aspects.

II. CELL BIOLOGY

One of the earliest types of dynamic processes analyzed in the EM was transcellular transport and release of secretory products, using radioactive labeling and autoradiography on the one hand, or cell fractionation combined with radioactivity measurements on the other hand.[4] Radioactivity can be applied best in pulse chase experiments, particularly in systems operating with appropriate synchrony. Conventional fixation and ultrathin sectioning techniques were fully adequate, e.g., to follow the main secretory pathways.[4]

This is no more the case with other "slow processes" mentioned in Table 1, because chemical fixation may act too slowly to prevent the occurrence of supravital artifacts (when cells are slowly "mummified", while they still can react to metabolic breakdown and ion shifts, etc.). Examples of such artifact phenomena in the context of membrane restructuring (rearrangements of IMPs = intramembranous particles), as well as possible ways to overcome such troubles, have been presented before.[5] The solution was mainly the application of fast freezing techniques (cryofixation) as physical fixation methods. For a long time this entailed the problem of unbearable ice crystal damage, until about 15 years ago, when sufficiently high cooling rates could be obtained. Otherwise ice damage not only obscures ultrastructural details, but it will also reduce the spatial resolution of analytical EM methods (like X-ray microanalysis, see Chapter 18 by M. Schmitz and K. Zierold, and Chapter 19 by M. Wendt-Gallitelli, in this book).

In the meantime, appropriate cryofixation methods are available for a variety of different types of samples (cell cultures in suspension or as monolayers, subcellular fractions and macromolecules). (For reviews see References 5 to 7; see also Table 3.) Only for bulk samples no solution has been found so far, since high pressure freezing, though yielding ice crystal-free preparations,[10] operates with low cooling rates,[11] even within small (<1 mm) tissue fragments (10^2 to 10^3 Ks^{-1} as opposed to $>10^4$ Ks^{-1}, obtainable with fast freezing techniques; see References 5,6). Such high cooling rates are just sufficient to reveal even very fast processes, such as membrane fusion[9] and some lipid phase transition phenomena (see Table 1 as well as Chapter 6 by G. Knoll and H. Plattner, and Chapter 5 by A. J. Verkleij). Of course, they "catch" also somewhat slower processes listed in Table 1.

Though the use of EM tracers also allowed investigation of the directionality of some dynamic processes like transcytosis a long time ago, important details of their dynamic sequence had to be corrected on the basis of later cryofixation studies,[12] which revealed transendothelial channels to be artifacts. Also dynamic changes occurring in thrombocytes

TABLE 1
Types of Dynamic Processes[a]

Very fast processes (time range: ms)
  Biological membrane fusion (e.g., during exocytosis), some lipid phase transitions;
  time resolution available: 0.1 ms. Follow up procedures: fast freezing → cryosections (e.g., for X-ray microanalysis), freeze drying (also with sections), freeze fracture, freeze substitution.

Fast processes (time range: 1 ms – or 10—100 ms)
  Muscle contraction (10—100 ms), ciliary beat (10—50 ms), exo-endocytosis (individual events: 50 ms, Ref. 2), transcytosis, various membrane vesiculation processes in cells, small scale self-assembly processes, ion shifts. Methods available: as indicated above.[b]

Slow processes (time range: s to min)
  IMP clustering (1—10 $\mu m \cdot s^{-1}$, Ref. 3) and other rearrangement of membrane components, cell and organelle shape changes, large scale rearrangement of cytoskeletal components, transcellular transport, large scale self-assembly processes, gap junction responses (see text and other chapters of this book). Methods available: as indicated above.[b]

Very slow processes
  Chromosome movement, cell division

[a] Numbers indicated are only rough key values.
[b] "Standard" preparation techniques (chemical treatments) might allow for the occurrence of artifacts with even slower processes.

TABLE 2
Alternative (Nonultrastructural) Methods

Vital observation of cells, eventually after microinjection,
  phase and differential interference contrast, confocal laser scanning microscopy image intensification, "nanovid tracking," etc.

Cell fractionation (yet: redistribution artifacts) and immunological methods

Fluorescence energy transfer for membrane fusion experiments

Nuclear magnetic resonance (NMR) Differential scanning calorimetry (DSC) X-ray diffraction (eventually at 77 k)	Analysis of phase changes in parallel to cryofixation and freeze fracturing (see A. J. Verkleij's Chapter 5 in this book).

Electrophysiology
  Analysis of gap junction function (see Chapter 3 by D. Hülser et al.)
  Surface capacitance measurements of secretory activity (see G. Knoll and H. Plattner, Chapter 6 in this volume, for a discussion).

during activation now look different in fast frozen samples, as compared to previous "standard" EM analyses (Chapter 7 by E. Morgenstern and L. Edelmann). This discrepancy is particularly serious for membrane fusions, for which we now can propose a general "focal fusion concept" (Chapter 5 by A. J. Verkleij, and Chapter 6 by G. Knoll and H. Plattner). These are just a few selected examples for the impact of cryofixation on the analysis of dynamic cellular processes on the ultrastructural level.

Beyond mere ultrastructural analyses it is now possible, e.g., for muscle contraction, to analyze (in conjunction with cryoultramicrotomy and X-ray microanalysis) ion shifts occurring during stimulus-contraction coupling[13] (see also Chapter 19 by M. Wendt-Gallitelli). The localization of a variety of different antigens in muscle cells by immuno-EM methods also has been very successful (see Chapter 16 by V. Small). What progress since Huxley's formulation of the sliding filament hypothesis on the basis of suggestive EM micrographs!

It is mandatory to check every newly developed method for possible new artifacts. We have done this, e.g., when cells were processed by spray or sandwich freezing (for a review see Reference 5). This postulate, of course, should be taken into consideration every time a new method is designed. Some of the methods listed in Table 2 appear appropriate for

TABLE 3
Cryofixation Methods Usable for Very Fast Processes

Required for vitrification of water: $>10^5$ Ks^{-1} (Ref. 8)
Required for cell biological problems: $>10^4$ Ks^{-1} (Ref. 9)
Goal:
"Fix" even very fast lipid phase transitions ($\leq$ms)
"Catch" membrane fusion events (ms)
Avoid redistribution of ions (in EM X-ray microanalysis) and possibly of soluble antigens (low temperature immunocytochemistry; see text), etc.
Type of method usable for dynamic processes ($>10^4$ Ks^{-1}):
Spray freezing, sandwich freezing, cold metal-block freezing (see Refs. 5,6).

such controls. Unfortunately most control methods, like vital observations in the light microscope (LM), provide only inferior resolution; but at least on this level, EM results should agree. For instance, organelle shape changes have been monitored by vital fluorescence observations with the use of organelle-specific fluorochromes, and in parallel by EM analysis.[14] Surprisingly, even such gross morphological details could not be reliably depicted in the EM after conventional chemical fixation techniques, but only after cryofixation. The goal and potential of cryofixation for EM studies are summarized in Table 3.

Important recent developments of control methods include different optical and fluorescent signal enhancement procedures (for a review see Reference 15) and confocal laser scanning microscopy.[16] This can eventually be combined with intracellular labeling via microinjected antibodies or via tagged molecular constituents, such as actin or tubulin, etc., which then become normally integrated into cell structures (fluorescent analogue labeling[17]). Microinjected antibodies, however, were meanwhile observed (particularly on the ultrastructural level with peroxidase as a tag) to cause clustering of antigens and sequestration of antigen-antibody complexes. Similar artifacts have also been realized with exogenous antibody labeling, e.g., in conjunction with membrane flow phenomena[18] (see also Reference 1 for a discussion). The use of monovalent Fab antibody fragments, therefore, appears advisable for parallel *in vivo* and EM analyses of dynamic events.

Other methods have been designed for the consecutive LM/EM labeling of cytoskeletal elements (microtubules), or transport phenomena mediated by them. Antibody coated colloidal gold particles are used.[19] Visualization in the LM is achieved by video enhanced interference contrast imaging.[15] New "nanovid tracking" methods are adapted to follow microinjected gold particles (20 to 40 nm) with a time resolution sufficient to register saltatory movements along microtubules.[20] Successive LM/EM localization is possible not only in conjunction with this method, but also with fluorochrome labeling, followed by EM localization with anti-fluorochrome-antibodies, detectable in turn by an EM tag.[21]

All this work illustrates various possibilities for controls applicable to the analysis of dynamic cellular processes. This not only provides a firm baseline to control any possible artifacts, but it also allows us to connect EM studies with the intravital observation of dynamic happenings in the living cell.

For a potential analysis of living cells by the newly developed X-ray microscope on a level of resolution between LM and EM, see section IV.

The potential of two extreme types of EM, operating either in the megavolt or in the kilovolt range, is presented in Chapter 17 by H. Ris and J. B. Pawley. These widely different EM types are valuable tools to investigate long range connections of dynamic ultrastructural details, including cytoskeletal elements. Immunolabeling procedures applicable to the contractile apparatus of the cell, but with the conventional EM, are illustrated by V. Small (Chapter 16).

Though resolution is less than desirable, i.e., only about 100 to 300 nm, with EM autoradiography radioactive labeling can operate with as few side effects as possible. Improved statistical evaluation methods allow one to follow rather precisely transcellular transport routes of secretory materials.[22] In Chapter 9 L. Thilo and C. de Chastellier give an example of ''metabolic'' labeling in conjunction with membrane internalization studies.

Other possibilities to follow the dynamics of cell surface components and their internalization are immunogold (Chapter 14 by W. Bohn) or ligand-gold affinity labeling (Chapter 8 by H. Robenek). In these chapters, labeling experiments are combined with surface replication methods.

Another quite new development is the registration of gold particles, down to approximately 10 nm diameter, via backscattered electrons on the surface of cells.[23] The relevance of this method for the study of cell surface dynamics is emphasized in Reference 24 and by E. DeHarven in Chapter 15.

A new type of labeling procedure has been introduced in the form of artificial glycoproteins with defined glycosyl moieties covalently attached to a protein carrier (see Chapter 13 by M. Monsigny). This now allows one to follow the fate of endogenous lectins, some of which are surface membrane components.

Unless *in vivo* labeling is used, intracellular antigens, receptors, or enzymes, etc., can also be localized efficiently under steady state conditions (i.e., with fixed materials). This can yield important clues to dynamic cellular processes, e.g., in the Golgi apparatus as a ''center stage for membrane traffic.''[25] In principle, pre-embedding methods (using permeabilized cells; Chapter 11 by M. Pavelka and A. Ellinger) or postembedding procedures are applicable. Among the latter, plastic sections of samples embedded in Lowicryl or similar materials (Chapter 12 by J. Roth and D. J. Taatjes) can be used. Ultrathin frozen sections are a third alternative (Chapter 10 by D. Brown). One or the other method might be preferred, depending on personal preference, equipment available, and the type of problem to be analyzed, etc. Although there might exist some differences in the ultrastructural preservation, antigen redistribution, accessibility of labeling sites, or sensitivity, many problems of cell dynamics have been solved in parallel by either one of these methods. Results were essentially identical, e.g., for microdomains in the Golgi apparatus (Chapter 10 vs. 11 and 12), or for identifying acidic compartments in cells,[26] to give just two examples pertinent to cell dynamics.

Intracompartmental pH can be differentiated in the EM with the use of permeable small indicator molecules which, after protonation, become trapped in certain trans-Golgi elements (stacks and vesicles), in lysosomes, and in endosomes, where they then can be visualized with antibodies (labeled with peroxidase for pre-embedding, or with colloidal gold particles for postembedding procedures) directed against the hapten used as a pH probe.[26]

Furthermore, postembedding methods have been shown to be a very powerful tool for localizing endogenous enzymes, e.g., certain glycosyl transferases in individual Golgi stacks (Chapter 12 by J. Roth and D. J. Taatjes). Either the enzyme can be offered in a gold labeled form, so that it binds to, and thus detects its substrate,[27] or an unlabeled enzyme is applied before the accomplished transglycosylation can be monitored by a lectin-gold conjugate.[28] Postembedding double labeling experiments enable one to find out diverging[29] or converging[30] vesicular transport routes through cells. Again, the labeling methods used may be widely different.

For cryosections, materials are slightly fixed, mainly with formaldehyde, before they are frozen; treatment with dimethylsulphoxide not only serves as a cryoprotectant impregnation, but also as a permeabilizing agent. This, however, might cause soluble antigens to undergo redistribution and fine morphological details to be altered (see Reference 1 for a discussion). Though the affinity- and immunolocalizations achieved with this method appear reliable (for the mannose-6-phosphate receptor see Reference 31 and Chapter 10 by D. Brown in this book), the observed ultrastructure of the subcellular compartments thus analyzed might not always be artifact-free. New low temperature-embedding media (see below) may be very helpful in this field.

TABLE 4
Localization of Subcellular Constituents Involved in Cell Dynamics

Problems
- Precursor → product relationships
- Steady state analyses or dynamic labeling (e.g., for receptor-ligand binding, intracellular sorting phenomena, microzonation formation, etc.)

Type of "primary" label
- Radioactive precursors → autoradiography
- Antibodies (immunocytochemistry)
  - New aspects: Exploitation of monoclonal antibodies and antibodies raised against synthetic peptides derived from cloned DNA sequences for EM localizations.[36] (See also Chapter 12 by J. Roth and D. J. Taatjes.)
- Affinity labels (pH measurements; lectins for glycoproteins and -lipids; synthetic glycoproteins for endogenous lectins, toxins, other ligands; see text for references.)

Type of EM tag
- Horseradish peroxidase (4.5 nm), microperoxidase (2 nm), ferritin (11 nm), colloidal gold (variable defined size classes 5—40 nm depending on purpose; see text), Ag grains (100—300 nm) in autoradiograms.[37,38]

Application of label
- *In vivo* (via membrane flow; microinjection)
- *In vitro* (fixed and/or embedded material) pre- or postembedding or cryosection labeling procedures

For soluble components, postembedding procedures appear optimal. With the use of older Lowicryl media, HM20, important progress already has been achieved in the microdomain localization of lipids by cryofixation, freeze substitution, and low temperature (238 K) embedding for EM autoradiography.[32] Currently, one can UV polymerize new low temperature embedding media, K11M (polar) or HM23 (apolar) Lowicryl resins, at temperatures around 193 K.[33] At this temperature ice crystal growth is not yet a serious problem;[5] therefore, attempts have been made to apply this embedding to native biological materials subjected to cryofixation, with or without freeze substitution in between.[34] Unexpectedly, antigen redistribution artifacts still may occur to a certain extent,[35] when fast frozen materials are embedded in such new media without chemical freeze substitution fixation.

Tables 3 to 5 are essentially summaries of what has been discussed so far. Table 3 underscores the fact that fast frozen materials should be preferred for a reliable analysis of many dynamic ultrastructural changes. (In ultramicrotomy freezing serves only to render materials so brittle that they can be cut, thus making intracellular labeling sites more easily accessible to immuno- or affinity-labeling.)

The potential of microwave fixation[39,40] as a supplementing or alternative, potentially fast "physical fixation" method, still requires more detailed investigations. Similar expectations exist for neutron flashes (see appendix discussion in Reference 1).

Table 4 presents a survey of the different labeling procedures relevant for cell dynamics. Depending on the type of problem, the "primary" label to be used is different; other variations consist in the types of EM tags, their application and instrumental evaluation. Possible pitfalls and perspectives related to the localization of subcellular constituents are put together in Table 5.

New methodical perspectives in the EM analysis of dynamic events arose with the advent of monoclonal antibodies (mAB) (Table 6). Frequently, it is impossible at the beginning of a project to isolate in pure form certain antigens from the bewildering jungle of subcellular components involved in dynamic processes. The mAB technique provides us with a new key to this kind of problem. Any of the above labeling methods may be used. Examples of successful use of mAB in cell dynamics are the detection of microdomains within cell membranes[41,42] and in the Golgi apparatus,[43] receptor dynamics,[44] and the precise localization of the proinsulin to insulin conversion site in acidic, clathrin-coated, Golgi-derived vesicles.[45] The rearrangement of deliberate (frequently yet unidentified) antigens can now be followed by EM immunocytochemistry on the way into and out of the cell (see for example, Reference 46).

TABLE 5
Pitfalls in the Analysis of Subcellular Constituents

Capping of membrane components (antigens) by bi- or multivalent ligands [lectins, IgG, $F(ab)_2$]
Altered targeting[18]
Clumping of soluble antigens after IgG microinjection →
 sequestration and lysosomal breakdown
Rearrangement of components:
 (A) Cellular level: IMP rearrangements during fixation and/or glycerol impregnation in older freeze fracture work; possibly antigen redistribution also after freeze substitution following cryofixation (even with new low temperature embedding media; see text)
 (B) Molecular level: Rearrangement of gap junction structural components during chemical fixation (see D. Hülser et al., Chapter 3)
Cationic or anionic tags (peroxidase, ferritin) may cause tag-specific targeting (see Ref. 1 for a discussion)

TABLE 6
Dynamic Cellular Processes to Be Studied with Monoclonal Antibodies[a]

Localization of various antigens (epitopes) before sufficient purification is possible
Analysis of the targeting of membrane vesicles
 Microzonations and membrane heterogeneities
 Cytoplasmic coats (clathrin vs. non-clathrin in Golgi region)
Proenzyme → enzyme conversion
Mapping of epitopes on macromolecules involved in dynamic processes (see e.g., cited work on myosin)

[a] For examples and references, see text.

TABLE 7
Examples of EM-Analyses of Dynamic Processes on the Molecular Level[a]

Muscle contraction
Membrane fusion (see Table 1 and section II)
Gap junctions during dynamic intercellular communication phenomena
Protein binding for transmembrane transport
Molecular events occurring during chromosome movement (spindle microtubule assembly and disassembly, dynein involvement)
Organelle transport (dynein, kinesin binding)
Self-assembly and polarity of actin filaments or of microtubules
2D/3D reconstruction of molecular assemblies or single particle averaging, applicable to dynamic molecular events[37,38]

[a] For references for most of these aspects, see text.

III. MOLECULAR BIOLOGY

This is only a marginal aspect in this volume on the EM analysis of dynamic events in cells. Again Tables 1 to 6 in subheading II should be consulted, as well as Table 7, which is more specifically devoted to dynamic aspects of molecular constituents.

Along these lines, EM methods have provided important contributions to the organization of contractile elements. With mAB one can now not only pinpoint distinct epitopes on individual macromolecules, e.g., along a myosin molecule,[47] but one can also at the same time study sites of functional interactions with other molecules of the contractile apparatus.[48] Recently, the first localization on the EM level of an active enzyme center was achieved, i.e., of the myosin head actin-activated ATPase activity via adenosinetriphosphate (ATP) analogue labeling (visualized by avidin binding, negative staining, minimal dose EM and image averaging).[49] So far, by image averaging, molecular electron microscopy provides a tool to analyze dynamic molecular interactions on the level of resolution of only about 1 to 2 nm.

As outlined by V. Small in Chapter 16, still much has to be learned about the arrangement and dynamic interaction of intermediate type filaments with other antigens within the contractile apparatus. The EM analysis of microtubular structure (dis-/assembly, polarity) and function (including associated proteins, dynein, and kinesin) have already been quite successful. Another example of the progress in this field is the work in Reference 50, for which fast freezing and low temperature EM analysis, without electron stains, were used.

For recent work on gap junction structure-function correlation, D. Hülser and his colleagues (Chapter 3) have combined cryofixation and freeze fracturing with electrophysiology. The results thus obtained have invalidated previous work with "conventional" chemical fixation and freeze fracturing. This again illustrates the potential of fast freezing techniques and verification by substantially different control techniques. With the same object, beautiful data were obtained for the transmembrane orientation of connexon subunits, by combining EM methods, including antibody labeling, with enzymatic attacks.[51]

An almost perfect match between EM data (obtained by fast freezing and freeze fracture analysis), and data obtained independently from electrophysiological capacitance change recordings, can now be registered for membrane fusion during exocytosis (see G. Knoll and H. Plattner, Chapter 6). Other examples of the use of cryomethods in molecular electron microscopy are contained in Reference 7. Controls appropriate for studying the precise molecular arrangement of lipids in membranes are listed in Table 2; however, none of them provides a time resolution comparable with fast freezing and freeze fracturing (see Chapter 5 by A. Verkleij). With regard to membranology it should be mentioned, furthermore, that the dynamic arrangement of membrane components such as IMPs can now be analyzed quite objectively by computer analysis, as shown by S. W. Hui in Chapter 4.

This book also contains a survey on the self-assembly of crystalline bacterial envelopes (Chapter 2 by U. B. Sleytr and P. Messmer). Freeze fracturing or negative staining, eventually combined with image processing by laser diffractometric evaluation of micrographs, can be applied. Some dynamic molecular aspects have already been analyzed on the 3-dimensional level, e.g., the dynamic conformational changes or gap junction connexon subunits during decoupling or coupling processes.[52] The requirement for low electron dose imaging, "sustains" or negative stains, high efficiency electron recording systems, and computer evaluation, as well as the restricted resolution (1 to 2 nm) in the context of image processing techniques, have been discussed elsewhere.[37,38]

IV. CONCLUSIONS AND OUTLOOK

In its first period of development, the EM has largely contributed to an understanding of cell dynamics, just by providing suggestive pictures, e.g., in support of the sliding filament hypothesis for muscle contraction or for secretory processes. When subsequently, labeling methods became available, they were still restricted to autoradiography, or to the application of visible EM markers on cell surfaces, or surface-connected compartments involved in vesicular transport. Only since about a decade ago can one address deliberate intracellular antigens; recently via mAB, labeling specificity also has increased further. In addition, cryomethods particularly have become of increasing importance for the EM analysis of dynamic processes (see recent monograph by Steinbrecht and Zierold, Reference 7).

In this context it is important that to an ever-growing extent, EM (with all its preparative and analytical developments) has matured to an analytical tool with quite diversified capabilities, though also with a variety of inherent pitfalls. Yet since many aspects can be analyzed (and thus, verified or falsified,) in parallel by widely different alternative methods, the EM provides us with increasingly reliable data on dynamic processes on the cellular and molecular level. A more comprehensive "survey of current problems, pitfalls and perspectives" of EM techniques has recently been published elsewhere.[1]

It remains to be seen to what extent methods and instruments other than the EM can solve some of the open questions. On the cellular level, the X-ray microscope could have the potential to visualize sub-LM details on a 5 nm scale in living cells.[53] The potential, although limited, of environmental chambers is addressed in S. W. Hui's Chapter 4 in this book. Future developments might render the scanning tunnelling microscope[54] or the atomic force microscope[55] (which requires no metal coating) valuable tools for analyzing dynamic structural aspects down to the atomic level, i.e., in the 0.1 nm range of resolution. If so, this would represent another important step in the history of microscopy on the molecular level.

On the cellular level, one can currently not recognize any alternative to the application of electron microscopy. For its applicability to the investigation of dynamic processes, preparative and analytical methods (as well as alternative controls) require permanent improvement. This volume is an attempt to present the current state in this field and to envisage future developments.

ACKNOWLEDGMENTS

Some of the work of the author's group cited herein was supported by SFB 156.

REFERENCES

1. **Plattner, H. and Knoll, G.,** Ultrastructural analysis of dynamic cellular processes: a survey of current problems, pitfalls and perspectives, *Scanning Microsc.*, 1, 1199, 1987.
2. **Plattner, H.,** Synchronous exocytosis in Paramecium cells, in *Cell Fusion,* Sowers, A. E., Ed., Plenum Press, New York, 1987, 69.
3. **Sowers, A. E. and Hackenbrock, C. R.,** Rate of lateral diffusion of intramembrane particles: Measurement by electrophoretic displacement and rerandomization, *Proc. Natl. Acad. Sci. U.S.A.*, 78, 6246, 1981.
4. **Palade, G. E.,** Intracellular aspects of the process of protein synthesis, *Science,* 189, 347, 1975.
5. **Plattner, H. and Bachmann, L.,** Cryofixation: A tool in biological ultrastructure research, *Int. Rev. Cytol.*, 79, 237, 1982.
6. **Plattner, H. and Knoll, G.,** Cryofixation of biological materials for electron microscopy by the methods of spray-, sandwich-, cryogen-jet- and sandwich-cryogen-jet-freezing: A comparison of techniques, in *The Science of Biological Specimen Preparation for Microscopy and Microanalysis,* Revel, J. P., Bardard, T., and Haggis, G. H., Eds., SEM Inc., AMF O'Hare IL, 1984, 139.
7. **Steinbrecht, R. A. and Zierold, K.,** Eds., *Cryotechniques in Biological Electron Microscopy,* Springer-Verlag, Berlin, 1987.
8. **Bachmann, L. and Mayer, E.,** Physics of water and ice: Implications for cryofixation, in *Cryotechniques in Biological Electron Microscopy,* Steinbrecht, R. A. and Zierold, K., Eds., Springer-Verlag, Berlin, 1987, 3.
9. **Knoll, G., Verkleij, A. J., and Plattner, H.,** Cryofixation of dynamic processes in cells and organelles, in *Cryotechniques in Biological Electron Microscopy,* Steinbrecht, R. A. and Zierold, K., Eds., Springer-Verlag, Berlin, 1987, 258.
10. **Moor, H.,** Theory and practice of high pressure freezing, in *Cryotechniques in Biological Electron Microscopy,* Steinbrecht, R. A. and Zierold, K., Eds., Springer-Verlag, Berlin, 1987, 175.
11. **Sitte, H., Edelman, L., and Neumann, K.,** Cryofixation without pretreatment at ambient pressure, in *Cryotechniques in Biological Electron Microscopy,* Steinbrecht, R. A. and Zierold, K., Eds., Springer-Verlag, Berlin, 1987, 87.
12. **Casley-Smith, J. R.,** Freeze-substitution observations on the endothelium, and the passage of ions, *Microcirculation,* 1, 12, 1981.
13. **Wendt-Gallitelli, M. F. and Wolburg, H.,** Rapid freezing, cryosectioning, and x-ray microanalysis on cardiac muscle preparations in defined functional states, *J. Electron Microsc. Techniques,* 1, 151, 1984.

14. **Robinson, J. R., Castello, J. J., and Karnovsky, M. J.,** Unusual lysosomes in aortic smooth muscle cells: presence in living and rapidly frozen cells, *J. Cell Biol.*, 102, 1615, 1986.
15. **Shotton, D. M.,** Review: Video-enhanced light microscopy and its application in cell biology, *J. Cell Sci.*, 89, 129, 1988.
16. **White, J. G., Amos, W. B., and Fordham, M.,** An evaluation of confocal versus conventional imaging of biological structures by fluorescence light microscopy, *J. Cell Biol.*, 105, 41, 1987.
17. **Taylor, D. L., Amato, P. A., McNeil, P., Luby-Phelps, K., and Tanasugarn, L.,** Spatial and temporal dynamics of specific molecules and ions in living cells, in *Applications of Fluorescence in the Biomedical Sciences,* Taylor, D. L., Waggoner, A. S., Lanni, F., Murphy, R. F., and Birge, R. R., Eds., Alan R. Liss, New York, 1986, 347.
18. **Mellman, I. S., Plutner, H., and Ukkonen,** Internalization and rapid recycling of macrophage F_c receptors ragged with monovalent antireceptor antibody: Possible role of a prelysosomal compartment, *J. Cell Biol.*, 98, 1163, 1984.
19. **DeBrabander, M., Nuydens, R., Geuens, G., Moeremans, M., and DeMey, J.,** The use of submicroscopic gold particles combined with video contrast enhancement as a simple molecular probe for the living cell, *Cell Motil.*, 6, 105, 1986.
20. **Geerts, H., DeBrabander, M., Nuydens, R., Geuens, S., Moeremans, M., DeMey, J., and Hollenbeck, P.,** Nanovid tracking: A new automatic method for the study of mobility in living cells based on colloidal gold and video microscopy, *Biophys. J.*, 52, 775, 1987.
21. **Luby-Phelps, K., Amato, P. A., and Taylor, D. L.,** Structural immunocytochemical detection of fluorescent analogs with antibodies specific for the fluorophore, *Cell Motil.*, 4, 137, 1984.
22. **Salpeter, M. M. and Farquhar, M. G.,** High resolution analysis of the secretory pathway in mammotrophs of the rat anterior pituitary, *J. Cell Biol.*, 91, 240, 1981.
23. **Walther, P., Kriz, S., Müller, M., Ariano, B. H., Brodbeck, V., Ott, P., and Schweingruber, M. E.,** Detection of protein A gold 15 nm marked surface antigens by backscattered electrons, *Scanning Electron Microsc.*, 3, 1257, 1984.
24. **Soligo, D., DeHarven, E., Naya, M. T., and Lambertenghi-Deliliers, G.,** Immuno-cytochemistry with backscattered electrons, in *The Science of Biological Specimen Preparation for Microscopy and Microanalysis 1985,* Müller, M., Becker, R. P., Boyde, A., and Wolosewick, J. J., Eds., SEM Inc., AMF O'Hare, IL, 1986, 289.
25. **Griffiths, G. and Simons, K.,** The trans Golgi network: Sorting at the exit site of the Golgi complex, *Science,* 234, 438, 1986.
26. **Anderson, R. G. W., Falck, J. R., Goldstein, J. L., and Brown, M. S.,** Visualization of acidic organelles in intact cells by electron microscopy, *Proc. Natl. Acad. Sci. U.S.A.*, 81, 4838, 1984.
27. **Bendayan, M.,** Protein A-gold electron microscopic immunocytochemistry: Methods, applications and limitations, *J. Electron. Microsc. Techniques,* 1, 243, 1984.
28. **Lucquoc, J. M., Berger, E. G., and Roth, J.,** Detection of terminal N-linked N-acetylglucosamine residues in the golgi apparatus using galactosyltransferase and endoglucosaminidase F/peptide N-glycosidase F: Adaptation of a biochemical approach to electron microscopy, *J. Histochem. Cytochem.*, 25, 175, 1987.
29. **Geuze, H. J., Slot, J. W., Strous, G. J. A., Peppard, J., Figura, K. v., Hasilik, A., and Schwartz, A. L.,** Intracellular receptor sorting during endocytosis: Comparative immunoelectron microscopy of multiple receptors in rat liver, *Cell,* 37, 195, 1984.
30. **Hedman, K., Goldenthal, K. L., Rutherfold, A. V., Pastan, I., and Willingham, M. C.,** Comparison of the intracellular pathways of transferrin recycling and vesicular stomatitis virus membrane glycoprotein exocytosis by ultrastructural double-label cytochemistry, *J. Histochem. Cytochem.*, 35, 233, 1987.
31. **Brown, W. J. and Farquhar, M. G.,** The distribution of 215-kilodalton mannose 6-phosphate receptors within cis (heavy) and trans (light) Golgi subfractions varies in different cell types, *Proc. Natl. Acad. Sci. U.S.A.*, 84, 9001, 1987.
32. **Knoll, G., Burger, K. N. J., Bron, R., VanMeer, G., and Verkleij, A. J.,** Fusion of liposomes with the plasma membrane of epithelial cells: Fate of incorporated lipids as followed by freeze-fracture and autoradiography of plastic sections, *J. Cell Biol.*, in press.
33. **Carlemalm, E., Villiger, W., Hobot, J. A., Acetarin, J. D., and Kellenberger, E.,** Low temperature embedding with Lowicryl resins: two new formulations and some applications, *J. Microsc.*, 140, 55, 1985.
34. **Humbel, B. and Mueller, M.,** Freeze substitution and low temperature embedding, in *The Science of Biological Specimen Preparation 1985,* Müller, M., Becker, R. P., Boyde, A., and Wolosewick, J. J., Eds., SEM Inc., AMF O'Hare, IL, 1985, 175.
35. **Stierhof, Y. D., Schwarz, H., and Frank, H.,** Transverse sectioning of plastic-embedded immunolabeled cryosections: morphology and permeability to protein A-colloidal gold complexes, *J. Ultrastruct. Molec. Struct. Res.*, 97, 187, 1986.
36. **Steiner, D. F., Michael, J., Houghten, R., Mathieu, M., Gardner, P. R., Ravazzola, M., and Orci, L.,** Use of a synthetic peptide antigen to generate antisera reactive with a proteolytic processing site in native human proinsulin: Demonstration of cleavage within clathrin-coated (pro)secretory vesicles, *Proc. Natl. Acad. Sci. U.S.A.*, 84, 6184, 1987.

37. **Plattner, H. and Zingsheim, H. P.,** Electron microscopic methods in cellular and molecular biology, in *Sub-cellular Biochemistry, Vol. 9,* Roodyn, D. B., Ed., Plenum Press, New York, 1983, 1.

38. **Plattner, H. and Zingsheim, H. P.,** *Elektronenmikroskopische Methodik in der Zell — und Molekularbiologie,* Gustav Fischer Verlag, Stuttgart, 1987.

39. **Login, G. R., Stavinoha, W. B., and Dvorak, A. M.,** Ultrafast microwave energy fixation for electron microscopy, *J. Histochem. Cytochem.,* 34, 381, 1986.

40. **Marti, R., Wild, P., Schraner, E. M., Mueller, M., and Moor, H.,** Parathyroid ultrastructure after aldehyde fixation, high-pressure freezing, or microwave irradiation, *J. Histochem. Cytochem.,* 35, 1415, 1987.

41. **Hubbard, A. L., Bartles, J. R., and Braiterman, L. T.,** Identification of rat hepatocyte plasma membrane proteins using monoclonal antibodies, *J. Cell Biol.,* 100, 1115, 1985.

42. **Drenckhahn, D. and Merte, C.,** Restriction of the human kidney band 3-like anion exchanger to specialized subdomains of the basolateral plasma membrane of intercalated cells, *Eur. J. Cell Biol.,* 45, 107, 1987.

43. **Dunphy, W. G., Brands, R., and Rothman, J. E.,** Attachment of terminal N-acetylglucosamine to asparagine-linked oligosaccharides occurs in central cisternae of the Golgi stack, *Cell,* 49, 463, 1985.

44. **Weissmann, A. M., Klausner, R. D., Rao, K., and Harford, J. B.,** Exposure of K562 cells to anti-receptor monoclonal antibody OKI9 results in rapid redistribution and enhanced degradation of the transferrin receptor, *J. Cell Biol.,* 102, 951, 1986.

45. **Orci, L., Ravazzola, M., Storch, M. J., Anderson, R. G. W., Vassalli, J. D., and Perrelet, A.,** Proteolytic maturation of insulin is a post-Golgi event which occurs in acidifying clathrin-coated secretory vesicles, *Cell,* 49, 865, 1987.

46. **Chicheportiche, Y. and Tartakoff, A. M.,** Monoclonal antibodies as markers of the endocytotic and secretory pathways, *Eur. J. Cell Biol.,* 44, 135, 1987.

47. **Claviez, M., Pagh, K., Maruta, H., Baltes, W., Fisher, P., and Gerisch, G.,** Electron microscopic mapping of monoclonal antibodies on the tail region of Dictyostelium myosin, *EMBO J.,* 1, 1017, 1982.

48. **Flicker, P. F., Peltz, G., Sheetz, M. P., Parham, P., and Spudich, J. A.,** Site-specific inhibition of myosin-mediated motility *in vitro* by monoclonal antibodies, *J. Cell Biol.,* 100, 1024, 1985.

49. **Tokunaga, M., Sutoh, K., Toyoshima, C., and Wakabayashi, T.,** Location of the ATPase site of myosin determined by three-dimensional electron microscopy, *Nature,* 329, 635, 1987.

50. **Mandelkow, E. M. and Mandelkow, E. R.,** Unstained microtubules studied by cryo-electron microscopy: substructure, supertwist, and disassembly, *J. Mol. Biol.,* 181, 123, 1985.

51. **Zimmer, D. B., Green, C. R., Evans, W. H., and Gilula, N. B.,** Topological analysis of the major protein in isolated intact rat liver gap junctions and gap junction-derived single membrane structures, *J. Biol. Chem.,* 262, 7751, 1987.

52. **Unwin, P. N. T. and Ennis, P. D.,** Two configurations of a channel-forming membrane protein, *Nature,* 307, 609, 1984.

53. **Feder, R., Banton, V., Sayre, D., Costa, D., Baldini, M., and Kim, B.,** Direct imaging of life human platelets by flash X-ray microscopy, *Science,* 227, 63, 1985.

54. **Binnig, G. and Rohrer, H.,** Scanning tunneling microscopy — from birth to adolescence, *Rev. Mod. Phys.,* 59, 615, 1987.

55. **Gould, S., Marti, O., Drake, B., Hellemans, L., Bracker, C. E., Hansma, P. K., Keder, N. L., Eddy, M. M., and Stucky, G. D.,** Molecular resolution images of amino acid crystals with the atomic force microscope, *Nature,* 332, 332, 1988.

Chapter 2

SELF-ASSEMBLY OF CRYSTALLINE BACTERIAL CELL SURFACE LAYERS (S-LAYERS)

Uwe B. Sleytr and Paul Messner

TABLE OF CONTENTS

I. INTRODUCTION

Many prokaryotic organisms and some green algae have regular arrays of (glyco)proteins on their outermost surface. These two-dimensional crystalline surface layers, termed S-layers,[1] are found in members of nearly every taxonomical group of walled eubacteria and represent an almost universal feature of archaebacterial cell envelopes (Figure 1) (for recent compilations see Sleytr and Messner,[2] Sleytr et al.,[3] Sleytr and Messner[4]). Due to the facts that S-layers possess a high degree of structural regularity and that S-layer proteins are the most abundant of all bacterial cellular proteins, these crystalline arrays are excellent models for studying the dynamic aspects of assembly of a supramolecular structure during cell growth.

It is particularly the surface location of S-layers which provides a clue for their biological significance.[2] Prokaryotic organisms possessing S-layers have been shown to be ubiquitous in the biosphere. Thus, like other bacterial surface components (e.g., glycocalyces), S-layers must have evolved as a consequence of quite diverse interactions between cells and their environment. S-layers apparently can provide organisms with an advantage of selection by functioning as protective coats, molecular sieves, and molecule and ion traps, as promotors for cell adhesion and surface recognition and frameworks, determining and maintaining cell shape or envelope rigidity (for reviews see Sleytr and Messner,[2-5] Smit,[6] and Koval and Murray[7]). As illustrated in Figure 1, S-layers are generally associated with quite complex supramolecular cell envelope structures, and may function only by specific interaction with these underlying structures.

This review will predominantly deal with the dynamic process of assembly of two-dimensional arrays of macromolecules on bacterial cell surfaces. For a more general introduction to the field the reviews[1,2,5-10] and the proceedings[11] of a recent workshop on S-layers are recommended for further reading.

II. LOCATION AND ULTRASTRUCTURE OF S-LAYERS ON INTACT CELLS

Our present knowledge of the molecular architecture of archaebacterial and eubacterial cell envelopes is based on results obtained by high resolution electron microscopy, and from biochemical and biophysical analysis primarily accomplished by subcellular fractionation. The location and ultrastructure of S-layers of a great variety of bacteria were investigated by electron microscopy of thin sections of freeze etched, freeze dried, shadowed, negatively stained, or frozen hydrated preparations (for review of methods see Sleytr et al.[12]). The most suitable technique for demonstrating crystalline surface layers on intact cells is freeze etching.[1,8,13,14] Analysis of freeze etched preparations of numerous archaebacteria and eubacteria have shown that S-layers completely cover the cell surface, leaving no gaps. S-layers of archaebacteria have predominantly hexagonal (p6) symmetry,[3,4] whereas in eubacteria, square (p4) and oblique (p2) lattices also are observed frequently.[4,5,15]

Whereas S-layer lattices on rod-shaped cells are generally uniform over large areas of the cylindrical part of the cells, at regions near the cell poles and at the septation sites where the curvature alters, the orientation of the pattern in adjacent areas is different[1,16-19] (see also section IV).

By evaluation of the local frequency of lattice faults at septation sites and cell poles, it was found that in rod-shaped cells crystallites covering the cell poles are generally larger. This provided first evidence for a dynamic process of assembly of the S-layer involving a continuous entropy-driven recrystallization of constituent subunits on a growing cell surface[14] (see also section IV).

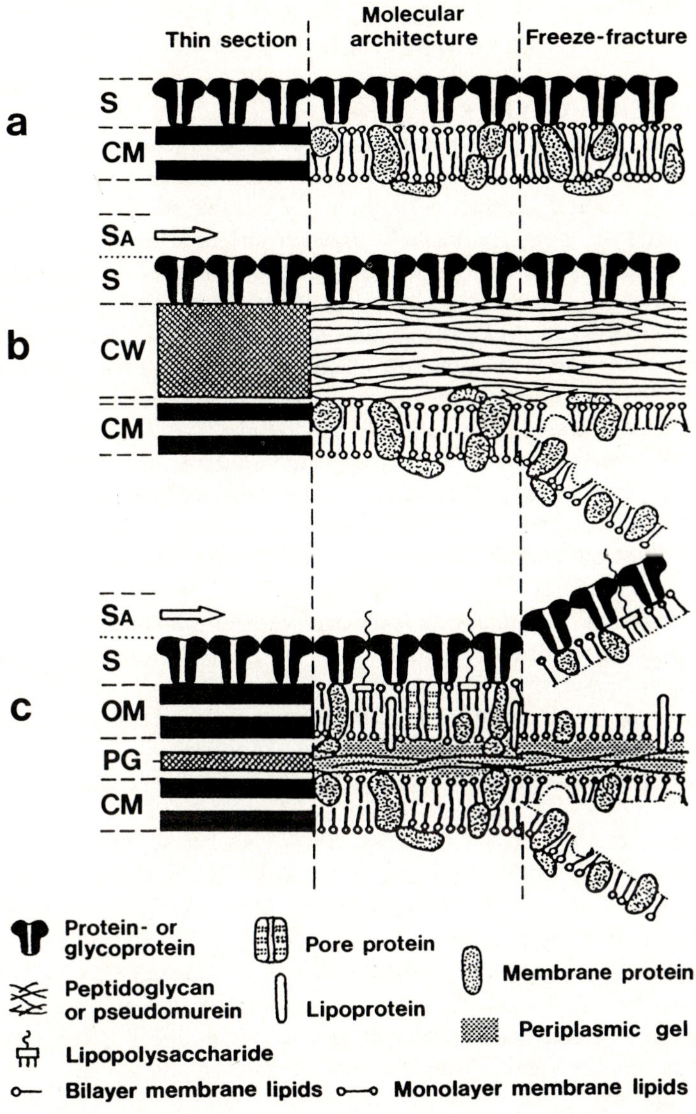

FIGURE 1. Schematic illustration of the three major classes of prokaryotic cell envelopes containing S-layers; left: thin section profiles as revealed by transmission electron microscopy; middle: molecular structure showing major components; right: freeze fracture behavior of cell envelopes. Cell envelope structure of Gram-negative archaebacteria (a) with crystalline surface layers (S-layers) as exclusive cell wall component; Gram-positive cell envelope (b), as observed in eubacteria and archaebacteria; Gram-negative cell envelope (c) as observed in eubacteria; CM, cytoplasmic membrane; CW, rigid cell wall layer (peptidoglycan in eubacteria, pseudomurein in archaebacteria); OM, outer membrane; PG, peptidoglycan; S, crystalline surface layer; S_A indicates location for possible additional S-layer. (Modified from Sleytr and Messner.[2])

Previously high resolution electron microscopical studies and computer image processing on negatively stained S-layers of *Aeromonas salmonicida* showed two different types of pattern on individual cells, both of which have p4 symmetry.[20] It was assumed that the two patterns reflect a structural transformation of the layer that may modulate its permeability.

III. ISOLATION, CHEMICAL CHARACTERIZATION AND SELF-ASSEMBLY OF S-LAYERS

A. ISOLATION OF S-LAYERS

S-layers vary considerably with respect to their mechanical stability and ease of disintegration into their subunits.[2,8,21] The crystalline arrays can be detached and disintegrated by a variety of procedures, each being particularly suitable for the individual organisms.[22] These data indicate that S-layers are held together and onto the supporting cell envelope component by noncovalent forces, including hydrophobic interactions, hydrogen bonds, and ionic bonds involving divalent cations or direct interaction of polar groups.[1,2,23-26]

Unfortunately, very little is known about the specific interactions between S-layers and the underlying layer.[1,2,6] Valuable information about differences in the nature of the chemical bonds between the S-layer subunits themselves and of the crystalline array to the supporting supramolecular envelope layer (Figure 1) can be obtained from consideration of the reagents and conditions that cause detachment and/or disintegration of the lattice. Data obtained on S-layers of both Gram-positive and Gram-negative organisms indicate that the bonds (or combination of bonds) holding together the S-layer constituent subunits in the two-dimensional array are stronger than those binding S-layer lattices to the underlying layer of the cell envelope (for review see Sleytr and Messner,[2] Smit,[6] and Masuda and Kawata[27]).

Further information on bonding properties have been obtained by studying conditions required for self assembly of S-layer subunits *in vitro*, both in the presence and absence of surfaces suitable for adhesion[1,24,28] (see section III. C and D).

A few organisms, particularly some archaebacteria, assemble crystalline surface layers, which are highly resistant to common denaturants, suggesting that most of the constituent subunits are covalently linked.[29-31]

B. CHEMICAL CHARACTERIZATION OF S-LAYER SUBUNITS AND RELATEDNESS

Chemical analysis of a variety of eubacterial and archaebacterial S-layers have shown that the crystalline arrays are composed of single homogeneous protein, or glycoprotein species (for review see Sleytr and Messner[2,4,5]). The molecular weight (M_r) of S-layer protomers varies from 40,000 to 220,000.[2,4] Carbohydrate chains covalently linked to the protein moieties have been demonstrated in both eubacterial[32-34] (for review see Messner and Sleytr[35]) and archaebacterial S-layers.[36-39] The S-layer proteins are acidic with a large proportion of hydrophobic amino acids, and contain little or no sulfur-containing amino acids.[2,26,29,38,40] Studies on the secondary structure of selected S-layers by circular dichroism and by infrared spectroscopy revealed a low level of α-helical structure.[26,41,42]

Diversity of S-layers of strains within a species has been reported for both Gram-positive and Gram-negative bacteria.[10,15,40,43-45] From the few comparative studies performed, it appears that in many organisms S-layers can be considered only a strain-specific taxonomical feature. Since the primary structure of only a few S-layer polypeptides has been elucidated, predictions on the evolutionary relationship of S-layers are not yet possible.[46-50]

C. SELF-ASSEMBLY OF S-LAYER SUBUNITS IN THE ABSENCE OF SUPPORTING LAYERS

S-layer subunits from many Gram-positive and Gram-negative eubacteria and archaebacteria have shown the ability to assemble into two-dimensional arrays identical to those observed in intact cells upon removal of the disrupting agent used for their isolation. This entropy-driven crystallization process may take place both in the presence or in the absence of supporting layers. All information for the assembly product is entirely contained in the individual protomeric units. Since S-layer proteins have a high proportion of non-polar amino

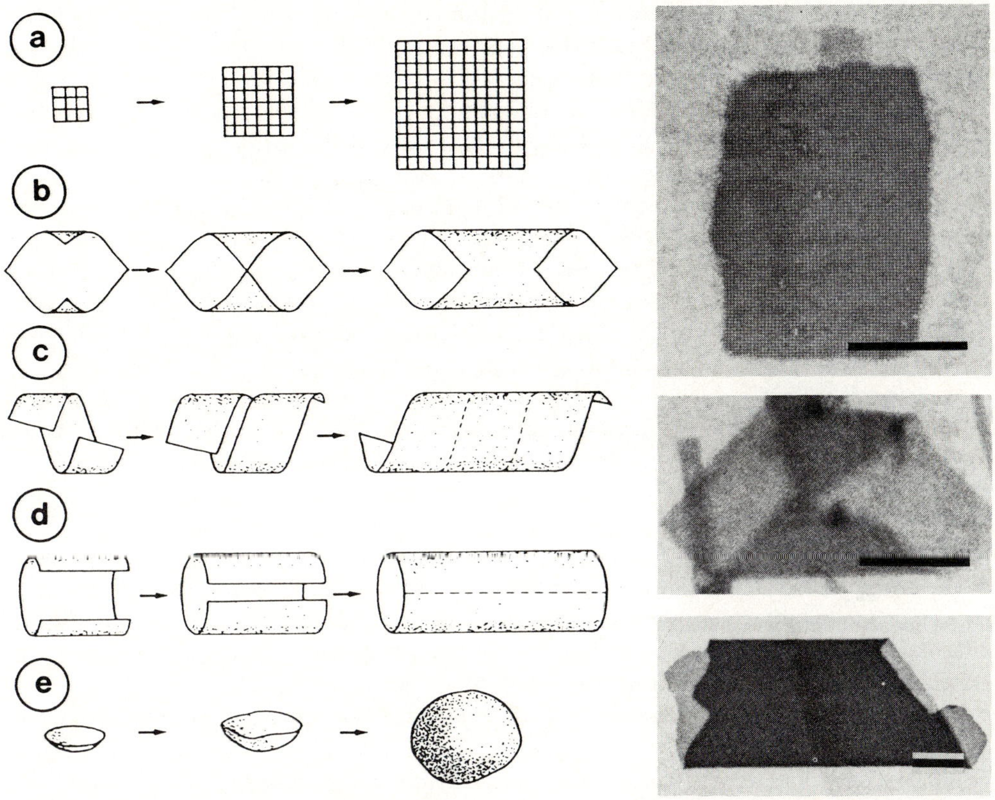

FIGURE 2. Diagram illustrating different self-assembly routes of S-layer subunits leading to the formation of sheets (a), cylinders (b,c,d), and spheres (e). The right hand panel shows from top to bottom representative micrographs of negatively stained (1% uranylacetate) self-assembly products corresponding to a, b and d. Bars = 0,5 μm (From Sleytr, U. B., *Cytomorphogenesis in Plants, Cell Biology Monographs,* Vol. 8, Kiermayer, O., Ed., Springer-Verlag, Vienna, 1981, 3. With permission.)

acids, most likely hydrophobic interactions are involved in the assembly process.[51] Other S-layers are stabilized by divalent cations interacting with acidic amino acids.[1,8,19,52] Studies on the distribution and functional significance of charged groups on both the outer and inner faces of S-layers of *Bacillaceae* have shown that on both faces free amino and carboxyl groups of adjacent protomers are arranged in proximity so as to contribute by electrostatic interactions to the cohesion of the protomers in the lattice.[53] Labeling experiments with polycationic ferritin (PCF) revealed that in many S-layers the two surfaces have a different net charge.[19,53]

Depending on the specific bonding properties of the S-layer subunits and their tertiary structure, protomers may aggregate into flat sheets, open ended cylinders, or closed vesicles (Figure 2). The latter are only formed by subunits with the ability of forming p6 lattices[54] (for review see Sleytr and Messner,[2] Beveridge,[8] and Sleytr[28]).

Few systematic studies have been done on the time course of the self-assembly of S-layer subunits. The crystallization process of the square S-layer lattice of *Bacillus stearothermophilus* NRS 1536/3c was monitored by light scattering as well as by chemical cross-linking and negative staining. It was demonstrated that temperature and protein concentration determine both the extent and rate of association. Obviously oligomeric precursors ($M_r > 10^6$) are involved in the assembly mechanism. The assembly kinetics are multiphasic, with a rapid initial phase and slow consecutive rearrangement steps, which finally lead to extended surface lattices similar to those observed on intact cells.[51]

Detailed studies on the time course of the self-assembly process of cylindrical structures have shown that different assembly routes can lead to the formation of S-layer cylinders (Figure 2b-d). In all cases the crystalline arrays will grow with a defined curvature and eventually fuse to form cylinders.[28] Depending on the type of S-layer species, the cylinders may have either a curvature identical or opposite to that of the lattice occurring on intact cells. With S-layers from *Bacillus sphaericus* P1, it was demonstrated that by reducing the molecular weight of the constituent subunits by specific proteolysis, subunits reassemble into cylinders instead of sheets.[55,56]

Frequently flat or cylindrical *in vitro* self-assembly products are composed of double-layers. A remarkably broad spectrum of mono- and double-layer self-assembly products could be obtained upon recrystallization of the S-layer subunits of *Bacillus stearothermophilus* strain NRS 2004/3a.[19]

In vitro self-assembly experiments with S-layers of different organisms have also shown that S-layer fragments, obtained either by mechanical disruption of self-assembly products or by detachment from envelope components, have a strong inclination to fuse and to recrystallize.[28]

D. RECONSTITUTION OF S-LAYERS ON SUPPORTING LAYERS

The ability of isolated S-layer subunits to reattach to cell surfaces from which they had been removed (homologous reattachment), or to those of other organisms (heterologous reattachment), was examined to elucidate the mechanisms involved in the development and maintenance of crystalline arrays of macromolecules on a growing cell surface.[54]

Isolated S-layer subunits or fragments from numerous Gram-positive and Gram-negative bacteria have shown to reattach to cell walls in various conditions, such as dialysis of a mixture of chaotropic agents with "naked" cell walls against buffer systems or distilled water.[27,54,57,58] Reattachment of S-layer subunits or fragments isolated from Gram-negative envelopes frequently requires the presence of bivalent cations or a defined ionic strength.[23,24,59,60] Subunits reattached to cell wall fragments generally form small crystallites at various orientations (Figure 3), but not the large regular arrays observed on intact cells. When mixtures of S-layer protomers from lattices with different geometry are used in reassembly studies, both types of lattices can develop on one cell (Figure 3b). Using the S-layers of two thermophilic *Clostridia* as a model system it was demonstrated that hexagonal and tetragonal S-layer crystallites are formed with equal frequency.[54] These observations demonstrated that the information for the formation of the crystalline array resides in the subunits themselves, but is not affected by the supporting cell envelope component. The fact that the crystallites are oriented at random indicates that their orientation is not determined by any pattern in the binding sites in the cell wall. When S-layers are reconstituted on "naked" cells which have maintained their cylindrical shape, the lattice orientation on the cylindrical part generally appears the same as observed by freeze etching of intact cells, showing few faults and good long range order. These results provide evidence that the curvature of the cylindrical part of the cell induces a recrystallization and/or an orientation of the array with least stress between adjacent subunits.[14,28]

IV. DYNAMIC PROCESS OF ASSEMBLY OF S-LAYERS DURING CELL GROWTH

A. SYNTHESIS AND SECRETION

With all the foregoing information in mind it becomes evident that the assembly and recrystallization properties contained within the individual S-layer protomers are sufficient for the maintenance of a crystalline surface layer with the dynamically lowest energy state during cell growth. An additional necessity for maintaining a closed surface layer is a

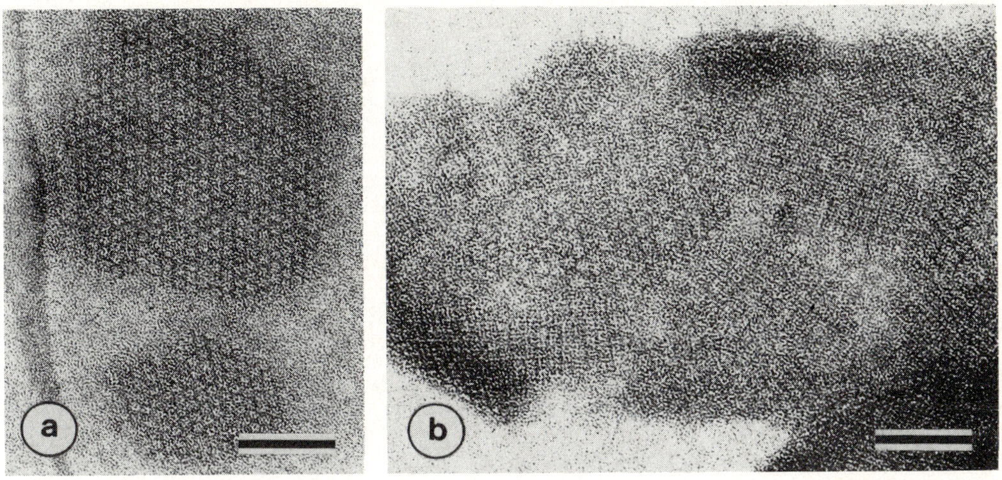

FIGURE 3. Electron micrographs of negatively stained preparations illustrating reattachment of S-layer subunits to peptidoglycan fragments. Homologous reattachment (a) of S-layer subunits of *Clostridium thermohydrosulfuricum* L111-69 to peptidoglycan fragments of this strain. Upon dialysis, randomly oriented crystallites are found on the peptidoglycan. (b) Heterologous reattachment. Murein layers of *Clostridium thermosaccharolyticum* D120-70 were incubated with a mixture of equal parts of S-layer subunits from *Clostridium thermosaccharolyticum* and from *Clostridium thermohydrosulfuricum*. After dialysis, randomly distributed crystallites with hexagonally and tetragonally arranged subunits can be seen. Bars = 100 nm.

continuous synthesis of the constituent protomers and their translocation to areas of insertion. For maintaining a crystalline array, an average size bacterial cell has to produce more than 400 copies of the S-layer subunits per second when growing with a generation time of 20 to 30 min.[2] In some organisms the rate of synthesis appears strictly controlled, in others a considerable excess of S-layer material is secreted into the medium.[2,6,61] With the exception of those Gram-negative archaeabacteria (Figure 1a) in which the S-layer is directly associated with the cytoplasmic membrane, the exported polypeptides pass through intermediate layers before reaching the areas of lattice growth. In Gram-positive eubacteria and archaebacteria, S-layer subunits or small oligomeric precursors will have to penetrate or even to diffuse laterally within the peptidoglycan or the pseudomurein layer before reaching their incorporation sites (Figure 1b). In Gram-negative eubacteria (Figure 1c) two alternative S-layer secretion pathways have to be considered: (1) over fusions between the cytoplasmic membrane and the outer membrane,[6,62,63] or (2) through the periplasmic space and the outer membrane.[64]

There is no evidence for pools of S-layer proteins in the cytosol. With some Gram-positive eubacteria (e.g., Bacillaceae), it was shown that, upon separation of the cytoplasmic membrane from the peptidoglycan layer or in the course of cell breakage, an additional S-layer assembles on the inside of the rigid wall layer. This strongly indicates that a surplus of S-layer subunits can be present in the open meshwork of the peptidoglycan layer.[1,17,65,66]

From the *in vitro* assembly experiments (see section III.C and D) it is evident that S-layers are highly anisotropic structures. Consequently, in the first instance, nonspecific interactions (e.g., charge-to-charge and/or hydrophobic interactions) may be responsible for selectively trapping the subunits in a proper orientation for their subsequent accurate incorporation into expanding lattices.[2] Several organisms, e.g., *Aquaspirillum serpens* strain MW5,[67-69] *Bacillus brevis* 47,[66] and *Nitrosocystis oceanus,*[70] have been described to possess a double S-layer composed of different subunit species. For maintaining such a two-layered crystalline array during cell growth, different but highly specific structural interactions are required. In *Aquaspirillum serpens* MW5 the two S-layers with hexagonal symmetry were

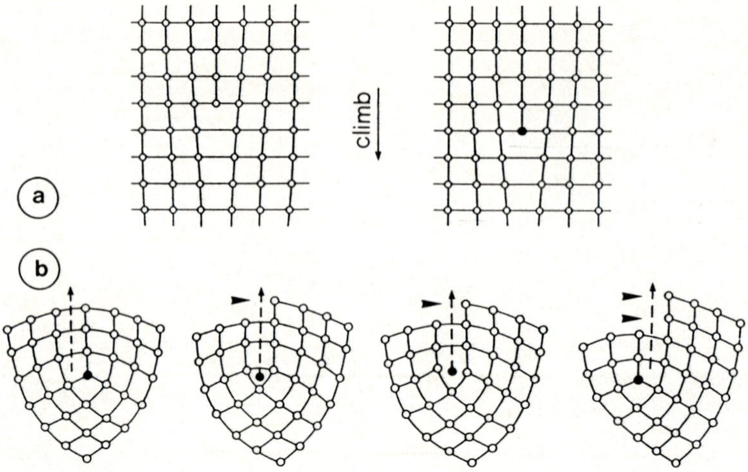

FIGURE 4. Schematic illustration of dislocations and disclinations as observed on S-layers. Intussusceptive growth of a square (p4) crystal by nonconservative climb of a dislocation (a). By adding a new subunit (filled dot), the dislocation climbs one lattice spacing to a new position. (b) Movement of a local wedge disclination in a square lattice. The disclination is shifted diagonally across a distorted square, and in the process it generates two edge dislocations (arrows). (Modified from Harris and Scriven.[73])

shown to assemble sequentially. The inner S-layer can self-assemble entropically by itself as well as assemble on the outer membrane, whereas formation of the other S-layer would require that array as a template for assembly.[68]

B. INCORPORATION OF NEW SUBUNITS INTO CLOSED SURFACE CRYSTALS

Topologically, most prokaryotic cells can be described as spheres, or as cylinders with two hemispherical ends. Theoretically, to cover such bodies with identical macromolecules to form closed surface crystals, no lattice faults would be required for the cylindrical parts. On the other hand, it is necessary to introduce lattice faults for covering the rounded surfaces at the cell poles, and the septation sites of a rod-shaped cell or the spherical surface of a coccus. The design of capsids of spherical (icosahedral) viruses represents a clear example for the geometrical necessity of lattice faults in a closed surface crystal.[71] Most of these capsids contain hexamers, as well as twelve pentamers held in a state of bonding termed "quasi-equivalent".[71,72] Quasi-equivalence in closed surface crystals can be defined as any small nonrandom variation in a regular bonding pattern which leads to a more stable structure than strictly regular bonding.[72] As in icosahedral shell designs in S-layer lattices covering spherical surfaces, quasi-equivalent bonding is a geometrical necessity. Most important, such an arrangement will have the greatest possible number of most stable bonds formed.

Different from the morphologically strictly predetermined icosahedral shell, S-layers cover living cells which undergo dimensional variations during cell growth and division. Consequently, S-layers have to be seen as "dynamic closed surface crystals" with the intrinsic tendency to assume a structure of the lowest free energy during growth.

As discussed in detail by Harris and co-workers,[73-76] dislocations (Figure 4a) can serve as sites for incorporation of new subunits in crystalline arrays which grow by "intussusception". Further, as a geometrical necessity, closed surface crystals must contain local wedge (screw) disclinations (Figure 4b), which themselves can act as sources of edge dislocations.[73] Thus, theoretically, the rate of growth of a closed surface crystal by the

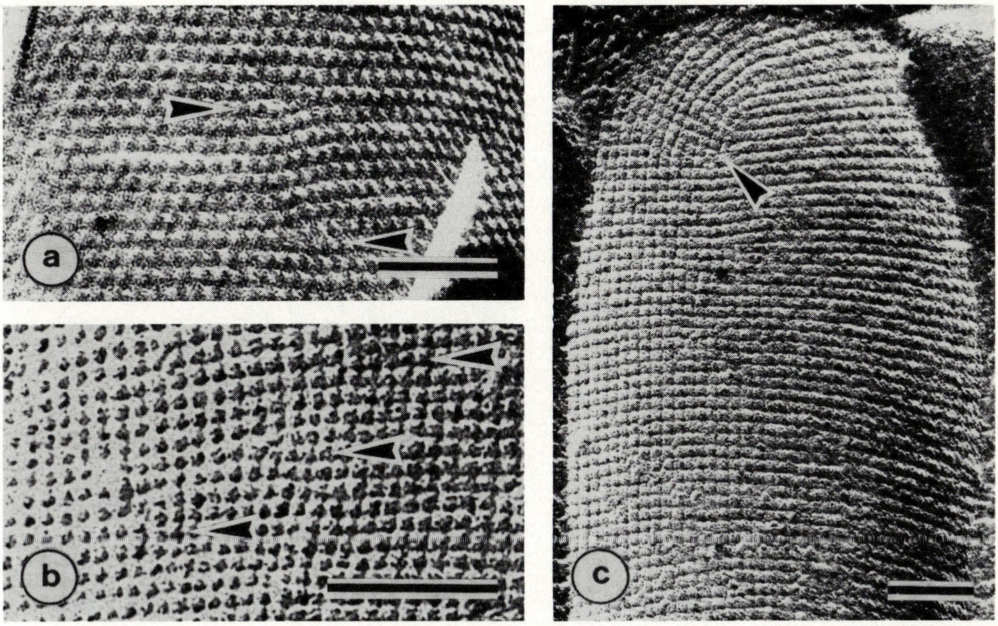

FIGURE 5. Freeze etched preparations of intact bacteria showing common lattice faults of S-layers. (a) Arrows indicate dislocations in the hexagonal lattice of *Clostridium thermohydrosulfuricum* L111-69. (From Sleytr, U. B. and Glauert, A. M., *J. Ultrastruct. Res.*, 50, 103, 1975. With permission.) (b) Dislocations in the square lattice of *Desulfotomaculum nigrificans* NCIB 8706 (arrows). (c) On the cell pole of *Bacillus stearothermophilus* NRS 106/lb$_2$ a local wedge disclination (arrow) can be seen in the square lattice (arrow). Bars = 100 nm.

mechanism of nonconservative climb of dislocations (Figure 5a,b) will depend on the number of dislocations present and the rate of incorporation of new subunits at these sites. Both dislocations (Figure 5a,b) and disclinations (Figure 5c), as well as grain boundaries, could be clearly demonstrated in high resolution freeze etching replicas of S-layers on intact cells (for example see Sleytr,[1] Holt and Leadbetter,[13] Sleytr and Glauert,[14] Kawata et al.,[16] and Nabarro and Harris[74]). Nevertheless, the actual proof for dislocations being sites of intussusceptive growth will require labeling procedures (see section IV. C) at the high resolution where individual rows of subunits can be detected. On the other hand, as discussed earlier, dislocations and disclinations may simply represent faults generated during recrystallization of S-layer crystallites, which are initially randomly incorporated during lattice growth.[1,5,14,28]

C. STUDIES OF THE EXTENSION OF S-LAYER LATTICES ON INTACT CELLS

1. Eubacteria

For localizing incorporation sites of new S-layer subunits during cell growth, both the indirect fluorescent antibody technique[77,78] and the protein A/colloidal gold marker method have been used.[78,79] From the limited number of available data it seems likely that, depending on the type of bacteria (Figure 1), different patterns of S-layer growth have to be considered. For the Gram-positive eubacteria (e.g., *Bacillus sphaericus, Bacillus stearothermophilus*), it was shown by fluorescent antibody techniques (Figure 6a-d) that the major area of S-layer deposition was a band at the sites of an incipient cell division. This band subsequently split and covered the new poles of each progeny cell. These pole areas resemble static lattice domains, since little or no S-layer material is inserted into pre-existing S-layers at the poles. Growth over the remainder of the cylindrical part of the cell occurs by insertions of multiple bands of new S-layer material.[77,78] Protein A/colloidal gold labeling (Figure 7a-c) confirmed

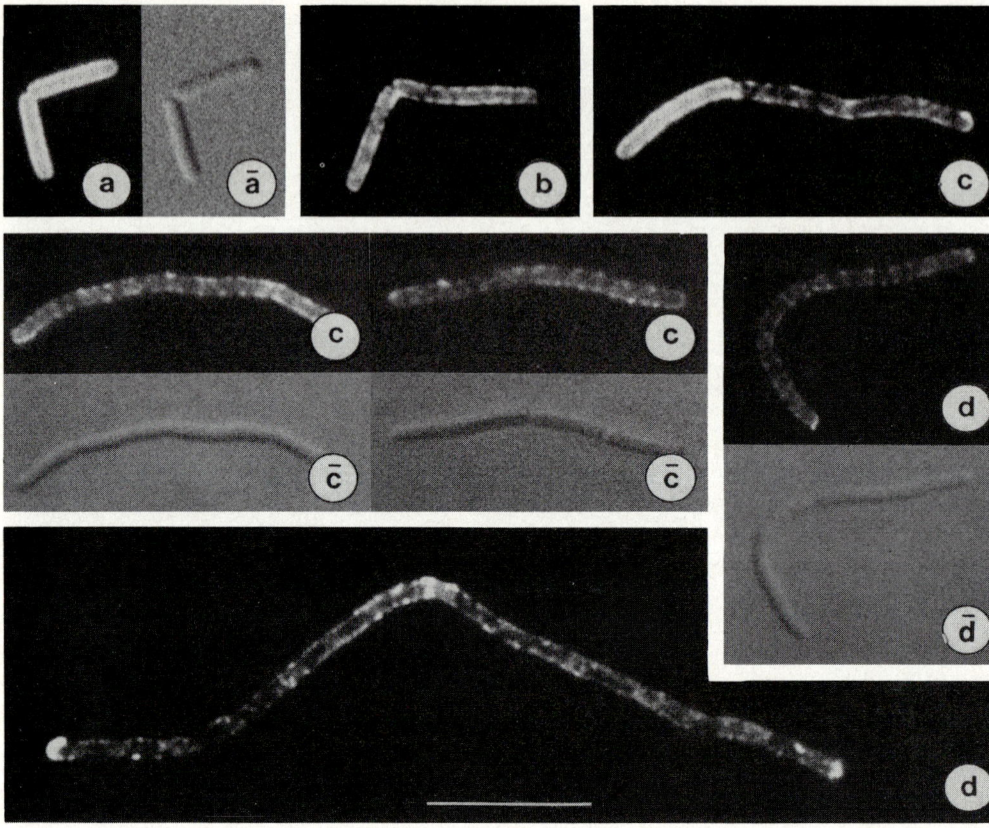

FIGURE 6. Localized insertion of new S-layer during growth of *Bacillus stearothermophilus* PV72. The old S-layer is labeled with indirect fluorescent antibodies at 0 time ($a,\bar{a}$); after 15 min (b); after 30 min ($c,\bar{c}$); after 60 min of growth ($d,\bar{d}$). Fluorescence illumination (a,b,c,d); phase interference contrast ($\bar{a},\bar{c},\bar{d}$); generation time approx. 20 min. Bar = 10 μm. (From Gruber, K. and Sleytr, U. B., *Arch. Microbiol.*, 149, 485, 1988. With permission.)

these results. Due to the higher spatial resolution of this technique, it could be clearly demonstrated that in two strains of *B. stearothermophilus* new S-layer material was deposited in helically arranged bands over the cylindrical surface of the cell at a pitch angle related to the orientation of the lattice vectors of the crystalline array.[78] As illustrated in the schematic drawing (Figure 7d), the geometry and distribution of these growth areas on the cylindrical part is explained by two mechanisms: (1) as discussed above, the entropy driven incorporation of new S-layer subunits could occur along conservative climbing dislocations, and (2) by the tensions within the S-layer lattice generated by the growth of the underlying peptidoglycan sacculus, which can be expected to occur along narrow bands perpendicular to the long axis of the cell.

A somewhat different pattern of surface growth was observed for the stalk-forming Gram-negative *Caulobacter crescentus*.[80] Similar to the Gram-positive bacilli, the region of incipient cell division is covered with new S-layer protein, but over the main cell body the S-layer enlarges by insertion of new protomers at diffuse sites, with no indication of growth zones or bands. Only the stalk region required for cell adhesion revealed entirely new S-layer material.

2. Archaebacteria

Whereas in Gram-positive eubacteria and archaebacteria (Figure 1b), and Gram-negative eubacteria (Figure 1c), the dynamic process of assembly and recrystallization of S-layers

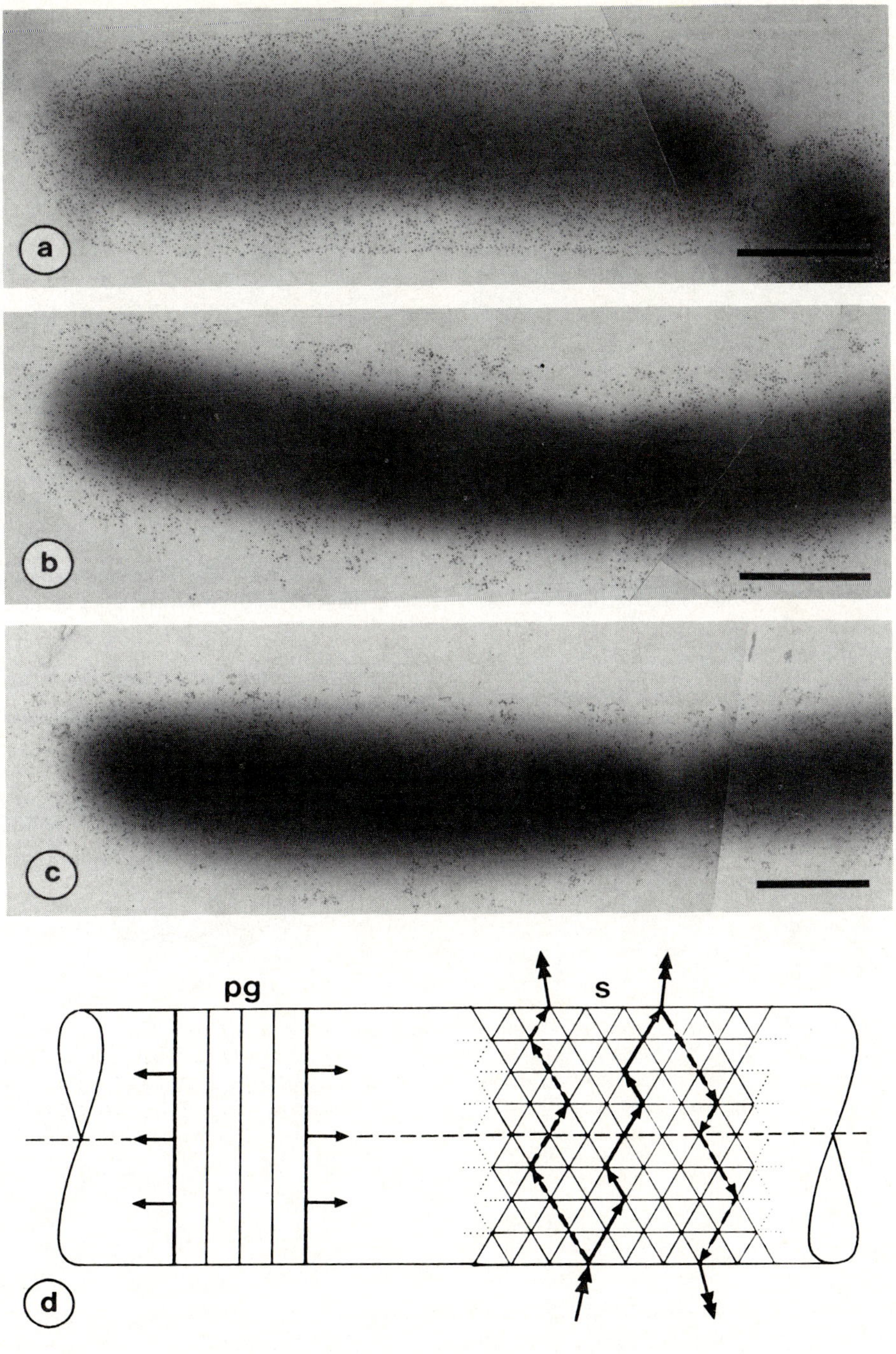

FIGURE 7. Localized insertion of new S-layer during growth of *Bacillus stearothermophilus* PV72. Labeling of the old S-layer with the protein A/colloidal gold technique. At 0 time (a); after 30 min of growth (b); after 60 min of growth (c). Diagrammatic representation (d) of the expansion of the peptidoglycan (pg) and the S-layer (s) of *Bacillus stearothermophilus* PV72. Localized insertion of new peptidoglycan material can be expected to occur in bands perpendicular to the longitudinal axis of the rod. Incorporation of new S-layer subunits can take place at sites of sliding dislocation (single-headed arrows). The pitch of the helical bands of new S-layer at growth-zones (double-headed arrows) is determined by the stress generated in the S-layer lattice by the expanding peptidoglycan. Bars = 0,5 μm. (From Gruber, K. and Sleytr, U. B., *Arch. Microbiol.*, 149, 485, 1988. With permission.)

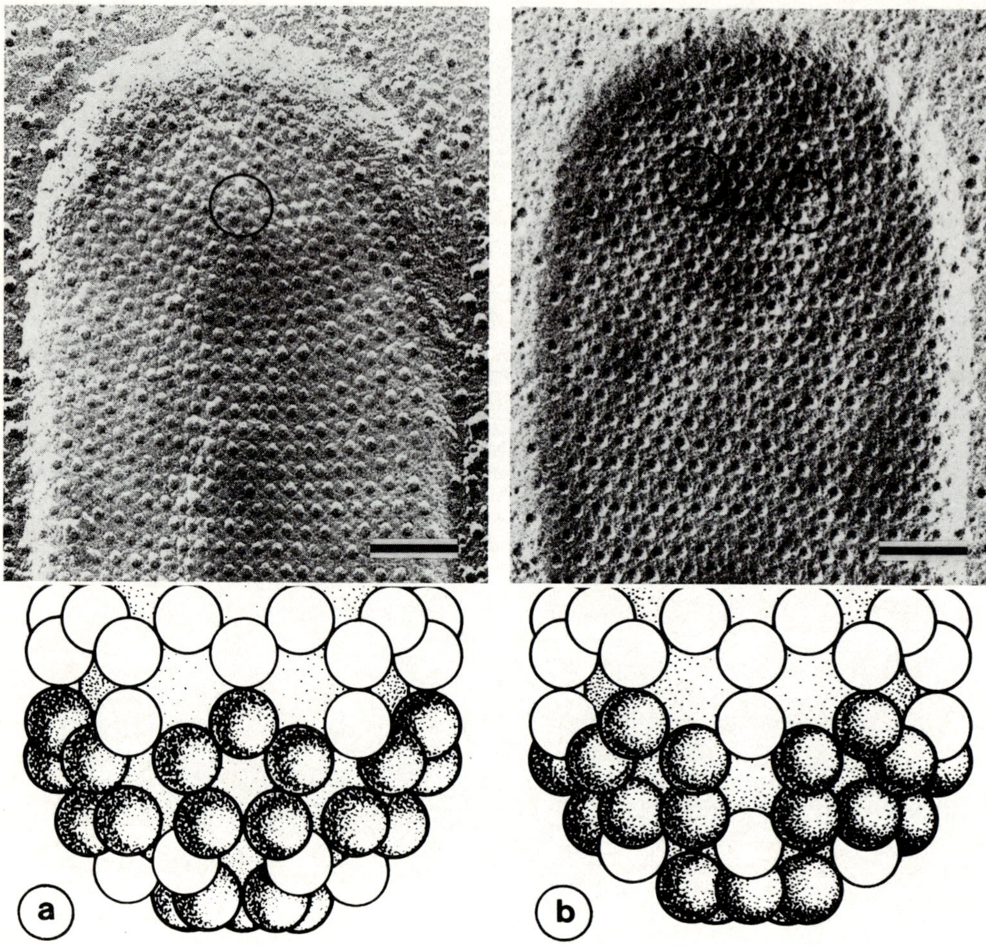

FIGURE 8. Freeze dried and shadowed preparation of envelopes of *Thermoproteus tenax* labeled at pH 5.7 with polycationic ferritin. The marker binds to the extracellular surface of the S-layer in a regular fashion. Depending on the orientation of the cell in the course of adsorption, one (a) or two (b) local wedge disclination ("pentamers") can be seen in the hexagonal lattice (circles). The lower panel demonstrates schematically that local wedge disclinations represent an essential structural element on hemispherical cell poles on a hexagonal lattice. Bars = 100 nm. (From Messner, P., Pum, D. S., Sára, M., Stetter, K. O., and Sleytr, U. B., *J. Bacteriol.*, 166, 1046, 1986. With permission.)

occurs on the surface of a rigid supramolecular envelope structure, in many archaebacteria S-layers represent the exclusive wall component (Figure 1a), and thus can be expected to have the intrinsic potential to fulfill cell shape maintaining and determining functions.[2,81,82]

As discussed earlier, the shape determining potential of a protein container composed of identical subunits is well known from viruses with icosahedral symmetry. Recently, the structural analysis of the extremely thermophilic archaebacterium *Thermoproteus tenax,* for the first time provided strong evidence that S-layers, as exclusive wall components, determine the morphology of a rod shaped cell.[83] The organism has a cylindrical shape with constant diameter but variable in length and hemispherical cell poles. By labeling the S-layer surface with polycationic ferritin, no dislocations were detected on the cylindrical part, but a distinct number of wedge disclinations could be visualized on the hemispherical caps (Figure 8). By evaluating the locations and distances of these wedge disclinations, the presence of 6 pentameric units per cell pole could be calculated, thus resembling the minimum of lattice

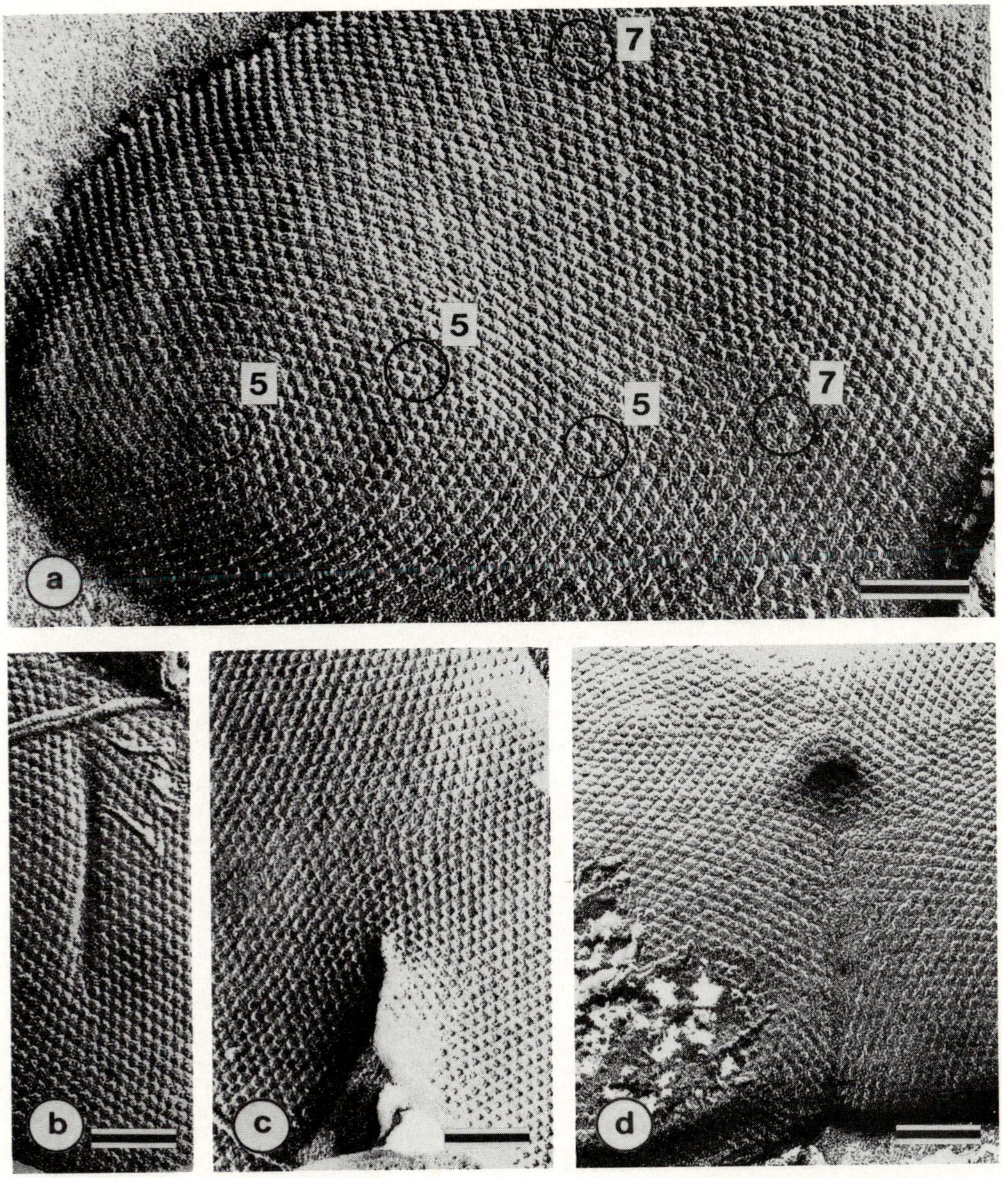

FIGURE 9. Freeze etching preparation of *Methanocorpusculum sinense* illustrating successive states of the division of a closed surface crystal. Numerous dislocations and disclinations (a) can be identified on the hexagonal S-layer lattice (five- and seven-coordinated units are encircled). An early stage (b) and advanced stages (c,d) of invagination. Bars = 100 nm.

faults required for covering the rounded surface. Consequently, elongation of the cylindrical part of the cell would require incorporation of S-layer material only at these lattice faults.[73,74,83,84]

In other archaebacteria with S-layers as exclusive wall component but less uniform morphology, the distribution and number of lattice faults is less well defined. For example, in high resolution freeze etching images of *Methanocorpusculum sinense*, S-layers reveal both dislocations and wedge disclinations (Figure 9a).[85] Upon close examination it can be clearly seen, that by the presence of local wedge disclinations, the original sixfold symmetry has been altered to fivefold and sevenfold, respectively. The necessity of fivefold and sevenfold vertices for producing the hexagonally arranged, highly lobed crystalline cell wall of *Sulfolobus acidocaldarius* has been discussed already by Deatherage et al.[86] Harris also drew attention to the fact that fissioning of cells can be represented by a model in which the essential process is the movement of a $-360°$ disclination on the closed surface crystal.[75,76] The disclination is localized at the end of the resulting fissure, and can be moved by extending the cuts and simultaneously resealing the edges. As schematically illustrated in the classical paper by Harris,[75] by this mechanism a closed surface crystal can be divided in half, leading to two separate cells without breaking the surface. Figures 9b to 9d show successive states of cell septation in *Methanocorpusculum sinense,* which apparently follows this mechanism. We assume that this mode of cell fission eventually will be verified for many other archaebacteria such as *Halobacteria,* thermo-acidophilic, and methano-genic bacteria which possess S-layers as exclusive wall components (for compilation see Sleytr et al.,[3] Sleytr and Messner,[4] and Kandler and König[38]).

Halophilic bacteria which maintain the shape of a flat rectangular box during the whole cell cycle (''square bacteria'') will represent particularly appealing models for studying this ''archaic cell fission process''.[87-89] In some *Halobacteria,* apparently, not only the S-layer protein, but also the covalently linked carbohydrate chains, are involved in determining the shape of the cell, since inhibition of glycosylation by bacitracin caused a change in cell shape from rods to spheres.[82,90]

V. S-LAYERS AND CRYSTALLINE OUTER MEMBRANE PROTEINS

2-D crystalline (regular) arrays of proteins (cOmp, rOmp) have also been detected as component of the outer membrane (Om) of Gram-negative eubacteria.[91-95] From the limited data available, it appears that these assemblies require a lipid bilayer for maintaining their structural integrity. With porins which represent a group of proteins forming channels across the outer membrane (Figure 1c), different crystalline packing arrangements were observed depending on the phospholipid-to-protein ratio.[96] Different from S-layer subunits which assemble into 2-D arrays (see section III. C), isolated porins have shown to form only 3-D crystals.[97] Since both S-layers[98,99] and porins[100-103] have the capability to function as accurate molecular sieves, detailed comparative studies may reveal phylogenetic relationships and common structural and functional principles.[4,95] Moreover, as schematically illustrated in Figure 10, ''intermediate states'' may be detected with respect to the degree of association and molecular interactions between S-layers and the cytoplasmic or outer membrane.

Structural analysis of isolated S-layers of archaebacteria have revealed significant topographical differences of the two surfaces. The inner surface often appears much more sculptured than the outer surface.[9,83] Thus, depending on the amphipathic structure, S-layers may only associate peripherically with the plasma membrane (Figure 10a), or penetrate into the lipid matrix (Figure 10b). In analogy, S-layers observed in Gram-negative eubacteria may only be attached to (Figure 10c), or partially (or even completely) penetrate (Figure 10d) the outer membrane. The latter situation closely resembles porins which generate an aqueous channel through the membrane (Figure 10e).[104] S-layers associated with a plasma membrane (Figure 10a, b) or an outer membrane (Figure 10c, d) can be expected to be involved in specific interactions with structural components of these layers.[6] Proteins are favorite candidates for such interactions, and may even lead to functionally ''cooperative assemblies'' (Figure 10c).

Finally, we would like to draw attention to the fact that in Gram-positive eubacteria and archaebacteria, S-layers have been suggested to be involved in preventing the release of macromolecules (e.g., cell wall associated enzymes).[2,28]

A similar ''periplasmic space'' could be generated by S-layers attached to the cytoplasmic membrane or the outer membrane, provided that the mass distribution of the S-layer protomers resemble those shown in Figures 10 a to 10 c.

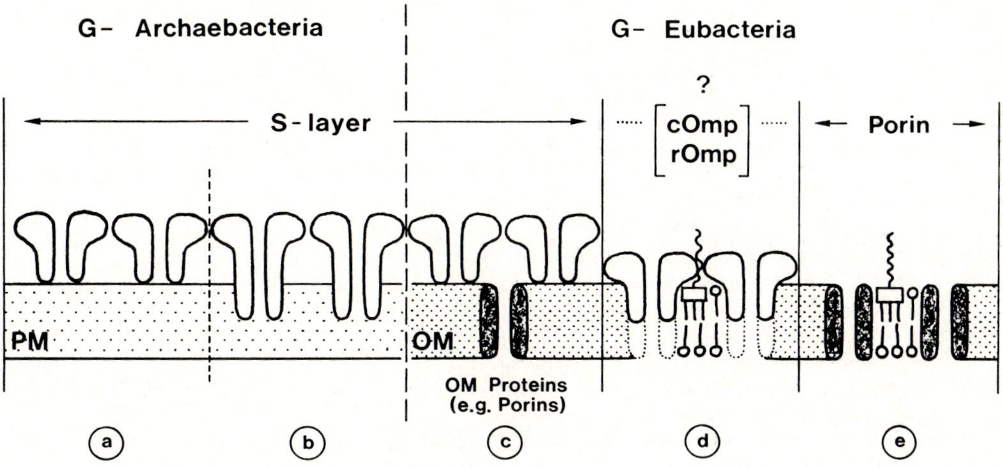

FIGURE 10. Diagrammatic representation of possible interactions of crystalline arrays of proteins with the cytoplasmatic membrane (PM) in Gram-negative archaebacteria (a,b), and the outer membrane (OM) in Gram-negative eubacteria (c,d,e). S layers may be superficially attached to (a,c) or anchored (b) in the lipid bilayer. Crystalline packing arrangements are also observed with porins which form channels across the outer membrane (e). Crystalline (regular) arrays of outer membrane proteins (cOmps, rOmps) (d) may represent structurally and functionally "intermediate states" between S-layers and porins (see also Figures 1a and c).

High resolution studies using advanced electron microscopical techniques, permeability studies, and analysis of sequence homologies will be required to provide a better insight into the functional and evolutionary relationship of crystalline arrays found in prokaryotic cell envelopes.

ACKNOWLEDGMENTS

The preparation of this review was supported, in part, by grants from the Fonds zur Förderung der wissenschaftlichen Forschung in Österreich, the Österreichisches Bundesministerium für Wissenschaft und Forschung and the Ludwig Boltzmann-Gesellschaft in Österreich.

We thank our colleagues at the Zentrum für Ultrastrukturforschung for their contributions incorporated into this review.

REFERENCES

1. **Sleytr, U. B.,** Regular arrays of macromolecules on bacterial cell walls: structure, chemistry, assembly, and function, *Int. Rev. Cytol.*, 53, 1, 1978.
2. **Sleytr, U. B. and Messner, P.,** Crystalline surface layers on bacteria, *Annu. Rev. Microbiol.*, 37, 311, 1983.
3. **Sleytr, U. B., Messner, P., Sára, M., and Pum, D.,** Crystalline envelope layers in archaebacteria, *Syst. Appl. Microbiol.*, 7, 310, 1986.
4. **Sleytr, U. B. and Messner, P.,** Crystalline surface layers on bacteria, in *Crystalline Bacterial Cell Surface Layers,* Sleytr, U. B., Messner, P., Pum, D., and Sára, M., Eds., Springer-Verlag, Berlin, 1988, 160.
5. **Sleytr, U. B. and Messner, P.,** Crystalline surface layers in procaryotes, *J. Bacteriol.*, 170, 2891, 1988.
6. **Smit, J.,** Protein surface layers of bacteria, in *Bacterial Outer Membranes as Model Systems,* Inouye, M., Ed., John Wiley & Sons, New York, 1987, 343.
7. **Koval, S. F. and Murray, R. G. E.,** The superficial protein arrays on bacteria, *Microbiol. Sci.*, 3, 357, 1986.

8. **Beveridge, T. J.,** Ultrastructure, chemistry, and function of the bacterial wall, *Int. Rev. Cytol.*, 72, 229, 1981.
9. **Baumeister, W. and Engelhardt, H.,** Three-dimensional structure of bacterial surface layers, in *Electron Microscopy of Proteins*, Vol. 6, *Membranous Structures*, Harris, J. R. and Horne, R. W., Eds., Academic Press, Orlando, 1987, 109.
10. **Koval, S. F.,** Paracrystalline protein surface arrays on bacteria, *Can. J. Microbiol.*, 34, 407, 1988.
11. **Sleytr, U. B., Messner, P., Pum, D., and Sára, M.,** Eds., *Crystalline Bacterial Cell Surface Layers*, Springer-Verlag, Berlin, 1988.
12. **Sleytr, U. B., Messner, P., and Pum, D.,** Analysis of crystalline bacterial surface layers by freeze-etching, metal shadowing, negative staining, and ultrathin sectioning, *Methods Microbiol.*, 20, 29, 1988.
13. **Holt, S. C. and Leadbetter, E. R.,** Comparative ultrastructure of selected aerobic spore forming bacteria: a freeze-etching study, *Bacteriol. Rev.*, 33, 346, 1969.
14. **Sleytr, U. B. and Glauert, A. M.,** Analysis of regular arrays of subunits on bacterial surfaces; evidence for a dynamic process of assembly, *J. Ultrastruct. Res.*, 50, 103, 1975.
15. **Messner, P., Hollaus, F., and Sleytr, U. B.,** Paracrystalline cell wall surface layers of different *Bacillus stearothermophilus* strains, *Int. J. Syst. Bacteriol.*, 34, 202, 1984.
16. **Kawata, T., Masuda, K., Yoshino, K., and Fujimoto, M.,** Regular array in the cell wall of *Lactobacillus fermenti* as revealed by freeze-etching and negative staining, *Jpn. J. Microbiol.*, 18, 469, 1974.
17. **Sleytr, U. B. and Glauert, A. M.,** Ultrastructure of the cell walls of two closely related *Clostridia* that possess different regular arrays of surface subunits, *J. Bacteriol.*, 126, 869, 1976.
18. **Masuda, K. and Kawata, T.,** Ultrastructure and partial characterization of a regular array in the cell wall of *Lactobacillus brevis*, *Microbiol. Immunol.*, 23, 941, 1979.
19. **Messner, P., Pum, D., and Sleytr, U. B.,** Characterization of the ultrastructure and the self-assembly of the surface layer of *Bacillus stearothermophilus* strain NRS 2004/3a, *J. Ultrastruct. Mol. Struct. Res.*, 97, 73, 1986.
20. **Stewart, M., Beveridge, T. J., and Trust, T. J.,** Two patterns in the *Aeromonas salmonicida* A-layer may reflect a structural transformation that alters permeability, *J. Bacteriol.*, 166, 120, 1986.
21. **Koval, S. F. and Murray, R. G. E.,** The isolation of surface array proteins from bacteria, *Can. J. Biochem. Cell Biol.*, 62, 1181, 1984.
22. **Messner, P. and Sleytr, U. B.,** Separation and purification of S-layers from Gram-positive and Gram-negative bacteria, in *Bacterial Cell Surface Techniques*, Hancock, I. and Poxton, I., Eds., John Wiley & Sons, Chichester, 1988, 97.
23. **Beveridge, T. J. and Murray, R. G. E.,** Dependence of the superficial layers of *Spirillum putridiconchylium* on Ca^{2+} or Sr^{2+}, *Can. J. Microbiol.*, 22, 1233, 1976.
24. **Thorne, K. J. I.,** Regularly arranged protein on the surface of Gram-negative bacteria, *Biol. Rev.*, 52, 219, 1977.
25. **Koval, S. F. and Murray, R. G. E.,** Solubilization of the surface protein of *Aquaspirillum serpens* by chaotropic agents, *Can. J. Microbiol.*, 29, 146, 1983.
26. **Bingle, W. H., Doran, J. L., and Page, W. J.,** Characterization of the surface layer protein from *Azotobacter vinelandii*, *Can. J. Microbiol.*, 32, 112, 1986.
27. **Masuda, K. and Kawata, T.,** Reassembly of the regularly arranged subunits in the cell wall of *Lactobacillus brevis* and their reattachment to cell walls, *Microbiol. Immunol.*, 24, 299, 1980.
28. **Sleytr, U. B.,** Morphopoietic and functional aspects of regular protein membranes present on prokaryotic cell walls, in *Cytomorphogenesis in Plants, Cell Biology Monographs*, Vol. 8, Kiermayer, O., Ed., Springer-Verlag, Vienna, 1981, 3.
29. **König, H. and Stetter, K. O.,** Studies on archaebacterial S-layers, *Syst. Appl. Microbiol.*, 7, 300, 1986.
30. **Beveridge, T. J., Stewart, M., Doyle, R. J., and Sprott, G. D.,** Unusual stability of the *Methanospirillum hungatei* sheath, *J. Bacteriol.*, 162, 728, 1985.
31. **Sprott, G. D., Beveridge, T. J., Patel, G. B., and Ferrante, G.,** Sheath disassembly in *Methanospirillum hungatei* strain GP1, *Can. J. Microbiol.*, 32, 847, 1986.
32. **Christian, R., Schulz, G., Unger, F. M., Messner, P., Küpcü, Z., and Sleytr, U. B.,** Structure of a rhamnan from the surface layer glycoprotein of *Bacillus stearothermophilus* strain NRS 2004/3a, *Carbohydr. Res.*, 150, 265, 1986.
33. **Messner, P., Sleytr, U. B., Christian, R., Schulz, G., and Unger, F. M.,** Isolation and structure determination of a diacetamidodideoxyuronic acid-containing glycan chain from the S-layer glycoprotein of *Bacillus stearothermophilus* NRS 2004/3a, *Carbohydr. Res.*, 168, 211, 1987.
34. **Messner, P. and Sleytr, U. B.,** Asparaginyl-rhamnose: a novel type of protein-carbohydrate linkage in a eubacterial surface-layer glycoprotein, *FEBS Lett.*, 228, 317, 1988.
35. **Messner, P. and Sleytr, U. B.,** Evidence for the glycoprotein nature of eubacterial S-layers, in *Crystalline Bacterial Cell Surface Layers*, Sleytr, U. B., Messner, P., Pum, D., and Sára, M., Eds., Springer-Verlag, Berlin, 1988, 11.

36. **Mescher, M. F. and Strominger, J. L.,** Purification and characterization of a prokaryotic glycoprotein from the cell envelope of *Halobacterium salinarium, J. Biol. Chem.*, 251, 2005, 1976.
37. **Wieland, F., Heitzer, R., and Schaefer, W.,** Asparaginylglucose: novel type of carbohydrate linkage, *Proc. Natl. Acad. Sci. U.S.A.*, 80, 5470, 1983.
38. **Kandler, O. and König, H.,** Cell envelopes of archaebacteria, in *The Bacteria,* Vol. 8, *Archaebacteria,* Woese, C. R. and Wolfe, R. S., Eds., Academic Press, Orlando, 1985, 413.
39. **Paul, G., Lottspeich, F., and Wieland, F.,** Asparaginyl-N-acetylgalactosamine. Linkage unit of halobacterial glycosaminoglycan, *J. Biol. Chem.*, 261, 1020, 1986.
40. **Sleytr, U. B., Sára, M., Küpcü, Z., and Messner, P.,** Structural and chemical characterization of S-layers of selected strains of *Bacillus stearothermophilus* and *Desulfotomaculum nigrificans, Arch. Microbiol.*, 146, 19, 1986.
41. **Baumeister, W., Karrenberg, F., Rachel, R., Engel, A., ten Heggeler, B., and Saxton, W. O.,** The major cell envelope protein of *Micrococcus radiodurans* (R1). Structural and chemical characterization, *Eur. J. Biochem.*, 125, 535, 1982.
42. **Phipps, B. M., Trust, T. J., Ishiguro, E. E., and Kay, W. W.,** Purification and characterization of the cell surface virulent A protein from *Aeromonas salmonicida, Biochemistry,* 22, 2934, 1983.
43. **Lapchine, L.,** Regularly arranged structures on the surface of some *Pseudomonas sp., FEMS Microbiol. Lett.*, 5, 223, 1979.
44. **Word, N. S., Yousten, A. A., and Howard, L.,** Regularly structured and non-regularly structured surface layers of *Bacillus sphaericus, FEMS Microbiol. Lett,* 17, 277, 1983.
45. **Kay, W. W., Phipps, B. M., Ishiguro, E. E., Olafson, R. W., and Trust, T. J.,** Surface layer virulence A-protein from *Aeromonas salmonicida* strains, *Can. J. Biochem. Cell Biol.*, 62, 1064, 1984.
46. **Smit, J. and Agabian, N.,** Cloning of the major protein of the *Caulobacter crescentus* periodic surface layer: detection and characterization of the cloned peptide by protein expression assays, *J. Bacteriol.*, 160, 1137, 1984.
47. **Belland, R. J. and Trust, T. J.,** Cloning of the gene for the surface array protein of *Aeromonas salmonicida* and evidence linking loss of expression with genetic deletion, *J. Bacteriol.*, 169, 4086, 1987.
48. **Lechner, J. and Sumper, M.,** The primary structure of a procaryotic glycoprotein. Cloning and sequencing of the cell surface glycoprotein gene of halobacteria, *J. Biol. Chem.*, 262, 9724, 1987.
49. **Peters, J., Peters, M., Lottspeich, F., Schäfer, W., and Baumeister, W.,** Nucleotide sequence analysis of the gene encoding the *Deinococcus radiodurans* surface protein, derived amino acid sequence, and complementary protein chemical studies, *J. Bacteriol.*, 169, 5216, 1987.
50. **Yamagata, H., Adachi, T., Tsuboi, A., Takao, M., Sasaki, T., Tsukagoshi, N., and Udaka, S.,** Cloning and characterization of the 5′ region of the cell wall protein gene operon in *Bacillus brevis* 47, *J. Bacteriol.*, 169, 1239, 1987.
51. **Jaenicke, R., Welsch, R., Sára, M., and Sleytr, U. B.,** Stability and self-assembly of the S-layer protein of the cell wall of *Bacillus stearothermophilus, Biol. Chem. Hoppe-Seyler,* 366, 663, 1985.
52. **Koval, S. F. and Murray, R. G. E.,** Effect of calcium on the *in vivo* assembly of the surface protein of *Aquaspirillum serpens* VHA, *Can. J. Microbiol.*, 31, 261, 1984.
53. **Sára, M. and Sleytr, U. B.,** Charge distribution on the S-layer of *Bacillus stearothermophilus* NRS 1536/3c and importance of charged groups for morphogenesis and function, *J. Bacteriol.*, 169, 2804, 1987.
54. **Sleytr, U. B.,** Self-assembly of the hexagonally and tetragonally arranged subunits of bacterial surface layers and their reattachment to cell walls, *J. Ultrastruct. Res.*, 55, 360, 1976.
55. **Aebi, U., Smit, P. R., Dubochet, J., Henry, C., and Kellenberger, E.,** A study of the structure of the T-layer of *Bacillus brevis, J. Supramol. Struct.*, 1, 498, 1973.
56. **Hastie, A. T. and Brinton, C. C.,** Isolation, characterization, and *in vitro* assembly of the tetragonally arrayed layer of *Bacillus sphaericus, J. Bacteriol.*, 138, 999, 1979.
57. **Hastie, A. T. and Brinton, C. C.,** Specific interaction of the tetragonally arrayed protein layer of *Bacillus sphaericus* with its peptidoglycan sacculus, *J. Bacteriol.*, 138, 1010, 1979.
58. **Masuda, K. and Kawata, T.,** Reassembly of a regularly arranged protein in the cell wall of *Lactobacillus buchneri* and its reattachment to cell walls: chemical modification studies, *Microbiol. Immunol.*, 29, 927, 1985.
59. **Beveridge, T. J. and Murray, R. G. E.,** Reassembly *in vitro* of the superficial cell wall components of *Spirillum putridiconchylium, J. Ultrastruct. Res.*, 55, 105, 1976.
60. **Buckmire, F. L. A. and Murray, R. G. E.,** Substructure and *in vitro* assembly of the outer, structured layer of *Spirillum serpens, J. Bacteriol.*, 125, 290, 1976.
61. **Tsukagoshi, N., Yamada, H., Tsuboi, A., and Udaka, S.,** Effects of phosphate in medium on protein secretion in a protein-producing bacterium, *Bacillus brevis* 47, *Appl. Environ. Microbiol.*, 42, 370, 1981.
62. **Bayer, M. E.,** Ultrastructure and organization of the bacterial envelope, *Ann. N.Y. Acad. Sci.*, 235, 6, 1974.
63. **Smit, J. and Nikaido, H.,** Outer membrane of Gram-negative bacteria. XVIII. Electron microscopic studies on porin insertion sites and growth of cell surface of *Salmonella typhimurium, J. Bacteriol.*, 135, 687, 1978.

64. **Belland, R. J. and Trust, T. J.,** Synthesis, export, and assembly of *Aeromonas salmonicida* A-layer analyzed by transposon mutagenesis, *J. Bacteriol.*, 163, 877, 1985.
65. **Masuda, K. and Kawata, T.,** Characterization of a regular array in the wall of *Lactobacillus buchneri* and its reattachment to the other wall components, *J. Gen. Microbiol.*, 124, 1, 1981.
66. **Tsuboi, A., Tsukagoshi, N., and Udaka, S.,** Reassembly *in vitro* of hexagonal surface arrays in a protein-producing bacterium, *Bacillus brevis* 47, *J. Bacteriol.*, 151, 1485, 1982.
67. **Stewart, M. and Murray, R. G. E.,** Structure of the regular surface layer of *Aquaspirillum serpens* MW5, *J. Bacteriol.*, 150, 348, 1982.
68. **Kist, M. L. and Murray, R. G. E.,** Components of the regular surface array of *Aquaspirillum serpens* MW5 and their assembly *in vitro, J. Bacteriol.*, 157, 599, 1984.
69. **Stewart, M.,** Computer analysis of ordered microbiological objects, in *Ultrastructure Techniques for Microorganisms,* Aldrich, H. C. and Todd, W. J., Eds., Plenum Press, New York, 1986, 333.
70. **Watson, S. W. and Remsen, C. C.,** Cell envelope of *Nitrosocystis oceanus, J. Ultrastruct. Res.*, 33, 148, 1970.
71. **Caspar, D. L. D. and Klug, A.,** Physical principles in the construction of regular viruses, *Cold Spring Harbor Symp. Quant. Biol.*, 27, 1, 1962.
72. **Caspar, D. L. D.,** Design principles in organized biological structures, in *Principles of Biomolecular Organization,* Wolstenholme, G. E. W. and O'Connor, M., Eds., J. & A. Churchill, London, 1966, 7.
73. **Harris, W. F. and Scriven, L. E.,** Function of dislocations in cell walls and membranes, *Nature,* 228, 827, 1970.
74. **Nabarro, F. R. N. and Harris, W. F.,** Presence and function of disclinations in surface coats of unicellular organisms, *Nature,* 232, 423, 1971.
75. **Harris, W. F.,** Disclinations, *Sci. Am.*, 237, 130, 1977.
76. **Harris, W. F.,** Dislocations, disclinations and dispirations: distractions in very naughty crystals, *S. Afr. J. Sci.*, 74, 332, 1978.
77. **Howard, L. V., Dalton, D. D., and McCoubrey, W. K.,** Expansion of the tetragonally arrayed cell wall protein layer during growth of *Bacillus sphaericus, J. Bacteriol.*, 149, 748, 1982.
78. **Gruber, K. and Sleytr, U. B.,** Localized insertion of new S-layer during growth of *Bacillus stearothermophilus* strains, *Arch. Microbiol.*, 149, 485, 1988.
79. **Smit, J. and Todd, W. J.,** Colloidal gold labels for immunocytochemical analysis of microbes, in *Ultrastructure Techniques for Microorganisms,* Aldrich, H. C. and Todd, W. J., Eds., Plenum Press, New York, 1986, 469.
80. **Smit, J. and Agabian, N.,** Cell surface patterning and morphogenesis: biogenesis of a periodic surface array during *Caulobacter* development, *J. Cell Biol.*, 95, 41, 1982.
81. **Mescher, M. F. and Strominger, J. L.,** Structural (shape-maintaining) role of the cell surface glycoprotein of *Halobacterium salinarium, Proc. Natl. Acad. Sci. U.S.A.*, 73, 2687, 1976.
82. **Wieland, F., Lechner, J., and Sumper, M.,** The cell wall glycoprotein of *Halobacteria:* structural, functional and biosynthetic aspects, *Zbl. Bakt. Hyg., I. Abt. Orig.*, C3, 161, 1982.
83. **Messner, P., Pum, D., Sára, M., Stetter, K. O., and Sleytr, U. B.,** Ultrastructure of the cell envelope of the archaebacteria *Thermoproteus tenax* and *Thermoproteus neutrophilus, J. Bacteriol.*, 166, 1046, 1986.
84. **Wildhaber, I. and Baumeister, W.,** The cell envelope of *Thermoproteus tenax:* three-dimensional structure of the surface layer and its role in shape maintenance. *EMBO J.*, 6, 1475, 1987.
85. **Zellner, G., Stackebrandt, E., Messner, P., Tindall, B. J., Conway de Macario, E., Kneifel, H., Sleytr, U. B., and Winter, J.,** Isolation and characterization of *Methanocorpusculum sinense* spec. nov., and *Methanocorpusculum bavaricum* spec. nov., and installation of a novel family of methanogens, the *Methanocorpusculaceae, Arch. Microbiol.*, in press.
86. **Deatherage, J. F., Taylor, K. A., and Amos, L. A.,** Three-dimensional arrangement of the cell wall protein of *Sulfolobus acidocaldarius, J. Mol. Biol.*, 167, 823, 1983.
87. **Walsby, A. E.,** A square bacterium, *Nature,* 283, 69, 1980.
88. **Stockenius, W.,** Walsby's square bacterium: fine structure of an orthogonal procaryote, *J. Bacteriol.*, 148, 352, 1981.
89. **Kessel, M. and Cohen, Y.,** Ultrastructure of square bacteria from a brine pool in Southern Sinai, *J. Bacteriol.*, 150, 851, 1982.
90. **Mescher, M. F. and Strominger, J. L.,** Bacitracin induces sphere formation in *Halobacterium* species which lack a wall peptidoglycan, *J. Gen. Microbiol.*, 89, 375, 1975.
91. **Chang, J. -J., Leonard, K., Arad, T., Pitt, T., Zhang, Y. - X., and Zhang, L. -H.,** Structural studies on the outer envelope of *Chlamydia trachomatis* by electron microscopy, *J. Mol. Biol.*, 161, 579, 1982.
92. **Kerosuo, E., Haapasalo, M., Ranta, H., and Lounatmaa, K.,** Hexagonal periodicity in the outer membrane of *Bacteroides buccae, J. Gen. Microbiol.*, 133, 2217, 1987.
93. **Gerbl-Rieger, S., Schneider, S., Engelhardt, H., Peters, J., and Baumeister, W.,** Characterization of the surface protein of *Comamonas acidovorans,* in *Crystalline Bacterial Cell Surface Layers,* Sleytr, U. B., Messner, P., Pum, D., and Sára, M., Eds., Springer-Verlag, Berlin, 1988, 65.

94. **Ranta, H., Haapasalo, M., Kerosuo, E., and Lounatmaa, K.,** Occurrence and characterization of S-layers in oral bacteria, in *Crystalline Bacterial Cell Surface Layers,* Sleytr, U. B., Messner, P., Pum, D., and Sára, M., Eds., Springer-Verlag, Berlin, 1988, 118.
95. **Baumeister, W., Wildhaber, I., and Engelhardt, H.,** Bacterial surface proteins: Some structural, functional and evolutionary aspects, *Biophys. Chem.,* 29, 39, 1988.
96. **Dorset, D. L., Engel, A., Häner, M., Massalski, A., and Rosenbusch, J. P.,** Two-dimensional crystal packing of matrix porin, *J. Mol. Biol.,* 165, 701, 1983.
97. **Garavito, R. M., Hinz, U., and Neuhaus, J. -M.,** The crystallization of outer membrane proteins from *Escherichia coli, J. Biol. Chem.,* 259, 4254, 1984.
98. **Sleytr, U. B. and Sára, M.,** Ultrafiltration membranes with uniform pores from crystalline bacterial cell envelope layers, *Appl. Microbiol. Biotechnol.,* 25, 83, 1986.
99. **Sára, M. and Sleytr, U. B.,** Molecular sieving through S-layers of *Bacillus stearothermophilus* strains, *J. Bacteriol.,* 169, 4092, 1987.
100. **Schindler, H. and Rosenbusch, J. P.,** Matrix protein from *Escherichia coli* outer membranes forms voltage-controlled channels in lipid bilayers, *Proc. Natl. Acad. Sci. U.S.A.,* 75, 3751, 1978.
101. **Benz, R.,** Structure and selectivity of porin channels, *Curr. Top. Membr. Transp.,* 21, 199, 1984.
102. **Engel, A., Massalski, A., Schindler, H., Dorset, D. L., and Rosenbusch, J. P.,** Porin channel triplet merge into single outlets in *Escherichia coli* outer membranes, *Nature,* 317, 643, 1985.
103. **Hancock, R. E. W.,** Model membrane studies of porin function, in *Bacterial Outer Membranes as Model Systems,* Inouye, M., Ed., John Wiley & Sons, New York, 1987, 187.
104. **Singer, S. J., Maher, P. A., and Yaffe, M. P.,** On the translocation of proteins across membranes, *Proc. Natl. Acad. Sci. U.S.A.,* 84, 1015, 1987.

Chapter 3

GAP JUNCTIONS: CORRELATED ELECTROPHYSIOLOGICAL RECORDINGS AND ULTRASTRUCTURAL ANALYSIS BY FAST FREEZING AND FREEZE-FRACTURING

Dieter F. Hülser, Dietmar Paschke and Joachim Greule

TABLE OF CONTENTS

I. INTRODUCTION

The evolution of complex organisms depends on the appropriate coordination of growth, differentiation, and function of the participating cells. Without intercellular communication, neither plants nor animals would have developed. In animal cells the exchange of information is maintained by three different mechanisms:

1. Indirect signal transfer, when signal substances such as neurotransmitters, hormones, growth factors or humoral antibodies are released into extracellular fluids and react with congruous receptors located in the plasma membranes of target cells. This signal is translocated across the membrane into the cytoplasm where second messengers transmit the information.
2. Direct signal transfer from cell to cell by plasma membrane bound receptors, as is the case in cell-mediated immune response or during sperm-egg-binding.
3. Direct signal transfer from cell to cell by proteinaceous channels, which provide hydrophilic paths across the plasma membranes of adjacent cells. These so-called gap junctions are bidirectionally permeable for both charged and neutral molecules, and also connect adjacent cells mechanically.

A communicating gap junction channel consists of two hemichannels, the connexons, to which both contacting cells contribute.[1,2] Each connexon can be subdivided into six subunits (for review see Revel et al.[3]). Up to several hundred channels are assembled to form a typical gap junction plaque (for review see Bennett and Goodenough[4]); the channel density in the plasma membrane may reflect their activity.[5-13] Gap junctions are ubiquitously found in the fauna, from mesozoa to vertebrates. Not only regulation of embryonic development, cell differentiation, and growth control depend on the existence of gap junctions, but also the synchronous beating of heart muscle cells and the coordinated contractions of smooth muscle cells in the intestine (for review see Pitts and Finbow[14]).

Gap junctions are very often established in permanently growing cell cultures, which adhere to the surface of plastic Petri dishes. These monolayer cultures facilitate investigations of gap junction properties, since individual cells can easily be discriminated under light-microscopical observation. They led to the detection of metabolic cooperation, which allows cells with certain enzymatic defects to proliferate without a substitution for the missing enzyme or its product, as long as they are coupled to wild type cells by gap junctions.[15] Using fluorescent dyes of different sizes, it has been detected that the upper limit for molecules which can pass intercellularly via gap junctions in vertebrate cells is about M_r = 900.[16] Since, in most cases, the intercellular signal which is exchanged between the cells is not known, artificial signals are used for the demonstration of open gap junctions. A favorite tracer is the fluorescent dye Lucifer Yellow, which can be injected iontophoretically or by pressure into a cell, and its spreading into adjacent cells can be followed under an epifluorescent microscope.[17] If Lucifer Yellow is retained in the injected cell, this block of dye coupling does not necessarily indicate a complete closure of gap junction channels, since ionic coupling still may be observed.[18,19] Thus, the gap junctional conductivity is not a constant value, but is part of the dynamic regulation processes in cells.

The opening and closing of gap junction channels is influenced by two endogenous signal pathways, using as second messengers cAMP or diacylglycerol.[20,21] The cAMP signal pathway opens gap junction channels, whereas the diacylglycerol signal pathway down-regulates the gap junction permeability. Besides a possible regulation by phosphorylation, the fine tuning may well be different in different cells, since many other substances with regulatory effects have been described.[22] However, not only physiological interactions by pH, Ca^{++} and retinoic acid, but also nonphysiological reactions by heptanol, benzhydrol, or glutardialdehyde cause a closing of gap junction.

The opening and closing characteristics of a single gap junction channel can be followed by patch clamp measurements, where the current across a channel can be determined. Depending on the type of experiment ("whole cell" recording with a cell pair, or "cell attached" measurements), only open and closed states have been determined,[23-25] or several states have been described for connexons.[26,27] In the latter case, the final proof for a gap junction channel—the test with blocking antibodies— is still missing.

Gap junctions can be demonstrated in the electron microscope after conventional fixation and staining procedures. On ultrathin sections of embedded cells they are characterized by a 2 to 3 nm wide gap between two contacting membranes, which is bridged by regularly arranged particles. When using lanthanum for membrane impregnation, the gap between the junctional proteins is filled with this electron dense material, and these proteins appear brighter than the lanthanum-filled space.[2] Larger areas of gap junction plaques can be visualized when cells are freeze-fractured and their membranes are replicated. Both the embedding procedure and the freeze-fracture techniques are conventionally performed with chemically fixed cells. In most cases, glutardialdehyde is used as a fixative, which, however, causes always irreversible uncoupling of gap junction channels.[9,12,28] It is, therefore, impossible to demonstrate open gap junction channels in such pretreated cells. The question remains, whether the *in situ* arrangement of gap junction plaques— which may consist of several hundred or thousand individual gap junction channels— is unaffected, or if the pattern of these plaques is also altered under these conditions.

Rapid freezing of biological material is considered as a method which may overcome the artifacts caused by chemical fixation and infiltration with glycerin. According to Moor[29] a freezing rate of 10^4 Ks^{-1} is necessary for the preservation of a structure in its native state. This vitrification, therefore, should be a preferential method when gap junction structures are to be investigated. Unfortunately, biological material has a poor temperature conductivity and—when cryoprotectives or high pressure cryofixation should be avoided—it can be vitrified only within a thickness of about 10 to 30 μm,[30,31] which depends on the preparation techniques (for review see Plattner and Zingsheim[32]). It is obvious that the biological probe must be very small and that a cryogen must be used which is characterized by high freezing rates.

II. MATERIALS AND METHODS

A. CELL CULTURES

Two permanently growing tumor cell lines have been used in this study; BICR/M1R_k, a neoplastic cell line in monolayer culture derived from a spontaneous mammary tumor of the Marshall rat,[33] and BT5C1, a neoplastic neurogenic cell line derived from cultures of fetal brain cells of BD IX rats.[34] These cells were transferred into monolayer culture at 90 h after an i.v. transplacental pulse of 75 μg of N-ethyl-N-nitrosourea per g body weight of the pregnant female (18th day of gestation). After an average period of 202 d, these cultures underwent malignant transformation.[34]

Cells were cultured in plastic flasks (Falcon 3024, Falcon Plastics, Oxnard, CA,) at 37°C, using modified[35] Eagle-Dulbecco medium supplemented with 10% calf serum, under an atmosphere of 5% CO_2 in air. The medium was renewed at 2- to 3-d intervals, and cultures were passaged at confluency by trypsinization [0.25% trypsin in Ca^{2+}-Mg^{2+}-free isotonic phosphate-buffered saline (PBS; pH 7.2 to 7.4)].

For the electrophysiological experiments, cells were grown in plastic Petri dishes (Falcon 3002, 60 × 15 mm), and fused by treating almost confluent monolayers with 40% (w/w) polyethylene glycol (PEG, molecular weight 1540, Koch Light Ltd., Haverhill, Suffolk, U.K.) for about 2 min at room temperature, followed by careful washing in medium and incubation at 37°C for about 4 h.[36] The homokaryons were trypsinized and transferred to

HEPES-buffered medium. Pairs of small homokaryons (about 5 nuclei/cell) were selected for the uncoupling experiments with glutardialdehyde. For the electron microscope preparations the cells were grown in Optical Tissue Culture Film Liners with Holders and Lids (Falcon 3006, 60 × 15 mm), or on plastic coverslips (Thermanox, Lux 5408, Miles Laboratories Inc., Naperville, IL,) in plastic Petri dishes (Falcon 3002).

B. ELECTROPHYSIOLOGY

Electrical coupling was determined by measuring the intercellular electrical signal transfer across the gap junctions with three glass microelectrodes.[37] They were pulled out of glass capillaries (outer diameter 1 mm, with inner filament, Hilgenberg Glas, Malsfeld, F.R.G.) in a vertical pipette puller (700 C David Kopf Instruments, Tujunga, CA, U.S.A.), and backfilled with 3 *M* KCl. A high-impedance amplifer (WPI 750, World Precision Instruments, New Haven, CT,) was used to measure the membrane potentials in two cells, and into one of them rectangular current pulses (generated by a GX 239 pulse generator, ITT-Metrix, Paris, France) were injected via a WPI 701 amplifier. These signal pulses V_1 were superimposed onto the cell's membrane potential and—when the cells were electrically coupled—signal pulses V_2 could be recorded in the neighboring cell, indicating the extent of coupling by the ratio V_2/V_1. These measurements were registered together with the membrane potentials pd_1 and pd_2 on a strip-chart recorder (Servokass 600, Gebr. Laumann, Selb, F.R.G.). In addition, this method also allows the determination of the cellular input resistance, which is the effective transmembrane resistance seen by the membrane potential recording electrode. The coupling capacity of the cells was also determined by iontophoretic injection of the fluorescent dye Lucifer Yellow (Sigma L-0259, St. Louis, MO; 4% in LiCl) into one cell, and observation of the dye spreading into the coupled cells.[17] These measurements were performed with electrically driven micromanipulators (Märzhäuser, Wetzlar, F.R.G.) under inverted microscopes (Epivert, Leitz, Wetzlar, F.R.G., or IM 35, Zeiss, Oberkochen, F.R.G.).

Patch clamp experiments[38] were carried out for the characterization of the connexon conductivity,[26] as well as for the measurement of intercellular coupling in isolated cell pairs.[23-25] Patch pipettes were pulled out of soft glass (Micro-Haematocrit-tubes No. 564, Bardram, Birkerod, Denmark) with a vertical electrode puller (L/M-3P-A, List, Darmstadt, F.R.G.), and filled with a patch solution (120 m*M* KCl, 5 m*M* EGTA, 10 m*M* HEPES, adjusted to pH 7.3 with KOH). Cell attached- as well as whole cell-recordings were measured with patch clamp amplifers (L/M-EPC 7, List). Observed channel activities were digitized with a pulse code modulator (PCM-501 ES, Sony, Japan), stored with a video recorder (Ultravideo 6A10 VHS, Saba, Villingen-Schwenningen, F.R.G.), and analyzed by an IBM-PC AT compatible computer with a program designed according to McCann et al.[39] using Asyst as a scientific software (Asyst, Keithley Instruments GmbH, München, F.R.G.).

C. ELECTRON MICROSCOPY

In situ freeze fracturing of cell monolayers was performed by a procedure modified after Pauli et al.,[40] including some advice given by Dahl[41] and Dermietzel.[42] Eighty ml PBS (pH 7.45) and 20 ml of 100% glycerol (Merck, Nr. 4093, Darmstadt, F.R.G.) were mixed at room temperature, and under continuous stirring, 20 g polyvinyl alcohol (15,000, Fluka, Buchs, Switzerland) were added slowly. The stirring was continued over night at room temperature till the mixture was centrifuged at 27,000 g for 30 min. The supernatant dimethylsulfoxide (DMSO, Serva, Nr. 20385, Heidelberg, F.R.G.) was added to a final concentration of 2%; fungal growth was prevented by a crystal of thymol. This polyvinyl alcohol solution was heated to 90°C for 40 min, portioned into Eppendorf tubes, frozen, and stored at −20°C.

For freeze fracturing a portion was slowly warmed to room temperature, and a droplet of polyvinyl alcohol was allowed to polymerize on a gold specimen carrier (3.0 mm diameter, Balzers, Liechtenstein) for about 5 min. Falcon Film Liners or Thermanox coverslips were cut to 2 × 2 mm squares, adhesive monolayer cells were washed carefully with PBS, the remaining fluid was drained with a filter paper, and the coverslip was placed with the cells on the highly viscous polyvinyl alcohol droplet. This sandwich was immediately plunged into liquid propane (about 90 K); after about 10 s the specimen carrier was mounted on a liquid nitrogen (LN_2) cooled object table (BB 172 058-U, Balzers), and transferred into a Balzers 301 instrument on the 123 K cold rotary stage (BB 176 312-T, Balzers).

When a vacuum of about 130 μPa was reached, the cooled (123 K) microtom knife was placed under an edge of the plastic square and lifted till the coverslip separated from the polyvinyl alcohol. Thus, preferentially, cell membranes sticking to polyvinyl alcohol were fractured and immediately replicated by shadowing with platinum/carbon under 45° to about 2 nm thickness by a high voltage evaporation device (EVM 052A, Balzers) under thickness control with a quartz crystal thin film monitor (QSG 201D, Balzers), and stabilized by perpendicular carbon evaporation (~20 nm). Replicas were separated from the polyvinyl alcohol by immersing the specimen carrier into 0.9% NaCl solution. They were cleaned overnight in 12% sodium hypochloride solution (Hedinger, Stuttgart, F.R.G.), washed three times in distilled water, and transferred onto copper grids (G 300 HS, SCI, Science Services, Munich, F.R.G.). Replicas were investigated in a Zeiss EM 10 transmission electron microscope; pictures were taken at magnifications between 20,000 and 40,000 on High Speed Clear Negative Film (HS-7 Cronalar Dupont Ltd., St. Neots, Huntington, U.K.).

For comparison, cells were also fixed with 2% glutardialdehyde (electronmicroscopy grade, Nr. 4239, E. Merck, Darmstadt, F.R.G.) in PBS for 30 min, washed several times in PBS, and infiltrated with glycerol solutions in PBS with increasing concentration to a final glycerol concentration of 30%. After infiltration over night at 4°C, the cells were either sandwiched with polyvinyl alcohol, as described above, or collected with a rubber policeman, pelleted, sandwiched between two gold specimen carriers (4.6 mm diameter, Balzers), frozen in LN_2, and fractured on a double replica table (BB 172 137-T, Balzers).

D. IMAGE ANALYSIS

For the analysis and quantitative evaluation of gap junction replicas, a computerized image processing system was used. These programs have been established at the Institut für Physikalische Elektronik (University of Stuttgart), and have been adapted for the pattern analysis of gap junction plaques.[12] The electron microscopical micrograph negatives were digitized by an A/D converter, resulting in images with a resolution of 256 gray levels, which were stored on tape. These data were then analyzed on a minicomputer (VAX, DEC, Maynard, MA). Local gray level maxima were determined in selected gap junctional areas. By shrinkage of these local maxima to single points, a matrix was formed which contained all spacings between these points, i.e., particles on PF-face leaflets and pits on EF-face leaflets. These parameters characterize the gap junction pattern in a computer-applicable mode. The center-to-center spacings between all skeletonization points were classified by 1 nm step, and spacings between 5 nm and 25 nm were included for subsequent determination of frequency distributions and statistical anaylsis, which revealed nearest neighbor distances as well as particle densities.

III. RESULTS

A. GAP JUNCTION PERMEABILITY

Under normal culture conditions it is difficult to increase the gap junctional conductivity of BICR/M1R_k- or BT5C1-cells by substances which are effective in primary cultures or *in vivo*.[19,43,44] These substances are known to interfere with the synthesis of the intracellular second messenger cAMP (3′,5′-cyclic adenosine monophosphate), e.g., forskolin or methylisoxanthin, or they substitute its function, as does dBcAMP (dibutyryl-cAMP). As monolayer cultures, these two cell lines seem to be completely upregulated; their gap junctional conductivity, however, can be downregulated by retinoic acid, intracellular pH-shifts to values lower than pH 6.8, and by unphysiological substances such as benzhydrol, heptanol, and glutardialdehyde.[19]

The extent of gap junctional coupling in BICR/M1R_k cells is demonstrated with a dye-spreading experiment, where cells were in contact only with some fine cellular protrusions, as is shown in Figure 1a. Two min after injection of Lucifer Yellow into the cell located in the center of this group, the dye had spread into all cells which were in direct contact with the injected one, and into some indirect neighbors, as is revealed from the same spot under epifluorescent illumination (Figure 1b).

As most other channels in the plasma membrane, individual gap junction channels are not permanently open, but may open or close in response to cellular regulation processes. Thus, *in situ,* an active gap junction plaque consists of channels which can be open or closed at a given time. It is difficult to follow this dynamic process by voltage measurements or by dye spreading experiments, as both determine the total conductivity of gap junction plaques. It can be detected, however, in the cell attached mode of the patch clamp technique, which resolves the current through one or few individual channels. Examples for channel characteristics are shown in Figure 2, where two different conductivity levels of an individual channel (connexon?) were measured with a BICR/M1R_k-cell (Figure 2a), and four different levels with a BRL-cell (Figure 2b). Their maximum conductivity of about 250 to 320 pS is of the same order as has been described for gap junction channels,[24,25] which consist of two hemichannels.

Conventional preparation methods for freeze fracturing or embedding in epon requires chemical fixation of the biological material. Very often glutardialdehyde is used, which, however, closes the gap junction channels irreversibly within 1 or 2 min. This uncoupling effect of glutardialdehyde is demonstrated with homokaryons of BICR/M1R_k-cells in Figure 3. The lower trace shows the rectangular current pulses which were fed into cell 1 (middle trace), where they caused voltage changes V_1 (proportional to the input resistance) superimposed to the cell's membrane potential (pd_1). Cell 2 (upper trace) was electrically coupled to cell 1, as can be seen by the voltage changes V_2 superimposed to the membrane potential of cell 2 (pd_2). The arrow indicates the timepoint when 0.5 ml of 2% glutardialdehyde was added to the dish so that a final concentration of about 0.5% was reached. The effect of glutardialdehyde can be seen about 1 min later (start of reaction: timepoint "0"), when it had diffused across the impaled cells. Due to a reaction of glutardialdehyde with the plasma membrane, the input resistances of both cells increased, which is indicated by higher amplitudes of both V_1 and V_2. The coupling ratio V_2/V_1, however, remains unaffected at about 0.65. Shortly thereafter (1.84 $\pm$ 0.21 min; mean $\pm$ standard error; n $=$ 70) the cells started to uncouple, as can be seen from the V_2-amplitudes, which decreased continuously because less current passed when gap junction channels closed. Consequently, this uncoupling from neighboring cells further increased the input resistance in cell 1, as is indicated by the growing V_1-amplitude. Finally, when the coupling ratio V_2/V_1 was below 0.08, the cell membranes suddenly became very leaky, as is indicated by the collapse of the membrane potentials.

The same uncoupling effect of glutardialdehyde is also demonstrated using the patch clamp technique for whole cell recordings with a pair of BT4C1-cells (Figure 4). The upper trace represents cell 1, where the voltage was pulsed between -5 and -15 mV; the lower trace represents cell 2 where the voltage was clamped to a constant value of -10 mV. For both cells only the currents necessary for maintaining the preset voltage clamp conditions are shown. The arrows indicate the time when the medium was replaced by 0.2% glutar-

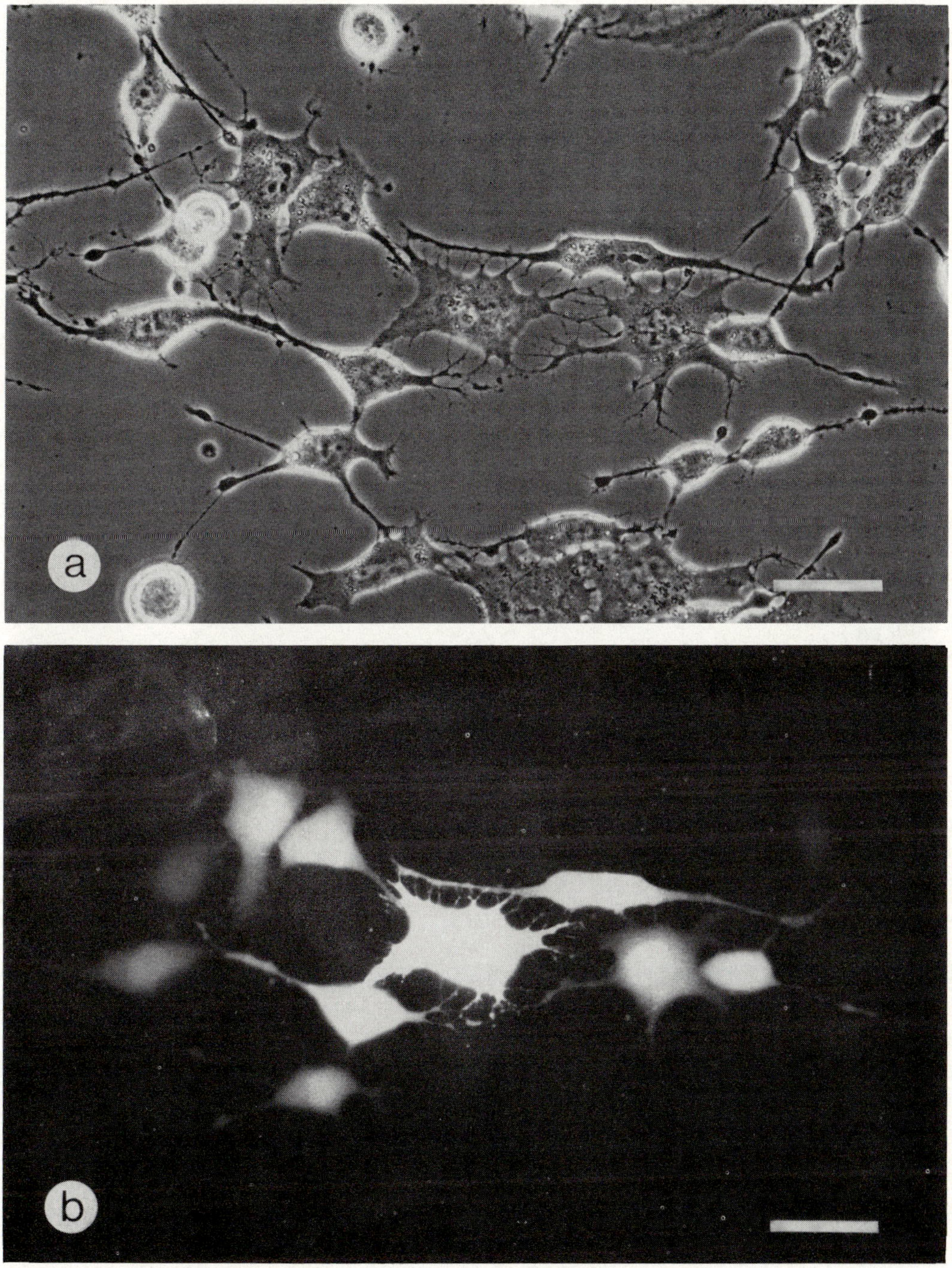

FIGURE 1. Intercellular communication between mammary tumor cells BICR/M1R$_k$ in culture demonstrated with Lucifer Yellow. The fluorescent dye was iontophoretically injected with a glass micropipette into the cell located in the center. Spreading of Lucifer Yellow through gap junctions into all directly and into some indirectly coupled cells was documented 2 min after the injection. (a) phase contrast, (b) epifluorescent illumination. Bar: 50 μm.

dialdehye in PBS. This led to a decrease in junctional coupling, as can be seen from the reduction of the current in cell 2. Assuming a conductivity of about 150 pS[24,25] for a fully open gap junction channel, these two cells were originally coupled by at least 220 open channels, whereas 3 min after the addition of glutardialdehyde, a conductivity of about 2

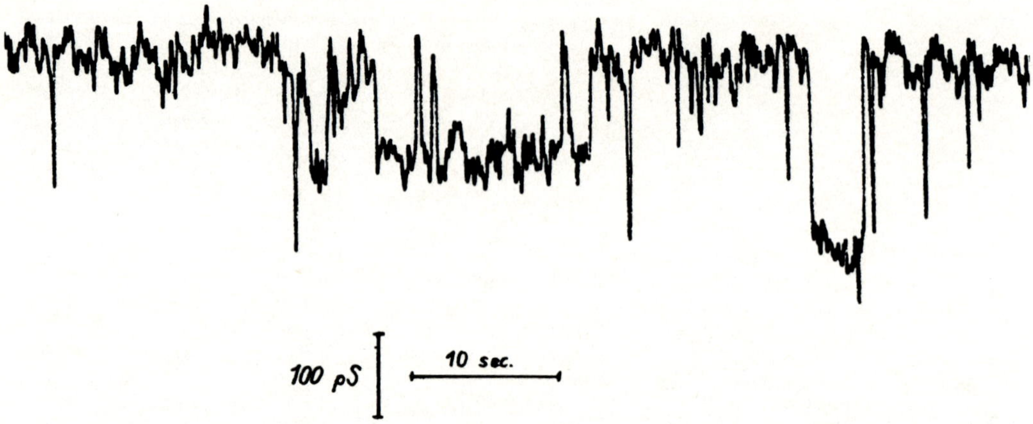

A

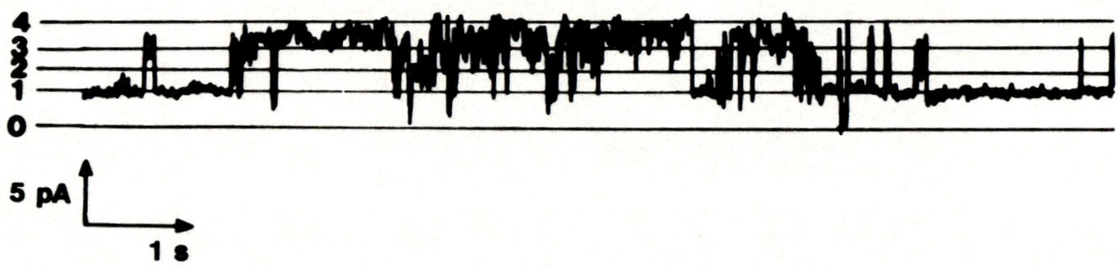

B

FIGURE 2. **Patch clamp measurements of channels (connexons?) with different conductivity levels. (A) Mammary tumor cell BICR/M1R$_k$ (B) buffalo rat liver (ATTC CRL 1442).**

nS was measured. This corresponds to about 10 to 15 fully open channels, or accordingly, more channels with reduced conductivities.

B. MORPHOLOGY OF GAP JUNCTION PLAQUES

Glutardialdehyde does not preserve gap junction channels in their open state, and it cannot be excluded that the gross morphology of gap junction plaques is also affected. Rapid freezing without preceding addition of chemical fixatives should, therefore, give some information about the *in situ* structure of gap junction plaques.

After conventional fixation with glutardialdehyde, and with 30% glycerol as a cryoprotectant, gap junctions between BICR/M1R$_k$-cells and between BT5C1-cells showed a typical clustered arrangement of tightly packed channels within a plaque (Figure 5). This crystalline-like structure normally identifies gap junctions, especially when both PF-face particles and EF-face pits are preserved and replicated at the same plaque. Nonjunctional membranes are characterized by a homogeneous distribution of membrane particles in PF-face leaflets, and by a low particle density in EF-face leaflets. When the cells were rapidly frozen by dipping a sandwich of plastic, polyvinyl alcohol, and gold specimen carrier into liquid propane, the fractured membranes were less smooth; PF-face leaflets revealed unevenly distributed particles, but also particle-free areas, and EF-face leaflets appeared rather particle-free. Under this condition, tightly packed gap junction plaques also have been found (Figures 6a and 8b), which, however, were always smaller than in fixed preparations. In addition, many loosely packed clustered gap junction plaques could be detected (Figures 6b, 7a, and 8a), where the particle density and cluster size varied considerably. Whereas all these structures

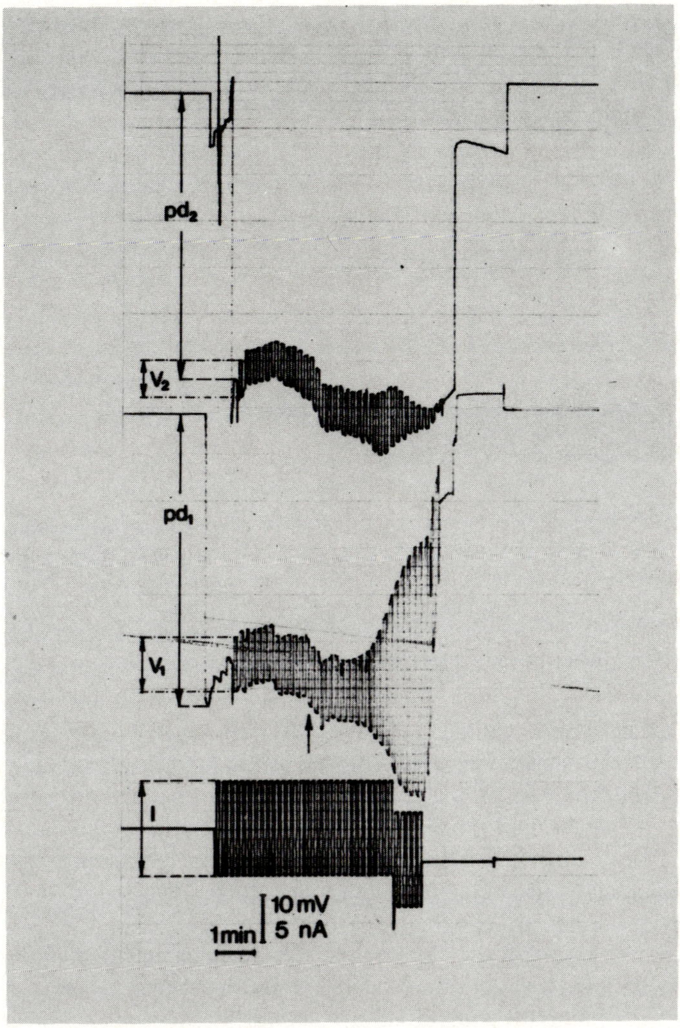

FIGURE 3. Uncoupling effect of glutardialdehyde. Electrically coupled homokaryons of BICR/M1R$_k$-cells were treated with 0.2% glutardialdehyde (arrow), which resulted in an increase of the membrane input resistance, an uncoupling of cell 2 (upper trace), and a collapse of the membrane potentials, within 3 min. For details see text. pd_1 and pd_2: membrane potentials of cell 1 and 2; V_1 and V_2: voltage changes in cells 1 and 2, due to the injected current pulse I in cell 1.

may well represent gap junctions, it is difficult to decide whether the particle aggregations shown in Figures 7b and 8c also consist of gap junction proteins.

A quantification of the different gap junction plaques has been performed by a pattern analysis for both fixed and unfixed BT5C1-cells. The mean particle density (± SE) measured on PF-replicas for conventionally treated and fixed cells is 5,408 ± 340 μm^{-2}, and 2,685 ± 340 μm^{-2} for unfixed cells. Comparison with frequency distributions of trigonal, tetragonal, and hexagonal patterns revealed a rather tetragonal packing of gap junction proteins with mean values for the nearest neighbor distance (± SE) of 7.08 ± 0.12 nm for the fixed, and 9.33 ± 0.49 nm for the unfixed preparations. BICR/M1R$_k$-cells have only been analyzed for fixed and conventionally frozen monolayer pellets, where the nearest neighbor distance was 6.70 ± 0.18 nm, and the particle density 6,990 ± 476 μm^{-2}.

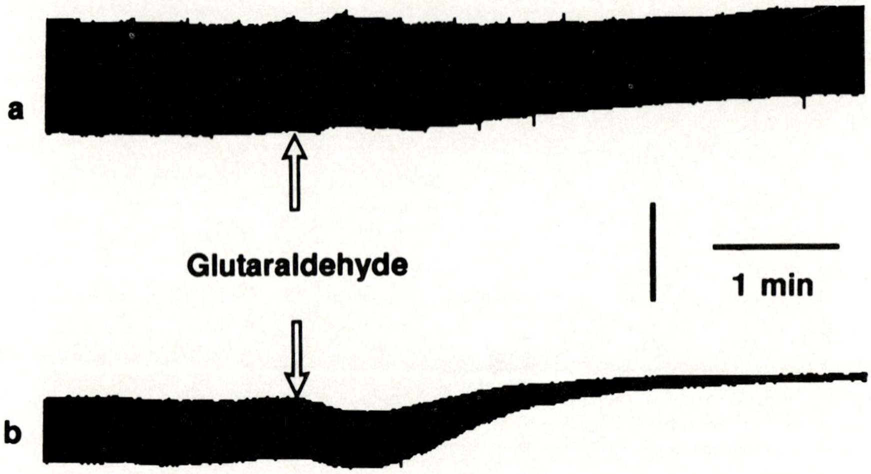

FIGURE 4. Uncoupling effect of 0.2% glutardialdehyde in BT5C1 cells measured with the patch clamp technique in whole cell configuration. Trace a: current in cell 1, trace b: current in cell 2; the amplitude is proportional to the degree of coupling. Horizontal bar: 1 min. Vertical bar: 1 nA (trace a), 0.5 nA (trace b).

IV. DISCUSSION

BICR/M1R_k- and BT5C1-cells are always very well coupled when grown as monolayer cultures (Figures 1, 3, and 4), but even with their gap junctional conductivity completely upregulated, individual channels may well vary in their conductivity levels. This cannot only be seen from patch clamp experiments where different conductivity levels of large channels are resolved (Figure 2), but also from simultaneous measurements of dye spreading and ionic coupling,[18,19] where ionic coupling can still be detected when dye spreading is already blocked.

With an open bore size between 1 and 2 nm[45] for a channel, it can be doubted that open channels in a plaque could be discriminated from closed ones when the membrane is replicated by Pt/C evaporation. From this point of view, it should be irrelevant whether fixed or unfixed material is used for the demonstration of gap junctions. However, when two cells initiate intercellular coupling, the gap junction conductivity increases with a cooperative characteristic, which indicates that more than one channel is involved in the formation of a stable contact.[46] These electrophysiological measurements correspond very well with the often documented ultrastructural morphology of freeze fractured gap junctions, where many individual channels are always arranged to aggregates.

The question arises whether the particle packing within a plaque may give some information on the coupling state of the cell, or of the conductivity of the gap junction plaques. Several experiments have been carried out to correlate the structure of freeze fractured gap junctions with their functional state;[5-13,47-50] however, the results were equivocal. In retrospect, this may be explained by the preparation techniques used before, which did not always preserve the *in situ* situation. Conventional preparation methods require fixation with glutardialdehyde, which causes irreversible uncoupling of the cells, as can be seen in Figures 3 and 4.

The morphology of unfixed gap junctions ranges from loosely to tightly packed plaques, whereas in fixed cells, loosely packed gap junction plaques are missing. The particle diameter is similar under both preparation techniques. Do these structures represent different dynamic states of gap junctional activity, or have artifacts been induced? The electrophysiological measurements revealed that it takes some time before the cell depolarizes after the addition

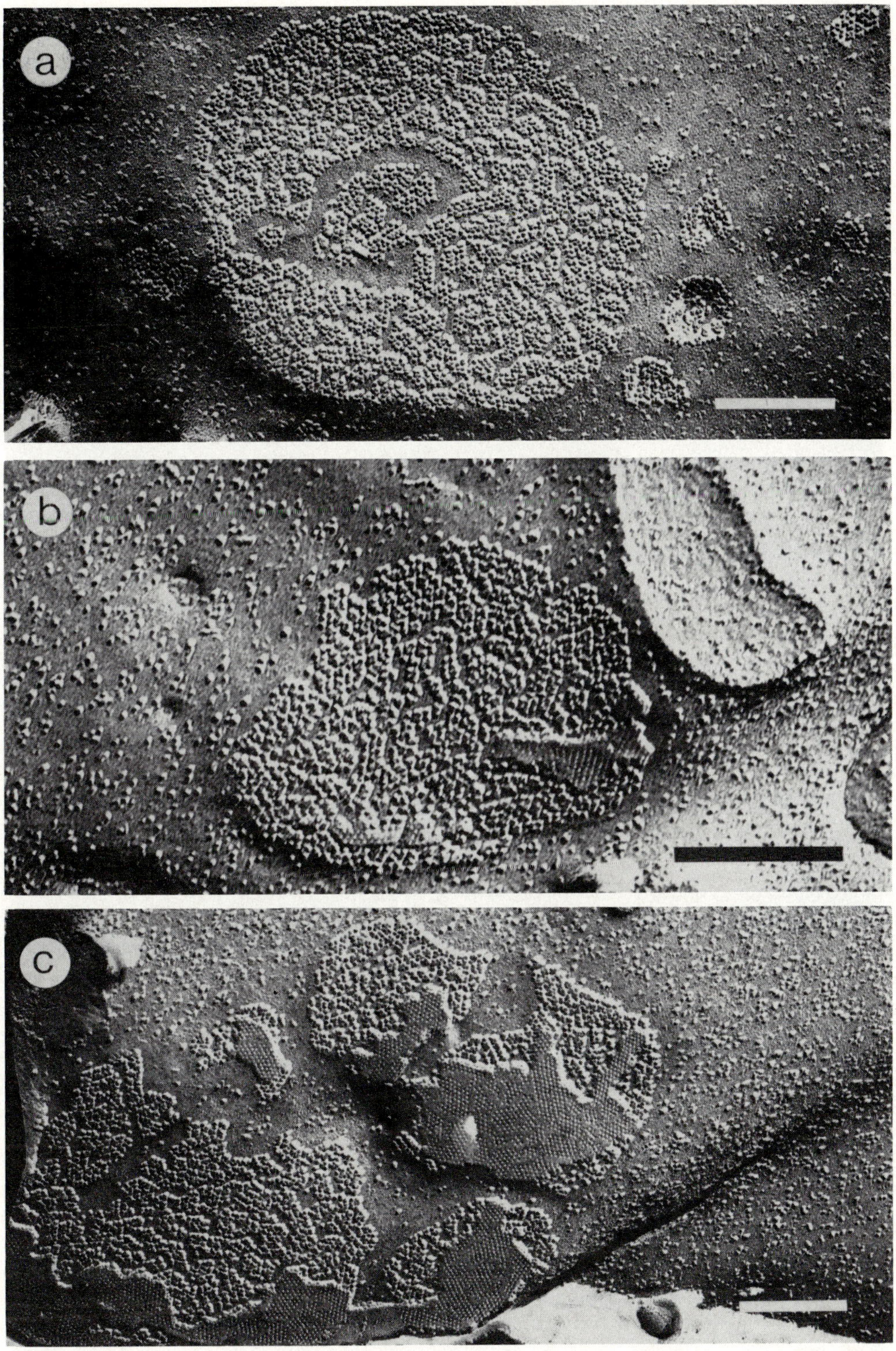

FIGURE 5. Gap junctions between glutardialdehyde-fixed cells. (a) BT5C1-cells, (b) and (c) BICR/M1R$_k$-cells. Bar: 0.2 μm.

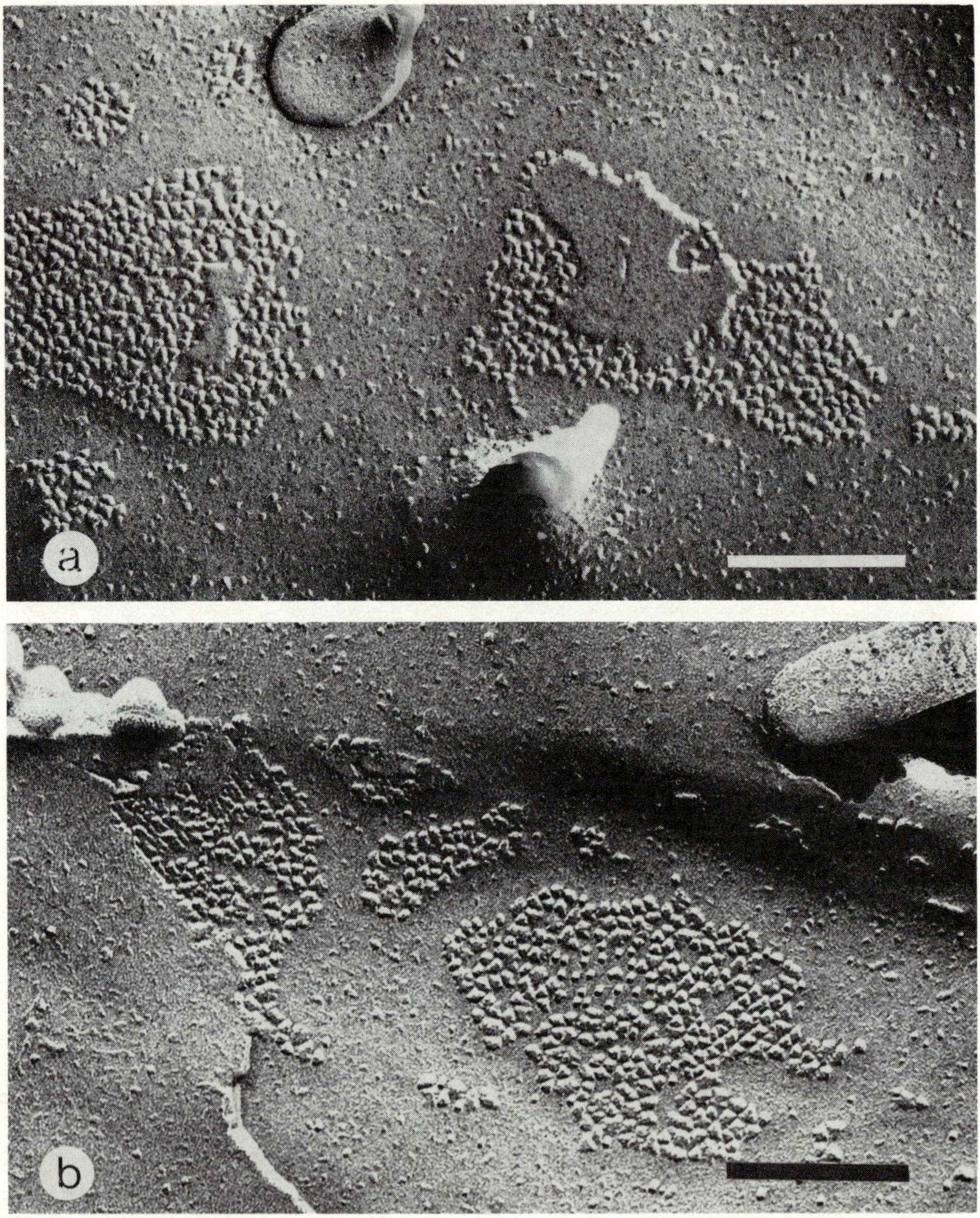

FIGURE 6. Gap junctions between unfixed BT5C1-cells, rapidly frozen by dipping a plastic-polyvinyl-sandwich in liquid propane. (a) tightly packed, (b) loosely packed and clustered. Bar: 0.2 μm.

of glutardialdehyde, even when some channels may close immediately. The depolarization may be accompanied by a collapse of electrical charges of proteins, which—in addition to the crosslinking property of glutardialdehyde— may result in a closer aggregation of the gap junction particles. On the other hand, a 5 min fixation with glutardialdehyde and freezing of BICR/M1R$_k$-cells by dipping them as a sandwich with 2 gold specimen carriers into liquid Freon® revealed gap junction patterns which were indistinguishable from propane jet preparations, i.e., loosely as well as tightly packed gap junction plaques. However, fixation

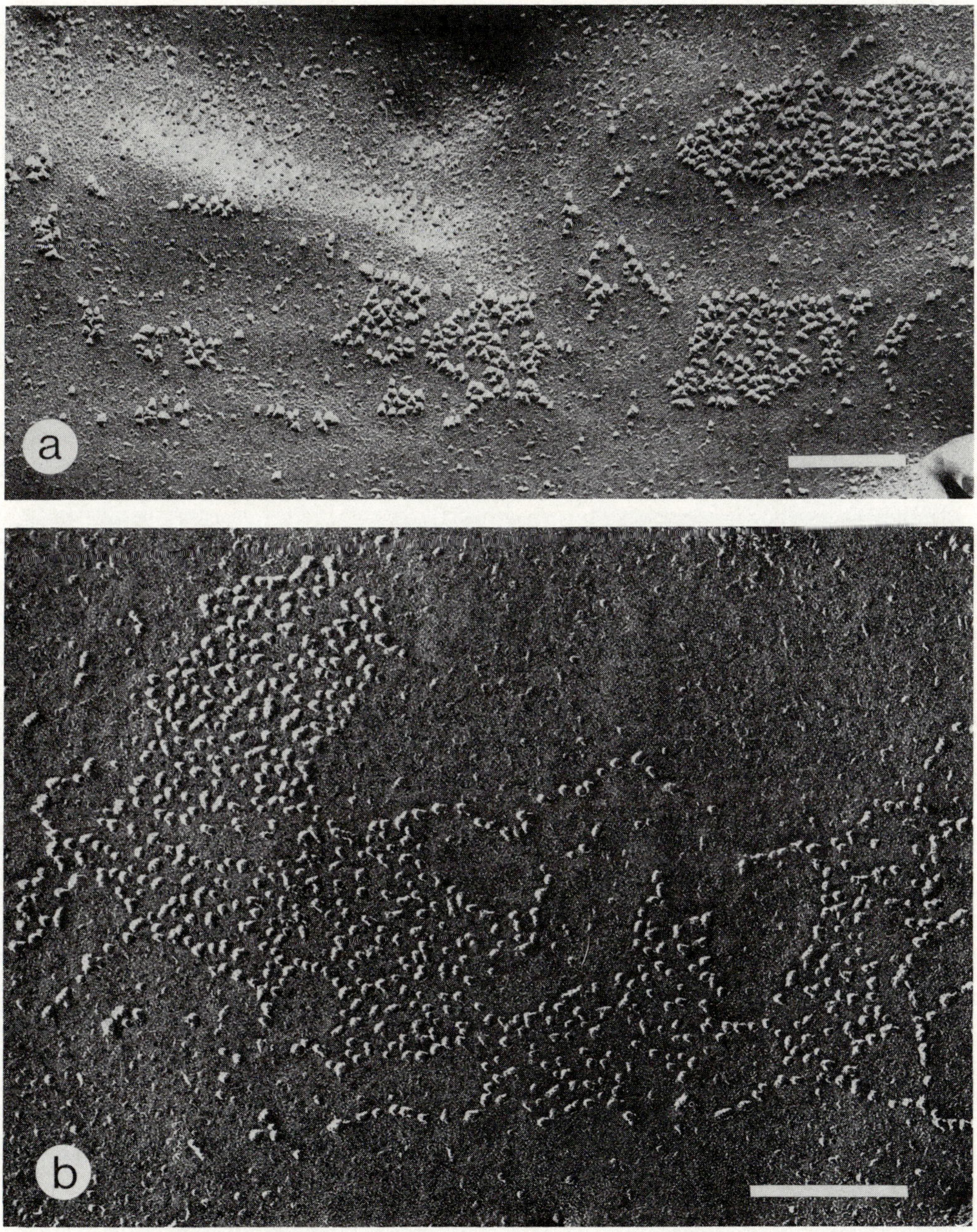

FIGURE 7. Gap junctions between unfixed BT5Cl-cells, rapidly frozen by dipping a plastic-polyvinol-sandwich in liquid propane. (a) Clustered and loosely aggregated particles, (b) loosely aggregated particles (gap junction?). Bar: 0.2 μm.

for 30 min and infiltration with glycerol increased the number of tightly packed gap junction plaques considerably.[12] Similar results have been reported for pancreatic acinar cells[47] and for heart cells.[7,49,50] Thus, a closing of most gap junction channels within 3 min is not necessarily accompanied by changes of the gap junction plaque morphology.

But why then are tightly packed gap junction plaques also found in unfixed cells? Interestingly, in thin sections of BICR/M1R_k- and BT5Cl-cells sometimes internalized gap junctions can be detected,[51] as have been found in many other tissues.[51,52] These internalized

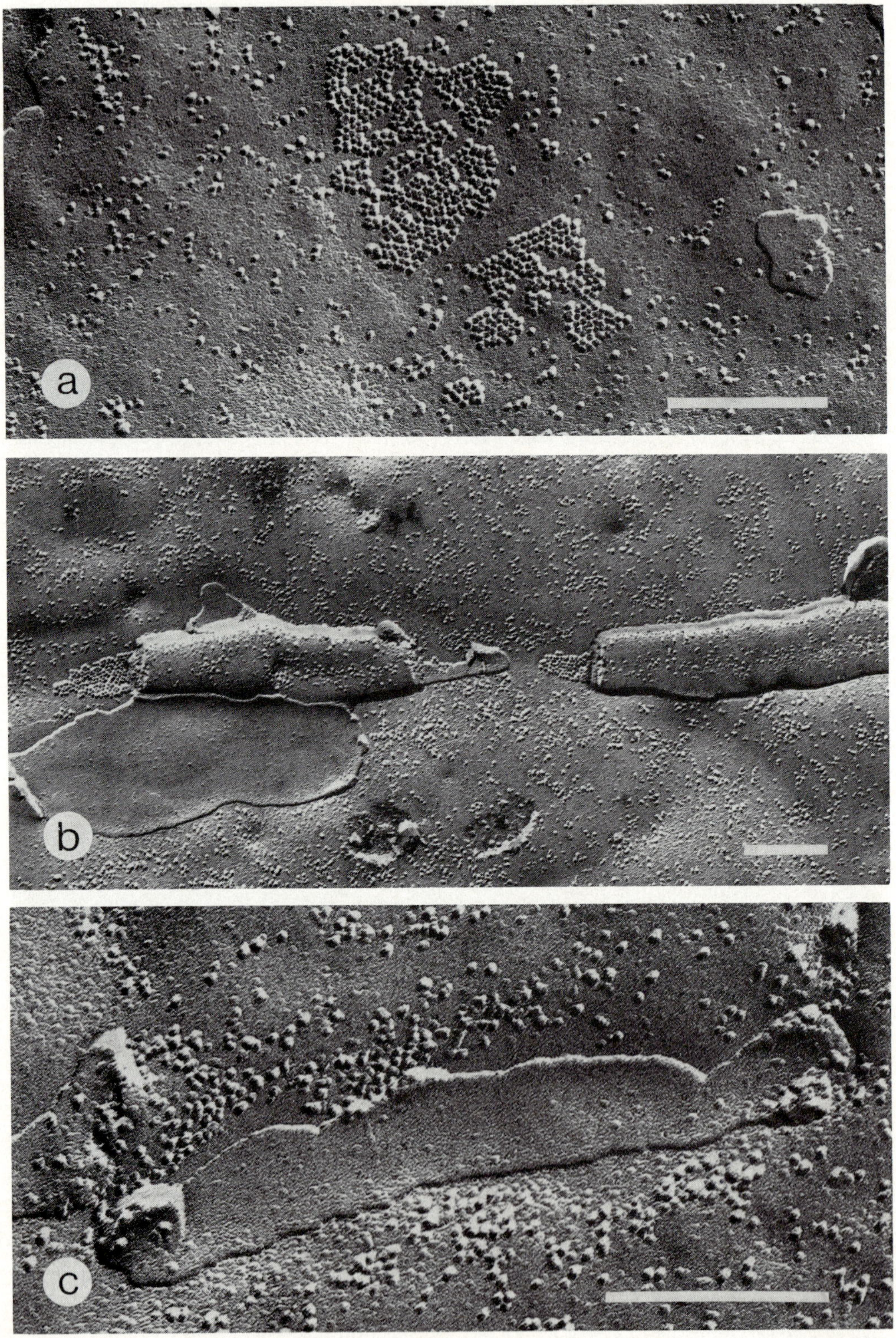

FIGURE 8. Gap junctions between unfixed BICR/M1R_k-cells, rapidly frozen by dipping a plastic-polyvinol-sandwich in liquid propane. (a) Loosely packed and clustered, (b) tightly packed, (c) loosely aggregated particles (gap junction?). Bar: 0.2 μm.

gap junctions might provide a clue for the interpretation of unfixed gap junction plaques: if one assumes that gap junction proteins have a limited life span and will be recycled, then the densely packed structure is the final state of a gap junction plaque before the proteins are internalized, as membrane receptors are collected in coated pits. Consequently, the loosely arranged particles may represent active gap junction plaques, where channels open and close to regulate the intercellular signal and metabolite transfer. Furthermore, it explains why both tightly and loosely packed gap junction plaques are found in unfixed material. Under these circumstances, i.e., with fluent transitions in the morphology of gap junction plaques, a pattern analysis cannot be very helpful in discriminating different functional states of gap junctional activities.

Rapid freezing should cause fewer alterations in biological membranes than fixation with glutardialdehyde and subsequent addition of glycerol as a cryoprotectant. According to Knoll et al.,[53] the cooling rate for a copper-Thermanox-sandwich is several times lower when dipped by hand into propane than with a propane jet application, which reaches a cooling rate of about 10^4 Ks^{-1} for monolayer thicknesses. This condition is certainly not fulfilled for the polyvinyl alcohol technique with monolayer cells grown on thin Film Liners or on Thermanox coverslips. The electron microscopic experiments show, however, that the obtained cooling rate is sufficient for the preservation of plasma membranes, since the formation of ice crystals is obviously suppressed to an extent that they do not interfere with the investigated gap junction structures. Our electrophysiological experiments reveal that this cooling rate is certainly fast enough to detect different functional gap junction plaques.

V. CONCLUSIONS

The effect of glutardialdehyde on the dynamic organization of gap junctions cannot only be seen by electrophysiological measurements where the uncoupling of cells occurs within 3 min, but also by freeze fracturing the cells. Gap junctions from unfixed cells rapidly frozen by dipping into liquid propane appear polymorphic; loosely packed and clustered plaques are found, as well as tightly packed aggregates, which are mainly found in fixed preparations. Whether these different structures correspond with different functional states, or whether they depend on the local configuration of the contacting membranes, is difficult to decide. The presented results, however, support the idea of active (coupling competent) gap junctions with loosely packed channels and nonactive (permanently closed) gap junctions where the channels are tightly packed.

ACKNOWLEDGMENTS

We wish to thank Prof. Dr.-Ing. W. H. Bloss and his colleagues from the Institut für Physikalische Elektronik (Universität Stuttgart) for providing assistance with their pattern analysis system. We also thank Susann Becker and Simone Rhein for technical assistance. This work has in part been supported by the Deutsche Forschungsgemeinschaft.

REFERENCES

1. **Goodenough, D. A. and Revel, J. P.,** A fine structural analysis of intercellular junctions in the mouse liver, *J. Cell Biol.*, 45, 272, 1970.
2. **Revel, J. P. and Karnovsky, M. J.,** Hexagonal array of subunits in intercellular junctions of the mouse heart and liver, *J. Cell Biol.*, 33, C7, 1967.
3. **Revel, J. P., Yancey, S. B., and Nicholson, B. J.,** The gap junction proteins, *Trends Biochem. Sci.*, 11, 375, 1986.

4. **Bennett, M. C. L. and Goodenough, D. A.,** Gap junctions, electrotonic coupling, and intercellular communication, in *Neurosciences Research Program,* Bulletin 16, MIT Press, Cambridge, MA, 372, 1978.
5. **Peracchia, C.,** Gap junctions—structural changes after uncoupling procedures, *J. Cell Biol.,* 72, 628, 1977.
6. **Raviola, E., Goodenough, D. A., and Raviola, G.,** Structure of rapidly frozen gap junctions, *J. Cell Biol.,* 87, 273, 1980.
7. **Hanna, R. B., Reese, T. S., Ornberg, R. L., Spray, D. C., and Bennett, M. V. L.,** Fresh frozen gap junctions: resolution of structural detail in the coupled and uncoupled states, *J. Cell Biol.,* 91, 125a, 1981.
8. **Green, C. R. and Severs, N. J.,** Gap junction connexon configuration in rapidly frozen myocardium and isolated intercalated disks, *J. Cell Biol.,* 99, 453, 1984.
9. **Berdan, R. C. and Caveney, S.,** Gap junction ultrastructure in three states of conductance, *Cell Tissue Res.,* 239, 111, 1985.
10. **Miller, T. M. and Goodenough, D. A.,** Gap junction structures after experimental alteration of junctional channel conductance, *J. Cell Biol.,* 101, 1741, 1985.
11. **Page, E., Karrison, T., and Upshaw-Earley, J.,** Freeze-fractured cardiac gap junctions: structural analysis by three methods, *Am. J. Physiol.,* 244, H525, 1983.
12. **Müller, A., Blanz, E. W., Laub, G., and Hülser, D. F.,** Pattern analysis of gap junction plaques with open and closed pores, *Stud. Biophys.,* 110, 185, 1985.
13. **Greule, J. and Hülser, D. F.,** Comparison of different freeze-fracture methods, *Eur. J. Cell Biol.,* 43(Suppl.17), 20, 1987.
14. **Pitts, J. D. and Finbow, M. E.,** The gap junction, *J. Cell Sci.,* Suppl.4, 239, 1986.
15. **Subak-Sharpe, H., Bürk, R. R., and Pitts, J. D.,** Metabolic cooperation between biochemically marked mammalian cells in tissue culture, *J. Cell Sci.,* 4, 353, 1969.
16. **Simpson, I., Rose, B., and Loewenstein, W. R.,** Size limit of molecules permeating the junctional membrane channels, *Science,* 195, 294, 1977.
17. **Stewart, W. W.,** Functional connections between cells as revealed by dye-coupling with a highly fluorescent naphthalimide tracer, *Cell,* 14, 741, 1978.
18. **Hülser, D. F. and Brümmer, F.,** Closing and opening of gap junction pores between two- and three-dimensionally cultured tumor cells, *Biophys. Struct. Mech.,* 9, 83, 1982.
19. **Brümmer, F.,** *Farbstoffausbreitung durch Gap Junctions in tierischen Zellkulturen,* Stöffler & Schütz, Stuttgart, 1988, 38.
20. **Flagg-Newton, J. L., Dahl, G., and Loewenstein, W. R.,** Cell junction and cyclic AMP: I. Upregulation of junctional membrane permeability and junctional membrane particles by administration of cyclic nucleotide or phosphodiesterase inhibitor, *J. Membrane Biol.,* 63, 105, 1981.
21. **Murray, A. W. and Fitzgerald, D. I.,** Tumor promoters inhibit metabolic cooperation in coculture of epidermal and 3T3 cells, *Biochem. Biophys. Res. Commun.,* 91, 395, 1979.
22. **Spray, D. C., White, R. L., Campos de Carvalho, A., Harris, A. L., and Bennett, M. V. L.,** Gating of gap junction channels, *Biophys. J.,* 45, 219, 1984.
23. **Spray, D. C., Ginzberg, R. D., Molralese, E. A., Gatmaintan, Z., and Bennett, M. V. L.,** Electrophysiological properties of gap junctions between dissociated pairs of rat hepatocytes, *J. Cell Biol.,* 103, 135, 1986.
24. **Neyton, J. and Trautmann, A.,** Single-channel currents of an intercellular junction, *Nature,* 317, 331, 1985.
25. **Veenstra, R. D. and DeHaan, R. L.,** Measurement of single channel currents from cardiac gap junctions, *Science,* 233, 972, 1986.
26. **Schwarze, W. and Kolb, H. -A.,** Voltage-dependent kinetics of an anionic channel of large unit conductance in macrophages and myotube membranes, *Pflügers Arch.,* 402, 281, 1984.
27. **Paschke, D. and Hülser, D. F.,** Voltage dependent behaviour of a large membrane channel with subconductance states, *Eur. J. Cell Biol.,* Suppl. 17, 41, 1987.
28. **Spray, D. C., Harris, A. L., and Bennett, M. V. L.,** Glutaraldehyde differentially affects gap junctional conductance and its pH and voltage dependence, *Biophys. J.,* 33, 108a, 1981.
29. **Moor, H.,** Theory and practice of high pressure freezing, in *Cryotechniques in Biological Electron Microscopy,* Steinbrecht, R. A. and Zierold, K., Eds., Springer-Verlag, Berlin, 1986, 175.
30. **Van Harreveld, A., Trubatch, J., and Steiner, J.,** Rapid freezing and electron microscopy for the arrest of physiological processes, *J. Microsc.,* 100, 189, 1974.
31. **Heuser, J. E., Reese, T. S., Dennis, M. J., Lan, Y., Jan, L., and Evans, L.,** Synaptic vesicle exocytosis captured by quick freezing and correlated with quantal transmitter release, *J. Cell Biol.,* 81, 275, 1979.
32. **Plattner, H. and Zingsheim, H. P.,** *Elektronenmikroskopische Methodik in der Zell- und Molekularbiologie,* Gustav Fischer, Stuttgart, 1987, 49.
33. **Rajewsky, M. F. and Grüneisen, A.,** Cell proliferation in transplanted rat tumors: influence of the host immune system, *Eur. J. Immunol.,* 2, 445, 1972.

34. **Laerum, O. D. and Rajewsky, M. F.,** Neoplastic transformation of fetal rat brain cells in culture after exposure to ethylnitrosourea *in vivo, J. Natl. Cancer Inst.*, 55, 1177, 1975.
35. **Frank, W., Ristow, H. J., and Schwalb, S.,** Untersuchungen zur wachstumsstimulierenden Wirkung von Kälberserum auf Kulturen embryonaler Rattenzellen, *Exp. Cell Res.*, 10, 390, 1972.
36. **Krähling, H., Schinkewitz, U., Barker, A., and Hülser, D. F.,** Electronmicroscopical and electrophysiological investigations on polyethylene glycol induced cell fusion, *Eur. J. Cell Biol.*, 17, 51, 1978.
37. **Hülser, D. F. and Webb, D. J.,** Relation between ionic coupling and morphology of established cells in culture, *Exp. Cell Res.*, 80, 210, 1973.
38. **Hamill, O. P., Marty, A., Neher, E., Sakmann, B., and Sigworth, F. J.,** Improved patch-clamp techniques for high-resolution current recording from cells and cell-free membrane patches, *Pflügers Arch.*, 391, 85, 1981.
39. **McCann, F. V., Stibitz, G. R., and Keller, Th. M.,** A computer method for the acquisition and analysis of patch-clamp single-channel currents, *J. Neurosci. Meth.*, 20, 45, 1987.
40. **Pauli, B. U., Weinstein, R. S., Soble, L. W., and Alroy, J.,** Freeze-fracture of monolayer cultures, *J. Cell Biol.*, 72, 763, 1977.
41. **Dahl, G.,** 1983, personal communication.
42. **Dermietzel, R.,** 1987, personal communication.
43. **Winterhager, E., Brümmer, F., Dermietzel, R., Hülser, D. F., and Denker, H. -W.,** Gap junction formation in rabbit uterine epithelium in response to embryo recognition, *Dev. Biol.*, 126, 203, 1988.
44. **Saez, J. C., Spray, D. C., Nairn, A. C., Hertzberg, E., Greengard, P., and Bennett, M. V. L.,** cAMP increases junctional conductance and stimulates phosphorylation of the 27-kDa principal gap junction polypeptide, *Proc. Natl. Acad. Sci. U.S.A.*, 83, 2473, 1986.
45. **Makowski, L., Caspar, D. L. D., Phillips, W. C., and Goodenough, D. A.,** Gap junction structures II. Analysis of the X-ray diffraction data, *J. Cell Biol.*, 74, 629, 1977.
46. **Hülser, D. F.,** Electrical characterisation of gap junction formation, *Stud. Biophys.*, 90, 197, 1982.
47. **Meda, P., Findlay, I., Kolod, E., Orci, L., and Peterson, O. H.,** Short and reversible uncoupling evokes little change in the gap junctions of pancreatic acinar cells, *J. Ultrastruct. Res.*, 83, 69, 1983.
48. **Baldwin, K. M.,** Cardiac gap junction configuration after an uncoupling treatment as a function of time, *J. Cell Biol.*, 38, 15, 1979.
49. **Sikerwar, S. S. and Malhotra, S. K.,** A structural characterisation of gap junctions isolated from mouse liver, *Cell Biol. Int. Rep.*, 7, 897, 1983.
50. **Délèze, J. and Hervé, J. C.,** Quantitative gap junction alterations in mammalian heart cells quickly frozen or chemically fixed after electrical uncoupling, *J. Membr. Biol.*, 93, 11, 1986.
51. **Bräuner, Th.,** *Interzelluläre Kommunikation und invasives Wachstum maligner Zellen,* Stöffler & Schütz, Stuttgart, 1987, 85.
52. **Larsen, W. J., Tung, H. N., Murray, S. A., and Swenson, C. A.,** Evidence for the participation of actin microfilaments and bristle coats in the internalisation of gap junction membrane, *J. Cell Biol.*, 83, 576, 1979.
53. **Knoll, G., Oebel, G., and Plattner, H.,** A simple sandwich-cryogen-jet procedure with high cooling rates for cryofixation of biological materials in the native state, *Protoplasma,* 111, 161, 1982.

Chapter 4

ANALYSIS OF IRREGULAR DISTRIBUTIONS AND DYNAMICS OF MEMBRANE COMPONENTS BY COMPUTER AIDED IMAGE RECOGNITION AND STATISTICAL EVALUATION

Sek Wen Hui

TABLE OF CONTENTS

I. INTRODUCTION

Intramembrane particles (IMPs) are the most prominent features in freeze fracture electron micrographs. The IMPs in cell membranes represent a wide variety of proteins with variations in molecular weights, aggregation, and structural details. Although this ubiquitous feature has been widely used to describe membrane morphology, quantitating the size, density and distribution of intramembrane particles requires the development of techniques which combine knowledge in pattern recognition, computer processing, and statistics.

The IMP distribution varies a great deal from membrane to membrane, as expected, although their sizes remain within a narrow range. The planar distribution is, in most cases, irregular, though not random (see below). For membranes containing a single protein species, such as the purple membrane of halobacteria, or artificially reconstituted membranes with a purified protein, the identification of IMPs is a relatively simple task. In these cases, the densities of IMPs have been correlated with protein concentrations, and individual IMPs have been shown to correspond to definite oligo-complexes of protein molecules.[1,2] In general, however, the complexity of integral membrane proteins and the irregularity of IMPs render the task of identifying IMPs exceedingly difficult. Only recently have labeling techniques been developed to identify these particles.[3,4] With increasing knowledge of the identity of IMPs, there are more calls for quantitation, both of their occurrence and planar distribution, and of their changes with time and environment.

IMPs have been found to change in varying physiological states and with cellular functions, i.e., cell cycle,[5] light excitation,[6] junctional communication[7] (see also Chapter 3 by Hülser et al.), membrane transport,[8] and cell fusion.[9,10] In some studies, rearrangement of the IMP distributions has been experimentally induced by pH,[11] temperature changes,[12] or cryoprotectants.[13] Differences in IMP density and distribution between normal and transformed or cancer cells have often been objects of ultrastructural study.[14-16]

There have been many attempts to describe and compare variations of the IMP distribution revealed by freeze fracture electron microscopy. Many studies have been restricted to qualitative description. Visual inspection and verbal assessment alone are limited to observations of obvious or large scale alterations to IMP distributions.[17] Additional problems of subjectivity and noncomparability between results occur. Manual measurement of features of the IMP distribution, such as IMP size or density,[5,18-21] are possible, but extremely time consuming, with a consequent potential for human error. The large sample sizes needed for analyses of statistical relevance are prohibitive.

A quantitative analysis of spatial distribution requires the application of statistical methods. Zernicke and Prins[22] studied the spatial distribution of finite size particles in a general random model. Early studies of molecular distribution in gases and liquids[23] laid the foundation of radial distribution analysis of irregular distributions. Later, these and other statistical methods have been applied to quantitate IMP and cell surface label distributions.[16,24-34] Semiautomatic approaches, using computers to process manually fed data, have enhanced the ability to quantitate the distribution of IMP, yet lengthy procedures are still needed to input IMP locational coordinates and/or size information, using digitizer tablets.[24-33] Discrimination of the IMP from the background is visual, and involves a stage of manual digitization. It is difficult to obtain the large sample sizes desirable for assessments of statistical relevance. This is particularly the case where the IMP population is not homogeneous.

A suitable method to analyze the IMP distribution should permit large numbers of IMP to be objectively and rapidly obtained and measured. Quantitative results will then be statistically relevant. Time-consuming stages of manual digitation, or manual discrimination of the IMP, with their consequent potential for human error, will be eliminated. In order to achieve full automation for speed and objectivity, image processing and recognition techniques have to be applied to replace human selection.[35-38]

The following is a synopsis of the computer approach to image processing and recognition, as well as distribution analysis problems.

II. AUTOMATIC RECOGNITION OF INTRAMEMBRANE PARTICLES

A. SAMPLE PREPARATION

A meaningful detection and analysis of IMP size, density and distribution hinges on the quality of sample preparation. Since IMP distribution is sensitive to the environment, especially upon fixation and cryoprotection, it is imperative to preserve the "native" condition as much as possible, with minimal artificial alteration. Rapid freezing is the preferred technique for preservation. The advantages of ultra-rapid freezing technology in avoiding chemical fixation or cryoprotectant procedures, and in preservation of cellular ultrastructure, have repeatedly been demonstrated.[13,39] This subject has been extensively reviewed in other chapters in this book. Of special interest is the methods to rapidly freeze cells grown on substrates. Some of these methods employ various chemical treatments of the substrates to render them more adhesive.[40,41] Others use propane jet or projection freezing of cells, which are sandwiched against copper supports[42-44] (see also Chapters 3, 6, and 7). While a good preservation is prerequisite for analysis, the detailed discussion of the relative merits of these experimental procedures is beyond this article. I will use as examples only cells preserved by this method.

The procedure in our laboratory for freezing cells in suspension (such as erythrocytes) is the following: an aliquot of 0.1 μl of sample is placed between two thin (gold plated) copper plates. For cells cultured on solid substrates (such as fibroblasts), the solid substrate is used as one side of the sandwich.[44] As long as the aqueous space between the two sides of the sandwich is kept below 20 μm, morphological damage by ice crystals can be kept at an undetectable level. The sandwich is then plunged into slush propane, using a spring-accelerated plunger.[45] The asymmetric sandwich is in fact, more advantageous than symmetric ones for rapid freezing.[42] A number of variations of the sandwich method have been reported, with equally satisfactory results.[40-43] Figure 1 shows the fracture face of the upper plasma membrane of a cultured 10 T-1/2 murine embryo fibroblast. The extreme flatness of these cells posts a challenge in freeze fracture.

B. DETECTION OF IMP

The first problem of the detection process is the undulation of the fracture face because of the natural undulation of the cell surface. The undulation causes the platinum coating of the replica to be uneven, as shown in Figure 2A. The gross variation in optical density due to membrane undulation usually exceeds those produced by the shadowing of individual IMP (Figure 2B). This effect is often not appreciated by visual examination, because the visual process can discriminate against background features. However, the slow but significant variation of the background precludes the use of gray level detection as the first criterion.

The second problem is the "grainy noise" produced by the platinum deposition, and the imperfections of the replica. Again, such high spatial frequency noise (henceforth all frequency and filtering are referred to in the spatial rather than in the temporal context) is often overlooked by the eye, but is an obstacle for machine vision.

The slowly varying background due to undulation may be removed by a high pass filter, whereas the "grainy noise" can be removed by a low pass filter, or by size discrimination. The effect is illustrated in Figures 2D and 2C. Spatial filtering is accomplished in real space by the use of kernal averaging in digital image process. Figure 3b represents a filtered image of Figure 3A, using a 3 × 3 unitary kernal as a low pass filter. The same figure is then filtered by using a 15 × 15 unitary kernal to average out the background undulation, as indicated in the area marked by an asterisk in Figure 3C.

FIGURE 1. The freeze fracture E-face of the upper plasma membrane of a 10T-1/2 fibroblast. Intramembrane particles (IMPs) are shown on the undulating background of the lipid bilayer. Circular cuts represent openings to microvilli and other cellular protrusions. Bar = 100 nm.

The detection of IMPs are based on both the gray level threshold and the optical density difference between each pixel and its immediate neighbors. An IMP is recorded as a contiguous group of detected pixels within certain size limits. Figure 3D shows an idealized image in which each IMP is constructed as a common size circular disc centering on the pixel of the highest optical density. Here we treat all IMPs in the fracture face of human erythrocyte membrane as the same size, an idea that is widely accepted.

The processing routine described above is commonly available in commercial image processors. As an example, the IMP distribution in the sample shown in Figure 1 is detected using the image processing capabilities of the Cambridge Instruments Quantimet 920 Image Analyzer. Negatives of freeze fracture micrographs are imaged with the macroviewer attachment. In the negatives, an IMP appears as a light, usually circular cap and a dark shadow, against a similarly shaded background. Separation of the IMP is achieved utilizing options of the Detection subroutine and other binary manipulative procedures available within version 5.0 of QUIPS (Quantimet Interactive Programming System).

The 2-D threshold detections are carried out first, selecting black areas darker than a certain threshold grey level. The detected binary image contains time-dependent fluctuating noise. Because each image is captured and stored instantaneously (in 1/30 s), the stored pattern of fluctuation is different in each video frame. A second black threshold detection redetects black areas above the same threshold. These two detected images retain the same shadow areas, but differ in noise patterns of the background. The Boolean AND operation of these two threshold-detected images eliminates the time-dependent fluctuating noise. A detection criterion called "White Tracking" detects abrupt shade changes from black to white, such as the change of peak intensity of an IMP to its black shadow. This acts also as a high pass filter. The Boolean AND of this detection and previous detections combines threshold and edge detections into a single binary image. The gradually undulating fracture plane is thus eliminated. The uneven IMP edges caused by the remaining "grainy noise" is smoothed by a one pixel dilation. Any remaining sporadic one pixel point noise is eliminated by a size criterion.

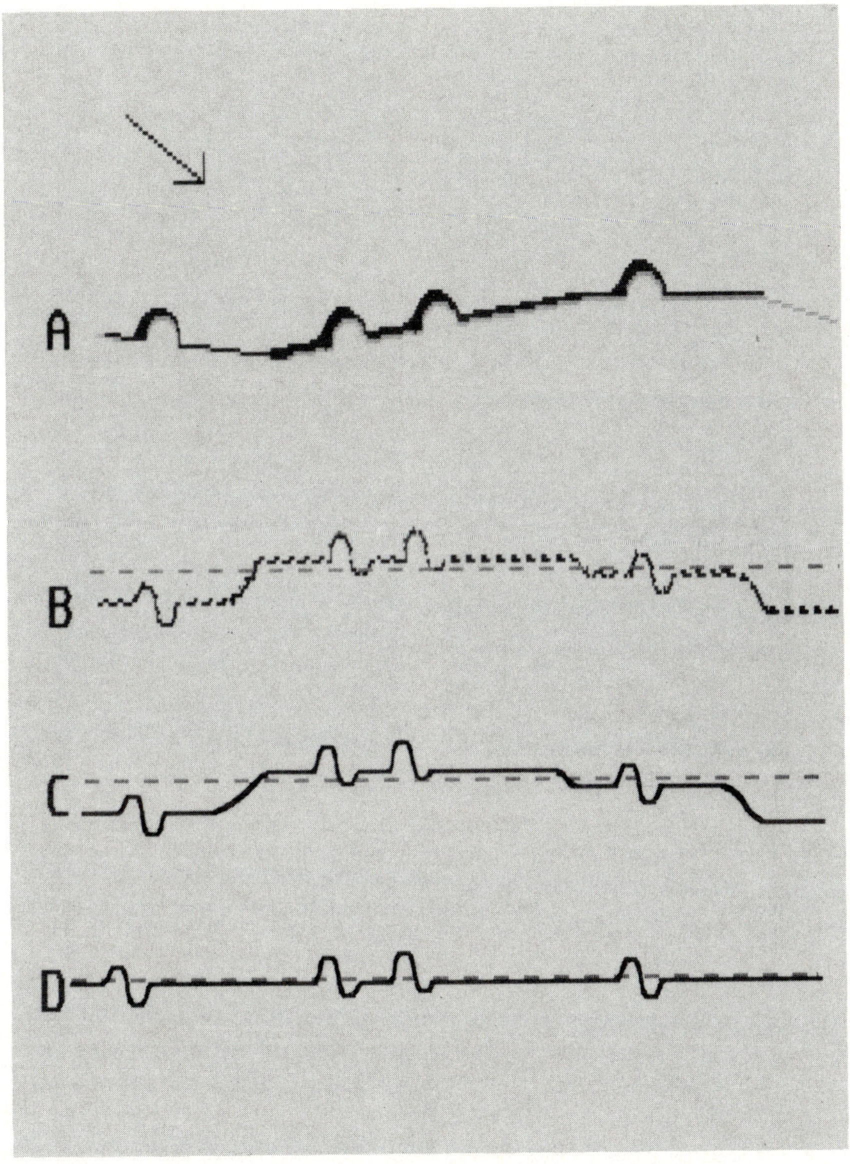

FIGURE 2. (A) The schematic representation of the cross section of a freeze fracture replica. The dark layer represents platinum shadowing material from the arrow direction. The lighter layer represents carbon. (B) The optical density tracing along the profile shown in A. (C) The optical density tracing of B after a low pass filter, and (D) that of C after a high pass filter application.

Various stages of these manipulations are shown starting in Figure 4A. The unprocessed image of the IMP distribution is seen at the point of scanner adjustment of illumination and sensitivity levels. Note the varying background grey levels. In Figure 4B, the 2-D Detector was selected to detect the dark thresholds of the IMP shadow. Unwanted overlapping background levels are noticable throughout the electron micrograph image. This noise is rapidly fluctating in the live detected image. It appears white in the figures. The third image detection, shown in Figure 4C, employs the white tracking enhancement of the edges. In Figure 4D, the final detected image is shown superimposed over the image of the original electron

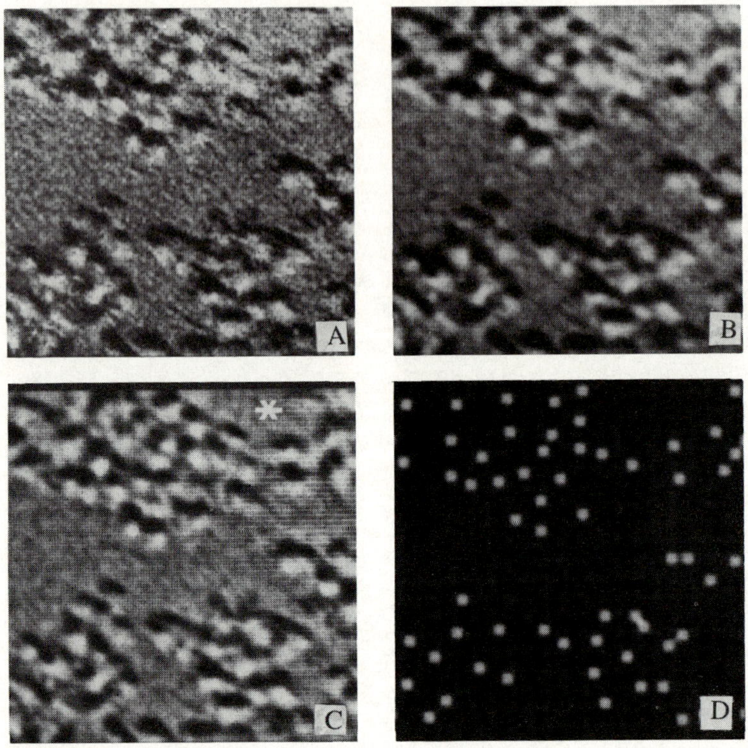

FIGURE 3. A portion of a freeze fracture electron micrograph of a human erythrocyte membrane. (A) Unprocessed image, (B) after a low pass filter of a 3 × 3 kernel, (C) after a high pass filter of a 15 × 15 kernel, and (D) an idealized image formed by expanding the highest density pixel of each detected IMP into a disc (appeared square because of limited pixel number per IMP). Length of squares = 100 nm. (From Hui, S. W. and Frank, J., *J. Microsc.*, 137, 293, 1985. With permission.)

micrograph. The final detected image was stored as a binary and recalled for quantitation. Comparison of the final binary image of the sample electron micrograph (Figure 4E) with the unprocessed freeze fracture electron micrograph image (Figure 4A) shows a good agreement and accurate identification of the IMP shadow.

The detected binary images sometimes require manual editing to separate accidentally adjoined IMPs. The feature is easy to recognize and constitutes only a minor part of the population. The entire detection process requires less than 4 min, including manual editing; therefore, a large quantity of micrographs can be processed for statistical analysis.

C. DENSITY AND SIZE MEASUREMENTS

The density and size measurement can be carried out using commonly available particle counting software packages in most image analyzers. The IMP density is measured and averaged for all binary images of the same sample. The IMP size parameter can be obtained in the same program. The width of the black shadow of the IMP is assumed to be representative of the parallel maximum dimension of the white cap. Both the maximum width of the IMP shadow perpendicular to the direction of the shadow,[5,19] and the area of the IMP shadow,[36,37] have been used to indicate IMP size. The size distribution of the IMPs in samples represented by Figure 1 is given in Figure 5 as frequency distribution histograms.

To facilitate comparison between data groups of different particle numbers, and summarize the data, a ΔS, or change in size parameter is defined.[44] The size distribution histograms of two data sets to be compared are normalized for differences in total numbers of IMP observed. The difference histograms are calculated using the LOTUS-123 routine. The absolute values of the deviations of the difference histogram are summed to give a one number parameter, the ΔS. The smaller the ΔS value, the smaller the difference between the two IMP size distributions.

III. SPATIAL DISTRIBUTION ANALYSES OF INTRAMEMBRANE PARTICLES

A. PRINCIPLE

Consider the fracture plane contains i-types of IMP, and each type has its own size, shape, and distribution pattern. The structure of a single IMP of the i-th type is given by $f_i(\mathbf{r})$. The distribution of this type of IMP is described mathmatically as $\delta(\mathbf{r}_O) + \rho_i P_i(\mathbf{r})$ where $\delta(\mathbf{r}_O)$ is a delta function of a test particle at $\mathbf{r}_O$ and $P_i(\mathbf{r})$ is the probability of finding a neighbor at vector displacement $\mathbf{r}$. The value ρ_i is the density of the i-type IMP. The structure of the entire fracture plane is then:

$$\phi(\mathbf{r}) = \sum_i \{f_i(\mathbf{r}) * [\delta(\mathbf{r}) + \rho_i P_i(\mathbf{r})]\} \tag{1}$$

where * denotes the convolution operation. In most cases, the freeze fracture technique does not preserve the structure of IMP other than the size. Therefore, the function $f(\mathbf{r})$ may be simplified into a disc of a given diameter, and will not be given further consideration. We shall focus on the characterization of the distribution function $P(\mathbf{r})$. For the sake of simplicity, only one type of IMP is considered in the following analysis. The principle may be applied to combinations of several types of IMPs, providing the overlapping problem is taken care of (see below).

The inter-IMP relationship may be characterized by the autocorrelation function;

$$\zeta(\mathbf{r}) = \int \phi(\mathbf{r}') \, \phi(\mathbf{r}-\mathbf{r}') \, d\mathbf{r}' \tag{2}$$

If there are preferred distances at preferred directions $\mathbf{r}$ between IMPs, the autocorrelation function will show a maximum at these $\mathbf{r}$, $2\mathbf{r}$, etc. values. Thus, the radial maxima of the 2-D autocorrelation function $\zeta(\mathbf{r})$ indicate favorable inter-IMP distance within a cluster, as well as inter-cluster distances. Angular relationships between IMPs, and between clusters of IMPs, are given in the azimuthal peaks of $\zeta(\mathbf{r})$. If the IMPs are packed in any regular or semiregular arrays, $\zeta(\mathbf{r})$ is also regular or semi-regular, in analogy to the Patterson function in crystallography. Figure 6 shows the autocorrelation function of Figure 3D, in which the IMPs form close but irregular (not random) clusters. The function $\zeta(\mathbf{r})$ is computed from the function $\phi(\mathbf{r})$, shown in Figure 3D, using a SPIDER program. The inner ring in Figure 6 indicates the minimal distance between IMPs in the close pack formation. The outer maxima are higher "order" correlation peaks of $\mathbf{r}_m$ and $\mathbf{r}_n$ of the distribution peaks of $P(\mathbf{r})$ at $\mathbf{r}_m$ and $\mathbf{r}_n$ etc.

Another way to characterize the distribution function $\phi(\mathbf{r})$ is by Fourier analysis. The Fourier transform of $\phi(\mathbf{r})$ is

$$\Phi(\mathbf{R}) = F(\mathbf{R})[1 + \rho Q(\mathbf{R})] \tag{3}$$

where $F(\mathbf{R})$ and $Q(\mathbf{R})$ are the Fourier transforms of $f(\mathbf{r})$ and $P(\mathbf{r})$, respectively. This operation can be performed either in a computer, or by the use of the analogous Abbe's diffraction theorem. The recorded diffraction pattern of $\phi(\mathbf{r})$:

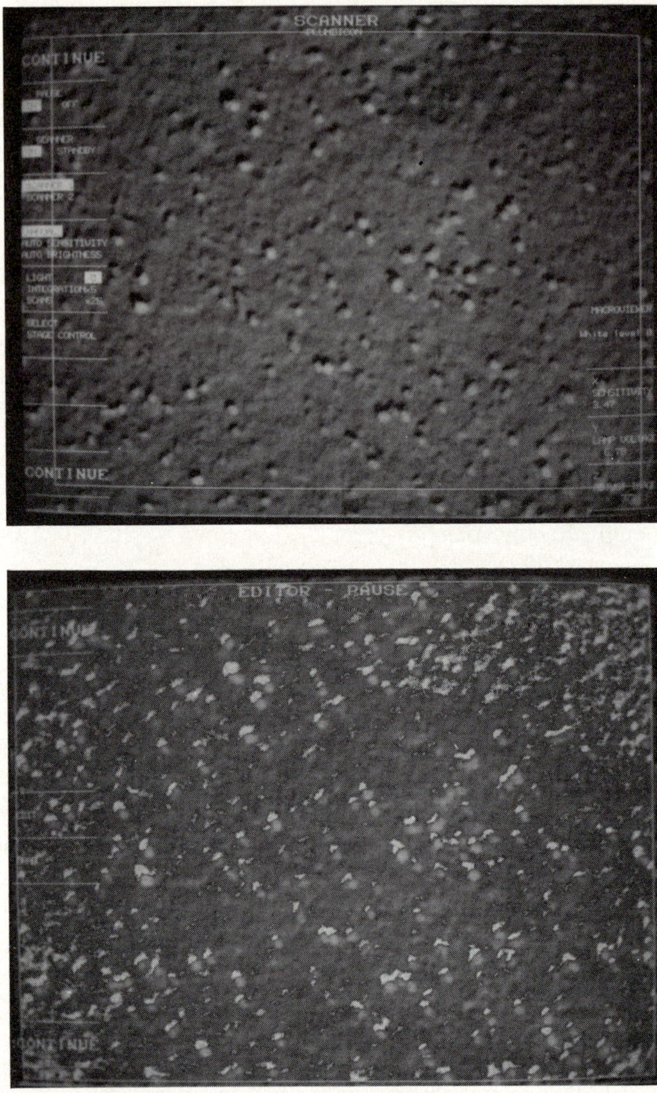

A

B

FIGURE 4. Sequential stages of IMP recognition in a Cambridge Quantimet-920 image analyzer. (A) Original image, (B) after the first 2-D DETECTION based on a grey level threshold. White spots indicate detected areas. (C) After WHITE TRACKING which detects abrupt changes of grey levels. (D) Result of Boolean AND of successive 2-D DETECTION and WHITE TRACKING. (E) The final binary image of the same field as A. Horizontal length of rectangles = 500 nm. (From Fassel, T. A. and Hui, S. W., *J. Microsc.*, 149, 37, 1988. With permission.)

$$I(\mathbf{R}) = [\Phi(\mathbf{R})\ \Phi(\mathbf{R})^*] \tag{4}$$

is in fact the Fourier representation of $\zeta(\mathbf{r})$. The correlation between IMP is expressed in the reciprocal space vector **R**. Figure 7 illustrates the expression of real space particle distribution (Figure 7 A-D) in terms of optical diffractions (Figure 8A-D). The regularity of particle arrangements is clearly reflected in their respective diffraction patterns. In relatively irregular distributions, such as in Figures 7B and 7C, the diffuse rings in Figures 8B and 8C show that there is a preferred radial interparticle distance, but no angular correlation. An entirely random distribution (Figure 7D) has no preferred radial nor angular correlation (Figure 8D).

C
D
E

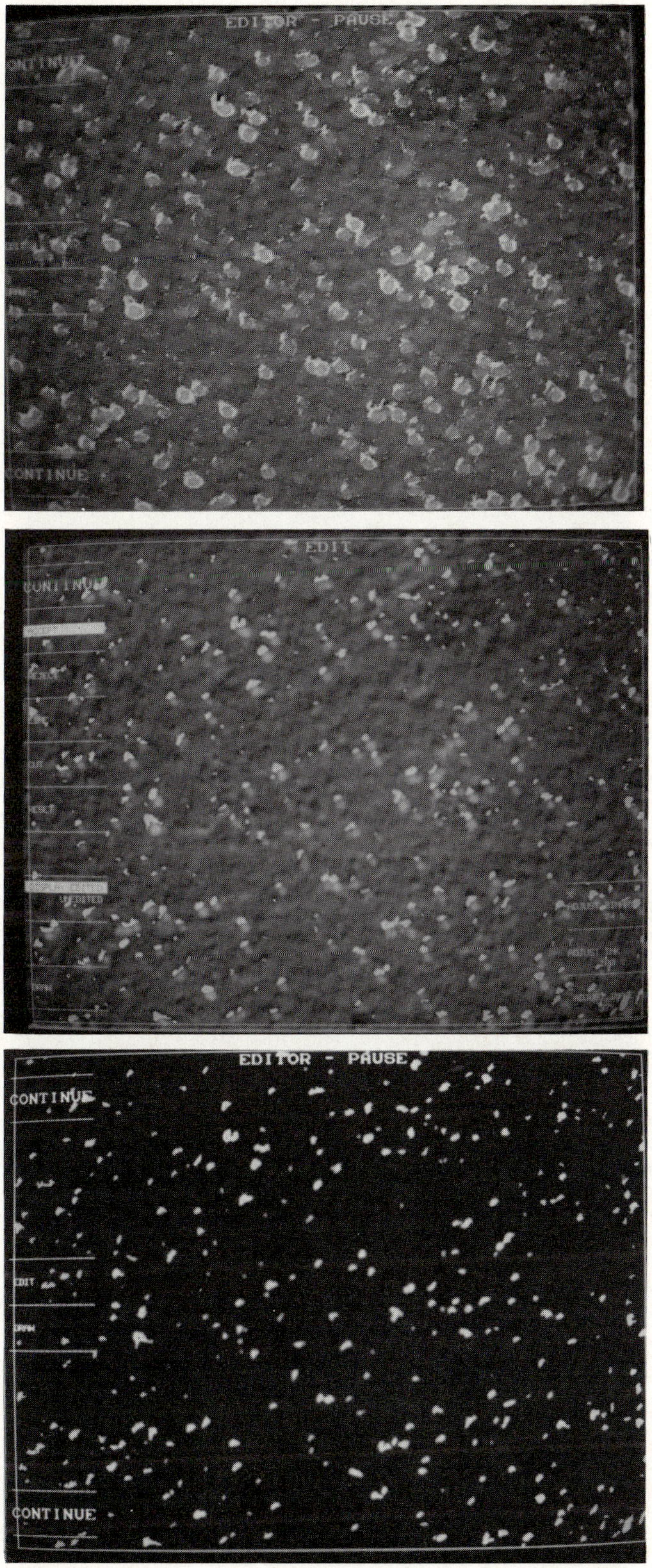

FIGURE 4 C, D, E.

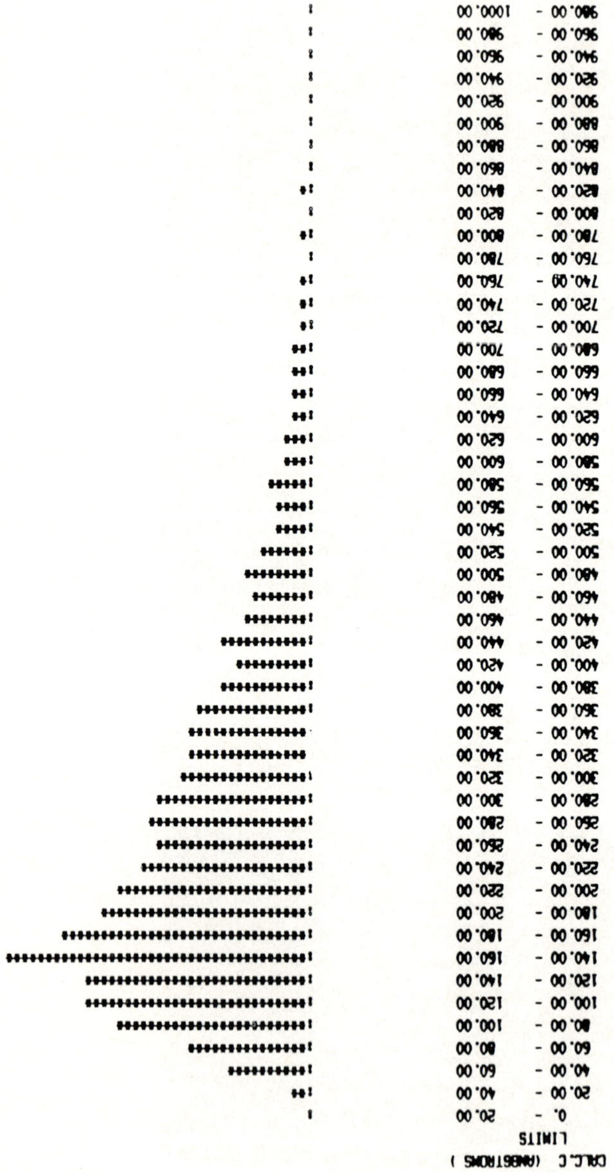

FIGURE 5. The size histogram of intramembrane particles from a sample represented by Figure 1.

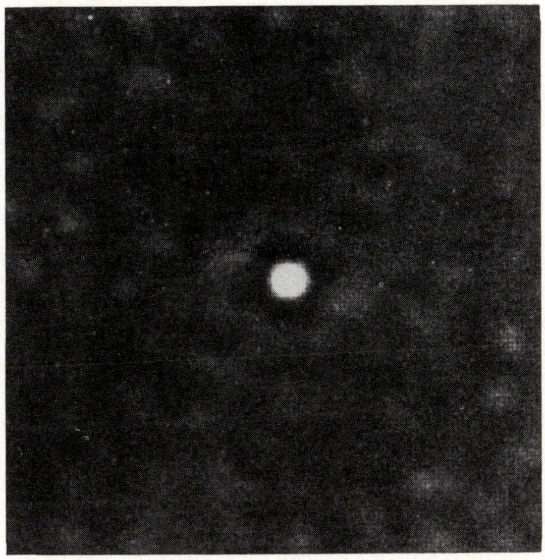

FIGURE 6. The 2-D autocorrelation function derived from Figure 3D, showing the preferred nearest neighbors as the pericentral "ring". Longer range ordering is indicated by outlying patches. Length of square = 100 nm. (From Hui, S. W. and Frank, J., *J. Microsc.*, 137, 293, 1985. With permission.)

B. CHOICE OF MODELS

Since P(**r**) is in general irregular for most biological membranes, a statistical approach is needed for quantitation. A standard distribution is required for comparison. One of the standard distributions is the random distribution. A truly random distribution obeys the Poisson statistics. Figure 7D is a 2-D random distribution. The x and y coordinates of the points were generated by an IBM Scientific Subroutine package RANDU. The points were generated and placed in sequence. Since the points are assumed to occupy no area, the chance of generating a pair of identical coordinates for a finite number of points is extremely small. The points can be placed extremely close to one another (random point model). This model, however, is not a good standard for a random distribution of IMPs, because IMPs occupy finite areas; therefore, there is a limit to how close each particle can be placed next to another.

In order to find a model to be the standard of randomness, conditions for particle size and density are imposed. We have generated a number of random models for different constraints.[32] The coordinate numbers in these models were generated by the linear congruential method,[46] and entered sequentially. Each coordinate pair was assigned a repulsion radius "d" within which no other points can be placed (hard disc model). Figure 7B represents one of these models, with the constraint $d^2\rho = 0.49$ where d is the diameter of the hard disc. The model seems to be much more regular in comparison to Figure 7D, where $d^2\rho = 0$. This is a more realistic "random" standard for this particular constraint. Figure 7C is an idealized hard disc map of Figure 11B. The distribution is equivalent to a model with constraint $d^2\rho = 0.22$. Therefore, the theoretical random statistics cannot be applied as a standard for IMP distribution analysis. A random model must be generated for each sample, with given constraints of IMP diameter and density.[32]

There are other models for IMP distribution, with various constraints. Potentials for interparticle attraction and adhesion have been considered.[25,34,47] These models help to simulate the interparticle interaction. Like all models, the more parameters one uses in the

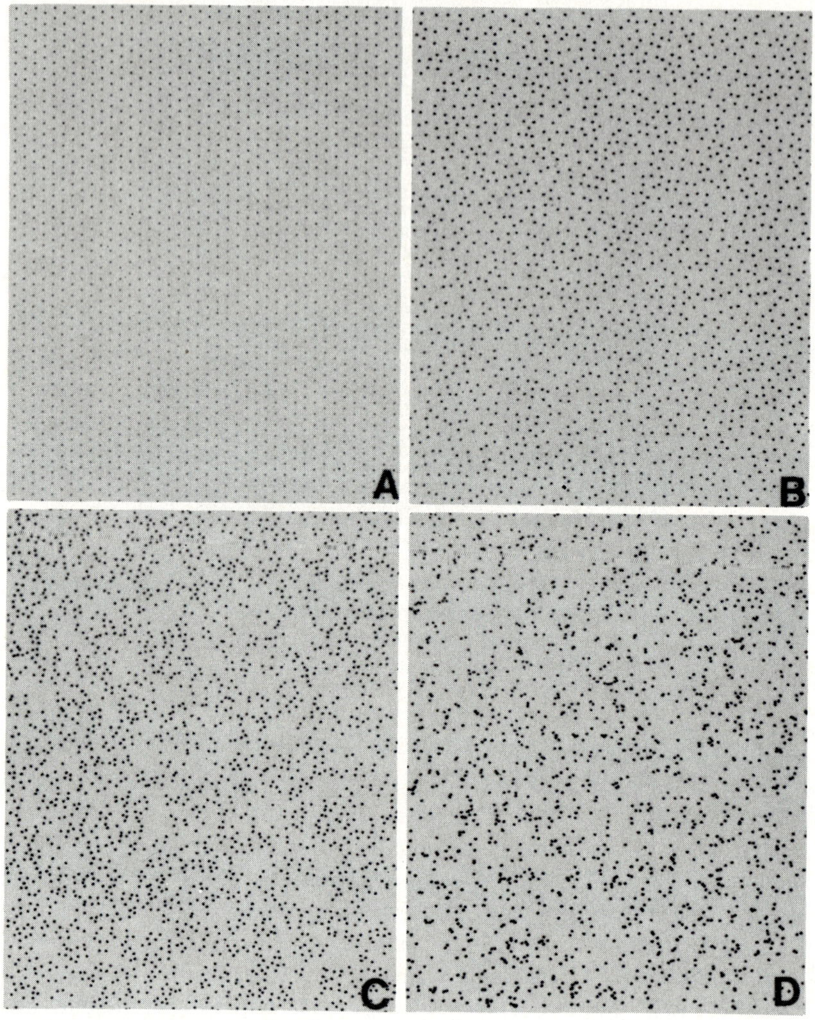

FIGURE 7. Particle distribution functions of various degrees of packing regularity. (A) Hexagonal close pack, (B) random model with constraint $d^2\rho = 0.49$ (random hard disc model). (C) IMP distribution as digitized from the micrograph Figure 11B. (D) Random model with constraint $d^2\rho = 0$ (unconstraint random point model).

model, the more information it provides, but it also is farther from being the uniquely fitted one.

C. ANALYSIS OF DENSITY DISTRIBUTION

A criterion for randomness against evenness (or regularity) of a 2-D distribution is the density distribution analysis (Figures 9 to 11). The 2-D field is divided by a grid so that the number of IMP in each grid square is sufficiently small ($n < 15$) for the Poisson condition to hold. A histogram is generated for comparison with those of a corresponding random model (Figure 9), and of a uniform model. By definition, the histogram for a uniform model is a delta function at the occupation number corresponding to ρ. The random model under our constraint is usually a bell-shaped curve. The absolute deviation χ^2 is the coefficient of dispersion.[5,16,31] The density distribution histogram D(n) for Figure 11B is plotted in Figure

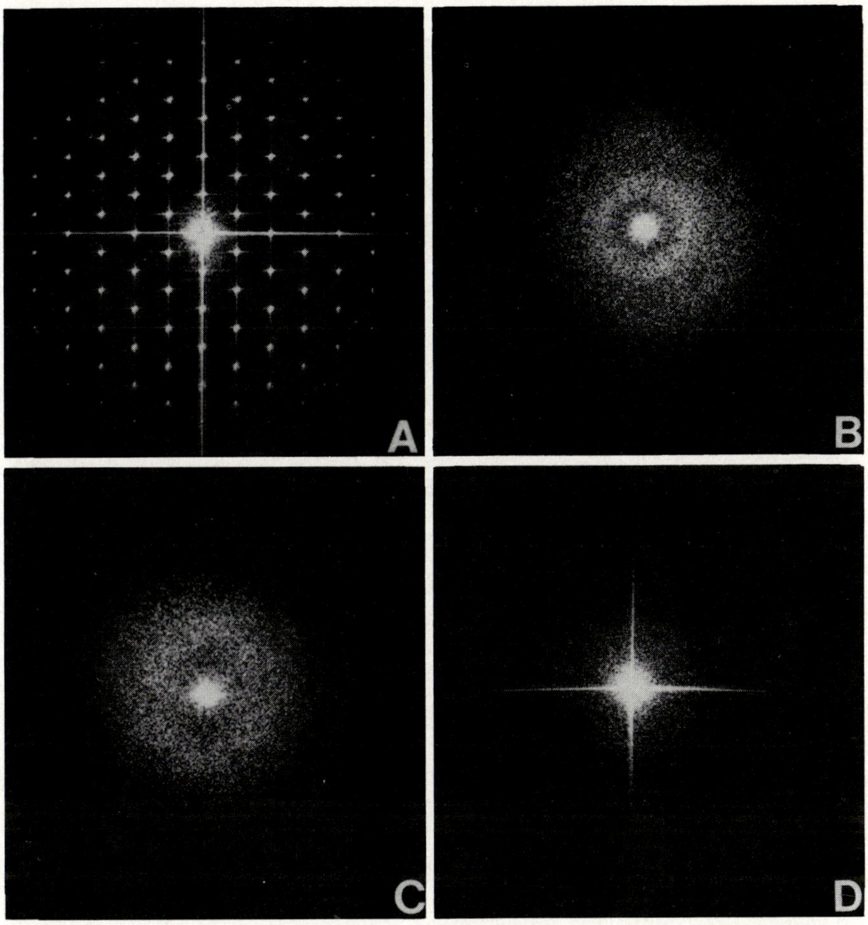

FIGURE 8. Optical diffraction patterns of respective distributions in Fig. 7. The crosses in the centers of (A) and (D) are overexposed optical artifacts due to the rectangular boundaries of the real space pictures.

9, together with those calculated from the corresponding random hard disc model and the random point model. The deviation from randomness is the difference function $\Delta(n)$ of the histograms. Since the density and the total number of grid squares applied to the constraint model must be the same as those applied to the sample field for a valid comparison, a model must be calculated for each micrograph field. Alternatively, the number of square and the mean density may be normalized to avoid repeating lengthy calculations.

The shape of the difference function $\Delta(n)$ contains information of the type of deviation. Since the histograms to be compared are normalized, the net difference is always zero. The curves of typical $\Delta(n)$ are plotted in Figure 10. The areas of each curve must satisfy

$$D1 + D3 = D2 \tag{5}$$

An even distribution has a large negative difference (D2) at the mean occupancy value (normalized to 1), and positive difference elsewhere. An aggregated distribution gives an opposite difference characteristic. All the irregular cases we studied are of the latter type; therefore, only this type of curve is discussed in detail.

The ratio D1/D3 is a sensitive indicator of the characteristics of density distribution.

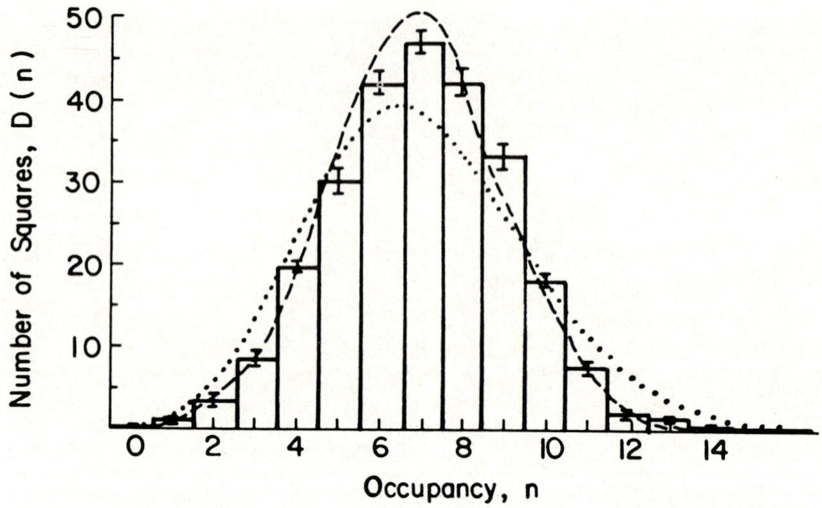

FIGURE 9. Particle density distribution histogram calculated from Figure 7C. The distributions calculated from constraint random models with $d^2\rho = 0$ and 0.22 are drawn as dotted and dashed curves, respectively. (From Pearson, R. P., Hui, S. W., and Stewart, T. P., *Biochim. Biophys. Acta,* 557, 265, 1979. With permission.)

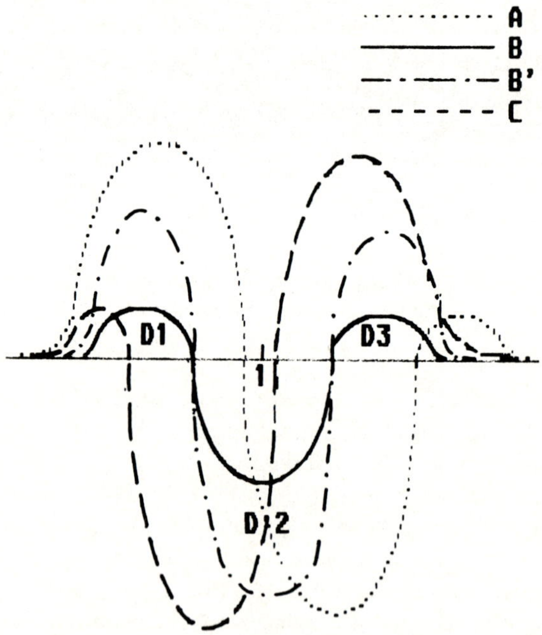

FIGURE 10. Typical differential density functions are depicted with their integral areas D1, D2, and D3, from the $\Delta(m) = 0$ axis. The point $m = 1$ represents the mean normalized occupancy number. The $\Delta(m)$ scale is arbitrary. The 3 cases D1 > D3, D1 = D3, D1 < D3 are represented by curves A, B, and C, respectively. B′ is a special case of B, with larger integral areas.

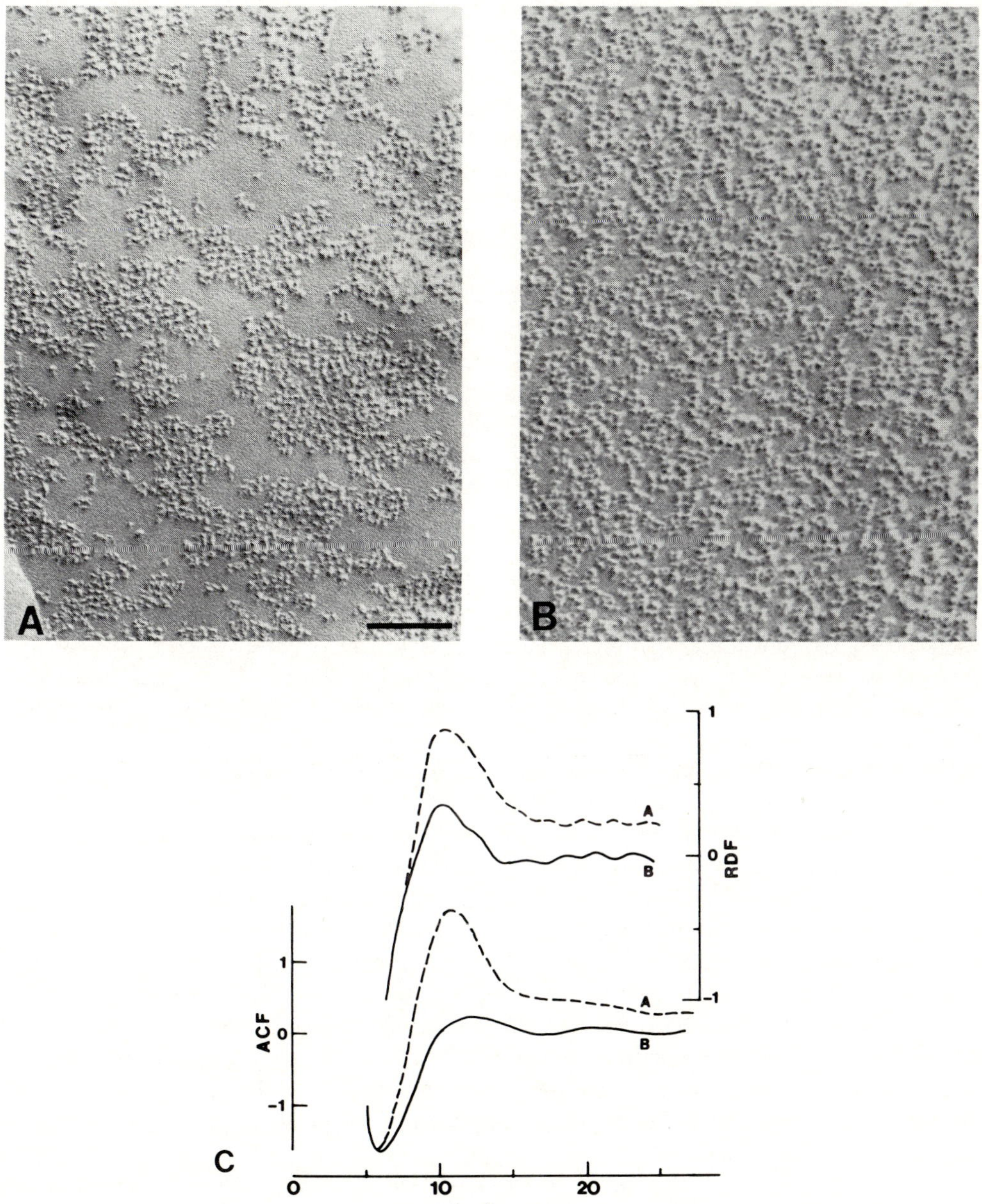

FIGURE 11. Freeze fracture electron micrographs of the P-face of human erythrocyte membranes (A) treated hypo-osmotically and at low pH to induce IMP aggregation, and (B) a control sample. Bar = 100 nm. (C) The radial distribution function (RDF) and the azimuthally averaged auto-correlation function (ACF) of the respective particle distributions in (A) and (B). (From Hui, S. W. and Frank, J., *J. Microsc.*, 137, 293, 1985. With permission.)

For instance, D1 ~ D3 (Figure 10, curve B) indicates that there are more high occupancy aggregates and local IMP-poor "holes" than those depicted by the random model. The unbalance is even greater in curve B′. A high D1/D3 ratio (curve A) indicates that there are more grid squares of low occupancy in comparison to the random model. In this case, the low density "holes" are created more at the expense of shifting the mean occupancy number; rather than creating more high occupancy squares. In other words, the particle-poor patches are formed by a slight increase the average density of the rest of the field, rather than forming high density IMP clusters. Similarly, in the case of small D1/D3 (curve C), aggregates are formed mainly by drawing IMPs globally. Such cases do occur in the kinetic progress of drug-treated cells. This analysis has been applied to differentiate the change in IMP distribution as functions of the length of drug treatment.[44,48]

Besides the test of randomness, the choice of grid size can indicate the characteristic size of aggregates and holes. If the size of these features matches that of the applied grid squares, the $\Delta(n)$ effect is at the maximum. Thus, a scan of grid size against the coefficient of dispersion usually gives the characteristic feature size.[31,48]

D. ANALYSIS OF INTER-PARTICLE DISTANCE DISTRIBUTION

Most of the IMP distributions we studied, apart from the regular arrays such as gap junctions and purple membranes, have weak of absent angular correlation. The distribution $P(\mathbf{r})$ is therefore circularly symmetric. Since $f(\mathbf{r})$ may be regarded as a circular disc, the 2-D distribution function is reduced to a 1-D distribution function. This simplification enables the quantitative analysis by the radial distribution method alone.

The one dimensional isotropic (azimuthal averaged) equivalence of equation (1) for a single IMP type is

$$\phi(r) = f(r) *[\delta(r) + \rho p(r)] = f(r) *[\delta(r) + \rho + \rho\{p(r) - 1\}] \qquad (6)$$

where p(r) is the probability of finding another IMP at any radial distance r. The function $\{p(r) - 1\}$ is generally known as the radial distribution function (RDF). A peak in the RDF indicates a preferred interparticle distance over others. For an even distribution, the RDF is a series of delta functions at the fundamental and higher harmonics of the unique nearest neighbor distance. For a truly random model, the RDF is a constant. This function is sometimes quoted as pair distribution function. [24,25,34]

For the cases of IMP distribution, including their corresponding hard disc models, the RDF is zero within the hard disc diameter due to the exclusion volume. The peaking of the function depends on the evenness and IMP density. In general, the peak height increases with $d^2 \rho$ of the model chosen. Since this value is an indication of nonrandomness, the lower and broader the peak in RDF, the closer it is to a true random distribution. An example is given in Figure 11. The RDF for the micrographs (Figures 11A and 11B) were obtained by counting the number of IMPs within a ring defined by two given radii, r_1 and r_2 from a chosen IMP. This was evaluated for incremental values of r to the edge of the field, and the process was repeated for all IMPs. The counting was done in a UNIVAC 90/60 computer, and plotted in Figure 11C. The highly aggregated sample gives a much more prominent peak in the corresponding RDF.

Interestingly, being a self convolution of $\phi(r)$, the azimuthal averaged autocorrelation function (ACF) also has its first correlation peak position coincide with the r value of the first peak of the RDF. The peak amplitudes should also correspond. This is indeed the case, as illustrated in Figure 11C. The ACF was calculated from the Fourier transform of the digitized micrographs, and inverse-transformed back to the real space, as discussed in section A. The computation utilized the SPIDER program designed for spatial correlation.[49]

The Fourier transform, or the corresponding optical diffraction pattern, can also be utilized to indicate preferential interparticle distances. The azimuthal averaged form of equation (3) contains the function $Q(\mathbf{R}) = q(R)$, where q(R) is the isotropic form of the Fourier transform of $P(\mathbf{r})$.

$$q(R) = \int p(r)\, r\, J_O(2\pi Rr)\, dr \qquad (7)$$

where J_O is the zeroth order Bessel function. The first order approximation of the diffraction intensity from equation (4) is

$$I(R) = F^2(R)[1 + 2\rho q(R)] \tag{8}$$

This is illustrated by the optical diffraction (Figure 8C) of Figure 7C, which is an idealized map of Figure 11B. The white ring in Figure 8C corresponds to a peak at $R_1 = 1/r_1$ where r_1 is the peak RDF value at Figure 11C, curve B. Therefore, the RDF, the ACF, and the Fourier or optical transform of the distribution, can equivalently indicate the state of aggregation of IMPs.

The interparticle distance analysis may also be accomplished by using commercially available particle counting routines. The Cambridge Quantimat QUIPS program has a nearest neighbor function, which measures the statistical distribution of all nearest neighbor distances. A nearest neighboring particle is found for each IMP considered, using the Pythagorean theorem. Once the nearest neighbor for a particular IMP is found, the measured nearest neighbor distance is stored, and that first IMP is excluded from further consideration in the image area. In this manner, all IMPs are tested, and the distributions of nearest neighbor distances are given. The frequency distribution histograms for nearest neighbors were normalized for differences in total numbers of distances measured. The peak of the nearest neighbor histogram also coincides with that of RDF, with corresponding width.[44,48] The difference is that all interparticle distances are considered in RDF, whereas only one nearest neighbor distance per IMP is registered in the latter method. At large r values, the IMP counts in RDF tend to average to ρ; the peak at small r is dominated by nearest neighbors. For a large sample size, the results from the two analyses is equivalent.

E. MULTIVARIANT ANALYSIS

There are many parameters to characterize the distribution of IMP. So far we have discussed the size, density, density distribution (coefficient of dispersion, D1/D3), preference of interparticle distance (peak positions in RDF or ACF, nearest and long range neighbors), packing regularity (peak height in RDF or ACF, amplitudes in $\Delta(n)$), and interparticle orientation. This list is by no means exhaustive. Even with a limited number of parameters, it is possible to classify the distribution using a second-tier, multivariant analysis.

The parameters listed above are not entirely independent. For instance, the density and the average interparticle distance has an inverse relationship.[48] These two parameters are, in turn, linked to the packing regularity via the particle size, because the evenness of packing increases with $d^2\rho$. These relations are not apparent, and may lead to trivial conclusions. However, their interdependent relationships may help us to identify subtle distribution characteristics.

The peak height of RDF and the integral value of $\Delta(n)$ were used together to reveal the difference between the actual IMP distribution in erythrocyte membranes and in various random models.[32] Figure 12 is an example which shows that increasing IMP aggregation from sample 1 to 10 is accompanied by the increase of both parameters, whereas increasing aggregation of points in the random models leads to the decrease of peak height in RDF. From the different trends we learned that the aggregation of IMP in high number samples is due to IMP grouping instead of randomization, and none of the samples 1 to 10 can be regarded as random, because none of them falls on the random model curve.

Recently, we have completed a study of the responses of culture fibroblasts to biomodifiers, such as retinoids and phorbol esters, and found that the D1/D3 vs. nearest neighbor peak height gave excellent characterization of IMP in response to drug treatment.[48] In this case, the value of ρ is very small; therefore, the packing regularity is almost independent of density.

As more statistical parameters are defined to describe the irregular planar distribution of IMP, the multivariant analysis is going to play a more important role in solving the ambiguity of single variant analysis.

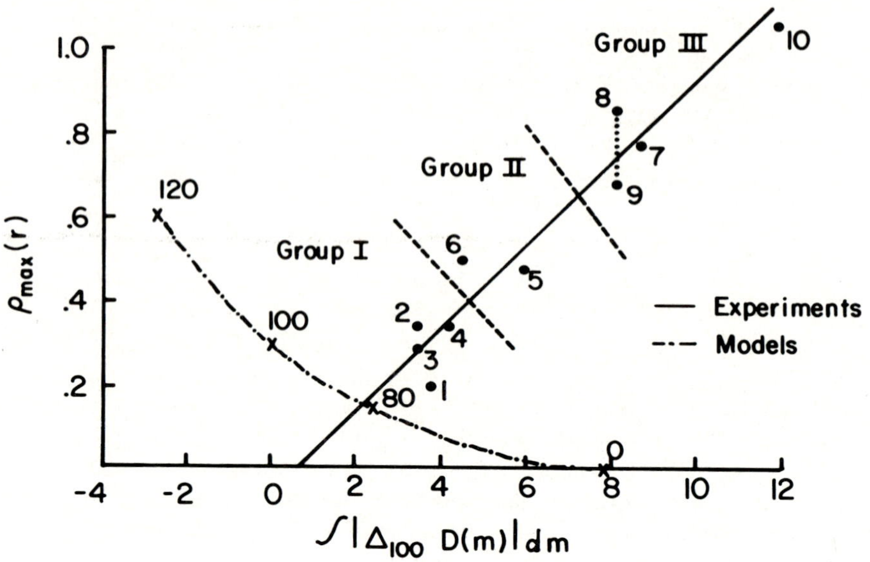

FIGURE 12. Multivariant analysis of the particle distributions of human erythrocyte membrane samples by the correlation of RDF [indicated by the peak height $\rho_{max}(r)$] and $\Delta(m)$ [indicated by the total integral difference area D1 + D2 + D3 = $\int\Delta_{100}D(m)$ dm]. Group I to III indicate samples with increasing degree of IMP aggregation. (From Pearson, R. P., Hui, S. W., and Stewart, T. P., *Biochim. Biophys. Acta,* 557, 265, 1979. With permission.)

IV. STATISTICAL MEASUREMENTS OF MEMBRANE DYNAMICS

With the advance of rapid freezing, kinetic events may be frozen at given time points to reveal the sequence of morphological changes.[50-52] So far, data derived from the majority of this type of studies are collected by visual examination and manual counting. It is expected that, as more time-dependent experiments are performed and a larger volume of data are collected, the need for automatic pattern recognition and statistical analyses will rise.

An alternative way to observe membrane dynamics is the study of *in vitro* membrane motion directly in the electron microscope, using an environmental chamber.[53] Studies of this type have been reported previously.[54,55] As an example of how statistical methods are applied to analyze these data, the following unpublished results are given. Human erythrocyte membrane was spread on the surface of a Langmuir trough, and reformed into a planar bilayer over small holes by the dipping method.[56] The reconstituted membrane contains essentially all the protein and lipid components of the native membrane, but for bands 5 and 7 proteins.[55] One side of the bilayer was labeled with concanavalin-A conjugated colloidal gold microspheres, which became attached to the mannose moiety of membrane glycoproteins. The components of the planar bilayer was free to move laterally, since there was no solid support. Figure 13 shows the movement of the labels in a 5 s exposure. The membrane within the field of view may be divided into several domains, according to the manner of the label movement. Some of the domains have stationary labels, while others contain labels which move in concert. Over 300 of such domains were counted, and the results were grouped according to the sample temperature and composition.

An attempt was made to correlate the size of the motional domains and their speed of motion. This is to answer the question whether the small domains are more mobile, and if the alteration of temperature and composition would change their dynamical states. The correlation plots are shown in Figures 14 and 15, where the statistics are given in terms of

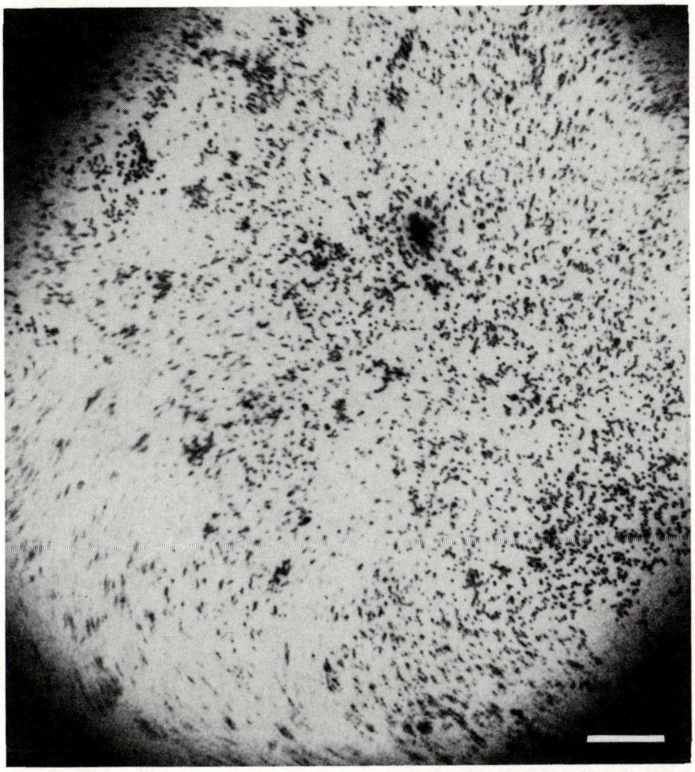

FIGURE 13. The motion of a free standing, reconstituted planar erythrocyte membrane labeled with concanavalin-A conjugated gold microspheres. The low dose micrograph was exposed for 5 s at 23°C(296 K) to trace the motion of individual labels. The different motional domains are readily discernible. Bar = 1 μm.

sample temperature and cholesterol/phospholipid (C/P) ratio, respectively (experimental details are given in Reference 55). Although the trend that the domains tend to be larger and less mobile at low temperature is apparent, the correlation between domain size and velocity at each temperature (or composition) is very weak. The highest correlation coefficient is only 0.59 for the sample temperature at 12.5°C (285.5 K). The data are, therefore, better reported in terms of mean size and mean velocity, independently.[55]

V. CONCLUSION

Electron microscopy has been a morphological tool mainly in the descriptive sense. Recent developments in cytochemical labeling and microanalysis have promoted it into an analytical tool. The application of morphometry enables quantitative data to be derived from pictorial results.

As all microscopists know, adequate sampling is always a major concern in electron microscopy. The advance of computers and image processing technology greatly fostered the development of quantitative analyses in electron microscopy. Many previously unsurmountable obstacles, because of either the immensity of data handling or the inevitable subjectivity in visual judgment, are now overcome. New techniques are developed to take advantage of these capabilities.

The examples I discuss above are but a small facet of the growing methodology. These examples are mainly centered on the analysis of IMP distribution, but the methods are

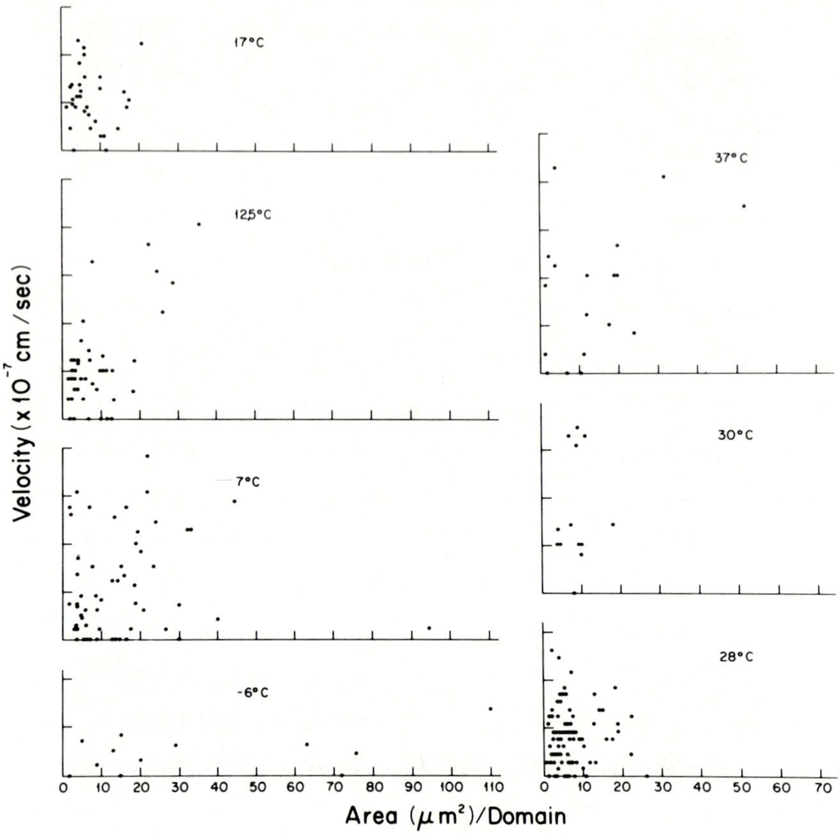

FIGURE 14. Correlations of domain size and velocity of samples described in Figure 13 at different temperatures.

equally applicable to analyzing distributions of cell surface labels, autoradiographic grains, and even cell or colony images in light microscopy. Many image analysis software packages (at ever lowering costs) may be adapted to perform the tasks which previously demanded professional programming. I expect that our knowledge of membrane structure and dynamics will be advanced more rapidly by the widespread use of the computer in electron micrograph analysis.

ACKNOWLEDGMENTS

I wish to express thanks for the efforts of Drs. R. H. Pearson, N. B. He, T. F. Fassel, and T. P. Stewart, who at various times have contributed to the various aspects of the above work in my laboratory. The collaboration with Dr. Joachim Frank is a key point in the analysis. The work at different stages is supported by grants GM-28120, GM-30969 from the National Institutes of Health. The use of the electron microscope and the Cambridge Quantimet 920 is through the shared instrument grants RR-01474 and RR-02347, respectively, from the National Institutes of Health.

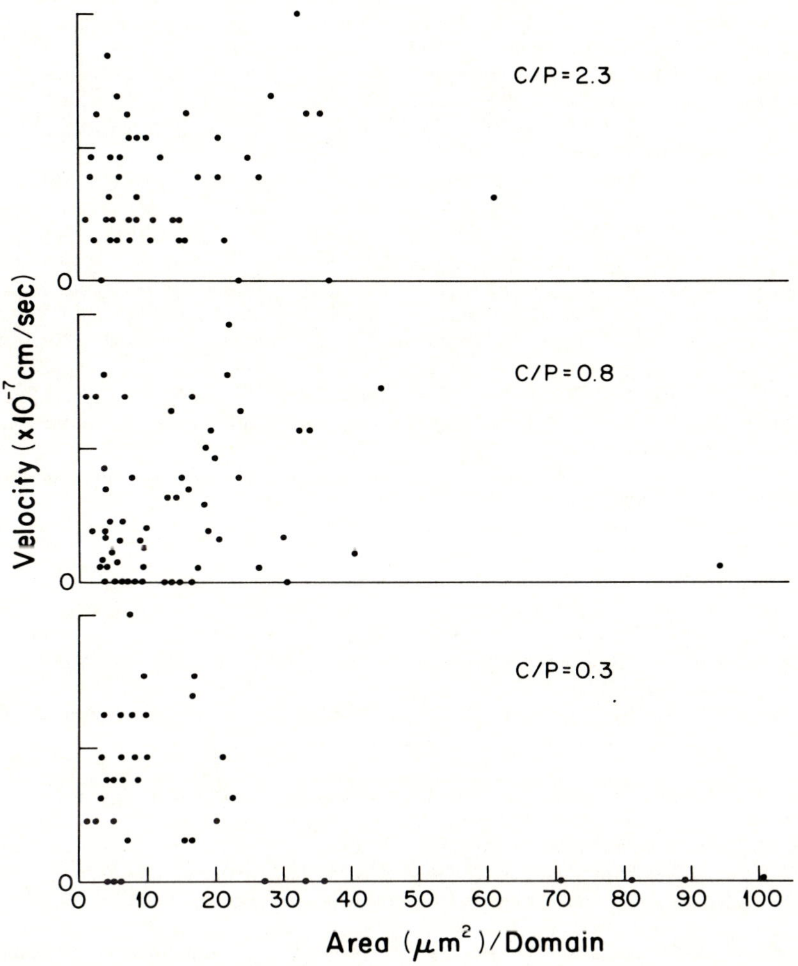

FIGURE 15. Correlations of domain size and velocity of samples described in Figure 13 for membranes containing different cholesterol/phospholipid (C/P) ratios, as modified by lipid exchange.[55]

REFERENCES

1. **Sase, S., Anraku, Y., Nagano, M., Osumi, M., and Kasahara, M.,** Random distribution of the glucose transporter of human erythrocytes in reconstituted liposomes, *J. Biol. Chem.*, 251, 11100, 1982.
2. **Hong, K. and Hubbell, W. L.,** Preparation and properties of phospholipid bilayers containing rhodopsin, *Proc. Natl. Acad. Sci. U.S.A.*, 69, 2617, 1972.
3. **Rash, J. E., Johnson, T. J. A., Dinchuk, J. E., and Levinson, S. R.,** Labeling intramembrane particles in freeze-fracture replicas, in *Freeze Fracture Studies of Membrane Structure*, Hui, S. W., Ed., CRC Press, Boca Raton, FL, in press.
4. **Pinto da Silva, P.,** Molecular cytochemistry of fractured cells: freeze etching fracture label, fracture permeation and label fracture, *Adv. Cell Biol.*, 1, 157, 1987.
5. **DeLaat, W., Tertoolen, L. G. J., van der Saag, P. T., and Bluemink, J. G.,** Quantitative analysis of modulations in numerical and lateral distribution of intramembrane particles during the cell cycle of neuroblastoma cells, *Eur. J. Cell Biol.*, 96, 1047, 1983.
6. **Staehelin, L. A.,** Reversible particle movements associated with unstacking and restacking of chloroplast membranes *in vitro*, *J. Cell Biol.*, 71, 136, 1976.

7. **Martinez-Palomo, A., Chavez, B., and Gonzalez-Robles, A.,** The freeze fracture technique: application to the study of animal plasma membranes, *Proc. 9th Int. Congr. Electron Microsc.*, 3, 1973, 503.
8. **Wade, J. B., Kachadorian, W. A., and Discala, V. A.,** Freeze-fracture electron microscopy: relationship of membrane structure features to transport physiology, *Am. J. Physiol.*, 232, F77, 1977.
9. **Roos, D. S., Robinson, J. M., and Davidson, R. L.,** Cell fusion and intramembrane particle distribution in polyethylene glycol-resistant cells, *J. Cell Biol.*, 97, 909, 1983.
10. **Hui, S. W., Isac, T., Boni, L. T., and Sen, A.,** Action of polyethylene glycol on the fusion of human erythrocyte membranes, *J. Membr. Biol.*, 84, 137, 1985.
11. **Pinto da Silva, P.,** Translational mobility of the membrane intercalated particles of human erythrocyte ghosts: pH-dependent, reversible aggregation, *J. Cell Biol.*, 53, 777, 1972.
12. **Speth, V. and Wunderlich, F.,** Membranes of tetrahymena II, direct visualization of reversible transitions in biomembrane structure induced by temperature, *Biochim. Biophys. Acta,* 291, 621, 1973.
13. **McIntyre, J. A., Gilula, N. B., and Karnovsky, J. J.,** Cytoprotectant-induced redistribution of intramembranous particles in mouse lymphocytes, *J. Cell Biol.*, 60, 192, 1974.
14. **Torpier, G., Montagnier, L., Biquard, J. -M., and Vigier, P. A.,** Structural change of the plasma membrane induced by oncogenic viruses: quantitative studies with the freeze fracture technique, *Proc. Natl. Acad. Sci. U.S.A.*, 72, 1695, 1975.
15. **Gilula, N. B., Eger, R. R., and Rifkin, D. B.,** Plasma membrane alteration associated with malignant transformation in culture, *Proc. Natl. Acad. Sci. U.S.A.*, 72, 3594, 1975.
16. **Weinstein, R. S.,** Changes in plasma membrane structure associated with malignant transformation in human urinary bladder epithelium, *Cancer Res.*, 36, 2518, 1976.
17. **Elgsaeter, A., Shotton, D. M., and Branton, D.,** Intramembrane particle aggregation in erythrocyte ghosts. II. The influence of spectrin aggregation, *Biochim. Biophys. Acta,* 426, 101, 1976.
18. **Arancia, G., Rosati, F., and Trovalusci, P.,** Density and size distribution of the intramembranous particles in the cytoplasmic membrane of human peripheral blood lymphocytes. II. Ultrastructural differences between T & B cells, *J. Electron Microsc.*, 31, 62, 1983.
19. **DeLaat, W., Tertoolen, L. G. J., and Bluemilk, J. G.,** Quantitative analysis of the numerical and lateral distribution of intramembrane particles in freeze-fractured biological membranes, *Eur. J. Cell Biol.*, 23, 273, 1981.
20. **Wold, V., Stockem, W., and Wohlfarth-Bottermann, K. -E.,** Influence of different freeze-fracture pretreatments on the fine structure of physarum polycephalum. A freeze-fracture and freeze-substitution study, *Eur. J. Cell Biol.*, 22, 667, 1980.
21. **Yoshimura, A., Yamashina, S., and Ohnishi, S. I.,** Mobilization and aggregation of integral membrane proteins in erythrocytes induced by interaction with influenza virus at acidic pH, *Exp. Cell Res.*, 160, 126, 1985.
22. **Zernicke, F. and Prins, J. A.,** Die Beugung von Röntgenstrahlen in Flüssigkeiten als Effekt der Molekulanordnung, *Z. Physik,* 41, 184, 1927.
23. **Bernal, J. D.,** The structure of lipids, *Proc. R. Soc.*, A280, 299, 1964.
24. **Pearson, L. T., Chan, S. I., Lewis, B. A., and Engelman, D. M.,** Pair distribution functions of bacteriorhodopsin and rhodopsin in model bilayers, *Biophys. J.*, 43, 167, 1983.
25. **Braun, J., Abney, J. R., and Owicki, J. C.,** How a gap junction maintains its structure, *Nature,* 310, 316, 1984.
26. **Pedro, N., Carmo-Fonseca, M., and Fernandes, P.,** Quantitative analysis of pore patterns on rat prostate nuclei using spatial statistics methods, *J. Microsc.*, 134, 271, 1984.
27. **Abbas, A. K., Ault, K. A., Karnovsky, M. J., and Unanue, E. R.,** Nonrandom distribution of surface immunoglobulin on murine B lymphocytes, *J. Immunol.*, 114, 1197, 1975.
28. **Gershon, N. D., Demsey, A., and Stackpole, C. W.,** Analysis of local order in the spatial distribution of cell surface molecular assemblies, *Exp. Cell Res.*, 122, 115, 1979.
29. **Knox, E. G.,** Distribution of rhesus antigens on red cell surfaces, *Br. J. Haematol.*, 37, 537, 1977.
30. **Markovics, J., Glass, L., and Maul, G. G.,** Pore patterns on nuclear membranes, *Exp. Cell Res.*, 85, 443, 1974.
31. **Niedermeyer, W. and Wilke, H.,** Quantitative analysis of intramembrane particle (IMP) distribution on biomembranes after freeze-fracture preparation by a computer-based technique, *J. Microsc.*, 126, 259, 1982.
32. **Pearson, R. P., Hui, S. W., and Stewart, T. P.,** Correlative statistical analysis and computer modelling of intramembraneous particle distributions in human erythrocyte membranes, *Biochim. Biophys. Acta,* 557, 265, 1979.
33. **Perelson, A. S.,** Spatial distribution of surface immunoglobulin on B lymphocytes, *Exp. Cell Res.*, 112, 309, 1978.
34. **Pearson, L. T., Edelman, J., and Chan, S. I.,** Statistical mechanics of lipid membranes: protein correlation function and lipid ordering, *Biophys. J.*, 45, 863, 1984.

35. **Hui, S. W. and Frank, J.,** Analysis of the irregular planar distribution of proteins in membranes, *J. Microsc.,* 137, 293, 1985.
36. **Usui, N., Takahamashi, I., Koga, T., and Kawaguchi, J.,** A quantitative study of freeze fractured biological membranes using an image analyzer, *J. Electron Microsc.,* 33, 363, 1984.
37. **Usui, N. and Takahashi, I.,** Quantitative analysis of modulations in intramembrane particles during maturation of starfish oocytes, *J. Ultrastruct. Res.,* 87, 86, 1984.
38. **Frey, H.,** Automatic measurement of intramembraneous particle distribution by digital image processing, *IEEE Trans. Biomed. Eng.,* 34, 553, 1987.
39. **Giley, J. C. and Straehelin, L. A.,** Advances in ultrarapid freezing for the preservation of cellular ultrastructure, *J. Electron Microsc. Techniques,* 3, 177, 1986.
40. **Pfenninger, K. H. and Rinderer, E. R.,** Methods for the freeze fracturing of nerve tissue cultures and cell monolayers, *J. Cell Biol.,* 65, 15, 1975.
41. **Nermut, M. V.,** The cell monolayer technique in ultrastructural studies of cell membranes, in *The Science of Biological Specimen Preparation for Microanalysis,* Revel, J. -P. and Haggis, G. H., Eds., SEM Inc., AMF O'Hare, IL, 1984, 181.
42. **Pscheid, P., Schudt, C., and Plattner, H.,** Cryofixation of monolayer cell cultures for freeze fracture without chemical pretreatment, *J. Microsc.,* 121, 149, 1981.
43. **Espevik, T. and Elgsaeter, A.,** *In situ* liquid propane jet freezing and freeze-etching of monolayer cell cultures, *J. Microsc.,* 123, 105, 1981.
44. **Fassel, T. A. and Hui, S. W.,** Quantitative analysis of intramembraneous particles in rapidly frozen 10T-1/2 cell monolayers, *J. Microsc.,* 149, 37, 1988.
45. **Costello, M. J., Fetter, R., and Corless, J. M.,** Optimum conditions for the plunge freezing of sandwiched samples in science of biological specimen preparation, *Scanning Electron Microsc.,* 2, 105, 1986.
46. **Knuth, D. E.,** The Art of Computer Programming II, Seminumerical Algorithms, Addison-Wesley, Reading, PA, 1969.
47. **Finegold, L. and Donnell, J. T.,** Maximum density of random placing of membrane particles, *Nature,* 27, 443, 1979.
48. **Fassel, T. A., Bertram, J., and Hui, S. W.,** Surface morphology and intramembranous particle changes in parental and initiated 10T-1/2 cells upon phorbol ester treatments, in press.
49. **Frank, J., Shimkin, B., and Dowse, H.,** SPIDER— a modular software system for electron image processing, *Ultramicrosc.,* 6, 343, 1984.
50. **Heuser, J. E., Reese, T. S., Dennis, M. J., Jan, Y., Jan, L., and Evans, L.,** Synaptic vesicle exocytosis captured by quick freezing and correlated with quantal transmitter release, *J. Cell Biol.,* 83, 91, 1979.
51. **Hui, S. W.,** Kinetics of artifically-induced membrane fusion, in *Freeze Fracture Studies of Membrane Structure,* Hui, S. W., Ed., CRC Press, Boca Raton, FL, in press.
52. **Stenger, D. A. and Hui, S. W.,** Kinetics of ultrastructural changes during electrically-induced fusion of human erythrocytes, *J. Membr. Biol.,* 93, 43, 1986.
53. **Hui, S. W., Hausner, G. G., and Parsons, D. F.,** A temperature controllable hydration of environmental stage for the Siemens Elmiskop IA, *J. Phys. E (Sci. Instrum),* 9, 67, 1976.
54. **Hui, S. W.,** Direct measurement of membrane motion and fluidity by electron microscopy, *Nature,* 262, 303, 1976.
55. **He, N. B. and Hui, S. W.,** Electron microscopic observation of domain movement in reconstituted erythrocyte membrane, *Proc. Natl. Acad. Sci. U.S.A.,* 82, 7304, 1985.
56. **Hui, S. W., Parsons, D. F., and Cowden, M.,** Electron diffraction of wet phospholipid bilayer, *Proc. Natl. Acad. Sci. U.S.A.,* 71, 5068, 1974.

Chapter 5

LIPID PHASE TRANSITIONS AND MICRO-DOMAINS IN BIOMEMBRANES AND THEIR POSSIBLE RELEVANCE FOR BIOLOGICAL PROCESSES

A. J. Verkleij

TABLE OF CONTENTS

I. INTRODUCTION

Lipids are commonly thought to be rather dull and less intriguing molecules than other biomolecules, such as nucleotides and proteins. They are merely seen as building blocks of biomembranes forming a bilayer continuum between the membrane proteins. Thus viewed, the role of lipids is simply to separate two aqueous compartments and to guarantee a semipermeable barrier. This rather static view of the role of lipids is in conflict with the astonishing variety of lipids in biomembranes, and the very dynamic properties which biomembrane phospholipids possess. The diversity of these biomolecules is extraordinary, and is not only expressed in the number of lipid classes having different head groups, but also in the number of lipid species having such a variety in fatty acid composition. The reason for this diversity may be fortuitous, but it would seem reasonable to assume that there is a method in such madness—that is to say, that each type of component fulfills a particular role in the successful functioning of a biological membrane.

More and more, crucial roles can be attributed to particular lipid classes during biological processes: polyphosphoinositides, diglycerides, and phosphatidylserines in signal transduction;[1,2] glycolipids in receptor function;[3] specific lipid classes in controlling and regulating membrane-enzyme activity;[4,5] and translocation of proteins through membranes.[6]

The dynamic state of membrane lipids is recognized (and is clearly expressed in the "fluid" nature of lipids) in that many can form nonbilayer structures under physiological conditions in isolated form (lipid polymorphism), and by their high metabolic turnover, e.g., polyphosphoinositides during signal transduction.[2]

Much knowledge concerning the physico-chemical properties of lipids has been obtained from model membranes using a variety of biophysical techniques, such as differential scanning calorimetry (DSC), X-ray diffraction, nuclear magnetic resonance (NMR), and other spectroscopic methods like fluorescence photo-bleaching, recovery (FPR), and freeze-fracture electron microscopy. These methods have proven that lipids in biomembranes are "fluid" in nature, and can diffuse laterally rather fast in the plane of the bilayer. They showed that under nonphysiological conditions lateral phase separation of solid lipid domains can occur, leaving "fluid" domains into which the membrane-spanning proteins are concentrated. This latter phenomenon has been visualized by freeze fracturing as intramembranous particles (IMP) aggregation.[7] Studies on model membranes also have taught us about the polymorphism of lipids, which has recently undergone a revival (for review see References[8,9]). This was fueled by the discovery that biomembranes also contain lipids, which under physiological conditions can form the hexagonal H_{II} phase. This revival has been made possible by a successful combination of ^{31}P NMR, X-ray diffraction and freeze-fracture electron microscopy. The visualization of the "lipidic particles" by freeze fracturing has stimulated many researchers to investigate the role of nonbilayer-preferring lipids in dynamic events, such as membrane fusion, contact sites between membranes in mitochondria, and chloroplasts, as well as ion and protein translocation (see below).

In this review I will describe and discuss the contributions to, and the value of electron microscopy in increasing our understanding of membrane structure and dynamics, with particular reference to membrane domains, lateral phase separation as a result of fluid-solid transitions, and lipid polymorphism, and their possible relevance for biological processes.

II. ELECTRON MICROSCOPY

Electron microscopy is a very powerful technique to unravel membrane architecture and dynamics. In the early days of electron microscopy, some expectations were not fulfilled because preparation procedures included chemical fixation, dehydration, and embedding methods, which induced many artifacts and imposed severe limitations upon sample choice and interpretation of results. Lipids are extracted by traditional dehydration techniques, and relocation of lipids, as well as of proteins, can occur.[10] Freeze fracture, however, has been proven to be a powerful method to study membrane architecture, since the samples under investigation are rapidly frozen. Physical fixation by rapid freezing is now seen as the most reliable method to trap membrane molecules in their native state (reviewed in Reference 11). Because of the ice crystallization during freezing, particularly in tissue samples, a compromise had to be reached initially, i.e., freezing was preceded by a slight chemical fixation in glutaraldehyde coupled with cryoprotection with glycerol. However, nowadays the technology is available to freeze the specimen rapidly, which excludes large ice crystal formation, and thus makes the use of chemical fixation and cryoprotectants unnecessary. Rapid freezing, which literally "traps" the living state within less than a millisecond, i.e., with estimated cooling rates of 5.5×10^{-4} Ks^{-1} at 40 μm to 10^{-6} Ks^{-1} at 10 μm from the contact with the freezing source.[12] This implies that dynamic events occurring in the millisecond range and longer can in principle, be visualized by electron microscopy. Such freezing rates have been found to be essential to prevent the lateral phase separation of lipids caused by fluid-solid transitions and hexagonal-lamellar transitions (see below), and permit the visualization of fusion intermediates.

Rapid freezing is now also used in combination with freeze substitution at low temperature, which gives us an alternative insight in membrane events. In addition to the gain in structural preservation by rapid freezing, both freeze fracturing and freeze substitution can be combined with immunogold labeling and autoradiography, greatly enhancing the extent of techniques available to the electron microscopist.

III. THE ARCHITECTURE OF THE BIOMEMBRANE

The generally accepted hypothesis regarding the structure of biological membranes has been the result of a variety of approaches, including monolayer studies, electron microscopy, biochemical analysis, X-ray diffraction, differential scanning calorimetry (DSC), nuclear magnetic resonance (NMR), and fluorescence microphotolysis, etc.

In the "fluid mosaic model" of Singer and Nicolson,[13] biomembranes are built up of (glyco)proteins and (glyco)lipids. The lipids are organized in a fluid bilayer in which the transmembrane proteins are freely mobile in the plane of the bilayer (Figure 1a). Membrane lipids diffuse very rapidly in the plane of the bilayer. Information concerning this property has been obtained by fluorescence photobleaching recovery.[14] In general, lipid diffusion is 3 to 10 times slower in biomembranes than in bilayer vesicles,[15] i.e., diffusion of lipid is 1.6 $\mu m^2 \cdot s^{-1}$ and 6.3 $\mu m^2 \cdot s^{-1}$ at 37°C in the biomembrane and the model system, respectively. Lipid diffusion has also been followed using a fluorescent lipid probe,[16] or ^3H labeled dipalmitoyl lecithin.[17] The labeled lipids were incorporated into liposomes and followed after fusion with epithelial cells. The lipids appeared to diffuse very fast (within minutes) from the apical to the basolateral membranes. The diffusion after fusion at the apical membrane is so fast that with freeze fracturing no smooth domains are visible, even after a few seconds of fusion of the liposomes with the membranes. Protein diffusion is generally more restricted.[14,15] A single exception is the value $D = 0.3$ $\mu m^2 \cdot s^{-1}$, which is measured for rhodopsin in the photoreceptor membrane.[18] This is likely because this protein is freely mobile in the bilayer, which is not the case for other proteins (see also above).

Lateral diffusion of membrane protein has also been studied by electron microscopy on the mitochondrial inner membrane, using the freeze fracture technique.[19] The mitochondria were first cooled down below the phase transition of the lipid. Upon warming up to physiological temperatures, this lateral phase separation is reversed and the former random distribution recovers. In order to determine the lateral diffusion coefficient of intramembrane particles (IMPs) in the mitochondrial inner membrane, electrophoresis and freeze-fracture

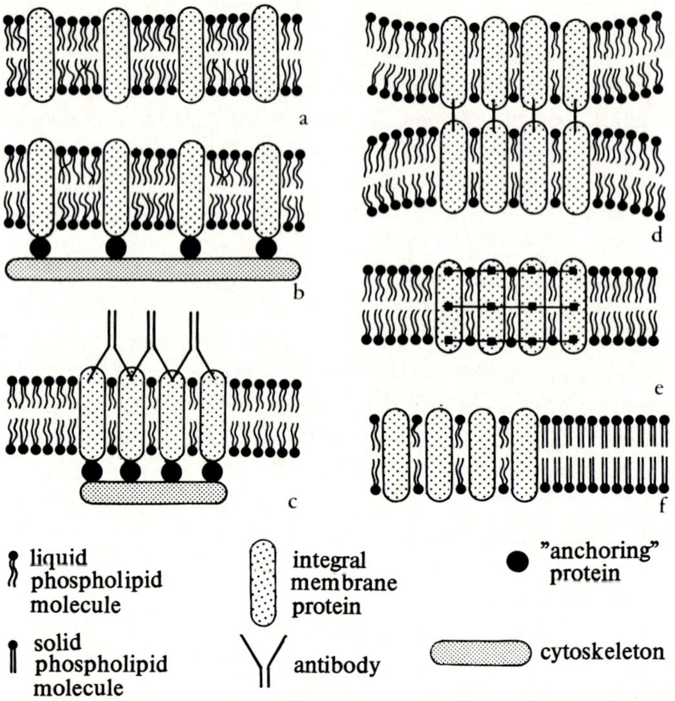

FIGURE 1. Schematic representation of the architecture of biological membranes. (a) Singer and Nicholson[13] concept: proteins are free to move. (b) Integral membrane proteins are connected with the membrane skeleton by means of an ankyrin protein, as in the erythrocyte. (c) Hormone or antibody-induced aggregation of receptors resulting in a coated pit. (d) Aggregation of proteins in contact sites between cells like the gap junction. (e) Paracrystalline protein structures in the plane of the bilayer, as in the "purple membrane" of *Halobacterium halobium.* (f) Phase separation of lipids resulting in the aggregation of proteins.

electron microscopy were combined.[20] An electric current was passed through a suspension of purified inner membrane vesicles, which caused an electrophoretic migration of IMPs in the membrane plane into a single crowded patch facing the positive electrode. The membrane vesicles were quenched at specific times after removing the electrophoretic force, while the particles were diffusing back into a random distribution. A diffusion coefficient of the IMPs at $8.3 \times 10^{-2}/\mu m^2/s$ at a temperature of 20°C. This diffusion coefficient for IMPs is consistent with values reported for a variety of proteins for various cellular membranes. In conclusion, lipids of protein (transmembrane) can diffuse rather fast, the former about 100 to 1000 times faster than the latter. This would predict a homogenous distribution of membrane molecules in the plane of the bilayer, consistent with the fluid-mosaic model.

However, there are many examples of proteins and lipids being restricted in lateral motion. A prime example is the tight junction in epithelial cells, which acts as a barrier preventing proteins moving from the apical to the basal-lateral membranes of the cell. It has been shown that a proportion of the lipids (the glycolipids) is not free to move through this barrier, whereas other lipid classes are. This has been attributed to the fact that the glycolipids are present in the outer leaflet (lipid asymmetry). So it is thought that the outer leaflet lipids are more restricted than the lipids in the inner leaflet.[21] The reason for this is unknown, but has to be sought in the structure of the tight junction itself. Alternatively, protein molecules can be noncovalently linked to the cyto- or membrane-skeleton, such as is the case in the erythrocyte membrane (Figure 1b). In that membrane the linkage between membrane proteins is known in some detail.[22] Part of the anion pump protein (''band 3'') is anchored to the membrane skeleton via ankyrin. In some cell types, binding of ligands (hormones, low density lipoproteins, growth factors) and antibodies with membrane receptor proteins can lead to noncovalent linkage to clathrin network (Figure 1c), a coated pit which will form coated vesicles by endocytosis.[23]

Clustering of receptors may also occur independently of coated pits. So it has been found that epithelial growth factor (EGF) receptors form microaggregates upon stimulation with EGF. This EGF receptor clustering is thought to be associated with the interaction with other cytoskeleton elements, most likely actin filaments.[24] Clustering of membrane receptors has also been found during capping of lymphocytes. Domain formation can be found in areas where two membranes make close contact (Figure 1d). The most prominent example is the gap junction where IMPs are aggregated at the contact area. (See Chapter 3 by Hülser.) Another example is the stacked membranes in the chloroplast, where, for instance, photosystem II is exclusively located in these membranes.[25] Lastly, a protein domain has been found in the membrane of *Halobacterium halobium,* where bacteriorhodopsin is organized in a paracrystalline manner (Figure 1e).

Comparable domains are not known for lipids, although lipids can form microdomains, where certain lipids are concentrated around a membrane protein. On the other hand, it is now clear that lipids are not homogeneously distributed over both leaflets of the bilayer; a phenomenon known as lipid asymmetry.[26] For example, it is known that the negatively charged phospholipids are exclusively present in the cytoplasmic leaflet of the erythrocyte[27] sarcolemma, and in the plasma membrane of cardiac muscle cells,[28] and that all lipopolysaccharides[29] are present in the outer leaflet of the outer membrane of *Escherichia coli.* This is also the case with glycolipids, which are selectively found in the outer leaflet of the apical membranes of epithelial cells.[30]

In summary, although proteins and lipids are free to diffuse extremely fast in the plane of the bilayer, there are many exceptions to this principle. Certain membrane proteins, and most likely, many lipid classes, are concentrated in particular stable domains with relatively long lifetimes.

IV. MEMBRANE FLUIDITY

A. FLUID-SOLID TRANSITION

An important feature of all membrane phospholipids is the existence of a temperature-dependent, reversible phase transition in which the hydrocarbon chains of the phospholipids undergo a transformation from an ordered (gel) state to a more disordered or ''fluid'' (lipid-crystalline) state. The gel to liquid-crystalline transition has been exhaustively examined in phospholipids by a wide variety of techniques.[31,32] Freeze fracture studies have shown that several synthetic phosphatidylcholines exhibit characteristic band patterns with defined periodicities at temperatures below the phase transition, whereas they show smooth fracture faces in the fluid state.[7] In addition, it has been found that the band patterns are confined to the $P\beta'$ gel phase, which exists at temperatures between the pretransition and the main transition. At temperatures below the pretransition, the $P\beta'$ gel state exhibits smooth fracture faces.[33,34] Figure 2 shows a compilation of these data. The band pattern of the $P\beta'$ gel phase has been used as a marker to investigate the miscibility properties of phosphatidylcholine (PC) species.

Mixture of PCs show cocrystallization by differential scanning calorimetry (DSC); this holds e.g., for an equimolar mixture of dilauroyl- and dimyristoyl-PC, exhibited a new specific $P\beta'$ phase with a new band pattern,[35] which suggested that the phospholipid molecules were homogeneously distributed below the phase transition temperature.

Mixtures of PCs which show monotectic behavior, when quenched from a temperature

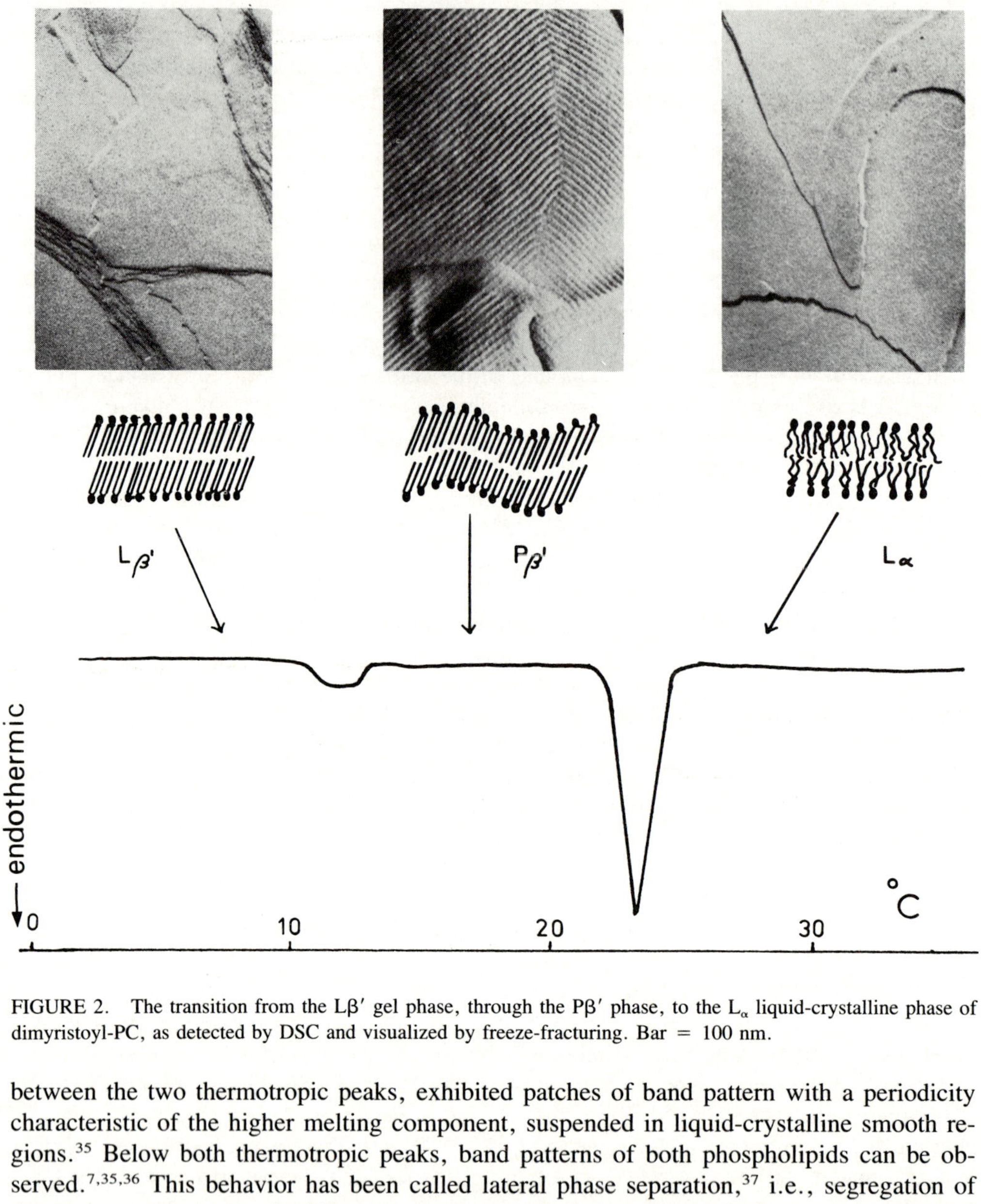

FIGURE 2. The transition from the Lβ′ gel phase, through the Pβ′ phase, to the L_α liquid-crystalline phase of dimyristoyl-PC, as detected by DSC and visualized by freeze-fracturing. Bar = 100 nm.

between the two thermotropic peaks, exhibited patches of band pattern with a periodicity characteristic of the higher melting component, suspended in liquid-crystalline smooth regions.[35] Below both thermotropic peaks, band patterns of both phospholipids can be observed.[7,35,36] This behavior has been called lateral phase separation,[37] i.e., segregation of the two components in the bilayer into separate domains. Figure 3 shows an example of such a domain formation, with crystalline areas floating in a "sea" of fluid lipids.

By this approach, information can also be gained about the distribution of cholesterol in the bilayer.[7] When the sterol is added to a cocrystallizing mixture, a homogenous band pattern with the same periodicity as in the pure mixture can be observed, but the intensity of the pattern fades away with increasing concentrations of cholesterol. From this observation we concluded that below the phospholipid phase transition the sterol is homogeneously distributed in the bilayer. When cholesterol is added to mixtures with a monotectic behavior at lower concentration, only the domains with the band pattern of lower melting component are affected, indicating that during the cooling process the sterol has preference for the liquid species.

Negatively charged phospholipids are of particular interest with respect to the gel-liquid-

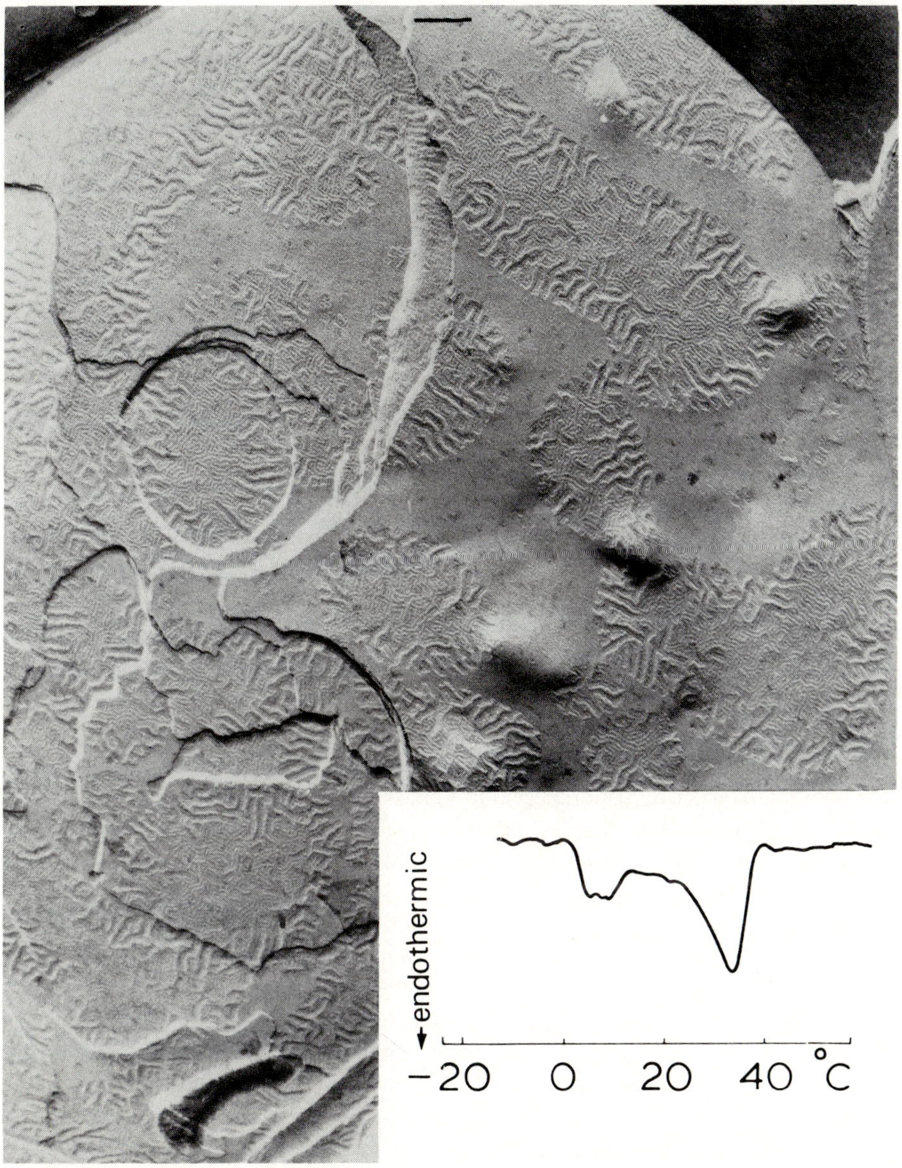

FIGURE 3. Lateral phase separation in a equimolar mixture of dipalmitoyl and 1-palmitoyl-2-oleoyl-PC, as measured by DSC and visualized by freeze-fracturing after quenching from a temperature between the two thermotropic peaks. Bar = 100 nm. (From Verleij, A. J., Ph.D. thesis, University of Utrecht, Utrecht, 1975.)

crystalline transition, as it has been found that such a transition can be induced isothermally by pH, Ca^{2+}, and ionic strength.[38-40] The effect of Ca^{2+} can be especially dramatic. It is found that Ca^{2+} can increase the transition temperature of phosphatidylglycerol and phosphatidylserine by more than 50°C, with an associated structural transformation into a different type of bilayer. Instead of liposomes composed of concentric spherical bilayers, a so-called cylindrical[39] or cochleated structure[41] is obtained. This structure has been extensively characterized.[42] These investigators also found that Ca^{2+}, but not Mg^{2+}, can induce phase separations in mixtures containing PC and phosphatidylserine (PS), or phosphatidic acid. Investigation of such mixtures with freeze fracturing could confirm the interpretation derived from DSC. It was found that in a cocrystallizing mixture of PS and PC, excess of Ca^{2+} (not Mg^{2+}) can induce a lateral phase segregation below a PS:PC molar ratio of 1:2, but that an additional structural phase segragation occurs at higher proportions of PS[43]. The latter segregation is associated with the formation of cochleated structures. Such behavior is not found in cocrystallizing mixtures of PC and phosphatidylglycerol.[39] Thus, the miscibility properties of a mixture of PC and a negatively charged phospholipid, upon the addition of Ca^{2+}, is strongly dependent on the type of negatively charged phospholipid involved.

B. IMP AGGREGATION

Lipids in biological membranes are fluid, which allows fast lateral diffusion. In several prokaryotic membranes[44] the lipids start to solidify below the growth temperature. This can be detected by DSC[45] and X-ray diffraction by a change from 4.6 to 4.2 reflection.[46] With freeze fracturing, one can visualize aggregation of IMPs upon lowering the temperature.[7] The membrane proteins and fluid lipids are squeezed out of the domains of solid lipids. Eventually, if all the lipids are solid, the IMPs are seen to be aggregated in large clusters (Figure 1f). It has to be said that fast freezing is required to detect a homogenous distribution of IMPs at 37°C. With conventional freezing in a mixture of liquid and solid nitrogen, IMP aggregation also is obtained if one freezes from 37°C (Verkleij, not published). Alternatively, the homogeneous distribution can be obtained at 37°C if the cells are fixed with glutaraldehyde before conventional freezing.

Lipid fluidity is essential for biological function. This is clear not only from the fact that cold shock alters membrane permeability, but also that, if one grows bacteria at lower temperature than 37°C, the lipid composition is corrected to maintain the appropriate fluidity. The membrane becomes enriched in unsaturated fatty acids.[44,47] IMP aggregation has been observed also in eukaryotic membranes, but it starts at lower temperatures 5 to 10°C.[7] This is consistent with the phase transition temperature of the lipids in these membranes. It is also found that below 10°C the IMPs of mitochondrial, alveolar, nuclear membranes undergo clustering.

Particle aggregation has also been found under pathological conditions in the sarcolemma membrane of cardiac muscle cells. This phenomenon is here induced isothermally by a Ca^{2+} overload in the cytoplasm to levels found in the blood (1.3 mM), after a period of ischemia, followed by Ca^{2+} rich perfusion (Figure 4).[48] This can be explained as follows. The membrane phospholipids are also asymmetrically distributed, with the negatively charged phospholipids in the cytoplasmic leaflet.[28] Upon Ca^{2+} entry, the negatively charged phospholipids start to solidify. Thus, solidification of phospholipids in one leaflet can induce lateral phase separation and IMP aggregation. Consequently, preference of the nonbilayer arrangement is thought to destabilize the bilayer configuration, and induces extensive blebbing (Figure 4).

A similar phenomenon has been found in the erythrocyte membrane, however, only when actin and spectrin have been removed. It was shown that in inside-out vesicles of actin-spectrin-stripped ghost, Ca^{2+} and low pH[49] can induce IMP aggregation. This phenomenon was initially explained as a result of the precipitation of the residual actin-spectrin.[49] However, this is more likely due to a lipid effect, i.e., lipid solidification, since this IMP aggregation also occurs after complete removal of actin and spectrin.[50] Moreover, IMP aggregation in these membranes can also be induced by lowering the temperature.[50] It is of interest to note that the cholesterol/phospholipid ratio is rather high, i.e., 0.9. This suggests that cholesterol is not able to obliterate phase transitions and thermally induced IMP patching. Thus, phase changes are also possible in the erythrocyte; however, they are obscured and no visible IMP aggregation is seen because of the interaction of IMPs (band 3) with the cytoskeleton elements spectrin-ankyrin-actin.

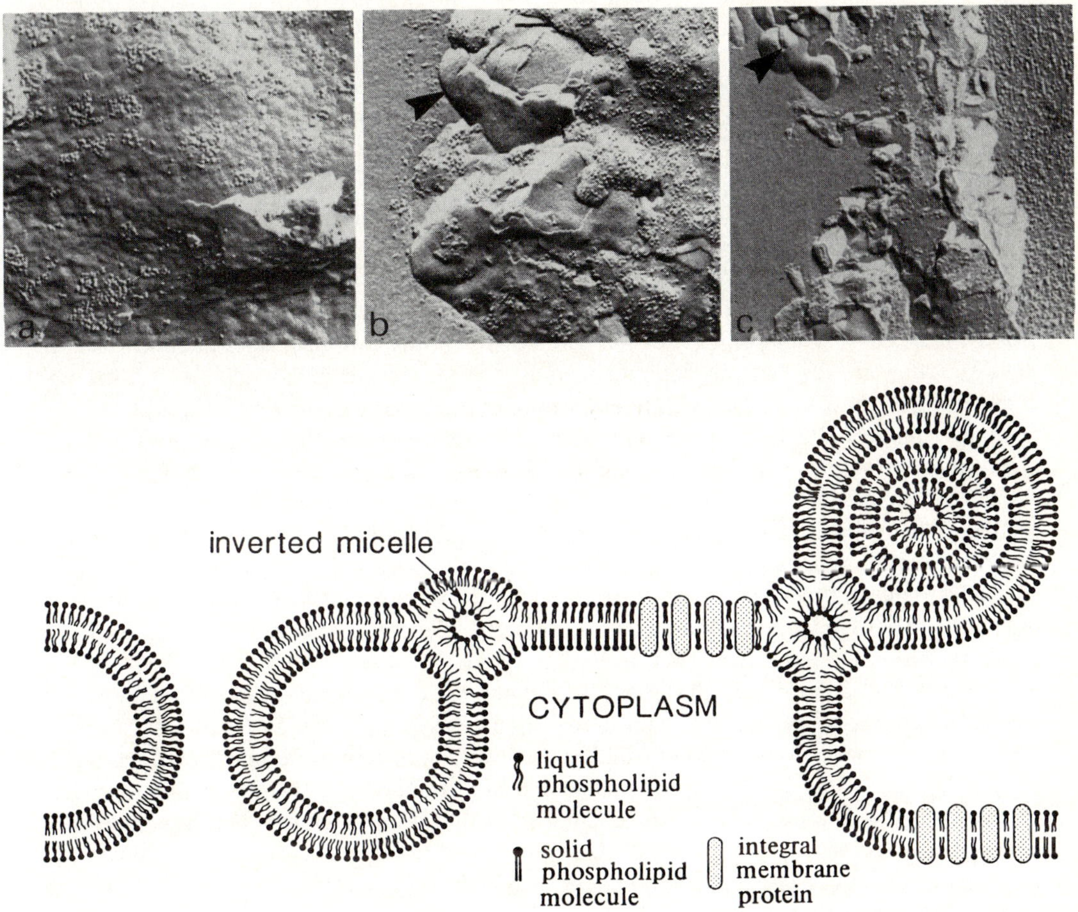

FIGURE 4. Freeze-fracture micrographs of the heart muscle cell sarcolemma after ischemia and reperfusion, showing severe aggregation of intramembraneous particles (a) and the extrusion of multilamellar structures (b and c). A schematic representation of this lipid extrusion and IMP aggregation as a result of lateral phase segregation in the cytoplasmic leaflet is shown in (d). Bar = 100 nm.

C. BIOLOGICAL RELEVANCE OF FLUID-SOLID TRANSITIONS

The fact that phospholipids in biomembranes are fluid at physiological temperatures indicates that this physicochemical state must be of relevance for membrane functioning. Biomembranes with solidified lipids, and exhibiting lateral phase segregation, become leaky for ions,[79,80] indicating that fluidity is of importance for the semipermeable properties of membranes. In eukaryotic cells exo- and endocytosis is prevented at low temperatures, indicating that fluidity is essential for fusion processes. The activities of many membrane enzymes have been shown to be strongly affected by the transition from the fluid to solid state. This all indirectly shows that fluidity is essential for membrane functioning. At present, there is no proof that a fluid-solid transition of lipids is of importance for membrane functions. It is of interest to note that alveolar membranes are very rich in dipalmitoylcholine, which has a transition at 41°C. It is conceivable that in these membranes the fluid-solid transition of this lipid plays a regulatory role in the compressibility of membrane surfaces. Moreover, it cannot be excluded that locally solid-fluid transitions may have a regulatory function in various membrane processes. In this way, it is reasonable to assume that Ca^{2+} at the cytoplasmic surface of cell membranes, where the negatively charged phospholipids are located (i.e., PS and PI), can solidify these lipids and may, in turn, regulate membrane enzyme activities.

In conclusion, fluidity is essential for membrane functioning. The fluid-solid transition has not yet been shown to be of direct biological relevance, except under pathological conditions. However, it cannot be excluded that in some membranes locally, fluid-solid transitions are important, for instance, for regulation of permeability properties and membrane enzyme activity.

V. LIPID POLYMORPHISM

A. HEXAGONAL H_{II} PHASE

The bilayer organization has been confirmed by a number of techniques: electron microscopy (negative staining, thin section and freeze fracture), and by other methods like X-ray diffraction and nuclear magnetic resonance (NMR). Despite the fact that it is a general consensus that lipids in biological membranes are organized in a bilayer configuration, it is now clear that in all biomembranes, up to 40% of lipids are found, which under physiological conditions do not form bilayers, but prefer the H_{II} phase.[9] Studies on micro-organisms clearly show that the amount of H_{II} preferring lipids, or better, the ratio H_{II}/bilayer lipid is strongly regulated, indicating a functional role for these lipids. So it has been shown for *Acholeplasma laidlawii* that the monoglucosyldiglyceride/diglucosyldiglyceride (the majority of the lipid) is strictly regulated in response to changes in the environment.[51] When growth conditions are such that the adoption of nonbilayer structures become more favorable, less of the nonbilayer lipid is synthesized, and vice versa.

The lipid structure of this phase, as elucidated by X-ray techniques, freeze fracture, and ^{31}P NMR, contrasts dramatically with the familiar bilayer organization (Figure 5).[8,9] First, the H_{II} phase is basically a liquid-crystalline hydrocarbon matrix, penetrated by hexagonally packed aqueous channels (of approximately 2.0 nm in diameter), towards which the polar groups of the lipid are oriented. Second, lipids in the H_{II} phase cannot provide a permeability barrier between internal and external environments. The observation that such lipids are major constituents of biological membranes raises questions about possible functions of these lipids in biomembranes.

Examples of lipids which adopt the H_{II} phase at physiological conditions are phosphatidylethanolamines, monogalactosyldiglycerides from chloroplasts and monoglucosyldiglycerides from *A. laidlawii.* The tendency of other naturally occurring lipids, like cardiolipin and phosphatidic acid, to form the H_{II} phase depends on pH and the presence of divalent cations. These negatively charged phospholipids can also adopt the H_{II} phase in the presence of the local presence of cytochrome c. Moreover, the H_{II} phase can be induced by gramidicin, can be modulated in the presence of cholesterol, and appears in complex mixtures of synthetic lipids and lipid extracts of *E. coli,* mitochondria, and rod outer segment membranes.[9,52] Worth mentioning in the light of the current discussions on phosphatidylinositol (PI) response and membrane fusion is that diacylglycerols can induce the H_{II} phase in PE, PC, and PS bilayers.[53] This seems to be particularly true for diglycerides with acyl chain substitution similar to that produced by PI hydrolysis *in vivo.*[54]

The transition from bilayer to hexagonal H_{II} phase in systems of purified phosphatidylethanolamine (PE) is remarkably abrupt and occurs within a temperature range of only a few degrees. The latter point can be deduced from the fact that one has to use fast freezing rates ($> 1000\ Ks^{-1}$) in order to preserve the H_{II} phase for freeze fracturing.[55] DSC also showed that the enthalpy change involved in the structural rearrangement is very small compared to the heat uptake needed to melt the solid bilayer (see Figure 5). Since the polymorphic transition occurs above the gel-liquid crytalline transition of the bilayers, it can be concluded that the H_{II} phase is a liquid-crystalline or fluid phase.[8]

Besides the pure H_{II} phase, an alternative analogous nonbilayer configuration can also be found, i.e., the inverted micelle. This inverted micelle is visible as a well-defined lipidic

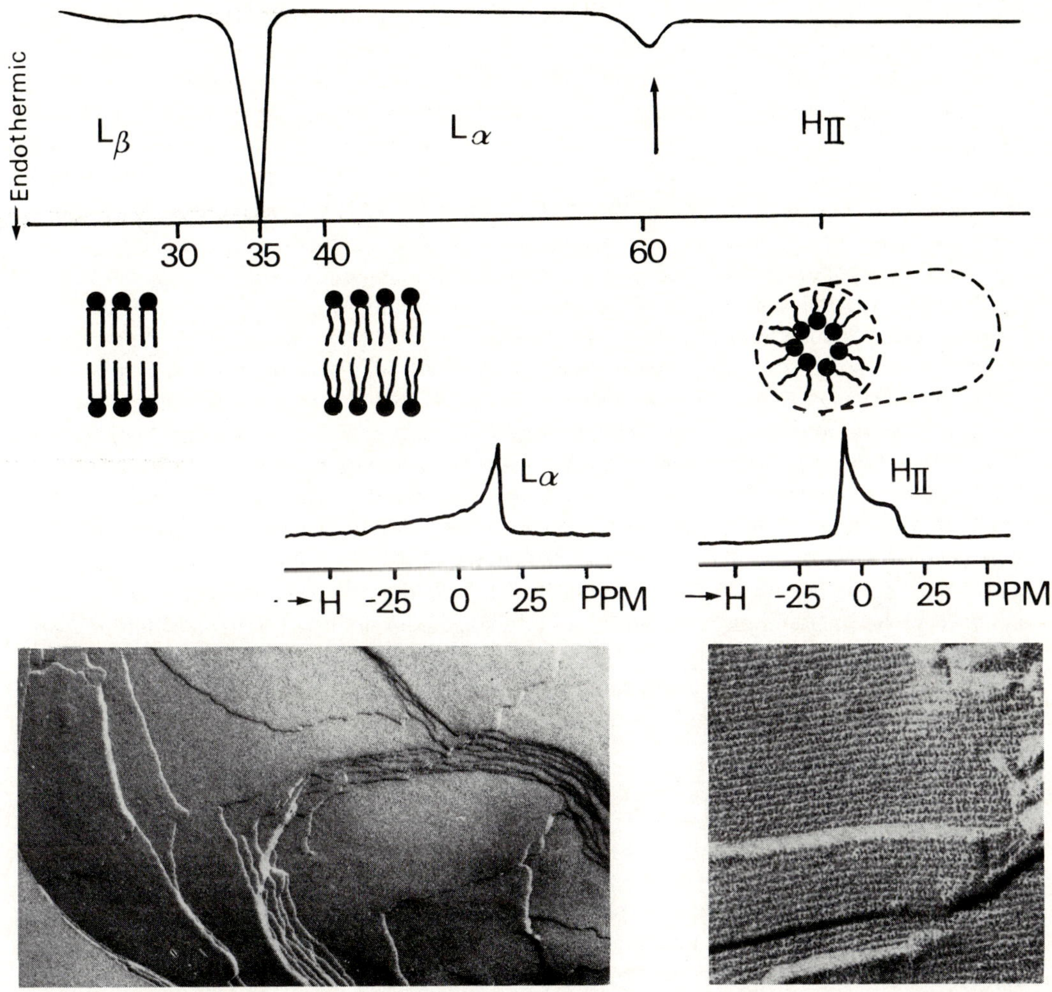

FIGURE 5. The transition from the L_β gel phase, through the L_α liquid-crystalline phase in dielaidoyl phosphatidylethanolamine, as detected by differential scanning calorimetry and visualized by freeze-fracturing. L_α and L_β phase both show smooth fractured bilayers. Bar = 100 nm.

particle.[52] Such lipidic particles have been found in pure PE and cardiolipin, as well as in lipid mixtures in which one of the lipids prefers the hexagonal H_{II} phase, e.g., PE-containing systems, cardiolipin/PC in the presence of Ca^{2+}, and cardiolipin in the presence of local anesthetics and peptides, like cardiotoxin or mellitin.[52,56-59] Inverted micelles or lipidic particles can in fact be considered as intermediary structures between bilayer and H_{II} phase and vice versa, and also as fusion intermediates in bilayer fusion.

B. LIPIDIC PARTICLES AND FUSION

The hypothesis that lipids preferring the hexagonal H_{II} phase and/or inverted micelle configuration play a crucial role in membrane fusion, has been strongly supported by experiments with model membrane systems. This was first demonstrated with vesicles composed of an equimolar mixture of cardiolipin and egg-PC. These vesicles, in fact, fuse in a non-leaky fashion.[60] More relevant for biological fusion events are mixtures of PE and negatively charged phospholipids. Ca^{2+}, which is well-known to promote biological membrane fusion, triggers bilayer to nonbilayer (H_{II} and/or inverted micelle) transitions isothermally in these mixtures.[9] An increase in temperature also promotes such a transition. It is therefore logical to suppose that the presence of Ca^{2+}, or an increase in temperature, allows the nonbilayer tendency of endogenous lipids to be expressed, thus promoting the fusion event. These predictions hold for a variety of mixtures when Ca^{2+} is added,[9,52] such as PE/PS, PE/PI, PE/phosphatidic acid, PE/cardiolipin, PE/PC/cholesterol/PS, and when the temperature is raised, PE/PC/cholesterol mixtures. In all these cases incubation results in the formation of larger liposome vesicles, showing that net fusion has occurred.

In all these mixtures undergoing fusion this event is associated with the appearance of lipidic particles sometimes localized in regions corresponding in the fusion interface. In a review,[52] it has been demonstrated that the different particle types, which confused the issue in the early days of the discovery of these lipidic particles, can be interpreted as a reflection of different stages during the fusion event.

In the "adhesion" stage, two neighboring bilayers form polar contact points requiring local dehydration and charge neutralization, which is characteristic for H_{II} preferring lipids. The bilayers are still intact but the contact points rise to deflections in the freeze-fractured membranes because of the local small curvature. Such deflections are most likely not homogeneous in size (diameter), and are not well defined. Cusp-like particles may reflect this stage, and indeed represent intermembrane attachment sites (IMAS model in Reference [61] and cusp model in Reference [62]). In the second stage, the bilayers "join", which enables intermixing of their lipids. This may either proceed with an inverted micelle or an extended micelle, i.e., an H_{II} cylinder as an intermediate. Fracturing of a "joining stage" reveals well-defined particles and complementary pits (Figure 6). In the last step there will be a fusion, which can give rise either to the two original vesicles, or to real fusion, by intermixing of the two aqueous compartments of both vesicles. In the former case, the bilayer will temporarily pass a stage similar to that of intermembrane attachment sites, whereas in the latter case, the fracturing passing through the newly formed aqueous channel will show a marked change from cusp shapes to volcanoes with flat tops. The flat tops arise as a result of fracturing through ice in aqueous pores.

Although all these lipidic particles have been found in systems which undergo fusion, these particles could not be found during the initial rounds of fusion, and are seen only at later stages (where these structures have a longer life time).[52,63] The reason for not seeing the lipidic particles during the initial fusion rounds is likely because fusion, which is not arrested at the joining stage but proceeds as expected, is so fast that the transient intermediates (the inverted micelles) escape detection by freeze-fracture electron microscopy. Structures occurring during biological membrane fusion are discussed by Knoll and Plattner in Chapter 6.

In PE-containing model membranes it has been found that at temperatures where the lipids are in the isotropic state between lamellar and H_{II} phase are relevant for nonleaky fusion.[64] The isotropic component seen by ^{31}P NMR is always associated with the presence of lipidic particles. This confirms that the intermediate structure is not the H_{II} phase,[64] as suggested by the inverted micelle model.[52] It was proposed that above TH_{II}— where the lipid is in the H_{II} phase—the number of inverted micelle intermediate (likely the lipidic particles) increase so fast that it leads to rapid collapse of the opposed bilayers towards H_{II} phase structures and H_{II} cylinders.[64] This is accompanied by leakage, and it would not be relevant for fusion. If the system exhibits isotropic ^{31}P NMR resonance (which represents the honeycomb network), and no H_{II} phase signal is formed, the inverted micelle remains isolated and can be intermediary in nonleaky fusion. Calculations show that inverted micelles can form new structures resulting in membrane fusion, but only if the inverted micelle intermediate remain isolated.[65] Thus, fusion has to be very fast and the inverted micelles must have a short lifetime if membrane collapse and consequent leakage is to be avoided.

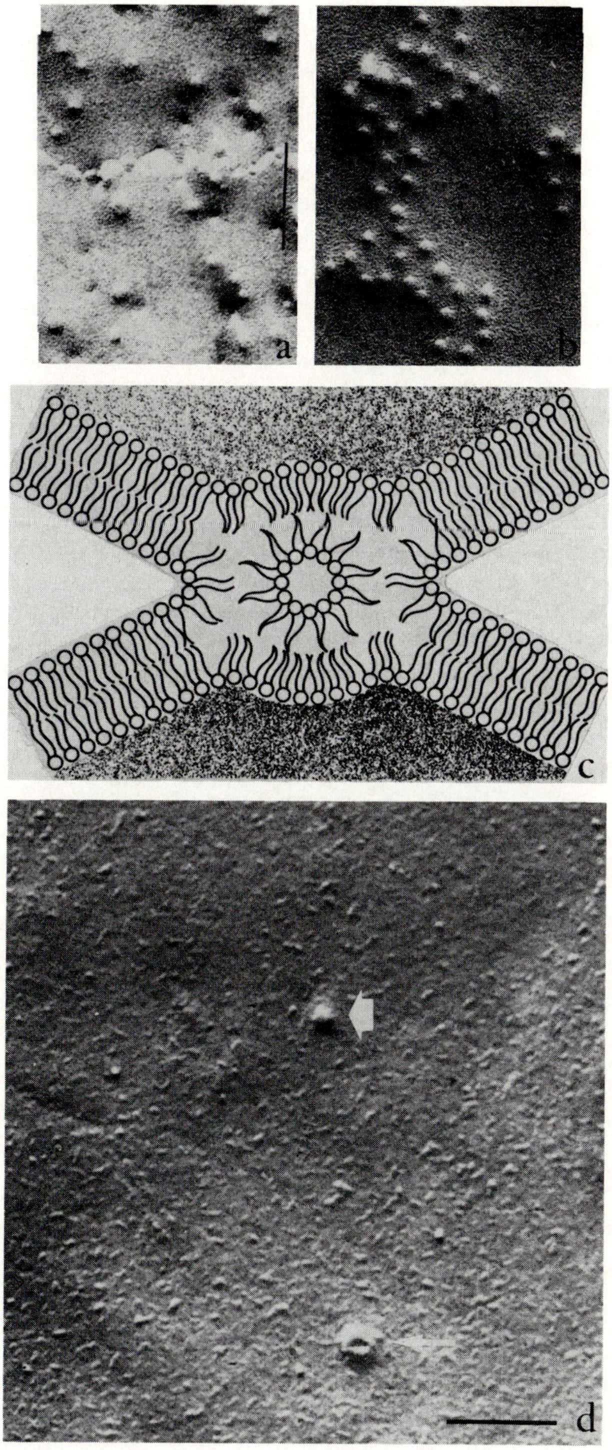

FIGURE 6. Inverted micelles as intermediate structures in membrane fusion. a and b: Fusion of model membranes, made of lipids extracted from biological membranes, showing lipidic particles. c: Schematical molecular interpretation of an inverted micel. d: Fusion-point during exocytosis of secretion granula in adrenal chromaffin cells. Bars = 100 nm. (Modified from Schmidt et al., *Eur. J. Cell Biol.*, 32, 31, 1983.)

C. BIOMEMBRANE FUSION

Initially, it was generally assumed that membrane proteins, both the extrinsic or membrane skeleton proteins, as well as the intrinsic membrane-spanning proteins, had to be cleared from the fusion site before actual fusion could start. This was deduced from freeze-fracture studies, where IMP-clearing and a so-called diaphragm at the site of fusion were seen. However, these studies were carried out using chemical fixation and cryoprotectants. Such diaphragm formation is not seen using fast freezing without these pretreatments.[66] Thus, only a small area about 10 nm in diameter is necessary for fusion, the so-called focal point fusion (see Chapter 6 by Knoll and Plattner).

The visualization of lipidic particles in artificial membranes[52] as possible fusion intermediates has led to further search for such structures during biological membrane fusion. In biological systems it is, in principle, difficult to trap fusion intermediates, because, (1) of the short life time, (2) the problem of synchronizing the fusion event, and (3) the amount of fusion events per surface. Even with exocytosis, only a few hundred vesicles per cell might be involved in fusion events. However, Schmidt et al.[67] have been able to capture exocytosis in chromaffin cells using fast freezing, freeze cleavage, and statistical analysis. After stimulation of the exocytotic activity in a controlled way by adding carbachol, the plasma membrane displayed structural features which strongly resemble, or better, which possibly are identical to the intermediates seen in model systems, i.e., lipidic particles and volcano-like protrusions (Figure 6d). Such structures have also been visualized using fast freezing during fusion of membranes in blood platelets upon stimulation.[68] Budding of viruses has been shown to be associated with well-defined particles (being fusion intermediates) at the nexus of the plasma membrane and the membrane of the budding influenza virions.[17] Membrane intercalated particles as fusion intermediates have also been found on liposomes and the apical membrane of MDCK cells during fusion of liposomes with these cells.[17]

In summary, well-defined particles have been seen to occur during fusion in different systems.

D. CONTACT SITES

On the basis of thin section electron micrographs, the occurrence of distinct contact sites between inner and outer membranes of isolated mitochondria already had been suggested in the sixties. These areas have been interpreted as sites of diminished negative charge, allowing close membrane opposition.[69] Freeze fractured, rapidly frozen mitochondria within intact cells exhibit frequent jumps or deflections of the outer to the inner membrane.[70,71] *In vitro* the frequency of these jumps can be increased by incubation of the isolated mitochondria with Ca^{2+}, Mn^{2+}, Mg^{2+}, and apocytochrome c, and by exposure to 37°C.[72] From these results it has been hypothesized that the fracture jumps represent contact points between inner and outer membrane, where the membranes are joined (semifusion) by local exhibition of a nonbilayer structure in membrane lipids.[72] This was assumed because it has already been demonstrated that divalent cations, increased temperature, and apocytochrome c are factors which induce nonbilayer configuration in lipids isolated from mitochondria.[9] Adriamycin, which competes with divalent cations for binding with negatively charged phospholipids and depresses the nonbilayer behavior of lipids in mitochondria, reduces the contact sites, as seen with freeze fracturing.[73] The frequency of fracture jumps has also been shown to be correlated with the metabolic state of the mitochondria, i.e., a higher frequency occurs with isolated well-coupled mitochondria during oxidative phosphorylation.[70] Only few deflections are observed with mitochondria frozen in other experimental states, namely without ADP but with substrate, or without both substrate and ADP. The phenomenon is totally abolished by chemical pretreatments or ''uncoupling'' of the mitochondria. Freeze substitution after rapid freezing shows that the membranes of mitochondria (exhibiting high frequencies of fracture plane jumps) are closely opposed, whereas a low frequency is correlated

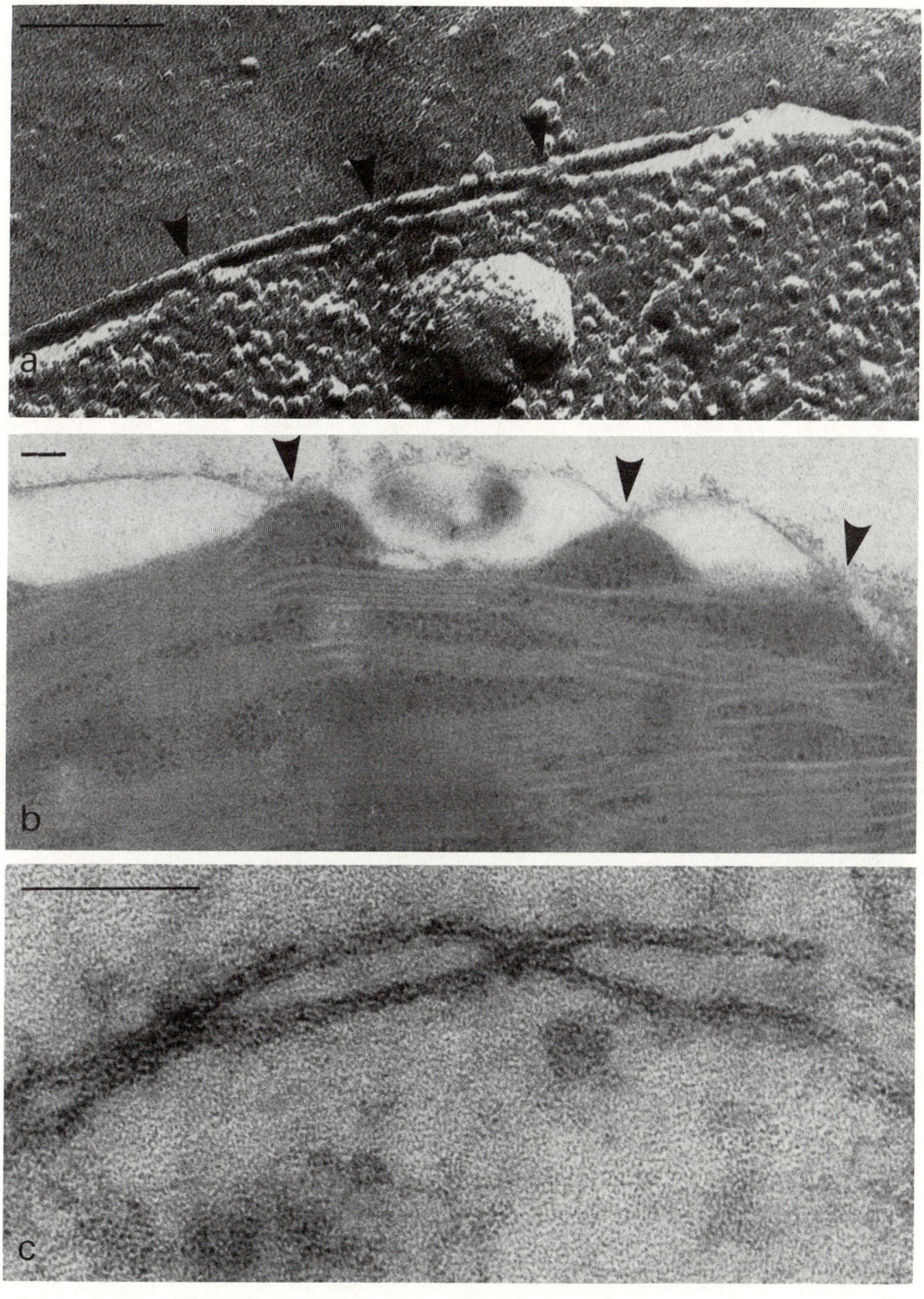

FIGURE 7. Contact sites: (a) cross-fractured chloroplast in isotonic buffer, (b) ultrathin secretion of freeze-substituted, rapidly frozen chloroplast under hypotonic conditions, (c) ultrathin section of a freshly isolated and rapidly frozen freeze-substituted mitochondrion. Arrow heads indicate contact sites. Bar = 100 nm.

with a pronounced intermembrane space.[70] These contact sites, also visible with freeze substitution (Figure 7c), are rather stable and can also be isolated by biochemical methods.[74] They appeared to be enriched in hexokinase and to possess a higher specificity for glutathione transferase.

Freeze-fracture and freeze substitution[75] indicate that attachments are also present between the outer and inner membrane of the chloroplasts. On cross fractures, one sees locally close appositions between the two membranes (Figure 7a). The existence of these contact sites seems to be real since under hypertonic conditions, when the outer and inner membrane are well-separated, the contact sites are still visible, both with freeze fracturing and freeze substitution, using rapid freezing (Figure 7b).

There are many reasons for the existence of contact sites between outer and inner membranes of both mitochondria and chloroplasts. It is assumed that ATP-ADP translocation occurs via such contact sites.[70] These contacts are also thought to play an important role in the transport of nuclear-encoded proteins from the cytoplasm into the chloroplast or mitochondria. Most chloroplast and mitochondrial proteins are encoded in the nuclear genome, and are synthesized as precursors, which are post-translationally taken up by these organelles. For the mitochondrial system, evidence for the role of contact sites in protein translocation has been presented.[77] It was shown that after binding of the precursor proteins at low temperature, processing could take place to the mature size, while a large part of the protein molecule still protruded from the outer membrane. Using immunocytochemical methods, it has been shown that these partly translocated precursor proteins are indeed located in areas where both membranes are in close proximity. Using immunogold labeling receptors for precursor proteins have been localized, too, at contact sites of chloroplasts.[78] Thus, evidence is growing that membrane contact sites are crucial in protein, and likely also in lipid translocation in mitochondria and chloroplasts.

VI. CONCLUSIONS

Electron microscopy has proven to be of great value in the analysis of membrane structure, dynamics, and function. This is due especially to the cryotechniques: freeze fracturing (etching) and freeze substitution, in combination with rapid freezing. Cryofixation reliably preserves structure (space resolution), and can be used to follow dynamic events (time resolution), such as lateral movement and fusion. Many concepts, such as lipid bilayer organization, membrane domains, membrane heterogeneity, lateral phase separation, lipid polymorphism, local point fusion, lipidic particle, and inverted micelle formation, etc., have been derived and/or supported by this morphological approach.

ACKNOWLEDGMENTS

I am grateful to José Leunissen-Bijvelt, Koert Burger, Fons Cremers, Ruud Hovius, Gerd Knoll, Jan Andries Post, Peter Thomas, and Wim Voorhout for their contributions to the work presented in this review.

REFERENCES

1. **Michell, R. H.,** Inositol phospholipids, and cell surface receptor function, *Biochim. Biophys. Acta,* 415, 81, 1975.
2. **Berridge, M. J. and Irvine, R. F.,** Inositol triphosphate, a novel second messenger in cellular signal transduction, *Nature,* 312, 315, 1984.
3. **Hakomori, S. and Melchior, D. L.,** How bilayer lipids affect membrane protein activity, *Trends Biochem. Sci.,* 11, 311, 1986.
4. **Carruthers, A.,** Glycolipids in cellular interaction, differentiation and oncogenesis, *Ann. Rev. Biochem.,* 50, 733, 1981.
5. **Hanahan, D. J. and Nelson, D. R.,** Phospholipids as dynamic participants in biological processes, *J. Lipid Res.,* 25, 1528, 1984.

6. **Rietveld, A. and De Kruijff, B.,** Phospholipids as a possible instrument for translocation of nascent proteins across biological membranes, *Biosci. Rep.*, 6, 775, 1986.
7. **Verkleij, A. J. and Ververgaert, P. H. J. T.,** Architecture of biological and artificial membranes, as visualized by freeze-etching, *Ann. Rev. Phys. Chem.*, 26, 101, 1975.
8. **Cullis, P. R. and De Kruijff, B.,** Lipid polymorphism and the functional roles of lipids in biological membranes, *Biochim. Biophys. Acta*, 559, 399, 1979.
9. **Cullis, P. R., Hope, M. J., De Kruijff, B., Verkleij, A. J., and Tilcock, C. P. S.,** Structural properties and functional roles of phospholipids in biological membranes, in *Phospholipids and Cellular Regulation*, Vol. 1, Kuo, J., Ed., CRC Press, Boca Raton, FL, 1985, 1.
10. **Weibull, C., Villiger, W., and Carlemaln, E.,** Extraction of lipids during freeze substitution of *Acholeplasma laidlawii* cells for electron microscopy, *J. Microsc.*, 134, 213, 1984.
11. **Knoll, G., Verkleij, A. J., and Plattner, H.,** Cryofixation of dynamic processes in cells and organelles, in *Cryotechniques in Biological Electron Microscopy*, Steinbrecht, R. A. and Zierold, K., Eds., Springer-Verlag, New York, 1987, 258.
12. **Kopstad, G. and Elgsaeter, A.,** Theoretical analysis of specimen cooling rate during impact freezing and liquid jet freezing of freeze etch specimens, *Biophys. J.*, 40, 163, 1982.
13. **Singer, S. J. and Nicolson, G. L.,** The fluid mosaic model of the structure of cells membranes, *Science*, 175, 720, 1972.
14. **Edidin, M.,** Rotational and lateral diffusion of membrane proteins and lipids: phenomena and function, *Curr. Top. Membr. Transp.*, 29, 91, 1987.
15. **Beck, K.,** Mechanical concepts of membrane dynamics: diffusion and phase separation in two dimensions, in *Cytomechanics*, Bereiter-Hahn, J., Anderson, O. R., and Reif, W. -E., Eds., Springer-Verlag, New York, 1987, 79.
16. **Van Meer, G. and Simons, K.,** The junction of tight junction in maintaining differences in lipid composition between the apical and the basolateral cell surface domains of MDCK cells, *EMBO J.*, 5, 1455, 1986.
17. **Knoll, G., Van Meer, G., Bron, R., Burger, K., and Verkleij, A. J.,** Fusion of liposomes with plasma membranes of epithelial cells. Fate of implanted lipids followed by freeze fracture and autoradiography of plastic sections, *J. Cell Biol.*, 107, 2511, 1988.
18. **Wey, C. and Cone, R. A.,** Lateral diffusion of rhodopsin in photoreceptor cells measured by fluorescence photobleaching and recovery, *Biophys. J.*, 33, 225, 1981.
19. **Höchli, M. and Hackenbrock, C. R.,** Fluidity in mitochondrial membranes. Thermotropic lateral translational motion of intramembraneous particles, *Proc. Natl. Acad. Sci. U.S.A.*, 73, 1636, 1976.
20. **Sowers, A. E. and Hackenbrock, C.,** Rate of lateral diffusion of intramembraneous particles: measurement by electrophoretic displacement and randomization, *Proc. Natl. Acad. Sci. U.S.A.*, 78, 6246, 1981.
21. **Van Meer, G., Gumbiner, B., and Simons, K.,** The tight junction does not allow lipid molecules to diffuse from one epitheleal cell to the next, *Nature*, 322, 639, 1986.
22. **Branton, D., Cohen, C., and Tyler, J.,** Interaction of cytoskeletal proteins on the human erythrocyte membrane, *Cell*, 24, 24, 1981.
23. **Goldstein, J. L., Anderson, R. G. W., and Brown, M. S.,** Coated pits, coated vesicles, and receptor-mediated endocytosis, *Nature*, 279, 679, 1979.
24. **Van Belzen, N., Rijken, P. J., Hage, W. J., De Laat, S., Verkleij, A. J., and Boonstra, J.,** Effect of epidermal growth factor on the lateral distribution of its receptors in A431 cells, as determined by label fracture, *J. Cell Phys.*, 134, 413, 1988.
25. **Miller, K. R.,** The photosynthetic membrane, *Sci. Am.*, 241, 102, 1979.
26. **Op den Kamp, J. A. F.,** Lipid asymmetry in membranes, *Ann. Rev. Biochem.*, 48, 47, 1979.
27. **Verkleij, A. J., Zwaal, R. F. A., Roelofsen, B., Comfurius, P., Kastelijn, D., and Van Deenen, L. L. M.,** The asymmetric distribution of phospholipids in the human red cell membrane. A combined study using phospholipase and freeze-etch electron microscopy, *Biochim. Biophys. Acta*, 323, 178, 1973.
28. **Post, J. A., Langer, G. A., Op den Kamp, J. A. F., and Verkleij, A. J.,** Phospholipid asymmetry in cardiac sarcolemma. Analysis of intact cells and ''gas-dissected'' membranes, *Biochim. Biophys. Acta*, 943, 256, 1988.
29. **Lugtenberg, B. and Van Alphen, L.,** Molecular architecture and functioning of the outer membrane of *E. coli* and other Gram-negative bacteria, *Biochim. Biophys. Acta*, 737, 51, 1983.
30. **Thompson, T. E. and Tillack, T. W.,** Organization of glycolipids in bilayers and plasma membranes of mammalian cells, *Ann. Rev. Biophys. Chem.*, 14, 361, 1985.
31. **Luzzati, V., Gulik-Krzywicki, T., and Tardieu, A.,** Polymorphism of lecithins, *Nature*, 218, 1031, 1968.
32. **Ladbrooke, B. D. and Chapman, D.,** Thermal analysis of lipids, proteins and biological membranes. A review and summary of some recent studies, *Chem. Phys. Lipids*, 3, 304, 1969.
33. **Janiak, M. J., Small, D. M., and Shipley, G.,** Nature of thermal pretransition of synthetic phospholipids: dimyristoyl and dipalmitoyllecithin, *Biochemistry*, 15, 4575, 1975.

34. **Lund, E. J. and McConnell, H. M.,** The intermediate monoclinic phase of PCs, *Biochim. Biophys. Acta,* 466, 381, 1977.
35. **Ververgaert, P. H. J. T., Verkleij, A. J., Elbers, P. F., and Van Deenen, L. L. M.,** Analysis of the crystallization process in lecithin liposomes: a freeze etch study, *Biochim. Biophys. Acta,* 311, 320, 1973.
36. **Grant, C. W. M., Hong-Wei-Wu, S., and McConnell, H.,** Lateral phase separation in binary lipid structures: correlation between spin label and freeze-fracture electron microscopic studies, *Biochim. Biophys. Acta,* 363, 151, 1974.
37. **Shimshick, E. J. and McConnell, H. M.,** Lateral phase separation in phospholipid membranes, *Biochemistry,* 12, 2351, 1973.
38. **Träuble, H. and Eibel, H.,** Electrostatic effects on lipid phase transitions: membrane structure and ionic environment, *Proc. Natl. Acad. Sci. U.S.A.,* 71, 214, 1974.
39. **Verkleij, A. J., De Kruijff, B., Ververgaert, P. H. J. T., Tocanne, J. F., and Van Deenen, L. L. M.,** The influence of pH, Ca^{2+} and protein on the thermotropic behaviour of the negatively charged phospholipid phosphatidylglycerol, *Biochim. Biophys. Acta,* 288, 326, 1974.
40. **Kimelberg, H. K. and Papahadjopoulos, D.,** Effects of phospholipid acyl chain fluidity, phase transitions, and cholesterol on $(N^+ + K^+)$-stimulated adenosine triphosphatase, *J. Biol. Chem.,* 49, 1071, 1974.
41. **Papahadjopoulos, D., Vail, W. J., Jacobson, K., and Poste, G.,** Cochleate lipid cylinders: formation by fusion of unilamellar lipid vesicles, *Biochim. Biophys. Acta,* 394, 483, 1975.
42. **Papahadjopoulous, D., Pontis, A., and Pangborn, W.,** Calcium-induced lipid phase transitions and membrane fusion, *Ann. N.Y. Acad. Sci.,* 308, 50, 1978.
43. **Van Dijck, P. W. M., De Kruijff, B., Verkleij, A. J., Van Deenen, L. L. M., and De Gier, J.,** Comparative studies on the effects of pH and Ca^{++} on bilayers of various negatively charged phospholipids and their mixtures with PC, *Biochim. Biophys. Acta,* 512, 84, 1978.
44. **Melchior, D. L.,** Lipid phase transition and regulation of membrane fluidity in prokaryotes, *Curr. Top. Membr. Transp.,* 17, 263, 1982.
45. **Steim, J. M., Tourtelotte, M. E., Reinert, J. C., McElhaney, R. N., and Rader, R. L. O.,** Calorimetric evidence for the liquid-crystalline state of lipids in a biomembrane, *Proc. Natl. Acad. Sci. U.S.A.,* 63, 104, 1960.
46. **Engelman, D. M.,** X-ray diffraction studies of phase transitions in the membrane of *Mycoplasma laidlawii, J. Mol. Biol.,* 47, 115, 1970.
47. **McElhaney, R. N.,** The effect of alteration in the physical state of the membrane lipids on the ability of *Acholeplasma laidlawaii B* to grow at various temperatures, *J. Mol. Biol.,* 84, 145, 1974.
48. **Post, J. A., Leunissen-Bijvelt, J., Ruigrok, T. J. C., and Verkleij, A. J.,** Ultrastructural changes of sarcolemma and mitochondria in the isolated heart during ischaemia and reperfusion, *Biochim. Biophys. Acta,* 845, 119, 1985.
49. **Elgsaeter, A., Shotton, D., and Branton, D.,** Intramembrane particle aggregation in erythrocyte ghosts II. The influence on spectrin aggregation, *Biochim. Biophys. Acta,* 426, 101, 1976.
50. **Gerritsen, W. J., Verkleij, A. J., and Van Deenen, L. L. M.,** The lateral distribution of intramembraneous particles in the erythrocyte membrane and recombinant vesicles, *Biochim. Biophys. Acta,* 555, 26, 1976.
51. **Wieslander, A., Rilfors, L., Johansson, L. B. A. and Lindblom, G.,** Lipid bilayer stability in membranes. Regulation of lipid composition in *Acholeplasma laidlawii* as governed by molecular shape, *Biochemistry,* 19, 3650, 1980.
52. **Verkleij, A. J.,** Lipidic intramembraneous particles, *Biochim. Biophys. Acta,* 779, 43, 1984.
53. **Das, S. and Rand, R. P.,** Modification by diacylglycerol of the structure and interaction of various phospholipid bilayer membranes, *Biochemistry,* 25, 2882, 1986.
54. **Siegel, D. P., Alford, P., Ellens, H., Lis, L., Quinn, P. J., Yeagle, P. L., and Bentz, J.,** A 2-arachidonyl diglyceride is particularly effective in inducing L/H_{II} phase transitions and membrane fusion, *Biophys. J.,* 51, 355, 1987.
55. **Van Venetië, R., Hage, W. J., Bluemink, J. G., and Verkleij, A. J.,** Propane jet-freezing. A valid rapid freezing method for the preservation of temperature dependent lipid phases, *J. Microsc.,* 123, 287, 1981.
56. **Gulik-Krzywicki, T., Balerna, M., Vincent, J. P., and Lazdunsky, M.,** Freeze-fracture study of cardiotoxin action on axonal membrane and axonal membrane lipid vesicles, *Biochim. Biophys. Acta,* 643, 101, 1981.
57. **Batenburg, A. M., Rochat, H., Verkleij, A. J., and De Kruijff, B.,** The penetration of a cardiotoxin into cardiolipin model membranes and its applications on lipid organization, *Biochemistry,* 24, 7102, 1985.
58. **Batenburg, A. M., Van Esch, J. H., Leunissen-Bijvelt, J., Verkleij, A. J., and De Kruijff, B.,** Interaction of melittin with negatively charged phospholipids: consequences for lipid organization, *FEBS Lett.,* 223, 148, 1987.
59. **Batenburg, A. M., Hibbeln, J. C. L., Verkleij, A. J., and De Kruijff, B.,** Melittin induces HII phase formation in cardiolipin model membranes, *Biochim. Biophys. Acta,* 903, 142, 1987.

60. **Wilschut, J., Holsappel, M., and Jansen, R.,** Ca^{++}-induced fusion of cardiolipin/phosphatidylcholine vesicles monitored by mixing of aqueous contents, *Biochim. Biophys. Acta,* 690, 297, 1982.
61. **Miller, R. G.,** Do "lipidic particles" represent intermembrane attachment sites? *Nature,* 287, 166, 1980.
62. **Hui, S. and Stewart, T.,** Lipidic particles are intermembrane attachment sites, *Nature,* 290, 427, 1981.
63. **Bearer, E. L., Düzgünes, N., Friend, D. S., and Papahadjopoulos, D.,** Fusion of phospholipid vesicles arrested by quick freezing. The question of lipidic particles as intermediates in membrane fusion, *Biochim. Biophys. Acta,* 693, 93, 1982.
64. **Ellens, H., Bentz, J., and Szoka, F. C.,** Fusion of phosphatidylethanolamine containing liposomes and mechanism of Lα-HII phase transition, *Biochemistry,* 25, 4141, 1986.
65. **Siegel, D. P.,** Inverted micellar intermediates and the transitions between lamellar, cubic and inverted hexagonal lipid phases. I. Mechanism of the L_{α}/H_{II} phase transition, *Biophys. J.,* 49, 1153, 1986.
66. **Chandler, D. E. and Heuser, J. E.,** Arrest of membrane fusion events in most cells by quick freezing, *J. Cell Biol.,* 86, 666, 1980.
67. **Schmidt, W., Patzak, A., Lingg, G., Winkler, H., and Plattner, H.,** Membrane events in adrenal chromaffin cells during exocytosis: a freeze-etching analysis after rapid cryofixation, *Eur. J. Cell Biol.,* 32, 31, 1983.
68. **Hols, H., Sixma, J. J., Leunissen-Bijvelt, J., and Verkleij, A. J.,** Freeze-fracture studies on human blood platelets activated by thrombin using rapid freezing, *Thrombosis and Haemostasis,* 54, 574, 1985.
69. **Hackenbrock, C.,** Chemical and physical fixation of isolated mitochondria in low energy and high energy states, *Proc. Natl. Acad. Sci. U.S.A.,* 61, 598, 1968.
70. **Knoll, G. and Brdiczka, D.,** Changes in freeze-fractured mitochondrial membranes correlated to their energetic state. Dynamic interactions of the boundary membranes, *Biochim. Biophys. Acta,* 733, 102, 1983.
71. **Klug, G. A., Krause, J., Östlund, A. -K., Knoll, G., and Brdiczka, D.,** Alterations in liver mitochondrial function as a result of fasting and exhaustive exercise, *Biochim. Biophys. Acta,* 764, 272, 1984.
72. **Van Venetië, R. and Verkleij, A. J.,** Analysis of the hexagonal II phase and its relation to lipidic particles and the lamellar phase. A freeze-fracture study, *Biochim. Biophys. Acta,* 645, 262, 1981.
73. **Nicolay, K., Timmers, R. J. M., Spoelstra, E., Van der Neut, R., Fok, J. J., Huigen, Y. M., Verkleij, A. J., and De Kruijff, B.,** The interaction of adriamycin with cardiolipin in model and rat liver mitochondrial membranes, *Biochim. Biophys. Acta,* 778, 359, 1984.
74. **Ohlendieck, K., Riesinger, J., Adams, V., Krause, J., and Brdiczka, D.,** Enrichment and biochemical characterization of boundary membrane contact sites from rat liver mitochondria, *Biochim. Biophys. Acta,* 860, 672, 1986.
75. **Cline, K., Keegstra, K., and Staehelin, L. A.,** Freeze fracture electron microscopic analysis of ultra-rapidly frozen envelope membranes in intact chloroplasts and after purification, *Protoplasma,* 125, 111, 1985.
76. **Cremers, F. F. M., Voorhout, W. F., Van der Krift, T. P., Leunissen-Bijvelt, J. J. M., and Verkleij, A. J.,** Visualization of contact sites between outer and inner envelope membranes in chloroplasts, *Biochim. Biophys. Acta,* 933, 334, 1988.
77. **Schwaiger, M., Herzog, V., and Neupert, W.,** Characterization of translocation contact sites involved in the import of mitochondrial proteins, *J. Cell Biol.,* 105, 235, 1987.
78. **Pain, D., Kanwar, Y. S., and Blobel, G.,** Identification of a receptor for protein import into chloroplasts and its localization to envelope contact zones, *Nature,* 331, 232, 1988.
79. **Haest, C. W. M., De Gier, J., Van Es, G. A., Verkleij, A. J., and Van Deenen, L. L. M.,** Fragility of the permeability barrier of *Escherichia coli, Biochim Biophys. Acta,* 288, 43, 1972.
80. **Blok, M. C., Van der Neut-Kok, E. C. M., Van Deenen, L. L. M., and De Gier, J.,** The effect of chain length and lipid phase transition on the selective permeability properties of liposomes, *Biochim. Biophys. Acta,* 706, 186, 1975.

Chapter 6

ULTRASTRUCTURAL ANALYSIS OF BIOLOGICAL MEMBRANE FUSION AND A TENTATIVE CORRELATION WITH BIOCHEMICAL AND BIOPHYSICAL ASPECTS

Gerd Knoll and Helmut Plattner

TABLE OF CONTENTS

I. INTRODUCTION

Membrane fusion is the temporal coalescence either of two membrane continua on the way to form one, or of one continuum to form two.[1,2] Membrane fusion is a general feature of all cells and membrane bounded organelles, irrespective of the high diversity of membranes involved.

Despite the dynamic nature of these events, much has been learned by electron microscopy (EM), although this method delivers only static snap shots. This is true of a variety of general concepts in cell biology, e.g., of exocytosis and endocytosis. Also, with regard to the development of ideas concerning molecular mechanisms of membrane fusion, EM analyses have provided a conceptual basis. Thus, the appearance of pure lipid areas in fusing membranes has led to the dogma that proteins would have to disappear from presumptive fusion areas (see Reference 3, for reviews). Considerable improvements of EM preparation techniques were then required to show this picture to be wrong (see e.g., References 4 to 7), and the resulting new "focal (point) fusion" concept corresponds well to the ever increasing evidence for a regulatory role of proteins in the fusion mechanism.[8]

Today there is a flood of information concerning new proteins or protein classes claimed to be involved in membrane fusion.[8-10] Only exceptionally, one or the other protein could be localized at fusion sites, e.g., calmodulin.[11] New EM methods, e.g., improved labeling techniques, will be helpful on the way of "the gene in search for a protein in search for a function" by structural correlation.

Concerning the impact of EM research on the development of concepts for membrane fusion, two things become apparent when one looks through recent extensive treatises on this subject (as summarized, e.g., in References 12, 13): (1) There is a great divergency of working hypotheses concerning the mechanism of biological membrane fusion, and (2) ultrastructural data are deliberately used as arguments in favor of different hypotheses, frequently irrespective of the validity of the methods used. In the meantime, methodical progress has, in fact, invalidated some interpretations of data obtained before. Therefore, the aim of this review is the following:

1. To determine the state of the art of ultrastructural methods with regard to possible consequences for the analysis of membrane fusion. This part will include a discussion of artifacts, of alternative methods, and of open methodical problems.
2. To categorize the diversity of biological membrane fusion and to exemplify where today we have valid data available, and where not (and where consequently acquainted views possibly have to be revised).
3. To summarize consistencies and diversities of data collected in the above paragraph, and to connect these data with recent views on biological membrane fusion obtained with complementary techniques.
4. To outline current problems as they could possibly be addressed by modern ultrastructural methods.

II. ASPECTS CONCERNING METHODS

A. THE PROBLEM: ARTIFACTS

"Almost every modification" (remark: of chemical fixation) "results in some change in cellular morphology, varying from size or 'evenness' of mitochondrial membrane and proportion of vesicular, as opposed to cisternal, endoplasmic reticulum to presence or absence of microtubules in the spindle . . . Under such conditions, personal opinion flowers extensively and the consequences are apparent, if unexpressed, in much of the literature on electron microscopy" (Rebhun).[14] We fully support this critical statement of a colleague who aimed at permanently improving EM methods himself.

After a flowering and very fruitful early era of EM applications to questions in cell biology, some disenchantment took place. The limitations and artifacts were soon recognized. However, "business went on as usual", but results were frequently questionable, or they were accepted only if illustrating what had been postulated before on the basis of other experiments.

The most obvious limitation of EM is the discrepancy between the dynamic nature of living cells and the static result of the necessary fixation. Membrane fusion is shown to be very rapid (see, e.g., References 15 to 17). Therefore, fixation procedures which take a longer time than the dynamic aspect under investigation are not suited for an analysis of such rapid events (see Reference 18 for a discussion). However, a bandwagon production of investigations on that theme resulted in a quite clear and consistent picture: all membrane fusion events in biological membranes seem to be accompanied by a local displacement of membrane proteins, and by formation of a "diaphragm" (for reviews see Reference 3). As shown below, this feature is an artifact of chemical fixation inherent to all studies performed on this basis.

But why is this artifact of such high reproducibility and specificity? An explanation has been presented by Chandler,[7] saying that membrane fusion is always accompanied by a local fluidization of the membranes. EM micrographs obtained by "standard methods", as discussed above, would then be an "exaggeration" of an otherwise much more restricted phenomenon (see also Plattner[6]). Indeed, an early fluidization of membranes in different biological fusion systems also has been stated with alternative methods (see Chandler[7], for references). Using stopped-flow fluorometry for the analysis of early transmembrane events in T-lymphocyte activation, in subsecond times a membrane fluidization was revealed preceding a release of Ca^{2+} from intracellular stores and extracellular Ca^{2+} influx.[19] The authors suggest that this sequence generally occurs during a variety of ligand-receptor interactions.

Besides the problem of stopping rapid events fast enough, chemical fixation is limited by its specificity; not all cellular components are immobilized in the same way and during the same time. Taking these drawbacks together, a faithful preservation, even of seemingly static cellular features, will not be possible, if these features depend on nonequilibrium steady states, which are susceptible to rapid changes, e.g., ion distributions, etc. A collection of examples of the failure of chemical fixation to preserve the native state of cellular structure and composition can be found in various reviews.[18,20-23]

B. THE ALTERNATIVE: RAPID FREEZING

As soon as the preparational limitations became apparent, cryofixation was propagated and elaborated as an alternative (e.g., References 24, 25). The development and state-of-the-art has been reviewed in References 18, 20-23, 26. Plattner and Bachmann[20] have discerned (1) spatial and (2) temporal resolution. (1) Refers to dislocation of the solutes by ice crystal formation; see Chapter 18 by Schmitz and Zierold. (2) Refers to dynamic cellular events. The time needed to freeze a suitably small specimen is in the range of a millisecond (for discussion see Reference 18); therefore, cryofixation seems much more suited to analyze membrane fusion phenomena than the much slower chemical fixation procedures. As a common result of a number of investigations using pure cryofixation, the absence of any clearing of intramembrane particles (IMPs), which in their majority are considered as integral membrane proteins (see Reference 27 for a discussion), and of diaphragm formation, has been found.[5,7,28-31] The induction of these phenomena by chemical pretreatments has also been demonstrated to be highly reproducible, though representing an artifact.[4,7] But why then do so many people stick uncritically to old results, even when they have unequivocally been unmasked as artifacts?

There seem to be two reasons (at least):

1. Due to the common observation of pure lipid patches correlated with biological membrane fusion, together with the perception of Ca^{2+} as "common denominator", studies

have concentrated on modeling membrane fusion with pure lipid systems by direct Ca^{2+} effects (e.g., References 32, 33). Though valuable information has resulted from these experiments, biological membrane fusion is, in fact, quite different according to our recent knowledge. Differences include nonleakiness, specificity and regulation, etc. (for discussions of this point, see e.g., References 15, 34, 35). Nevertheless, views evolved in this direction still have considerable impact on work with cells.

2. It is not so unusual to hear in discussions that every electron microscopist is enthusiastic toward his own method merely on the basis of personal taste. One will vote for glutaraldehyde, another one for osmium, and there are some people who like to freeze all "stuff". But EM methodology is no black magic, and one should not close his eyes even for seconds (or fractions thereof) when samples are prepared for EM analyses. (One of us [H.P.] still recalls the comments of a reviewer of an outstanding cell biology journal on a paper introducing a new fast freezing technique, about 15 years ago, which said that one could achieve equally smooth pictures just by adding some glycerol before freezing. This attitude certainly has changed since then.) It is one purpose of the authors to provide the reader with a brief appreciation of methodical developments.

Today cryofixation is the most reliable method known; results fit well with data obtained with so widely different alternative methods, such as electrophysiology (see section IV). Of course, the reliability or possible hazards of fast freezing methods also have to be investigated.[20,36,37] Some criticism has been raised concerning deformations during "slam freezing",[36] but with new developments in this field[39] and with the possibility of using alternative cryofixation techniques for most objects,[40] one can overcome this problem.*

C. RAPID FREEZING: THE PROBLEMS

Even with the use of rapid freezing (cryofixation) techniques one faces considerable difficulties in the analysis of rapid events, such as membrane fusion. (1) The first aspect concerns the chance to "catch" the event at all: the fixation method has to be more rapid than the event itself. (2) On the other hand, the events must have a high frequency and synchrony, and in addition they must conincide in time with the short moment of freezing. These difficulties have been discussed in detail.[18,29,38] Some prospects are discussed in section V.

III. BIOLOGICAL MEMBRANE FUSION

The general two possibilities of biological membrane fusion are (a) "planar" (the joining of formerly separated membranes), or (2) "annular" (or "tubular", i.e., the fission of a former continuum).[1,2] As examples of these complementary types one can think of (1) cell fusion (e.g., during fertilization) and exocytosis, or (2) cell division and endocytosis, respectively. Either the exoplasmic or the protoplasmic surface of a membrane (nomenclature according to Branton et al.,[43]) is involved in making contact, and finally in fusion (ES- or PS-fusion; see Figure 1). To achieve membrane fusion, in all cases the lipid bilayer structure has to be abandoned temporarily and locally. Hypotheses on fusion mechanisms have to take into account the asymmetry of biological membranes involved,[44] both for the partition of proteins and for lipids, to one half of the bilayer or the other.

For the sake of clarity, we will use this classificaton further on (Figure 1). It is not the aim to give a complete survey here (see e.g., References 12, 13 for more details), but to

* In fact, no specific artifacts of rapid freezing have been described till today, if there is no chemical pretreatment engaged. This is not the case, however, with deep etch studies after "quick freeze",[41] when samples have to be fixed before, and impregnated with methanol to allow deep etching, or with vitrification of biological bulk samples impregnated with cryoprotectants.[42]

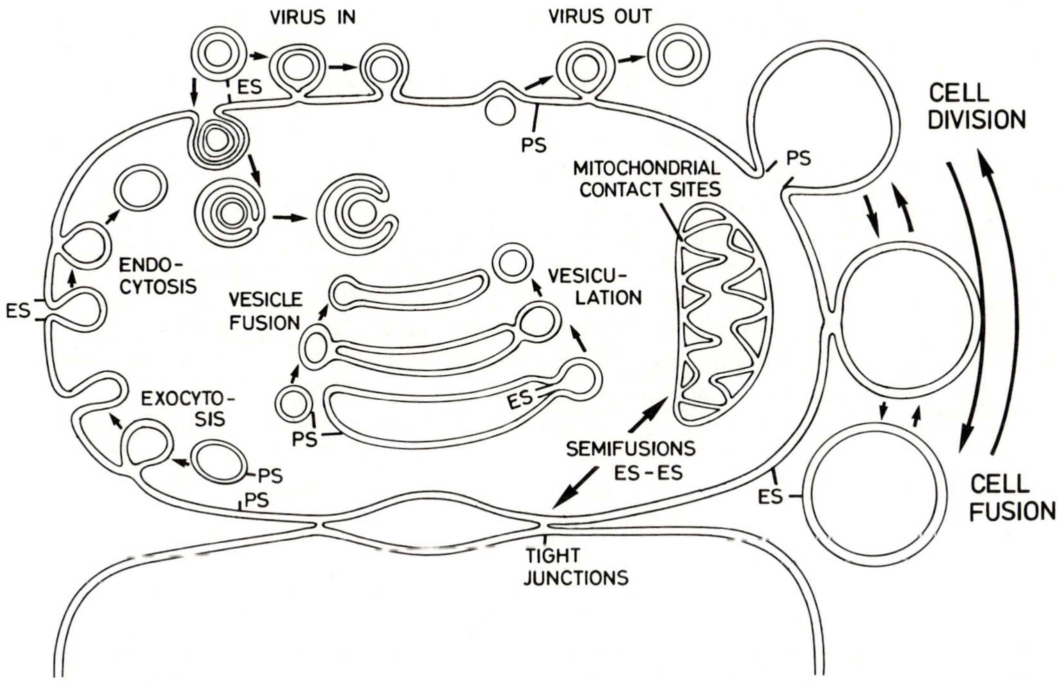

FIGURE 1. Types of biological membrane fusion. To depict the different types of biological membrane fusion a eukaryotic cell (nucleus omitted) is schematized. It is connected to a neighbor cell via tight junctions, and, during division and fusion, with a daughter and partner cell, respectively. The membrane surfaces involved in making the initial contact (followed by the intermediate nonbilayer stage of lipids and the final bilayer restabilization) are indicated as PS (protoplasmic surface) and ES (exoplasmic surface), following the terminology of Branton et al.[43] The principal mirror-like congruence of planar and annular fusion events is suggestive. For classification, examples, and details see text.

exemplify what has been learned from EM studies, what now can be judged as reliable data, to connect these data with current views on membrane fusion in a general frame, and to pinpoint arising questions that possibly could be addressed now by recent EM methods.

A. CELL FUSION

1. Cell-Cell Fusion

Cell-cell fusion is a highly specific event, and only a few cell types become fusion competent during the course of their development in a very well regulated way. Therefore, a primary question (and a possible key for an understanding of the mechanism) is the mode of regulation.

a. Fertilization

Recent hypotheses on the fusion mechanism point to the involvement of fusogenic proteins.[45,46] The regulation of the event concerns competence, both of the sperm and of the egg. Sperm must become fusion competent; related membrane differentiations have been analyzed by freeze fracture techniques (see References 45, 47 for reviews). When an egg cell is penetrated by a sperm cell, the egg loses its fusion competence by rapidly switching the membrane potential, which may lead to conformational changes of either fusogenic proteins or of receptors in the target membrane.[48] Exactly timed rapid freezing followed by freeze fracture would possibly be helpful to test the hypothesis that transbilayer movements of proteins are involved.[48] In fact, the transbilayer orientation of mellitin is influenced by the transmembrane voltage,[49] and vertical particle displacements have been observed in nerve endings and synaptosomes upon stimulation,[50-52] as well as in plasma membranes of *Tetrahymena*, in response to local anesthetics.[53]

In cases where cell to cell fusion competence is restricted to a localized site, this is easily recognized by dense material and IMP clusters in the fusogenic zone, e.g., with *Chlamydomonas reinhardtii* cells. Fusion starts in that zone densely studded with IMPs.[54]

b. Myoblast Fusion

EM studies combined with electrophysiology have allowed correlation of the time points for the onset of intercellular connections, and the structural appearance of individual cell couples at the moment of fixation. Analysis by serial sectioning revealed gap junction connections to be responsible for low efficiency coupling, and rapid disappearance of the surface membranes between fused cells after the onset of high efficiency coupling.[55]

In a study employing freeze fracture and thin sectioning (after chemical pretreatments), IMP-clearing, formation of diaphragms and of IMP-free vesicles in/at the plasma membranes of fusion competent cells, have been described.[56] The authors were aware of the artifact hazard, but they argued that "fixation 'reality' is difficult to define" before they asked, "Is the 'optimal' fixation condition one that preserves or eliminates certain structures?" One now can assume reasonably that the described structures are not destroyed by rapid freezing, but rather that they are evoked by the chemical pretreatments as consequence of membrane fluidization events (e.g., Reference 57).

But we also would like to mention that the use of rapid freezing has a low chance to capture events of such low frequency in time and space.[18,29,38] Indeed, an effort to elaborate the structural features of myoblast fusion (using sandwich freezing as described in References 37, 58, and thereby circumventing the hazard of cold metal block freezing thought to destroy sensitive structures[38]) was skipped by us for these reasons. However, some rare clear-cut instances have given us a different picture of the fusion event: we observed several local point fusions at a limited area of membrane contacts between myoblasts under exceptionally well synchronized fusion conditions (elaborated by Wakelam and Pette,[59] Figure 2). Another feature revealed by freeze fracture under these conditions was a very high density of coated pits (as already described in serial thin sections[60]), which might correspond to ligand-receptor interactions postulated to occur.[61] In each case, membrane retrieval during the joining of the two fusing cells has to take place to accomodate the surplus of membrane material (in contrast to the situation during cell division, see below); therefore, vesicles occurring in the vicinity of fusing plasma membranes are more likely endo- rather than exocytotic.

Together with the observation of rapidly increased phosphatidylinositol turnover and rise in intracellular Ca^{2+} (as observed also for fusing plant protoplasts[62]), the cell-cell fusion competence seems to follow the activation of intracellular events, as the activation of metalloendoproteases.[63] Finally, given the glycokalix as a contact barrier, these intracellular events have to influence some components on the outer surface of the membrane. A protein kinase has been suggested to be directly linked to membrane fusion.[64,65] Extracellular Ca^{2+} has been found necessary for the activation step, but not for the following membrane fusion itself.[61]

c. Tight Junctions = Semifusions?

On the basis of structural features a semifusion mechanism (linking adjacent plasma membranes) has been proposed for cells connected by tight junctions.[66,67] One prediction of this model, the continuity of the outer monolayers of the plasma membranes of connected cells, has been refuted.[68] Instead a proteinaceous nature is postulated. The recent identification of tight junction-specific proteins ZO-1[69] and cingulin[70] seems to support this view. The properties of these proteins, however, point to peripheral association to the membranes. (This localization is supported by immunoelectron microscopy.) Therefore, the molecular architecture of tight junctions remains obscure. The elucidation of the mechanism which induces these "focal regions of apparent fusion between the adjoining cell membranes"[70]

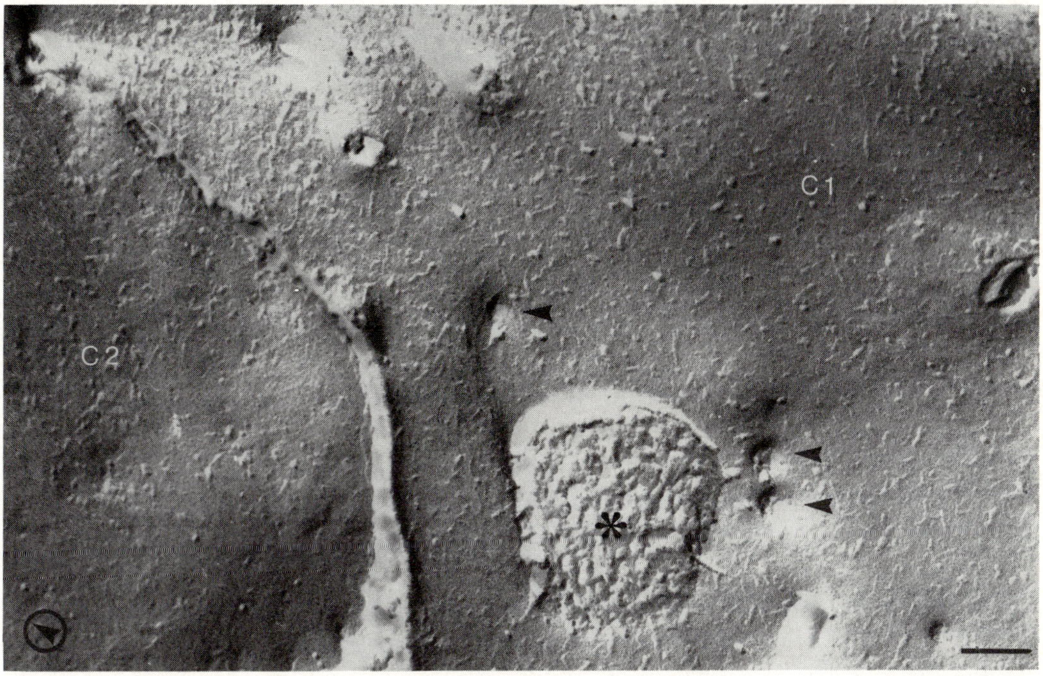

FIGURE 2. Myoblast fusion. Myoblasts cultured as overlapping monolayer (for details see Reference 163) were rapidly frozen on their substrate by a sandwich-propane-jet method[37,58] 5 min after calcium in the medium had been raised to induce massive and rather synchronous cell-cell fusion.[61] The freeze etch replica shows the EF face of a cell (marked as C1) on top of a second cell (C2), with the PF face partly exposed. A continuous aqueous space (asterisk) between the cytoplasm of both cells has a granular texture due to the "etching" applied. Otherwise, a smooth surface (which would be produced by merely fracturing the frozen cytosol) could be misinterpreted as a diaphragm. An artificial rupture of such a diaphragm, considered as a risk of slam freezing,[38] is not likely to occur because cells remained undisturbed on their growing substrate before and during (protective) sandwich-jet freezing.[37,58] Depressions (marked by arrowheads) in the EF face of C1 pointing to C2 can be reasonably interpreted as early fusion stages. Thus, membrane fusion between myoblasts seems to start at multiple points and without diaphragm formation. Also, no particle clearing or formation of smooth lipid patches can be recognized. Arrowhead in circle indicates shadowing direction. Bar = 100 nm. (From Knoll and Wakelam, unpublished.)

as relatively stable structures will be of high interest for an understanding of membrane fusion in general.

2. Cell Division

Immunocytochemical approaches have allowed the localization of specific proteins, most prominently cytoskeletal elements and calmodulin, in the region of the cleavage furrow.[71] Another focus of interest is the question of how and where new membrane components are provided to compensate the increase in surface area during the constriction of the parent cell by the contractile ring. Membrane discontinuities have been described by freeze fracture studies.[72] The topography of new membrane material insertion has been revealed by autoradiography.[73] New membrane structures destined for insertion seem to be very sensitive to chemical fixation;[74] rapid freezing helped to overcome this problem with growth cones.[75] Thus, valid data on fine structural detail of the membranes engaged in membrane fusion during cell division can be expected only from rapid freezing analyses.

B. CELLS AND VIRUSES

1. Entry of Enveloped Viruses

Enveloped viruses may fuse either with the plasma membrane (e.g., Sendai virus, a paramyxovirus, see References 2, 76) or with the endosomal membrane, by the action of specific viral fusion proteins.[2,77-80] The low pH-induced conformational change of the influenza viral hemagglutinin spike protein,[2,79,80] was suggested to induce "local perturbations of the membranes, which result in a lipid phase transition leading to coalescence and separation,"[2] is well characterized. While the general pathways have been elucidated successfully with conventional EM (see references above for reviews), detailed analyses of the fusion event itself are scarce and show contradictory results. While Knutton[81,82] described specific membrane distortions and changes in IMP size and distribution during fusion of Sendai virus with erythrocytes, others failed to observe these intermediate structures.[83] Since both studies were performed using "conventional" freeze fracture (i.e., with chemical pretreatments), we are not yet on solid ground. Using exclusively rapid freezing techniques, only membrane fusion occurring between influenza virus and liposomes has been investigated so far. The findings were focal attachment and fusion, followed by dispersal of viral membrane components.[84]

In a complementary way, liposomes are able to fuse with the plasma membrane of infected cells due to the fusogenic activity of the influenza viral hemagglutinin.[85,86] Freeze fracture after rapid freezing again revealed focal (point) fusion, with well defined IMPs in the otherwise smooth liposomal fracture faces.[87] These IMPs have tentatively been interpreted as inverted lipid micelles, representing fusion intermediates (Verkleij[88]; see Chapter 5 by Verkleij), evoked by the activity of the viral fusogenic protein (which *per se* is probably not recognized).[87]

2. Budding of Enveloped Viruses

Most enveloped viruses leave the cell by wrapping a piece of plasma membrane (containing the viral envelope proteins) around the nucleocapsid. This is a continuous process which finally results in the release of virus particles.[78,89] The last step, the pinching off, involves an annular fusion event which is not yet characterized very well. In the simplest way, the strain of high curvature could lead to the pinching off process and spontaneous resealing of the bilayers.[78,89] However, for Sindbis virus a definite "semifusion" stage has been described,[90] with the inner lipid leaflets of the bud and the plasma membrane already separated, but with the outer leaflet still continuous. A similar situation was found for influenza virus budding from epithelial cells.[87] Just at the nexus of the joined membranes a well defined particle was shown (Figure 3) which may be interpreted as an inverted lipid micelle.[87]

C. PLASMA MEMBRANE AND VESICULAR ORGANELLES

1. Exocytosis

Since this is a general capacity of all eukaryotic cells, membrane fusion during exocytosis has been the subject of a large number of investigations, either by "conventional" or by "advanced" preparation techniques. Therefore, exocytosis may exemplify best the developments towards our recent understanding of the structural features of membrane fusion. This development has been extensively reviewed (see Plattner[6,8,91-93]). In summary: whereas chemical pretreatments rather consistently lead to IMP clearing and diaphragm formation (see e.g., Reference 94), all studies performed until now with fast freezing techniques (regardless of the follow-up procedures) never revealed any of these features (see examples in Figures 4 and 5). On the contrary, the causal relationship between chemical pretreatments and the occurrence of these features has been demonstrated.[4,7]

The common picture emerging now out of the analyses of a variety of exocytosis systems endorses the focal fusion scheme proposed before.[6] As a first step, a close contact between the membranes is established, then a small pore opens and rapidly enlarges. Sometimes the contact is observed as an invagination of the plasma membrane (Figures 4, 5; see Reference 5, 93), interpreted as being induced by a force generating machinery.[95] But in this context

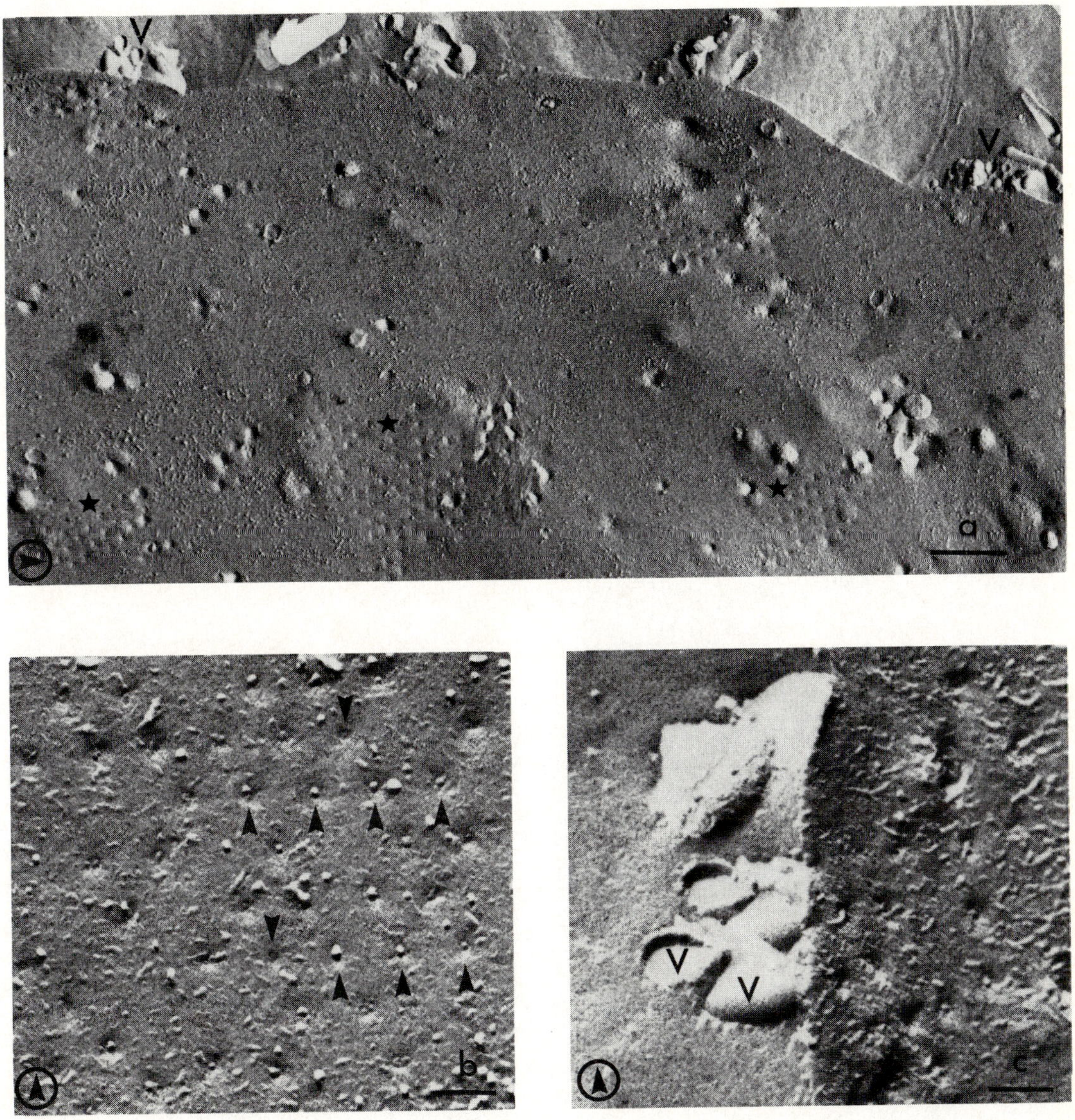

FIGURE 3. Budding influenza virus. Epithelial cells of a kidney cell line (MDCK strain II) were grown on plastic substrate[85,86] and rapidly frozen by a sandwich-propane-jet method[37,58] 4.5 h after infection with influenza virus. At that time the first virions start to bud from the apical surface of the cells.[85] The freeze fracture replica exhibiting the EF face of the apical membrane (a) shows clusters of depressions (asterisks), with central particles or pits [arrowheads at higher magnification in (b)]. These membrane structures are caused by viruses attached to the plasma membrane (c) in a late budding stage just before pinching off. The particles are suggested to represent inverted micelle-like structures of lipids, representing annular fusion intermediates. Arrowhead in circle indicates shadowing direction. Bars = 500 nm (a), 100 nm (b,c). (From Knoll et al., *J. Cell Biol.*, 107, 2511, 1988. With permission.)

it is interesting that mere binding (at neutral pH) of influenza virus to ganglioside-receptor carrying liposomes induces considerable protrusion in the liposomal bilayer.[84] The pore appears as volcano-like protrusions in freeze fractured membranes,[30] and the diameter is below 10 nm, which hampers visualization in thin section preparations.[5] In addition, some particulate protrusions without a central opening are visible on replicas (see Reference 30), which led to the interpretation of these structures as representing lipidic particles as fusion intermediates.[96] The lifetime of the pore seems to be in the millisecond range, followed by rapid enlargement.[4,5,17,30]

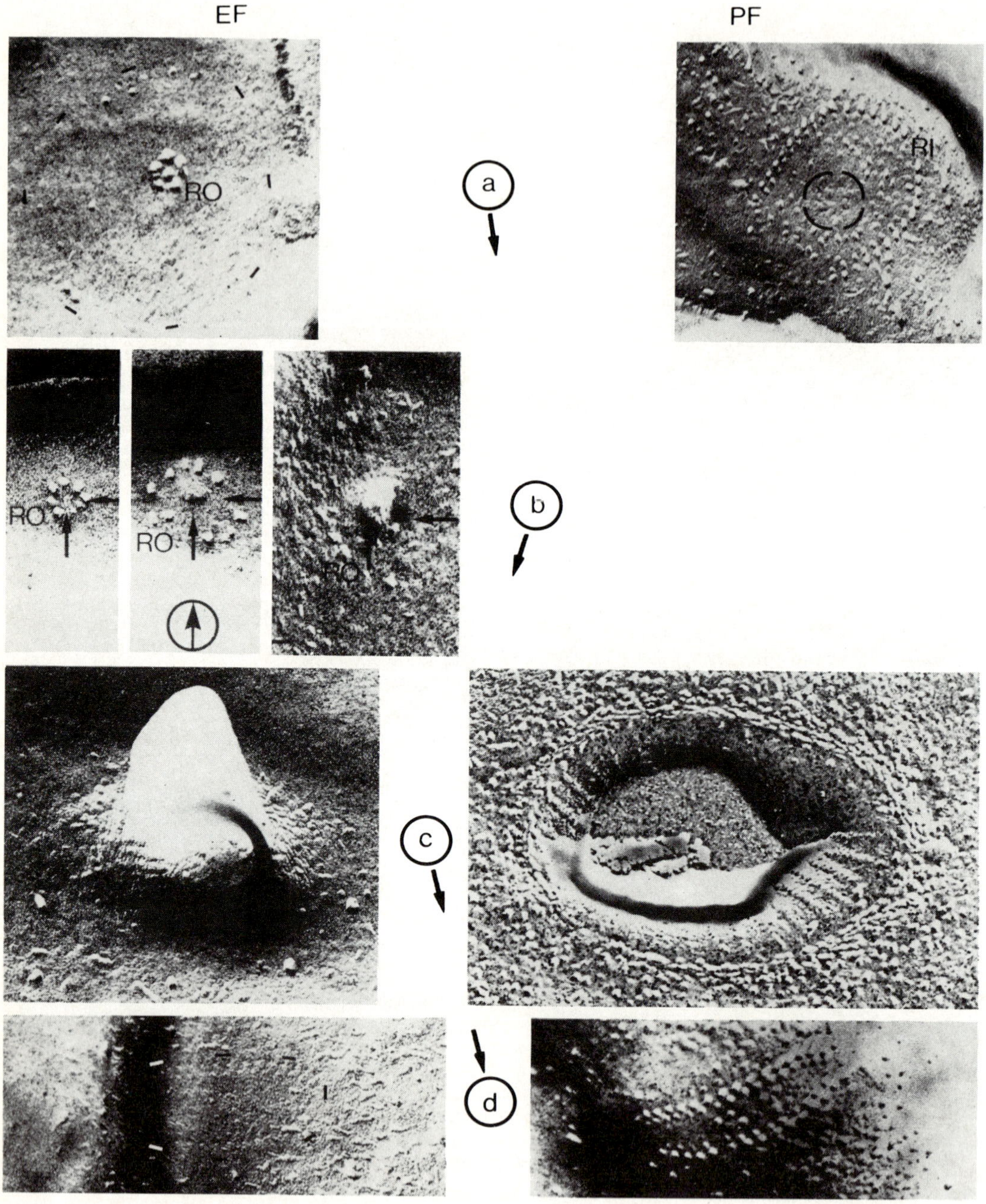

FIGURE 4. Preformed sites of exocytotic membrane fusion in *Paramecium* cells. Left: EF face, right: PF face, showing preferably rosette (RO) or ring (RI) subpopulations of IMPs, respectively. Rings delineate fusogenic sites, whereas only rosette IMPs are directly correlated with fusion capacity.[91] Stages (a-d) represent the resting stage; (a), (b) beginning of membrane fusion within the rosette IMP population; (c) release of secretory products, and (d) "resealing" (exocytosis-coupled endocytosis). Note that presumable early fusion stages [three examples shown in (b)] (1) are of a range of 10 nm, (2) are positioned on a funnel-like depression and (3) dispersal of rosette IMPs occurs only when membrane fusion is initiated. Also remarkable is the absence of rosette IMPs after resealing, suggesting their irrelevance for endocytotic membrane fusion, which, therefore, might require other regulation mechanisms than exocytosis. Pt/C Shadowing from bottom to top. Bar = 100 nm. (From Olbricht et al., *Exp. Cell Res.*, 151, 14, 1984. With permission.)

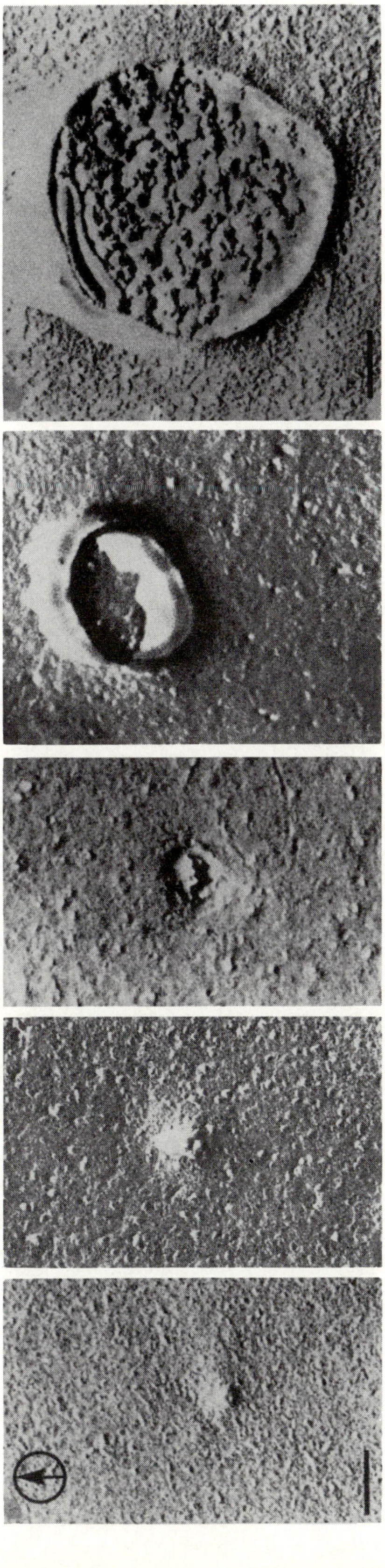

FIGURE 5. EF face of the cell membrane of a bovine chromaffin cell isolated from the adrenal medulla. Cells were exposed for 1.5 min to the trigger agent carbachol (for inducing exocytosis of catecholamines, etc.) and fast frozen (starting at a temperature of 37°C) by the sandwich-propane-jet method[37,58] for freeze fracturing and Pt/C replication (direction indicated by arrow). Since exocytosis is not synchronous in this system, the ultrastructurally detected increase of the number (per unit cell surface area) of *bona fide* exocytosis sites (for which a potential "morphogenetic sequence"—from left to right—has been assembled) had to be quantitatively correlated with a simultaneous increase in catecholamine output. (For details see Reference 30). Note the absence of IMP rearrangements around fusion sites, the occurrence of a ~10 nm large pit on a funnel-like depression as the first presumable fusion stage (left), and its subsequent expansion to a size corresponding to the diameter of a secretory vesicle (right). Bars = 100 nm. (From Schmidt et al., *Eur. J. Cell Biol.*, 32, 31, 1983. With permission.)

The fusion of isolated secretory vesicles (thus mimicking compound exocytosis) has been seen as a promising model to address general questions of exocytosis.[9] Applying conventional preparation techniques, IMP displacements and diaphragm-formation have been stated after the addition of millimolar Ca^{2+}.[97] (However, this might be due to extensive membrane contacts by unphysiologically high Ca^{2+} concentrations, since similar effects could be produced by forced physical contact during preparation.[98]) Rapid freezing analysis confirmed fusion, but not IMP displacements.[99] This study, in addition, revealed fusion to occur in a more physiological range of Ca^{2+} concentrations, as opposed to the fusion of liposomes prepared from the lipids of the same vesicles. The concluding assumption that proteins are responsible for this higher Ca^{2+} affinity has been affirmed, and meanwhile expanded to the description of the (allegedly fusogenic) group of Ca^{2+}-regulated phospholipid-binding proteins.[9,10]

By the use of recent methodical developments (cell permeabilization, microinjection, patch clamp, *in vivo* Ca^{2+} measurements by fluorescent dyes, isolated cell cortex fragments as *in vitro* systems) it was possible to define more clearly some requirements of the fusion of secretory vesicles with the plasma membrane. While for sea urchin cortices the role for Ca^{2+} as the only[100] and "proximal physiological stimulus"[101] could be demonstrated, many other systems investigated showed a strong dependency on still other factors. This was revealed either by varying the effective concentration of Ca^{2+}, or by measuring the actual *in vivo* Ca^{2+} concentration. The results imply that exocytosis occurs even at resting levels (see e.g., References 35, 102-105), or an apparent discrepancy in timing between Ca^{2+}-rises and exocytotic events was registered.[105] As modulating (or alternative) factors, diacylglycerol (one product of stimulated polyphosphatidylinositide cleavage[106]) and guanosinetrisphosphate (via GTP-binding proteins) have been identified in many systems.[35,102-104,107] See also section IV for further proteins possibly involved in membrane fusion during exocytosis.

2. Endocytosis

The mechanism by which a cell takes up extracellular material (and retrieves pieces of plasma membrane), is endocytosis. The membrane becomes locally constricted and two new separate membranes arise. The fusion intermediate to be postulated, however, remains unknown. This holds also for the question of how membrane contact can be achieved on this side of the cell membrane with the glycokalyx attached.

EM investigations have contributed a substantial part of our present knowledge about different aspects of endocytosis, including receptor-mediated[108] and exocytosis-coupled endocytosis.[109-112] This comprises evaluation of the compartments involved (e.g., endosomes, receptosomes, but see below), elucidation of the pathways of receptor recycling and sorting specificity during exo-/endocytosis coupling. Only some EM analyses on the kinetics of these events, even on a very short time scale,[113,114] were performed with fast freezing, whereas most other studies are based on "conventional" techniques (starting with chemical fixation). This will not compromise results with regard to the identification of general pathways. But conclusions concerning certain fine structural details (e.g., the controversy about "endosomes"[115] or "receptosomes"[116]) are confronted with all the uncertainties which are discussed in section II. In our eyes extensive morphological elaborations as presented, e.g., in Reference 117, should be repeated with "state of the art" preparational procedures. (see also section III.D.1.)

The fusion event itself, or an intermediate, has not yet been shown, and we know little about possible mechanisms. Morphological details observed during frustrated endocytosis suggest a final closing of the clathrin basket to be necessary,[118] maybe by applying sufficient strain (by a small curvature) on the membrane to pinch off. This might be a simplistic view of annular fusion that can be questioned, however, by the existence of mutants where coated pits are trapped, indicating that pinching off is a specifically blockable step.[119] For endocytosis, where the external leaflet of the plasma membrane undergoes annular fusion, extracellular Ca^{2+} has been found not to be essential.[120] This is opposite to previous data.[110]

D. INTRACELLULAR MEMBRANES

1. Vesicle Traffic

The intracellular vesicle traffic between the different compartments needs permanent fusion events, comprising both planar and annular. Specificity and mechanisms are difficult to address by biochemical methods. An important tool is provided by the use of well characterized enveloped viruses.[78] Traffic routes can conveniently be followed by immunocytochemical methods.[121]

The sensitivity of intracellular vesicular systems to chemical fixation procedures has been demonstrated repeatedly, and may be exemplified by recent studies on endothelial cells. Rapid freezing revealed a population of small vesicles "in various connective associations" which are "suggestive of imminent fission or fusion events."[122] In glutaraldehyde-fixed preparations most of these vesicles have disappeared, usually by artificial fusion with larger vesicles and the plasma membrane, which consequently exhibit a dramatic increase in membrane area. Artificially induced fusion and fission of intracellular membrane systems, which change the morphology of the organelles, not only hampers proper structural description and identification, but may require reassessment of functional data of EM tracer and label studies.

First attempts to reconstitute intracellular fusion systems[123-126] point to a need for ATP; increased Ca^{2+} does not seem to be necessary.[125] On the other side there is increasing evidence for the involvement of GTP-binding proteins,[127,128] known to increase Ca^{2+} sensitivity during exocytosis.[35,103] An intriguing feature of intracellular vesicle traffic between organelles is the abrupt stop at the onset of mitosis, which led Warren[129] to propose a common feature of the budding and fusion reactions, and a general inhibition of a common mechanism. Because the disassembly of the nuclear lamina involves the phosphorylation of specific proteins (see Reference 130 for review and newer data), Warren[129] speculates "that membrane traffic is inhibited in a similar way, perhaps by the phosphorylation of vesicle fusion proteins."

2. Mitochondria and Chloroplasts

Both organelles have to divide, and mitochondria seem to undergo frequent fusion and fission cycles.[131] In the light of two boundary membranes involved, the situation appears rather complex.

A recent focus of interest is contacts between the inner and outer membranes, which have been shown to be involved in protein import[132-134] and metabolic regulation;[135] they are correlated with dedifferentiation of cancer cells[136] (for summaries of different aspects see References 137-139). Contacts in mitochondria have been interpreted as "semifusions" of the membranes.[140] (For the analogy with tight junction structure see above and Reference 88.) Contacts in chloroplasts exhibit a specific particle at the contact spot.[141]

While conventional EM techniques allowed for localization of peptides to be imported at defined contact points of isolated mitochondria,[133] EM analysis after rapid freezing shows the close apposition of the mitochondrial membranes *in situ*.[142] These "semifusions" are dynamic and metabolically regulated,[143] and under hormonal control *in vivo*.[144]

IV. CONSEQUENCES FOR POSSIBLE FUSION MECHANISMS

A. CORRELATION WITH ALTERNATIVE APPROACHES

The common feature of all EM observations on biological membrane fusion (after rapid freezing) is the occurrence of a focal event without any restructuring of membrane components around the fusion point. Therefore the view that membranes " . . . form a pentalaminar junction from the two trilamellar membranes" followed by "monolayer fusion where the adjacent monolayers of the plasma membrane and the vesicular membrane fuse" with "subsequent breakage of the semifused membranes" (citations from two recent review articles, [9,33]) is certainly not the state of our current knowledge. Hypotheses on biological membrane fusion, which are based on fit with the cited scheme, have to be rejected today on the ground of opposite results obtained by advanced EM methods.

What is known today about induction of biological membrane fusion? The only well characterized system is represented by enveloped viruses with fusogenic spike proteins, e.g., the hemagglutinin (HA) of the influenza virus.[2,79,80] The conformational change of this protein is responsible for the induction of membrane fusion; this is triggered by low pH in endosomes (and is independent of Ca^{2+}, in contrast to other viral fusion proteins, see Reference 2). Interestingly, this concentration of protons ($5 < pH < 6$) correlates with the concentration of Ca^{2+} in the cytoplasm effective in triggering exocytosis in most systems.[15] Of course, it is suggestive to speculate on analogies, insofar as Ca^{2+} could transfer a hydrophilic target (inhibiting membrane fusion) into a hydrophobic (fusogenic) state, as it occurs with the HA protein when the pH drops.

Are there candidates for such functions? With regard to lipids, the hydrolysis of polyphosphoinositides (very often observed accompanying biological membrane fusions) would eliminate the highly charged headgroups at the membrane surface, and the remaining cleavage product, diacylglycerol, has been shown to induce inverted hexagonal lipid phases.[145] With regard to proteins, the observation of proteolytic induction of fusion[146] and the inhibition by specific protease inhibitors,[63] point to the controlled activation of specific proteases in some fusogenic systems. Other relevant enzymes, which can be activated by Ca^{2+}, include protein kinases and phosphatases, and both fusion related phosphorylation and dephosphorylation have been reported for a variety of systems. In line with the picture suggested above, a phosphorylated protein could inhibit fusion (though interpretations in the literature are usually the opposite), while the dephosphorylation would remove this block, and possibly, in addition, create a hydrophobic fusogenic peptide. This hypothetic mechanism[8] would provide a very rapid execution of the fusion if the membranes are already closely apposed, as is the case e.g., with trichocysts in *Paramecium* cells. In this system exocytosis is very rapid and synchronous, and accompanied by dephosphorylation of a 65 kDa protein.[147] A Ca^{2+}-calmodulin-dependent calcineurin-like phosphatase has been shown to be responsible.[148] All these proteins were shown to occur in the cortex of *Paramecium* cells.[8] Since rephosphorylation takes place within some seconds,[147] this may be a reason for the observation of phosphorylation (rather than dephosphorylation) in other systems, where the fusion is not so synchronous. Of course, there may take place totally other (de)phosphorylation events in systems where transport and other processes have to occur before fusion proper can start. This (and the rapid reversibility of 65 kDa phosphoprotein dephosphorylation) would impede similar analyses with other (nonsynchronous) systems.

What would happen with the lipid bilayers in such a scenario? For the influenza viral HA protein, White et al.[2] suggested a local phase transition, whereby the lipids temporarily leave the bilayer configuration. On the basis of morphological data, inverted micelles have been identified as probable candidates.[87] Arguments based on biophysical facts, in favor of the occurrence of phase transition intermediates (between liquid crystalline and inverted hexagonal phases for appropriate lipids) as fusion intermediates, have been presented, e.g., in References 15, 34, 88, 96, 100, and 149. Direct observation of these structures at membrane joining spots is problematic, probably due to the high speed of the transition. More or less random lipid rearrangement (rather than strict inverted micelles) also appears possible.[15] A much more convenient situation with respect to the speed of transition has been found during the budding of influenza virus, an annular fusion event.[87] Here both membranes seem to remain joined for a relatively long time, and an IMP (situated just in the nexus of the joined membranes, see Figure 3) may be interpreted as a relatively stable inverted micelle.[87] This situation, in addition, contradicts the assumption that annular fusion is an unspecific pinching off due to strain produced by the high curvature (resulting from the interaction of the viral membrane with the nucleocapsid). Corresponding annular fusion events occur during endocytosis with strain imposed by the clathrin basket[150] or during cell division and phagocytosis, where the contraction is executed by actomyosin filaments.[151]

Why should a fusion intermediate stay longer in one case (during virus budding) than in other situations (exocytosis)? By applying patch clamp techniques for the measurement of capacitance and resistance changes over fusing membranes an electrical intermediate was observed during membrane fusion, showing characteristics of a pore.[152,153] The lifetime of this pore proved to be very variable for different systems, ranging from μs in artificial systems up to 100 ms in some biological systems.[16] This pore exhibits a "flicker" phenomenon indicating open and closed states. The nature of the intermediate is not known; on the basis of the pore characteristics an analogy to gap junctions is suggested.[16] But at the moment we do not even know whether this "fusion pore" is proteinaceous or lipidic.[154]

Why is the lifetime of that intermediate so variable, and what is the driving force for completion of fusion? A likely candidate is osmotic force, long correlated with membrane fusion.[155] Osmotic forces as inducers of membrane fusion have been refuted by combining patch clamp measurements with imaging of the swelling of individual secretory granules.[152,153] Clearly, membrane fusion was shown always to precede granule swelling. Recent work corroborates this result, because permeabilization of secretory granules in oocytes in the presence of media with varying ion and sugar concentrations did not affect membrane fusion.[101] Unfortunately, it was not investigated by EM methods whether fusion pores expand. If not as inducers of membrane fusion, osmotic forces could very well drive fusion to completion (see Refs. 16, 100). It appears to us that the "decondensation" of secretory materials (see also Morgenstern and Edelmann, Chapter 7 in this book) might be only the driving force for the expulsion of secretory materials *after* membrane fusion. Both these steps could, in fact, be dissociated from each other in *Paramecium* cells.[156] While an elevation of intracellular Ca^{2+} was sufficient for membrane fusion, extracellular Ca^{2+} was necessary for decondensation.[156]

B. CONCLUSIONS

- Viral fusogenic proteins have shown that peptides are capable of inducing biological membrane fusion; a variety of proteins are candidates for an involvement in membrane fusion, but a non-viral "fusogenic protein" and its mode of action are not yet known.
- Locally, the bilayer lipids have to adopt a nonbilayer structure; EM studies at the moment are in line with the idea of inverted lipid micelles.
- Ca^{2+} seems to exert a regulatory function, rather than directly to effect fusion of biomembranes (though the charge screening and surface dehydration potency may provoke fusion in some systems, e.g., Reference 15); calmodulin and GTP binding-proteins seem to be involved in regulation, possibly by increasing Ca^{2+} sensitivity.
- The ultrastructural equivalent of these proteins may be represented by a more or less pronounced mass of electron dense "connecting material" between membranes to be fused, as observed in exocytotic systems.[6,92,93] (The protein equivalent of IMPs present in the fusion zone is not yet known in any particular case.)

V. PROSPECTS

As shown in section II, the main difficulty in the analysis of biomembrane fusion by rapid freezing is poor frequency and synchrony of these rapid events. There are two principal strategies to overcome this problem:

1. To apply a suitable experimental setup to biological systems with extraordinary high frequency and synchrony. Examples for such a strategy are the "Heuser-machine"[28] in combination with the precisely timed stimulation of frog neuromuscular junctions,[28,113,114,157] or sandwich freezing applied to the synchronous exocytosis of secretory materials in *Paramecium* cells.[17,31,91]
2. On the other side, there may be a chance for finding extremely long-lasting fusion events (as is seemingly the case with viral budding[87]), or to prolong the lifetime of such events (as may be possible after finding the forces driving fusion events to completion; see above). Along these lines, complementary patch clamp analyses appear useful.[16,152,153]

A current challenge is to define the structural identity of the pores which have been found to display widely different lifetimes in artificial and biological systems.[16] It is encouraging, though, that data obtained for exocytosis by such methods correlate well with data obtained by fast freezing and freeze fracturing, i.e., fusion pores of $\leq$10 nm within millisecond times (see sections III and IV).

Another line of work will focus on improving labeling techniques for the intracellular localization of candidates claimed to be involved in membrane fusion regulation. Efforts should fulfill two needs: (1) reliable structural preservation, and (2) sufficient temporal resolution during possible rearrangements.

1. The limitations of chemical fixation techniques in fine structural analyses have been discussed above. But there are biological objects where chemical fixation fails at all or leads to severe structural reorganization. This is known for some plant[158] or fungus cells.[159] The use of rapid freezing in combination with freeze substitution and low temperature embedding methods provides an alternative.[160,161] In addition to better preservation and more reliable localization of proteins, this method has been shown to reliably preserve also phospholipids[87] and lipophilic drugs.[162] This will allow to follow the traffic of lipidic constituents of membranes and their possible involvement in membrane fusion regulation. The now routinely applicable analytical technique of electron energy loss spectroscopy (EELS) makes it possible to visualize membranes by virtue of their phosphorus content (see Figure 6), and only a few dozens of P atoms are sufficient to be detected.
2. Membrane fusion phenomena are rapid, not only the lifetime of the intermediate itself. Structural changes and changes in distribution of relevant components could be followed by appropriate labeling procedures only if the time resolution of the fixation procedure is appropriate. Here again rapid freezing appears to be mandatory.

ACKNOWLEDGMENTS

Some of the authors' work summarized here was supported by SFB 156 and an EMBO long-term fellowship for G. K.

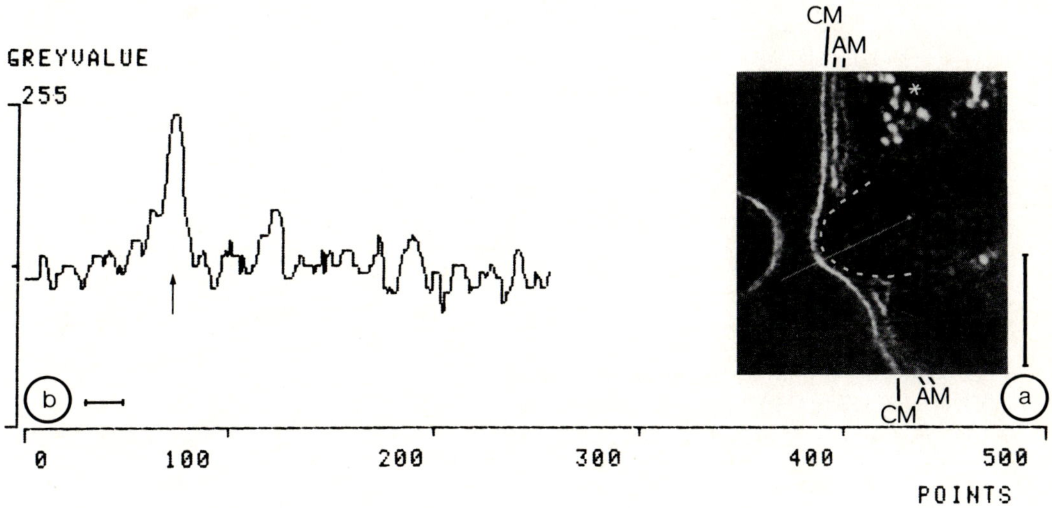

FIGURE 6. (a) Electron spectroscopic imaging (ESI, obtained from phosphorus signals, E = 150 eV, background subtracted, 80 kV acceleration voltage) of a median section through a secretory site in a *Paramecium* cell (''standard'' chemical fixation with glutaraldehyde and OsO_4 and epoxide resin embedding). It shows abundant P-content in the cell membrane (CM) and closely apposed flat ''alveolar sac membranes'' (AM), as well as in bona fide glycogen particles (asterisk), but none in the secretory organelle (trichocyst) membrane (dashed outlines). (b) Is a density line scan profile indicated in (a), from left to right; arrow indicates the position of the cell membrane; other peaks do not exceed background fluctuations. The picture is thought as a demonstration of (1) chemical heterogeneity of phospholipid contents in fusable membranes (though it could also be due to selective extraction from trichocyst membranes with the procedures used, which also would indicate, indirectly, inherent chemical differences in the phospholipid composition;* but see Reference 164 for summarized data that point to identical lipid compositions of plasma membranes and the membranes of other secretory organelles), and (2) the potential of this new analytical method. Unpublished data obtained with an analytical Zeiss EM 10 instrument, with the kind help of R. Bauer. Bars = 1 μm (a), 0.1 μm (b).

* An alternative explanation of low P signals from trichocyst membrances could be enhanced background due to 0s impregnation (a problem for which correction programs are currently developed) which again would indicate, indirectly, a peculiar lipid composition in this membrane.

REFERENCES

1. **Stossel, T. P., Bretscher, M. S., Ceccarelli, B., Dales, S., Helenius, A., Heuser, J. E., Hubbard, A. L., Kartenbeck, J., Kinne, R., Papahadjopoulos, D., Pearse, B., Plattner, H., Pollard, T. D., Reutter, W., Satir, B. H., Schliwa, M., Schneider, Y. -J., Silverstein, S. C., and Weber, K.,** Membrane dynamics group report, in *Transport of Macromolecules in Cellular Systems,* Life Sciences Research Report II, Dahlem Konferenzen, Silverstein, S. C., Ed., Berlin, 1978, 503.
2. **White, J., Kielian, M., and Helenius, A.,** Membrane fusion proteins of enveloped animal viruses, *Q. Rev. Biophys.,* 16, 151, 1983.
3. **Poste, G. and Nicolson, G. L.,** Membrane fusion, in *Cell Surface Reviews,* Vol. 5, North-Holland, New York, 1978.
4. **Chandler, D. E. and Heuser, J.,** Arrest of membrane fusion events in mast cells by quick-freezing, *J. Cell Biol.,* 86, 666, 1980.
5. **Ornberg, R. L. and Reese, T. S.,** Beginning of exocytosis captured by rapid-freezing of Limulus amebocytes, *J. Cell Biol.,* 90, 40, 1981.
6. **Plattner, H.,** Membrane behaviour during exocytosis, *Cell Biol. Int. Rep.,* 5, 435, 1981.
7. **Chandler, D. E.,** Comparison of quick-freeze and chemically fixed sea-urchin eggs: structural evidence that cortical granule exocytosis is preceded by a local increase in membrane mobility, *J. Cell Sci.,* 72, 23, 1984.

8. **Plattner, H., Lumpert, C. J., Gras, U., Vilmart-Seuwen, J., Stecher, B., Höhne, B., Momayezi, M., Pape, R., and Kersken, H.,** Enzymatic regulation of membrane fusion during synchronous exocytosis in *Paramecium* cells, in *Molecular Mechanisms of Membrane Fusion,* Ohki, S., Doyle, D., Flanagan, T., Hui, S. W., Mayhew, E., Eds., Plenum Press, New York, 1988, 477.
9. **Creutz, C. E., Zaks, W. J., Hamman, H. C., and Martin, W. H.,** The roles of Ca^{2+}-dependent membrane binding proteins in the regulation and mechanism of exocytosis, in *Cell Fusion,* Sowers, A. E., Ed., Plenum Press, New York, 1987, 45.
10. **Geisow, M. J., Walker, J. H., Boustead, C., and Taylor, W.,** Annexins-New family of Ca^{2+}-regulated-phospholipid binding protein, *Biosci. Rep.,* 7, 289, 1987.
11. **Momayezi, M., Kersken, H., Gras, U., Vilmart-Seuwen, J., and Plattner, H.,** Calmodulin in Paramecium tetraurelia: Localization from the *in vivo* to the ultrastructural level, *J. Histochem. Cytochem.,* 34, 1621, 1986.
12. **Düzgünes, N.,** Membrane fusion, in *Subcellular Biochemistry,* Vol. 11, Roodyn, D. B., Ed., Plenum Press, New York, 1985, 195.
13. **Sowers, A. E.,** *Cell Fusion,* Plenum Press, New York, 1987.
14. **Rebhun, L. I.,** Freeze-substitution: fine structure as a function of water concentration in cells, *Fed. Proc. (Suppl. 15),* 24, 217, 1965.
15. **Blumenthal, R.,** Membrane fusion, *Curr. Top. Membr. Transp.,* 29, 203, 1987.
16. **Zimmerberg, J.,** Molecular mechanisms of membrane fusion: steps during phospholipid and exocytotic membrane fusion, *Biosci. Rep.,* 7, 251, 1987.
17. **Momayezi, M., Girwert, A., Wolf, C., and Plattner, H.,** Inhibition of exocytosis in Paramecium cells by antibody-mediated crosslinking of cell membrane components, *Eur. J. Cell Biol.,* 44, 247, 1987.
18. **Knoll, G., Verkleij, A. J., and Plattner, H.,** Cryofixation of dynamic processes in cells and organelles, in *Cryotechniques in Biological Electron Microscopy,* Steinbrecht, R. A. and Zierold, K., Eds., Springer-Verlag, New York, 1987, 258.
19. **Utsunomiya, N., Tsuboi, M., and Nakanishi, M.,** Early transmembrane events in alloimmune cytotoxic T-lymphocyte activation as revealed by stopped-flow fluorometry, *Proc. Natl. Acad. Sci. U.S.A.,* 83, 1877, 1986.
20. **Plattner, H. and Bachmann, L.,** Cryofixation: A tool in biological ultrastructure research, *Int. Rev. Cytol.,* 79, 237, 1982.
21. **Gilkey, J. C. and Staehelin, L. A.,** Advances in ultrarapid freezing for the preservation of cellular ultrastructure, *J. Electron. Microsc. Techn.,* 3, 177, 1986.
22. **Menco, B. P. M.,** A survey of ultra-rapid cryofixation methods with particular emphasis on applications to freeze-fracturing, freeze-etching, and freeze-substitution, *J. Electron. Microsc. Techn.,* 4, 177, 1986.
23. **Plattner, H. and Knoll, G.,** Ultrastructural analysis of dynamic cellular processes: a survey of current problems, pitfalls and perspectives, *Scanning Microsc.,* 1, 1199, 1987.
24. **Fernandez-Moran, H.,** New approaches in correlative studies on biological ultrastructure by high-resolution electron microscopy, *J. R. Microsc. Soc.,* 83, 183, 1964.
25. **Van Harreveld, A. and Crowell, J.,** Electron microscopy after rapid freezing on a metal surface and substitution fixation, *Anat. Rec.,* 149, 381, 1964.
26. **Sitte, H., Edelmann, L., and Neumann, K.,** Cryofixation without pretreatment at ambient pressure, in *Cryotechniques in Biological Electron Microscopy,* Steinbrecht, R. A. and Zierold, K., Eds., Springer-Verlag, New York, 1987, 87.
27. **Plattner, H. and Zingsheim, H. P.,** Electron microscopic methods in cellular and molecular biology, in *Subcellular Biochemistry,* Vol. IX, Roodyn, D. B., Ed., Plenum Press, New York, 1983, 1.
28. **Heuser, J. E., Reese, T. S., Dennis, M. J., Jan, Y., Jan, L., and Evans, L.,** Synaptic vesicle exocytosis captured by quick freezing and correlated with quantal transmitter release, *J. Cell Biol.,* 81, 275, 1979.
29. **Chandler, D. E. and Heuser, J.,** Membrane fusion during secretion. Cortical granule exocytosis in sea urchin eggs as studied by quick-freezing and freeze-fracture, *J. Cell Biol.,* 83, 91, 1979.
30. **Schmidt, W., Patzak, A., Lingg, G., Winkler, H., and Plattner, H.,** Membrane events in adrenal chromaffin cells during exocytosis: a freeze-etching analysis after rapid cryofixation, *Eur. J. Cell Biol.,* 32, 31, 1983.
31. **Olbricht, K., Plattner, H., and Matt, H.,** Synchronous exocytosis in Paramecium cells. II. Intramembranous changes analysed by freeze-fracturing, *Exp. Cell Res.,* 151, 14, 1984.
32. **Papahadjopoulos, D., Poste, G., and Vail, W. J.,** Studies on membrane fusion with natural and model membranes, *Methods Membr. Biol.,* 10, 1, 1979.
33. **Szoka, F. C.,** Lipid vesicles: Model systems to study membrane-membrane destabilization and fusion, in *Cell Fusion,* Sowers, A. E., Ed., Plenum Press, New York, 1987, 209.
34. **Rand, R. P. and Parsegian, V. A.,** Mimicry and mechanism in phospholipid models of membrane fusion, *Ann. Rev. Physiol.,* 48, 201, 1986.
35. **Baker, P. F. and Knight, D. E.,** Theme and variation in the control of exocytosis, in *Cell Calcium and the Control of Membrane Transport,* Mandel, L. J., Eaton, D. C., Eds., Rockefeller University Press, New York, 1987, 1.

36. **Plattner, H., Schmitt-Fumian, W. W., and Bachmann, L.,** Cryofixation of single cells by spray freezing, in *Freeze-Etching Techniques and Applications,* Benedetti, E. L., Favard, P., Eds., Soc. Franc. Micr. Electronique, Paris, 1973, 81.
37. **Pscheid, P., Schudt, C., and Plattner, H.,** Cryofixation of monolayer cell cultures for freeze-fracturing without chemical pretreatments, *J. Microsc.,* 121, 149, 1981.
38. **Pinto da Silva, P. and Kachar, B.,** Quick freezing vs. chemical fixation: Capture and identification of membrane fusion intermediates, *Cell Biol. Int. Rep.,* 4, 625, 1980.
39. **Phillips, T. E. and Boyne, A. F.,** Liquid nitrogen-based quick freezing: experiences with bounce-free delivery of cholinergic nerve terminals to a metal surface, *J. Electron Microsc. Techn.,* 1, 9, 1984.
40. **Plattner, H. and Knoll, G.,** Cryofixation of biological materials for electron microscopy by the methods of spray-, sandwich-, cryogen-jet- and sandwich-cryogen-jet-freezing: A comparison of techniques, in *The Science of Biological Specimen Preparation for Microscopy and Microanalysis,* Revel, J. P., Barndard, T., and Haggis, G. H., Eds., AMF O'Hare, IL, SEM, Inc., 1984, 139.
41. **Heuser, J. E.,** Quick-freeze-deep-etch preparation of samples for 3-D electron microscopy, *Trends Biochem. Sci.,* 6, 64, 1981.
42. **Dubochet, J., Adrian, M., Chang, J.-J., Lepault, J., and McDowall, A. W.,** Cryo-electron microscopy of vitrified specimens, in *Cryotechniques in Biological Electron Microscopy,* Steinbrecht, R. A., Zierold, K., Eds., Springer-Verlag, New York, 1987, 114.
43. **Branton, D., Bullivant, S., Gilula, N. B., Karnovsky, M. J., Moor, H., Mühlethaler, K., Northcote, D. H., Packer, L., Satir, B., Satir, P., Speth, V., Staehelin, L. A., Steere, R. L., and Weinstein, S.,** Freeze-etching nomenclature, *Science,* 190, 54, 1975.
44. **Op den Kamp, J. A. F.,** Lipid asymmetry in membranes, *Ann. Rev. Biochem.,* 48, 47, 1979.
45. **Vacquier, V. D.,** Activation of sea urchin spermatozoa during fertilization, *Trends Biochem. Sci.,* 11, 77, 1986.
46. **Primakoff, P., Hyatt, H., and Tredick-Kline, J.,** Identification and purification of a sperm surface protein with a potential role in sperm-egg membrane fusion, *J. Cell Biol.,* 104, 141, 1987.
47. **Friend, D. S.,** Plasma-membrane diversity in a highly polarized cell, *J. Cell Biol.,* 93, 243, 1982.
48. **Jaffe, L. A. and Cross, N. L.,** Electrical regulation of sperm-egg fusion, *Ann. Rev. Physiol.,* 48, 191, 1986.
49. **Kempf, C., Klausner, R. D., Weinstein, J. N., Van Renswoude, J., Pincus, M., and Blumenthal, R.,** Voltage-dependent trans-bilayer orientation of melittin, *J. Biol. Chem.,* 257, 2469, 1982.
50. **Israel, M., Manaranche, R., Morel, N., Dedieu, J. C., Gulik-Krzywicki, T., and Lesbats, B.,** Redistribution of intramembrane particles related to acetylcholine release by cholinergic synaptosomes, *J. Ultrastruct. Res.,* 75, 162, 1981.
51. **Garcia-Segura, L. M., Muller, D., and Dunant, Y.,** Increase in the number of presynaptic large intramembrane particles during synaptic transmission at the Torpedo nerve-electroplaque junction, *Neuroscience,* 19, 63, 1986.
52. **Egea, G., Esquerda, J. E., Calvet, R., Solsona, C., and Marsal, J.,** Structural changes at pure cholinergic synaptosomes during the transmitter release induced by A-23187 in Torpedo marmorata. A freeze-fracture study, *Cell Tissue Res.,* 248, 207, 1987.
53. **Sekiya, T. and Nozawa, Y.,** Evidence for a vertical displacement if intramembranous particles on the plasma membrane of *Tetrahymena* cells by a local anesthetic, *Cell Struct. Funct.,* 8, 185, 1983.
54. **Weiss, R. L., Goodenough, D. A., and Goodenough, U. W.,** Membrane differentiations at sites specialized for cell fusion, *J. Cell Biol.,* 72, 144, 1977.
55. **Rash, J. E. and Fambrough, D.,** Ultrastructural and electrophysiological correlates of cell coupling and cytoplasmic fusion during myogenesis in vitro, *Dev. Biol.,* 30, 166, 1973.
56. **Kalderon, N. and Gilula, N. B.,** Membrane events involved in myoblast fusion, *J. Cell Biol.,* 81, 411, 1979.
57. **Herman, B. A. and Fernandez, S. M.,** Changes in membrane dynamics associated with myogenic cell fusion, *J. Cell Physiol.,* 94, 253, 1978.
58. **Knoll, G., Oebel, G., and Plattner, H.,** A simple sandwich-cryogen-jet procedure with high cooling rates for cryofixation of biological materials in the native state, *Protoplasma,* 111, 161, 1982.
59. **Wakelam, M. J. O. and Pette, D.,** The breakdown of phosphatidylinositol in myoblasts stimulated to fuse by the addition of Ca^{2+}, *Biochem. J.,* 202, 723, 1982.
60. **Lipton, B. H. and Konigsberg, I. R.,** A fine-structural analysis of the fusion of myogenic cells, *J. Cell Biol.,* 53, 348, 1972.
61. **Wakelam, M. J. O. and Pette, D.,** Myoblast fusion and inositol phospholipid breakdown: causal relationship or coincidence? in *Cell Fusion,* Ciba Found. Symp. 103, Evered, D. and Whelan, J., Eds., Pitman, London, 1984, 100.
62. **Boss, W. F.,** Fusion-permissive protoplasts: A plant system for studying cell fusion, in *Cell Fusion,* Sowers, A. E., Ed., Plenum Press, New York, 1987, 145.

63. **Strittmatter, W. J., Couch, C. B., and Mundy, D. I.,** Role of metalloendoprotease in the fusion of biological membranes, in *Cell Fusion,* Sowers, A. E., Ed., Plenum Press, New York, 1987, 99.
64. **Lognonne, J. L. and Wahrmann, J. P.,** Spontaneous myoblast fusion is mediated by cell surface Ca^{2+}-dependent protein kinase(s), *Exp. Cell Res.,* 166, 340, 1986.
65. **Lognonne, J. L. and Wahrmann, J. P.,** A cell surface phosphoprotein of 48 kDa specific for myoblast fusion, *Cell Differ.,* 22, 245, 1988.
66. **Kachar, B. and Reese, T. S.,** Evidence for the lipidic nature of tight junction strands, *Nature,* 296, 464, 1982.
67. **Pinto da Silva, P. and Kachar, B.,** On tight-junction structure, *Cell,* 28, 441, 1982.
68. **Van Meer, G., Gumbiner, B., and Simons, K.,** The tight junction does not allow lipid molecules to diffuse from one epithelial cell to the next, *Nature,* 322, 639, 1986.
69. **Stevenson, B. R., Siliciano, J. D., Mooseker, M. S., and Goodenough, D. A.,** Identification of ZO-1: a high molecular weight polypeptide associated with the tight junction (zonula occludens) in a variety of epithelia, *J. Cell Biol.,* 103, 755, 1986.
70. **Citi, S., Sabanay, H., Jakes, R., Geiger, B., and Kendrick-Jones, J.,** Cingulin, a new peripheral component of tight junctions, *Nature,* 333, 272, 1988.
71. **Sellitto, C. and Kuriyama, R.,** Distribution of a matrix component in the midbody during the cell cycle in Chinese hamster ovary cells, *J. Cell Biol.,* 106, 431, 1988.
72. **Bluemink, J. G., Tertoolen, L. G. J., Ververgaert, P. H. J. T., and Verkleij, A. J.,** Freeze fracture electron microscopy of preexisting and nascent cell membrane in cleaving eggs of Xenopus laevis, *Biochim. Biophys. Acta,* 443, 143, 1976.
73. **Byers, T. J. and Armstrong, P. B.,** Membrane protein redistribution during Xenopus first cleavage, *J. Cell Biol.,* 102, 2176, 1986.
74. **Spornitz, U. M.,** Lamellar bodies in the oocytes of *Xenopus laevis* and their relation to the mode of fixation, *Experientia,* 29, 589, 1973.
75. **Cheng, T. P. O. and Reese, T. S.,** Polarized compartmentalization of organelles in growth cones from developing optic tectum, *J. Cell Biol.,* 101, 1473, 1985.
76. **Poste, G. and Pasternak, C. A.,** Virus-induced cell fusion, in *Membrane Fusion,* Poste, G. and Nicolson, G. L., Eds., Elsevier/North-Holland, New York, 1978, 305.
77. **Marsh, M.,** The entry of enveloped viruses into cells by endocytosis, *Biochem. J.,* 218, 1, 1984.
78. **Simons, K. and Warren, G.,** Semliki Forest virus: A probe for membrane traffic in the animal cell, *Adv. Protein Chem.,* 36, 79, 1984.
79. **Doms, R. W., White, J., Boulay, F., and Helenius, A.,** Influenza virus hemagglutinin and membrane fusion, in *Cellular Membrane Fusion: Fundamental Mechanisms and Applications of Membrane Fusion Techniques.* Wilschut, J. and Hoekstra, D., Eds., Marcel Dekker, New York, in press.
80. **Wiley, D. C. and Skehel, J. J.,** The structure and function of the hemagglutinin membrane glycoprotein of influenza virus, *Ann. Rev. Biochem.,* 56, 365, 1987.
81. **Knutton, S.,** Changes in viral envelope structure preceding infection, *Nature,* 264, 672, 1976.
82. **Knutton, S.,** Studies of membrane fusion. II. Fusion of human erythrocytes by Sendai virus, *J. Cell Sci.,* 28, 189, 1977.
83. **Pinto da Silva, P., Shimizu, K., and Parkison, C.,** Fusion of human erythrocytes induced by Sendai virus: freeze-fracture aspects, *J. Cell Sci.,* 43, 149, 1980.
84. **Burger, K. N. J., Knoll, G., and Verkleij, A. J.,** Influenza virus-model membrane interaction. A morphological approach using modern cryotechniques, *Biochim. Biophys. Acta,* 939, 89, 1988.
85. **Van Meer, G. and Simons, K.,** An efficient method for introducing defined lipids into the plasma membrane of mammalian cells, *J. Cell Biol.,* 97, 1365, 1983.
86. **Van Meer, G., Davoust, J., and Simons, K.,** Parameters affecting low-pH-mediated fusion of liposomes with the plasma membrane of cells infected with the influenza virus, *Biochemistry,* 24, 3593, 1985.
87. **Knoll, G., Burger, K. N. J., Bron, R., Van Meer, G., and Verkleij, A. J.,** Fusion of liposomes with the plasma membrane of epithelial cells: fate of incorporated lipids as followed by freeze-fracture and autoradiography of plastic sections, *J. Cell Biol.,* 107, 2511, 1988.
88. **Verkleij, A. J.,** Lipidic intramembranous particles, *Biochim. Biophys. Acta,* 779, 43, 1984.
89. **Simons, K. and Garoff, H.,** The budding mechanism of enveloped animal viruses, *J. Gen. Virol.,* 50, 1, 1980.
90. **Brown, D. T., Waite, M. R. F., and Pfefferkorn, E. R.,** Morphology and morphogenesis of Sindbis virus as seen with freeze-etching techniques, *J. Virol.,* 10, 524, 1972.
91. **Plattner, H.,** Synchronous exocytosis in *Paramecium* cells, in *Cell Fusion,* Sowers, A. E., Ed., Plenum Press, New York, 1987, 69.
92. **Plattner, H.,** Ultrastructural aspects of exocytotic membrane fusion, in *Cellular Membrane Fusion: Fundamental Mechanisms and Applications of Membrane Fusion Techniques,* Wilschut, J. and Hoekstra, D., Eds., Marcell Dekker, New York, in press.

93. **Plattner, H.,** Freeze-fracture(-etching) analysis of exo-endocytosis, in *Freeze Fracture Studies of Membranes,* Hui, S. W., Ed., CRC Press, Boca Raton, FL, in press.
94. **Orci, L. and Perrelet, A.,** Ultrastructural aspects of exocytotic membrane fusion, in *Membrane Fusion,* (Cell Surface Reviews, Vol. 5), Poste, G. and Nicolson, G. L., Eds., North-Holland, New York, 1978, 630.
95. **De Lisle, R. C. and Williams, J. A.,** Regulation of membrane fusion in secretory exocytosis, *Ann. Rev. Physiol.,* 48, 225, 1986.
96. **Verkleij, A. J.,** Role of lipids during fusion of model and biological membranes, in *Phospholipid Research and the Nervous System. Biochemical and Molecular Pharmacology,* (Fidia Research Series, Vol. 4), Horrocks, L. A., Freysz, L., and Toffano, G., Eds., Livinia Press, Padua, Italy, 1986, 207.
97. **Schober, R., Nitsch, C., Rinne, U., and Morris, S. J.,** Calcium-induced displacement of membrane-associated particles upon aggregation of chromaffin granules, *Science,* 195, 495, 1977.
98. **Schuler, G., Plattner, H., Aberer, W., and Winkler, H.,** Particle segregation in chromaffin granule membranes by forced physical contact, *Biochim. Biophys. Acta,* 513, 244, 1978.
99. **Ekerdt, R., Dahl, G., and Gratzl, M.,** Membrane fusion of secretory vesicles and liposomes. Two different types of fusion, *Biochim. Biophys. Acta,* 646, 10, 1981.
100. **Whitaker, M.,** How calcium may cause exocytosis in sea urchin eggs, *Biosci. Rep.,* 7, 383, 1987.
101. **Zimmerberg, J. and Liu, J.,** Ionic and permeability requirements for exocytosis *in vitro* in sea urchin eggs, *J. Membr. Biol.,* 101, 199, 1988.
102. **Baker, P. F.,** Multiple controls for secretion? *Nature,* 310, 629, 1984.
103. **Gomperts, B. D.,** Calcium shares the limelight in stimulus-secretion coupling, *Trends Biochem. Sci.,* 11, 290, 1986.
104. **Neher, E.,** The influence of intracellular calcium concentration on degranulation of dialysed mast cells from rat peritoneum, *J. Physiol.,* 395, 193, 1988.
105. **Neher, E. and Almers, W.,** Fast calcium transients in rat peritoneal mast cells are not sufficient to trigger exocytosis, *EMBO J.,* 5, 51, 1986.
106. **Berridge, M. J.,** Inositol trisphosphate and diacylglycerol: two interacting second messengers, *Ann. Rev. Biochem.,* 56, 159, 1987.
107. **Burgoyne, R. D.,** Control of exocytosis, *Nature,* 328, 112, 1987.
108. **Brown, M. S. and Goldstein, J. L.,** A receptor-mediated pathway for cholesterol homeostasis, *Science,* 232, 34, 1986.
109. **Koike, H., Tanaka, Y., Rez, G., and Meldolesi, J.,** Membrane interactions in pancreatic acinar cells: Exocytosis, recycling and autophagocytosis, in *Biology of Normal and Cancerous Exocrine Pancreatic Cells,* INSERM Symp. 15, Ribet, A., Pradayrol, L., and Susini, C., Eds., Elsevier/North Holland, New York, 1980, 215.
110. **Koike, H. and Meldolesi, J.,** Post-stimulation retrieval of luminal surface membrane in parotid acinar cells is calcium-dependent, *Exp. Cell Res.,* 134, 377, 1981.
111. **Pape, R. and Plattner, H.,** Synchronous exocytosis in Paramecium cells. V. Ultrastructural adaptation phenomena during re-insertion of secretory organelles, *Eur. J. Cell Biol.,* 36, 38, 1985.
112. **Patzak, A. and Winkler, H.,** Exocytotic exposure and recycling of membrane antigens of chromaffin granules: Ultrastructural evaluation after immunolabeling, *J. Cell Biol.,* 102, 510, 1986.
113. **Miller, T. M. and Heuser, J. E.,** Endocytosis of synaptic vesicle membrane at the frog neuromuscular junction, *J. Cell Biol.,* 98, 685, 1984.
114. **Torri-Tarelli, F., Grohovaz, F., Fesce, R., and Ceccarelli, B.,** Temporal coincidence between synaptic vesicle fusion and quantal secretion of acetylcholine, *J. Cell Biol.,* 101, 1386, 1985.
115. **Helenius, A., Mellman, I., Wall, D., and Hubbard, A.,** Endosomes, *Trends Biochem. Sci.,* 8, 245, 1983.
116. **Pastan, I. and Willingham, M. C.,** Receptor-mediated endocytosis: coated pits, receptosomes and the Golgi, *Trends Biochem. Sci.,* 8, 250, 1983.
117. **Marsh, M., Griffiths, G., Dean, G. E., Mellman, I., and Helenius, A.,** Three-dimensional structure of endosomes in BHK-21 cells, *Proc. Natl. Acad. Sci. U.S.A.,* 83, 2899, 1986.
118. **Tooze, J.,** Blocked coated pits in AtT20 cells result from endocytosis of budding retrovirions, *J. Cell Biol.,* 101, 1713, 1985.
119. **Kosaka, T. and Ikeda, K.,** Reversible blockage of membrane retrieval and endocytosis in the garland cell of the temperature-sensitive mutant of Drosophila melanogaster, shibire^{ts1}, *J. Cell Biol.,* 97, 499, 1983.
120. **von Grafenstein, H. V., Roberts, C. S., and Baker, P. F.,** Kinetic analysis of the triggered exocytosis/endocytosis secretory cycle in cultured bovine adrenal medullary cells, *J. Cell Biol.,* 103, 2343, 1986.
121. **Griffiths, G. and Simons, K.,** The trans Golgi network: Sorting at the exit site of the Golgi complex, *Science,* 234, 438, 1986.
122. **Wagner, R. C. and Andrews, S. B.,** Ultrastructure of the vesicular system in rapidly frozen capillary endothelium of the rete mirabile, *J. Ultrastruct. Res.,* 90, 172, 1985.

123. **Beckers, C. J. M., Keller, D. S., and Balch, W. E.,** Semi-intact cells permeable to macromolecules: use in reconstitution of protein transport from the endoplasmic reticulum to the Golgi complex, *Cell,* 50, 523, 1987.
124. **Braell, W. A.,** Fusion between endocytic vesicles in a cell-free system, *Proc. Natl. Acad. Sci. U.S.A.,* 84, 1137, 1987.
125. **Davey, J., Hurtley, S. M., and Warren, G.,** Reconstitution of an endocytotic fusion event in a cell-free system, *Cell,* 43, 643, 1985.
126. **Davey, J.,** A cell-free analysis of the endocytic pathway, *Biosci. Rep.,* 7, 299, 1987.
127. **Melançon, P., Glick, B. S., Malhotra, V., Weidman, P. J., Serafini, T., Gleason, M. L., Orci, L., and Rothman, J. E.,** Involvement of GTP-binding "G" proteins in transport through the Golgi stack, *Cell,* 51, 1053, 1987.
128. **Segev, N., Mulholland, J., and Botstein, D.,** The yeast GTP-binding YPT1 protein and a mammalian counterpart are associated with the secretion machinery, *Cell,* 52, 915, 1988.
129. **Warren, G.,** Membrane traffic and organelle division, *Trends Biochem. Sci.,* 10, 439, 1985.
130. **Newport, J. W. and Forbes, D. J.,** The nucleus: structure, function, and dynamics, *Ann. Rev. Biochem.,* 56, 535, 1987.
131. **Tzagaloff, A.,** *Mitochondria,* Plenum Press, New York, 1982.
132. **Schleyer, M. and Neupert, W.,** Transport of proteins into mitochondria: translocation intermediates spanning contact sites between outer and innermembranes, *Cell,* 43, 339, 1985.
133. **Schwaiger, M., Herzog, V., and Neupert, W.,** Characterization of translocation contact sites involved in the import of mitochondrial proteins, *J. Cell Biol.,* 105, 235, 1987.
134. **Paln, D., Kanwar, Y. S., and Blobel, G.,** Identification of a receptor for protein import into chloroplasts and its localization to envelope contact zones, *Nature,* 331, 232, 1988.
135. **Brdiczka, D., Knoll, G., Riesinger, I., Weiler, U., Klug, G., Benz, R., and Krause, J.,** Microcompartmentation at the mitochondrial surface. Its function in metabolic regulation, in *Myocardial and Skeletal Muscle Bioenergetics,* Brautbar, N., Ed., Plenum Press, New York, 1985, 55.
136. **Denis-Pouxviel, C., Riesinger, I., Bühler, C., Brdiczka, D., and Murat, J.-C.,** Regulation of mitochondrial hexokinase in cultured HT29 human cancer cells. An ultrastructural and biochemical study, *Biochim. Biophys., Acta,* 902, 335, 1987.
137. **Verkleij, A. J. and Knoll, G.,** Structure and regulation of contact sites between outer and inner membrane of mitochondria, *Acta Biol. Hung.,* 37(Suppl.), 88, 1986.
138. **Brdiczka, D., Riesinger, I., and Adams, V.,** Functions in metabolic regulation of contact sites between mitochondrial boundary membranes, *Acta Biol. Hung.,* 37(Suppl.), 89, 1986.
139. **Schleyer, M., Schwaiger, M., Hartl, F. -U., and Neupert, W.,** Contact sites and protein translocation in mitochondria, *Acta Biol. Hung.,* 37(Suppl.), 90, 1986.
140. **Van Venetie, R. and Verkleij, A. J.,** Possible role of nonbilayer lipids in the structure of mitochondria. A freeze-fracture electron microscopy study, *Biochim. Biophys. Acta,* 692, 397, 1982.
141. **Cline, K., Keegstra, K., and Staehelin, L. A.,** Freeze-fracture electron microscopic analysis of ultrarapidly frozen envelope membranes on intact chloroplasts and after purification, *Protoplasma,* 125, 111, 1985.
142. **Malhotra, S. K. and Van Harreveld, A.,** Some structural features of mitochondria in tissues prepared by freeze-substitution, *J. Ultrastruct. Res.,* 12, 473, 1966.
143. **Knoll, G. and Brdiczka, D.,** Changes in freeze-fractured mitochondrial membranes correlated to their energetic state. Dynamic interactions of the boundary membranes, *Biochim. Biophys. Acta,* 733, 102, 1983.
144. **Riesinger, I., Knoll, G., and Brdiczka, D.,** Electronmicroscopic analysis of mitochondrial membranes *in situ, Eur. J. Cell Biol.,* 10(Suppl.), 60, 1985.
145. **Das, S. and Rand, R. P.,** Diacylglycerol causes major structural transitions in phospholipid bilayer membranes, *Biochem. Biophys. Res. Commun.,* 124, 491, 1984.
146. **Lucy, J. A.,** Do hydrophobic sequences cleaved from cellular polypeptides induce membrane fusion reactions *in vivo?, FEBS Lett.,* 166, 223, 1984.
147. **Zieseniss, E. and Plattner, H.,** Synchronous exocytosis in *Paramecium* cells involves very rapid (≤1 s), reversible dephosphorylation of a 65-kD phosphoprotein in exocytosis-competent strains, *J. Cell Biol.,* 101, 2028, 1985.
148. **Momayezi, M., Lumpert, C. J., Kersken, H., Gras, U., Plattner, H., Krinks, M. H., and Klee, C. B.,** Exocytosis induction in *Paramecium* tetraurelia cells by exogenous phosphoprotein phosphatase *in vivo* and *in vitro:* possible involvement of calcineurin in exocytotic membrane fusion, *J. Cell Biol.,* 105, 181, 1987.
149. **Siegel, D. P.,** Membrane-membrane interactions via intermediates in lamellar-to-inverted hexagonal phase transitions, in *Cell Fusion,* Sowers, A. E., Ed., Plenum Press, New York, 1987, 181.
150. **Pearse, B. M. F. and Crowther, R. A.,** Structure and assembly of coated vesicles, *Ann. Rev. Biophys. Biophys. Chem.,* 16, 49, 1987.
151. **White, J. G. and Borisy, G. G.,** On the mechanisms of cytokinesis in animal cells, *J. Theor. Biol.,* 101, 289, 1983.

152. **Breckenridge, L. J. and Almers, W.,** Final steps in exocytosis observed in a cell with giant secretory granules, *Proc. Natl. Acad. Sci. U.S.A.,* 84, 1945, 1987.

153. **Zimmerberg, J., Curran, M., Cohen, F. S., and Brodwick, M.,** Simultaneous electrical and optical measurements show that membrane fusion precedes secretory granule swelling during exocytosis of beige mouse mast cells, *Proc. Natl. Acad. Sci. U.S.A.,* 84, 1585, 1987.

154. **Almers, W. and Breckenridge, L. J.,** Early steps in the exocytosis of secretory vesicles in mast cells, in *Molecular Mechanisms of Membrane Fusion,* Ohki, S., Doyle, D., Flanagan, T., Hui, S. W., Mayhew, E., Eds., Plenum Press, New York, 1988, 197.

155. **Finkelstein, A., Zimmerberg, J., and Cohen, F. S.,** Osmotic swelling of vesicles: Its role in the fusion of vesicles with planar phospholipid bilayer membranes and its possible role in exocytosis, *Ann. Rev. Physiol.,* 48, 163, 1986.

156. **Matt, H. and Plattner, H.,** Decoupling of exocytotic membrane fusion from protein discharge in *Paramecium* cells, *Cell Biol. Int. Rep.,* 7, 1025, 1983.

157. **Heuser, J. E. and Reese, T. S.,** Structural changes after transmitter release at the frog neuromuscular junction, *J. Cell Biol.,* 88, 564, 1981.

158. **Fernandez, D. E. and Staehelin, L. A.,** Structural organization of ultrarapidly frozen barley aleurone cells actively involved in protein secretion, *Planta,* 165, 455, 1985.

159. **Hippe, S. and Hermanns, M.,** Improved structural preservation in freeze-substituted sporidia of Ustilago avenae—a comparison with low temperature embedding, *Protoplasma,* 135, 19, 1986.

160. **Humbel, B. and Müller, M.,** Freeze substitution and low temperature embedding, in *The Science of Biological Specimen Preparation 1985,* Müller, M., Becker, R. P., Boyde, A., and Wolosewick, J. J., Eds., Scanning Electron Microscopy, AMF O'Hare, Chicago, 1985, 175.

161. **Humbel, B. and Schwarz, H.,** Freeze-substitution for immunocytochemistry, in *Immuno-Gold Probes in Cell Biology,* Verkleij, A. J., Ed., Boca Raton, FL, CRC Press, in press.

162. **Hippe, S.,** Combined application of low temperature preparation and electron microscopic autoradiography for the localization of systemic fungicides, *Histochemistry,* 87, 309, 1987.

163. **Wakelam, M. J. O.,** Inositol phospholipid metabolism and myoblast fusion, *Biochem. J.,* 214, 77, 1983.

164. **Plattner, H.,** Fusion of cellular membranes, in *Transport of Macromolecules in Cellular Systems,* Silverstein, S. C., Ed., Berlin, Dahlem Konferenzen, Life Sciences Research Report 11, 1978, 465.

Chapter 7

ANALYSIS OF DYNAMIC CELL PROCESSES BY RAPID FREEZING AND FREEZE SUBSTITUTION

Eberhard Morgenstern and Ludwig Edelmann

TABLE OF CONTENTS

I. METHODOLOGY OF RAPID FREEZING AND FREEZE SUBSTITUTION TECHNIQUES

Fast dynamic processes captured by rapid freezing (see Knoll and Plattner, Chapter 6) have been mainly investigated with freeze fracturing in the past.[1] However, with this technique the cellular material is lost for any further investigation. A promising alternative method is the freeze substitution technique,[2] which yields ultrathin sections suited for heavy metal staining, immuno-labeling, and analysis of ion distribution.[3-10] Hence, even the localization of subcellular components and the reconstruction of dynamic cellular processes is rendered possible, e.g., by examination of serial sections.[11-13]

A. SHORTCOMINGS OF CONVENTIONAL CHEMICAL FIXATION

Up to now, the procedures which are mostly used to investigate biological material by electron microscopy include chemical fixation followed by dehydration in an organic solvent, embedding in a resin, and polymerization. Complex chemical reactions take place; shifts of ions and water, and thus changes in volume of cellular or subcellular compartments, are inevitable.[14-16] Rapid physiological structure-changes cannot be captured at all because chemical fixation does not arrest cell functions — even under the most favorable conditions — in less than 1 s. It is mainly due to findings obtained with cryo-techniques that many artifacts caused by chemical fixation have already been identified. Some of the characteristic artifacts should be mentioned here: cells or compartments of cells with high water content cannot be captured with conventional fixation, as has been shown with nervous tissue[3] or muscle cells.[9] Swollen secretory granules get a flattened appearance,[13,17] and labile cell structures are shrunken after fixation.[8] On the other hand, during chemical fixation, cell organelles might be deformed or new cell structures are generated.[18-20] The consequences, induction of membrane fusion or alteration of cytoskeletal components, lead to misinterpretations of the functional morphology.[14,21] Serious artifacts are provoked on membranes in labile states, e.g., during exocytosis. The formation of abundant and polymorphic vesicles in sea urchin eggs on sites of fusion with exocytic membranes has been attributed to aldehyde fixation in comparative studies with different fixation modes.[22] In rapidly frozen and freeze substituted platelets this phenomenon is not observed[13] (cf. section II. A and Figures 1 and 2). Further examples and a detailed description of artifacts caused by chemical fixation are given in recent studies[23,24] and reviews.[15,25,26]

B. ARTIFACTS RESULTING FROM RAPID FREEZING

Artifacts due to sampling and specimen pretreatment were reviewed recently by Robards and Sleytr.[25] Potential artifacts which may occur during rapid freezing on a metal mirror (impact freezing) will be discussed here, since the results we present were obtained with this method. When cells in suspension or in compact tissue blocks are catapulted against the surface of a solid metal a deformation of the specimen may result, and a shock wave possibly may be induced.[25,27,28] Splashing of suspensions by the same method causes shear forces which additionally damage the cells.[11] In spite of many efforts to suppress detrimental effects such as specimen bouncing or thrust by instrument improvement — as discussed by Sitte et al.[27] — the occurrence of artifacts cannot be ruled out completely, and they have to be recognized by the investigator during the observation in the electron microscope. Particularly, it is still under discussion whether the slam-provoked shock wave and a transmitted front of higher hydrostatic pressure induce "fusion artifacts" in secretory cells.[21] This precariousness necessitates an exact demonstration of the fusion site, i.e., the demonstration of membrane continuity in Ω-shaped structures and an absence of other signs of cell alterations. Under these conditions, it is possible to identify fusion events in the first 3 μm of depth of the rapidly frozen specimens. Whether the use of helium instead of liquid nitrogen (LN_2) as a coolant improves the structure preservation in the first micrometers of depth, and whether other freezing methods (jet freezing[29]) yield better preservation of fusion processes remains to be determined by future comparative investigations.

C. FREEZE SUBSTITUTION TECHNIQUE

This technique includes dehydration of the frozen specimen in an organic liquid at low temperatures (approximately 193 K), infiltration with a resin at temperatures between 203 and 293 K, and polymerization at low (203 K[30,31]) or higher temperatures (up to 343 K). Freeze substitution units are commercially available today,[10,32,33] which allow one to select and control the times and temperatures during substitution and low temperature embedding.[30,31,33]

For studying the morphology of fast dynamic processes in cells, the substitution in acetone, as has already been employed by Van Harreveld,[3] and the addition of OsO_4 to the medium, has been preferred. Although the capacity of acetone to take up water (<2% at 183 K and 3.5% at 203 K) is one tenth of that of methanol (for data see Reference 8), and even lower than that of ethanol; however, this property seems to guarantee a more gentle removal of water from cellular structures, thus preventing alterations. The addition of OsO_4 (up to 13%[34]) leads to a stabilization during the warming-up process, and facilitates the observation of cell structures by a satisfactory staining. The substitution of cell suspensions in acetone without additional OsO_4 leads to poor results, and similar phenomena were observed on solid tissues (mice lung[35]).

Methanol or ethanol are used as substitution media because of their high capacity to take up water. Therefore, the times required for freeze substitution may be shortened. It is also possible to add water — and aqueous fixatives, notably aldehydes — to the substitution media, up to a certain extent (<5% of the volume). The solubility of staining salts like uranyl acetate is also higher in alcohol than in acetone. Thus, a complex medium introduced by Müller[32] contains a mixture of 3% glutaraldehyde, 1% OsO_4 and 0.5% uranyl acetate in methanol, with 3% of water. Generally, it was found that substitution of alcohols causes a greater shrinkage and more intracellular alterations than of acetone. The addition of glutaraldehyde seems to raise problems, too.[35] Our investigations with cell suspensions have shown that the proteinaceous suspension medium (blood plasm) is not well preserved; the cells reveal empty halos and may be shrunken. These alterations are observed regardless of whether methanol or ethanol are used, with or without additional glutaraldehyde, or also with the complete Müller's medium during substitution until the warming-up. Even OsO_4 in alcohol does not seem to preserve the structural integrity as well as OsO_4 added to acetone.[35]

In view of these findings, we propose freeze substitution in acetone containing 4% OsO_4 for morphological studies. The staining — particularly of cytoskeletal components — can be enhanced by hafnium chloride.[11] For better staining of membranes, an osmification in water after substitution with acetone is possible.[36]

Freeze substituted specimens can be evaluated by structural criteria summarized in Table 1 for materials after rapid freezing and freeze substitution in acetone containing OsO_4, according to Van Harreveld and Crowell,[37] and as reported for different biological specimens.[8,11,25,27,37-40] It has been pointed out that the appearance of segregation artifacts in the nucleoplasm is a particularly sensitive indicator of insufficient cryofixation. Other very sensitive structures are membranes in pentalaminar apposition or during exocytotic fusion (see section II. A. 1 and 2). It must be pointed out that the given criteria are not necessarily applicable when other freeze substitution procedures are employed. For instance, poorly distinguishable cell membranes are obtained when substitution media without OsO_4[35] are used.

D. SPECIMEN PROCESSING SCHEDULES

Human platelets from healthy donors and bovine platelets from slaughter cattle were

TABLE 1
Criteria of Structure Preservation After Rapid Freezing and Freeze Substitution

Cell shape	Smooth-lined, cross-sectioned filopodia, microvilli, cilia, or axones round
Cell membranes	Sharply distinguished, smooth in contour, continuous and not wrinkled
Cytosol	Fine granular and homogenously dispersed, no segregation artifacts ("ice crystal ghosts")
Microfilaments, microtubules	Well preserved with no wrinkled lines, no aggregates, not shrunken and (cross sectioned) round
Ribosomes	Individually distributed, no aggregates
Nuclear envelope and endoplasmic reticulum	Smooth-contoured, no irregular dilated cisternae
Chromatin	Fine granular or filamentous components of the euchromatin and larger heterochromatin granules homogenous in dispersion (zone near the nuclear membrane is most critical)
Mitochondria	Smooth-contoured, no clear zones around the outer membrane, cristae membranes sharply distinguished

obtained after venous puncture. The procedures of platelet preparation are described in more detail elsewhere.[13,41,42] In some cases we used Fragmin for anticoagulation, a low molecular heparin (Deutsche KabiVitrum, Munich, F.R.G.), 5 U per ml of platelet rich plasm, instead of trisodium-citrate or EDTA. Stimulation of platelets was induced with either thrombin (Behringwerke AG, Marburg, F.R.G.) or collagen from equine tendon (Hormonchemie, Munich/F.R.G.). Cationized ferritin (Miles-Yeda, Frankfurt, F.R.G.) was added as described in detail elsewhere.[43] The specimens were rapidly frozen with the metal-mirror attachment MM 80 to the KF 80 cryofixation unit (Reichert-Jung). Using plastic spacers with 3 mm of inner diameter and 0.4 mm of thickness (filling capacity 6 μl), the platelet suspensions were slammed against an LN_2-cooled copper block (80 K) with a speed of approximately 0.2 $m \cdot s^{-1}$ (see Reference 27). For freeze substitution we used the CS-auto cryosubstitution unit (Reichert-Jung, Nussloch, F.R.G.). Freeze substitution was programmed for 48 h at 193 K and for subsequent continuous warming up at 10 $K \cdot h^{-1}$ up to room temperature. If not mentioned otherwise, we used acetone with 4% OsO_4 as substitution medium. Hafnium chloride (Ventron-Alpha Produkte, Karlsruhe, F.R.G.), 0.1% in acetone, was used in some cases. The specimens were embedded in araldite or Spurr's resin and sectioned on an Ultracut E (Reichert-Jung). Section staining was carried out with uranyl acetate and lead citrate.

Experimental results obtained from resting and contracting frog sartorius muscles are described in section II. B. 2; experimental procedures are described in the original papers (see Reference 9). The method used in the freeze substitution experiments is the following: frog sartorius muscles were isolated in a sterile fashion from North American leopard frogs (*Rana pipiens pipiens*), and used either with a normal content of K^+, or after a major portion (≥80%) of cell K^+ had been replaced by Cs^+ or Tl^+ in a reversible process.[44,45] The muscles were frozen by contact with a falling LN_2-cooled copper block. Freezing of contracting muscles was carried out in such a way that electrical stimulation (rate: 50 s^{-1}) started at a given time (e.g., 100 ms) before freezing. A detailed description of the method will be published elsewhere. By using the CS auto (Reichert-Jung) the muscles were freeze substituted for 7 d at 193 K in pure acetone. Embedding in Lowicryl K11M and polymerization by UV irradiation at 213 K is carried out in the same apparatus as described by Sitte et al.[10]

II. DYNAMIC PROCESSES CAPTURED WITH RAPID FREEZING AND FREEZE SUBSTITUTION

A. DYNAMICS DURING EXOCYTOSIS

Despite the fact that specific information can be obtained from the freeze substituted specimens, it was the freeze fracture technique that was preferred in the years following the

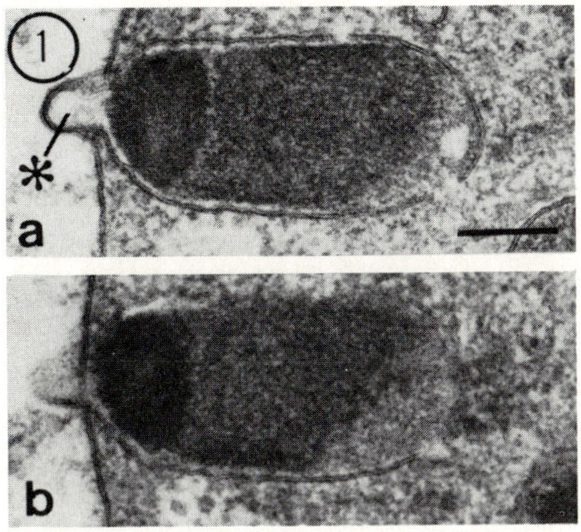

FIGURE 1. Membrane apposition in stimulated platelets after chemical fixation; (40 μg collagen per ml citrated blood plasm, 10 min at 37°C); for details see Reference 42. The membrane of an elongated α-granule in apposition with the plasmalemma is shown in two subsequent sections (a and b). Characteristic artifacts (the small bleb (asterisk in a) at the apposition site and the space between the organelle matrix and membrane) are never seen after rapid freezing (cf. Figure 2)

first successful demonstration of fusion processes by freeze substitution.[2] This has two main reasons: certain changes at the beginning of exocytosis (e.g., the lateral movement of membrane constituents) could easily be seen in replicas, and the detection of the rare fusion events in ultrathin sections is very laborious and requires careful preparation of serial sections.

During the last few years we investigated blood platelet exocytosis by means of the serial-section technique after freeze substitution. In this section we compare our findings with the observations on other cell types during exocytosis.

1. Prefusion States

The initial step of exocytosis is the formation of contacts of the secretory organelle membrane with the target membrane (plasmalemma, membranes of other organelles or membrane systems). The adjacent membranes adhere to each other and the small cytoplasmic space between them is bridged by fine electron dense connections,[11] which represent modulatory proteins in exocytosis.[46] Subsequently, both membranes form an apposition (Figures 1, 2, 3b, 4 and 5e) with a pentalaminar structure of the membrane layers. Appositions are relatively stable (lifetime: several minutes).[47] The distance between the apposed membranes in cells was estimated to be ten times as large as in liposome preparations.[48] In sections of freeze substituted secretory cells a dense intermediate layer is seen in pentalaminar contacts. Cells with large secretory organelles (*Limulus* amebocytes) showed several punctate pentalaminar contacts in apposition with the plasmalemma.[11] In goblet cells, apposition areas with diameters of more than 0.5 μm with a continuous pentalaminar structure were observed.[20]

The prefusion states in platelets widely resemble the phenomena described in other cells. Platelets contain two types of secretory organelles. The so-called α-granules store coagulation factors, platelet-specific proteins and cationic proteins. Another type of secretory organelle in platelets (dense bodies) contains calcium, phosphorus, and biogenic amines, as well as adenine nucleotides in a very electron-dense matrix.[49-51] The formation of protrusions and the elongation of the secretory organelles in stimulated platelets, as seen in Figures 1, 2 and 4, is a characteristic phenomenon at the beginning of the secretory pathway. These changes can even be seen in discoid platelets prior to shape change or to aggregation of platelets.[43]

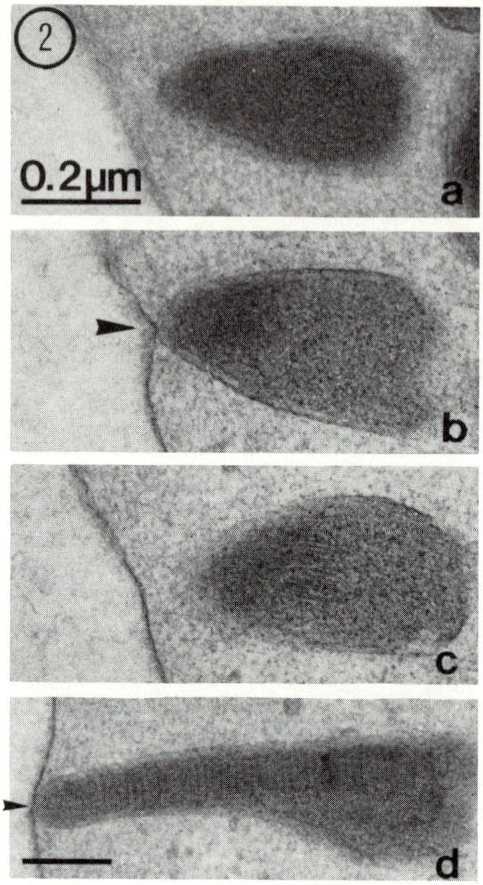

FIGURE 2. Punctate membrane appositions of elongated α-granules in thrombin-stimulated (1U per ml EDTA-plasm, 35 s at 37°C) rapidly frozen and freeze substituted bovine platelets. The sections seen in (a to c) show the apposition site with the plasmalemma (arrowhead) and the dimpled surface in its vicinity. In (d) the section of an elongated α-granule in apposition with the plasmalemma reveals the typical striation (23 nm period) of the granular matrix. The apposition (arrowhead) is situated at a distance of 0.6 μm from the metal mirror.

Similar shape changes of secretory organelles were observed on other cells when secretion started.[41,52] The tips of protrusions — often seen as coated membranes[43] — may indicate a membrane retrieval process, as described for different secretory cells.[53-55] Until now, the significance of the coated membranes in platelets has been unknown.[41,51] The characteristic periodic 2.3 nm striation — seen in elongated α-granules (Figure 2d) — has been interpreted as polymerized fibrin from fibrinogen within the organelles.[49] Influx of water into the α-granules during platelet activation — a prerequisite for the enzyme-mediated polymerization process — could cause the shape change of these organelles. The α-granules of platelets elongate in an undirected manner, and thus enable the granule membranes to be apposed to the plasmalemma or other membranes.[41] The elongated secretory organelles contact the plasmalemma in single very small pentalaminar apposition zones (Figures 1, 2, 4; see also References 13, 41). Pentalaminar appositions with an area of 0.1 up to 0.2 μm in diameter and a thickness of approximately 13 nm are seen between the membranes of platelet organelles (Figure 5d). As seen in other cells, the intermediate dense layer is about 4.5 nm thick and thus twice as thick as the dense layer of a platelet membrane (approximately 2.25 nm). The thickness of two membranes in a pentalaminar apposition is smaller than the sum of thickness of two single membranes (each 10.5 ± 0.8 nm thick). This characterizes

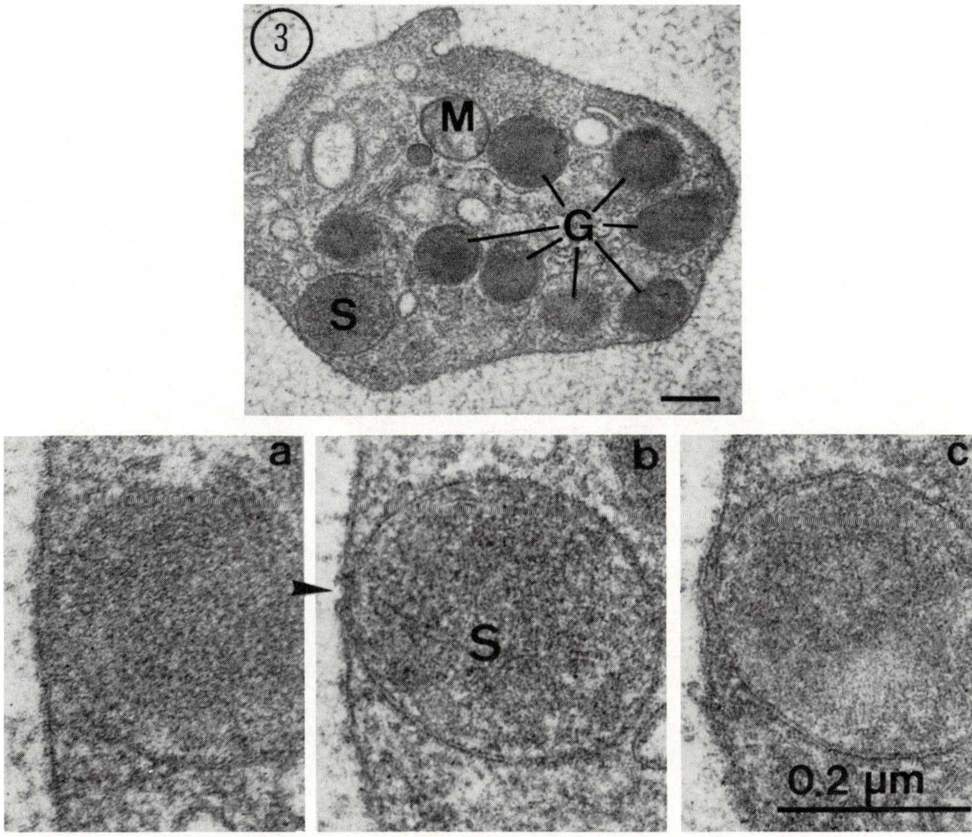

FIGURE 3. Membrane apposition and formation of a fusion pore in a collagen-stimulated (40 μg per ml Fragmin-plasm, 150 s at 37°C) rapidly frozen and freeze substituted human platelet. The whole section profile shows free individual α-granules (G) with dense matrix. An α-granule (S) has been apposed to the plasmalemma and shows moderate swelling and dispersal of the matrix (indicating the formation of a pore). In (b) the apposition (0.9 μm distant from the metal mirror) is seen to be impressed (arrowhead), but the pore itself is not visible. In the adjacent sections (a) and (c) both membranes are separated from each other. M = mitochondrium.

the absence of membrane-associated material in the zone of a pentalaminar contact after freeze substitution with OsO_4 in acetone.

The phases that have been described hitherto are relatively stable and proceed slowly (cf. Table 2). Shape changes of the granules are also visible in conventionally fixed platelets. The observation of appositions with a pentalaminar structure is very rare and critical to evaluate (cf. Figure 1 with Figure 2). Formation of blebs or artificial vesiculation during chemical fixation is probably caused by the lability of the lipids in these membrane areas.[22] In rapidly frozen blood platelets, pentalaminar structures were only clearly visible in the best frozen zone of our specimens (up to 3 μm in depth).

2. Membrane Fusion (Pore Formation)

The rearrangement of lipid molecules in two apposed membranes during the formation of a fusion pore proceeds very rapidly. Intermediates of this process (inverted micellar lipids or other changes in the bilayer structure,[14,26,48] as discussed in Chapters 15 and 16 in this book) are small, and their half-life is estimated to be 0.45 ms.[56] From measurements on phospholipid vesicles fusing with planar phospholipid membranes, a time range of up to 50 μsec for fusion is expected, and for cell membranes this period is calculated to be markedly

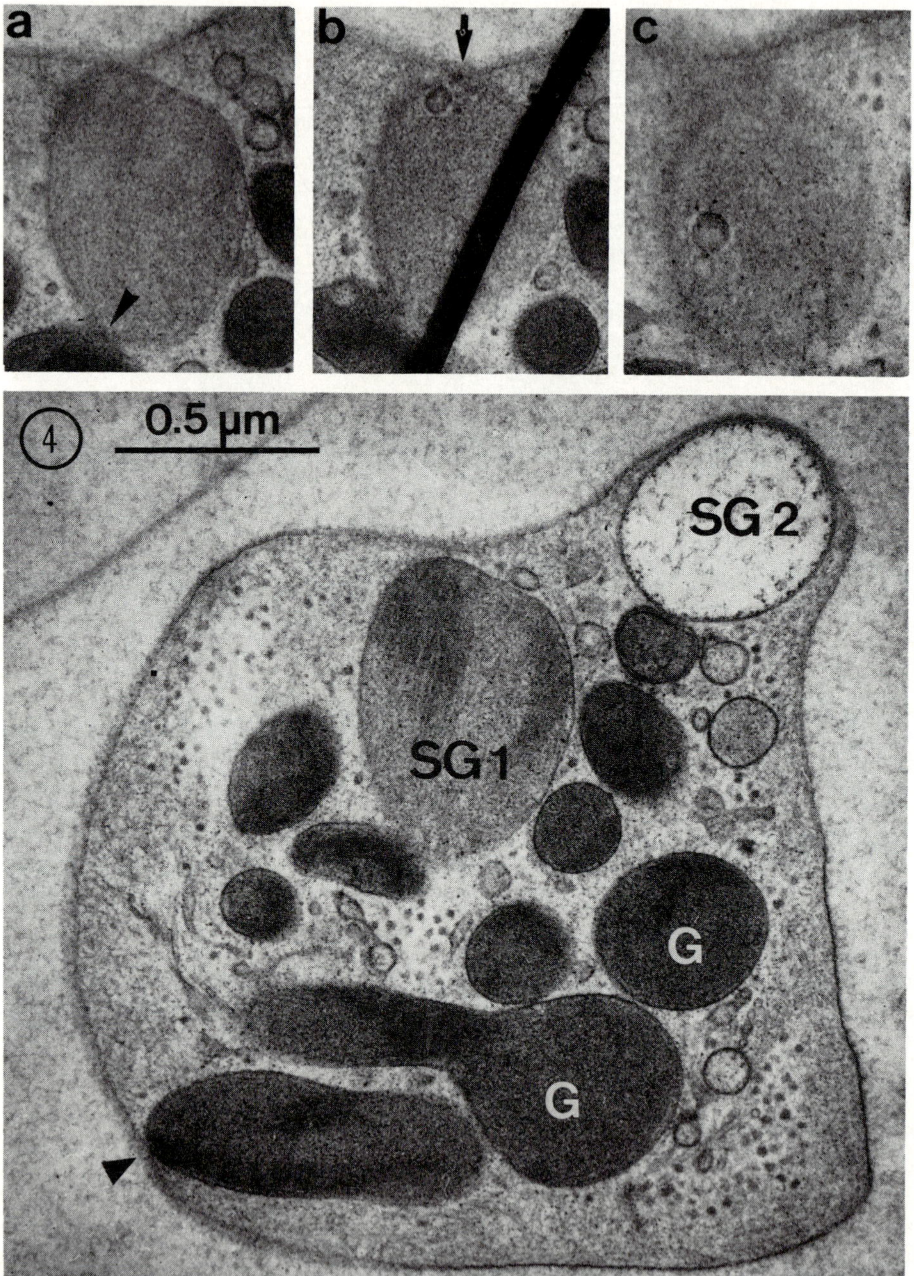

FIGURE 4. Initial steps of exocytosis in thrombin-stimulated, rapidly frozen and freeze substituted bovine platelets. (1.0 U per ml EDTA-plasm, 35 s at 37°C). The whole section profile shows α-granules (G) with protrusions or with an elongated shape in apposition to the plasmalemma (arrowhead). The swollen α-granule (SG 1) has a dispersed matrix. Another strongly swollen organelle is seen (SG 2). In (a), (b) and (c) the pore is invisible (arrow in (b) indicates an electron dense dot). Swelling and matrix dispersal of SG 1 (0.6 μm distant from the metal mirror) indicate influx of water after formation of a nonvisible pore. In (a) an apposition of the SG 1 membrane with another granule membrane is seen (arrowhead).

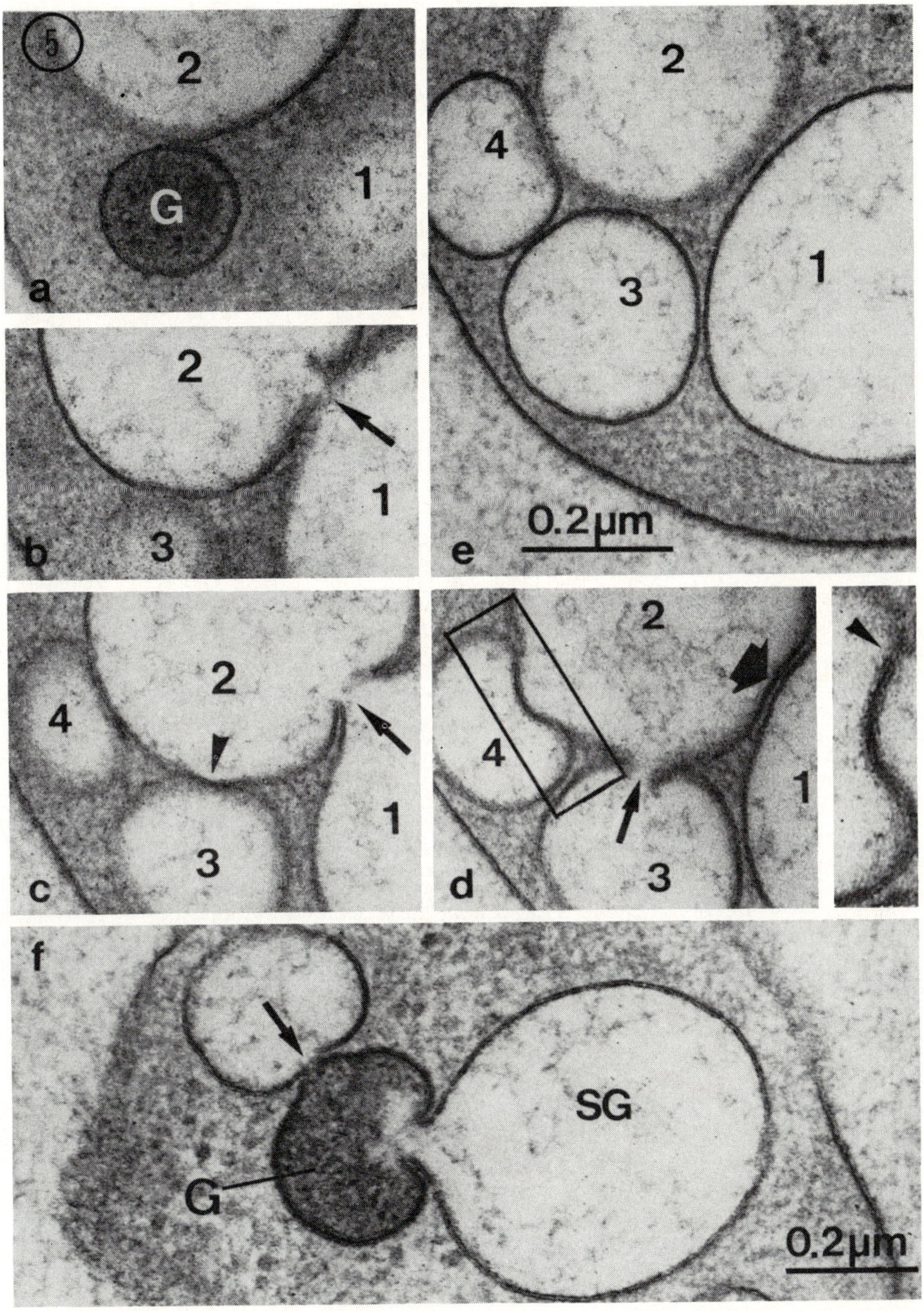

FIGURE 5. Compound exocytosis in a thrombin-stimulated (0.1 U per ml EDTA-plasm, 3 min at 37°C) rapidly frozen and freeze substituted human platelet. One section of the series (a to e) between (a) and (b) is not demonstrated. Swollen secretory organelles numbered 1 to 4 in (e) show adjacent membranes (thick arrow in (d), appositions (arrowheads in (c) and inset from (d)) and fusion pores (arrows in (b to d)). Organelle 2 is 1.7 μm distant from the metal mirror. In (a) an unswollen α-granule (G) is apposed to the membrane of the swollen organelle 2. The section in (f) shows a platelet stimulated in a similar manner and displays a fusion pore at the very moment of influx of water in an unswollen α-granule (G) that has fused with the swollen organelle (SG). The formation of a further fusion is seen (arrow).

TABLE 2
Reaction Times of Platelets After Stimulation

Reaction	Morphology	Time
Shape change	Disc-sphere transformation surface enlargement and pseudopodia formation at the expense of the surface connected system (SCS)	ms (?)-s
Internal contraction	Formation of contractile filaments, centralization of organelles	s-min (?)
Internalization exocytosis	Membrane invaginations	s-min
Alterations of α-granules	Formation of protrusions, elogation; transformation of the granular matrix (periodic striation)	s-min
Membrane apposition	Formation of pentalaminar contacts of granular membranes and plasmalemma or SCS	s-min
Membrane fusion	Pore formation	μs-ms

less than 1 ms[48] (see also Chapter 6 by Knoll and Plattner). Torri-Tarelli[34] observed the first fused synaptic vesicles 2.5 ms after their triggering with an electric shock. It appears that capturing of the formation of a pore poses problems even for rapid freezing. Minute particles were demonstrated in replicas of rapidly frozen and freeze fractured artificial membranes, and interpreted as inverted micellar intermediates.[57,58] Similar phenomena were demonstrated on freeze fractured plasmalemmata of rapidly frozen carbachol-triggered chromaffin cells.[59] In sections with a minimal thickness of 20 nm, inverted micellar intermediates, as well as small pores, can escape observation. The smallest example of a fission at a site of apposition between plasmalemma and organelle membrane was shown in stereo pairs of micrographs of an ultrathin section, by Ornberg and Reese.[11] Minute pores in a pentalaminar apposition zone were demonstrated on secretory granules of mucus-secreting cells in ultrathin sections.[20] Fusion pores of <100 nm diameter were observed in platelets.[13] In Figure 5 (d and f) small pores with a diameter of approximately 30 nm are shown. Indirect evidence for the existence of small pores is indicated by "swelling" of the organelle, or by electron dense markers in the secretory organelle (cf. section II. A. 3 and Figures 3, 4, and 6, respectively). The opening of a pore rapidly expands after fusion. In this phase cytoplasmic fragmentations are observed in sea urchin eggs[22] or in goblet cells,[20] which appear as small spheric or large elongated exclaves of cytoplasm. In other cells mentioned in this chapter, cytoplasmic fragmentation was not observed during exocytosis.

Generally, the analysis of ultrathin sections of cells rapidly frozen during exocytosis has confirmed the observations of Ornberg and Reese,[11] who found that the pore formation is a direct but infrequent outcome of the formation of a pentalaminar contact between membranes. Since evidence for semifusion states[28] was not obtained,[14,60] the occurrence of a "diaphragm" or a trilaminar intermediate phase seems very improbable.[13,20,61] Considering these findings, the formation of a micropore from inverted micellar intermediates between apposed membranes (reviewed in Reference 75) appears to remain as a logical model whose morphologic evidence ("lipidic particles") was taken exclusively from freeze fractured specimens.[1,57,62] On the other hand, the comprehensive studies on *Paramecium* cells during exocytosis by Plattner and his group could not provide evidence of micellar intermediates.[46] Instable focal rearrangements in the lipid molecules within the apposed membranes are considered to cause the formation of a small pore ("focal fusion model",[14,21,46,59]). These aspects are discussed in more detail in Chapters 5 by Verkleij and Chapter 6 by Knoll and Plattner.

3. Attendant Phenomena of Pore Formation: Swelling and Dispersal of Organelle Contents

The reasons for the most striking phenomenon accompanying exocytosis (swelling of the secretory organelles) have been controversially discussed hitherto.[48,63] As was shown in

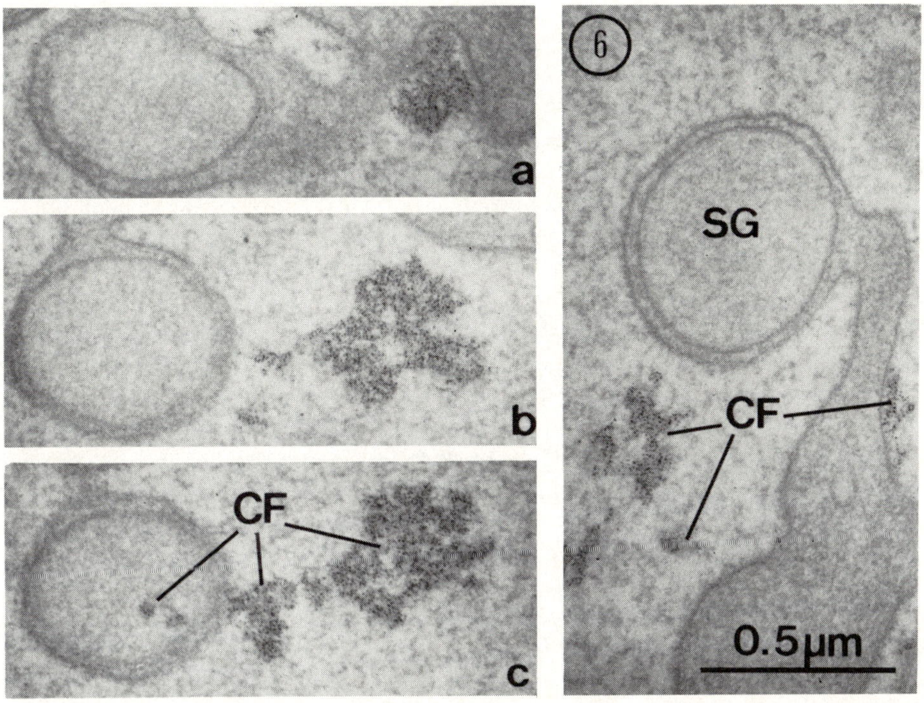

FIGURE 6. Indirect evidence for a fusion pore; cationized ferritin (40 mg final concentration/ml (for detail see Reference 43) was added to platelet rich citrated plasm 5 s prior to stimulation with thrombin (0.5 U/ml) for 20 s at 37°C) and rapidly frozen thereafter. Particles (CF) are seen only in (c) within the swollen organelle (SG), shown in the adjacent sections (a to c), indicating the existence of a hidden pore. SG is 1.8 μm distant from the metal mirror. (Block-staining with 0.1% hafnium chloride, see section I. C, no post-staining).

studies with liposomes, an osmotic swelling (caused by the Ca^{++} dependent increase in the influx of water into the organelle across the apposed membranes) seemed to be the driving force for membrane fusion and pore formation.[64] More recently, electrophysiological measurements of membrane capacitance and of the simultaneously observed swelling of secretory granules during exocytosis, gave strong evidence for the fact that the formation of a pore is the initial event and triggers the swelling.[65,66] The stepwise increase in membrane capacitance during secretion indicated the existence of a fusion pore before swelling. Morphological findings also argue against osmotic swelling. No swelling of the secretory organelles (trichocysts) was found with the light microscope in stimulated living *Paramecia* prior to expulsion.[46] Recent findings on stimulated platelets, whose secretory organelles swell strongly during exocytosis,[13] also indicate the existence of a pore prior to swelling. We found particles of cationized ferritin (11 nm in diameter) — added before triggering exocytosis with thrombin — in swollen secretory organelles (Figure 6). Yet individual pores could not be shown on ultrathin sections (about 30 nm) — not even in section series — because of their minor dimension, or because of an oblique section plane through a pore in the apposed membranes.

It has to be assumed that swollen organelles live very briefly (reference 48, see Table 2). Only rapid freezing can preserve the extent of swelling. Water is not retained during the long period needed for arrest when chemical fixation is used (see section II. A). The interpretation of the secretory pathway of cells (i.e., platelets) is problematic when swelling is missing. Resting discoid human platelets reveal a surface connected membrane system, which is evaginated after stimulation to render possible surface enlargement during shape change.[13,67] The flattened membranes of chemically fixed secretory organelles may be misinterpreted as remnants of this canalicular system (for controversial discussion see References 13, 68, 69). Additionally, stimulated platelets simultaneously start internalization of surface membranes,[41,43,51,70] which may also be mistaken for membranes of secretory organelles.

A remarkable extent of swelling was observed with mast cell granules[71] and platelet granules.[13] Such organelles contain high concentrations of proteins and other molecules, or of minerals.[51,72,73] In amebocytes[11] or goblet cells[20] the secretory granules show less swelling. A swelling of synaptic vesicles was not observed.[2,34] The membranes of swollen organelles appose to the membranes of numerous other swollen and unswollen organelles (Figures 4 and 5), thus obviously enhancing the sequential fusion of organelles during "compound exocytosis" in platelets.

The degree of matrix dispersal seen in ultrathin sections of secretory organelles is assessable by their density and their quality. High concentrations of electrolytes in the dense bodies of platelets[50] may cause stronger influx of water, and subsequently, a very lucent interior of the swollen secretory organelle results (SG 2 in Figure 4). Some structural parts of secretory granules (e.g., the "nucleoids" in platelet α-granules; Figure 7) resist the dispersal longer than others.[13] The matrix of mast cell granules or of dense bodies of platelets often remains condensed even after extrusion.[13]

4. Driving Forces for the Discharge of Secretory Products

The rapid enlargement of the organelle volume after the formation of a pore, and the resulting pressure inside the organelle for a short time, as well as the simultaneous expansion of the pore, support the expulsion of organelle contents in some of the examples mentioned here. The secretory organelles (trichocysts) of ciliated protozoa (*Paramecium*) are expelled in the range of milliseconds by this mechanism ("decondensation", i.e., stretching of their contents), independently of assistance by the contractile system.[46] Similarly, in stimulated *Limulus* amebocytes cytochalasin D did not prevent degranulation.[11] In platelets the role of contractile forces for discharging secretory products is not yet clear. Only a portion of the secretory products was discharged when phorbol myristate acetate-stimulated platelets in citrated blood plasm were treated with high concentrations of cytochalasin B.[74] However, an increase of serotonin release was measured on washed platelets stimulated with thrombin or other strong stimulants, and treated with lower doses of cytochalasin B.[75] We found the formation of fusion pores in discoid (resting) platelets and swelling of secretory platelet granules in response to the weak stimulator dihexanoylglycerol. It was measured biochemically that only 7 to 20% of the serotonin contents were released from the platelet organelles (dense bodies) 12 min after stimulation, instead of approximately 60% after strong stimulation.[76] Further investigations are necessary to determine whether the forces of the contractile system in platelets increase — as in other cells[46,77] — the extent of discharge of organelle contents.

5. Compound Exocytosis

The sequential fusion of secretory organelles with one another after the fusion of the most proximal organelle with the plasmalemma in the set of fusing organelles was observed in secretory cells.[60,78] This type of exocytosis exists in goblet cells,[20] in mast cells,[79] as well as in platelets,[13] in addition to the fusion of single organelles with the plasmalemma. In several cells one can find the secretory organelles in the resting cell separated from each other by the cytoplasmic matrix. In stimulated cells organelles can fuse subsequently at sites where effective membrane contact is established and certain conditions are fulfilled (e.g., adequate concentration of Ca^{++}, and presence of intermediates of phospholipid turnover as fusogenes; reviewed in References 46, 60, 77). The apposition of the organelle membranes occurs by the formation of protrusions (see above section II. A. 1) or — as seen in platelets[13] — by an attachment which is supported by constricting a contractile gel. Compound granules develop very quickly (Table 2), and as their swollen shape is not preserved after chemical fixation they escape estimation in ultrathin sections. After rapid freezing, compound granules are characterized by their more or less electron dense contents and their swollen shape. In serial sections of stimulated human platelets, tapering of the shape of compound granules can be reconstructed; this indicates the participation of the former organelles.[13] Tapered compound granules are also demonstrated in freeze fracture replicas of neutrophilic granulocytes.[80]

6. Concluding Interpretation of the Secretory Pathway in Blood Platelets

One of the first observable structural alterations in stimulated blood platelets are protrusions of the secretory granules (α-granules and dense bodies) and the apposition of the elongated granules to the plasmalemma or to surface connected membranes (Table 2). Granules in membrane apposition (already after formation of a pore (?) as indicated in Figure 3b) seem to maintain this position during subsequent shape change and internal contraction, reactions which progress in the time range of ms or s. Granules which are apposed to the surface connected membranes, which are evaginated for surface enlargement (shape change), are moved to the platelet surface. Under platelet stimulation as described here and after rapid freezing, no evidence was found for fusion of α-granule membranes with membranes of the surface connected system,[13,17] as discussed by others.[69] The simultaneous commencement of concentric internal contraction of microfilaments and microtubules drives (mobile) platelet organelles (including mitochondria) to the platelet center and into contact with each other, thus supporting further apposition. There is strong evidence (Figures 2, 3, 4, and 6, Reference 13) that secretory organelles which are apposed to the surface membrane (and arrested) fuse and induce sequential fusion (compound granules). Membranes of the former surface connected system which were not evaginated may be involved in this compound. However, as shown in the section series in Figure 7, the surface connected membranes are absent within the contractile sphere, and this strongly suggests that their presence is not a prerequisite for human platelet exocytosis. The findings from bovine platelets which lack surface connected membranes[81,82] support this interpretation. The action of the contractile gel in platelets may obstruct the discharge of secretory products by squeezing the organelles (Figure 7) during the process of secretion.[13,83] Therefore, the necessity of contractile forces for complete release is not yet proven. The fundamental phenomena of platelet exocytosis described in this work correspond to the findings on other cells. The interaction of the concentric moving contractile gel with the secretion process may be a particular functional property of the platelets.

B. WATER MOVEMENT DURING DYNAMIC CELLULAR PROCESSES

Most dynamic cell processes can be characterized by the following events: Conformation change of macromolecules, redistricution of solid particles, and redistribution of water. The swelling of subcellular compartments during exocytosis is an example of the redistribution of cellular water. Another example illustrating this process can be found in the contracting muscle: during shortening, water is redistributed within the cytoplasm (see below). The problem of water movement during dynamic cell processes is far from being understood. But since it is now possible to visualize water movements (volume changes) by the freeze substitution technique, it is to be expected that in the future increasing attention will be paid to this problem. In this section, we discuss theoretical models which could be used and further developed to explain intracellular water movements. Results obtained on resting and contracting muscles, which are consistent with a theoretical explanation of water movement during muscle contraction, are presented.

1. Theoretical Considerations

Swelling of a subcellular membrane-enclosed compartment may be explained by a change in membrane properties. For instance, if the osmotic activity inside the compartment is

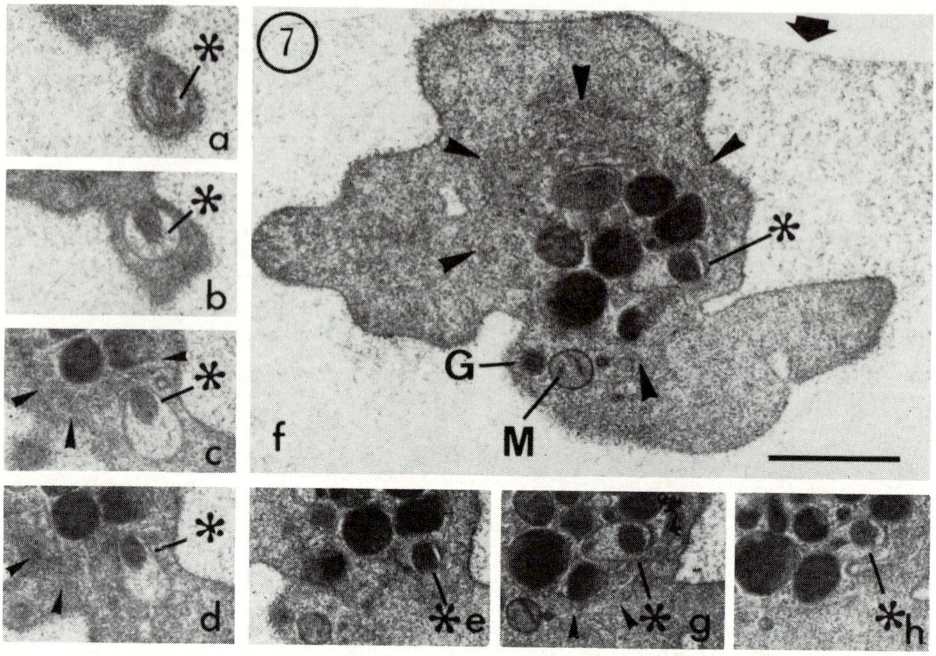

FIGURE 7. Interactions of the contractile gel with the exocytosis in a collagen-stimulated (40 μg per ml citrated plasm, 20 s at 37°C) rapidly frozen and freeze substituted human blood platelet. In eight sections of a series (a to h) the mass of the α-granules is located within the sphere of the contractile gel, indicated by arrowheads. No surface connected membrane system is seen. The swollen granule (asterisks) shows a persisting nucleoid and is situated outside the contractile sphere in (a to d), seen to be squeezed by the contractile gel in (e) and located inside the contractile gel in (f to h). A pore is not visible. The thick arrow in (f) indicates the metal mirror. Bar = 0.5 μm, as in Figure 6.

constantly higher than outside, a sudden increase in the water permeability of the membrane, or puncturing of the membrane, will lead to swelling. Assuming that the membrane has a high water permeability, and assuming that the osmotic activity is equal inside and outside the compartment, a rapid swelling must be explained by a rapid increase in osmotic activity (decrease in the water activity), either by a decomposition of the compartment compounds into many smaller units, or by binding of water to macromolecules.

Water movement within the cytoplasm of a cell could be caused by a shape change of the cell, or by a movement of macromolecules carrying a large amount of water, or by local changes in the osmotic activity. The latter event could be caused either by a local liberation (binding) of many particles, or by a sudden binding (release) of water molecules to (from) macromolecules.

The idea of a rapid local change in the water activity during cellular dynamic processes has not yet been considered very often in the literature. This is due to the dominance of the membrane pump theory in biological research, and the belief that membrane properties are mainly responsible for changes of volume in cells and subcellular compartments. For instance, it is commonly assumed that free solutes in the cell — including K^+ which is supposed to be free — help to balance the osmotic pressure of the cell interior to that of the outside solution; a Na^+ pump is necessary to prevent swelling of cells under normal conditions, despite the colloido-osmotic pressure of intracellular macromolecules.[84]

According to the association-induction hypothesis[85-87] — an alternative view of the cell — the cytoplasm and most subcellular compartments are seen as protein-ion-water complexes. In a resting cell, most of the cellular anions and cations are weakly bound to cellular proteins, and most of the cellular water is under the influence of extended proteins in such a way that it suffers both translational and rotational restriction, and thus has a reduced solubility for complex molecules, including the large hydrated Na^+ ion. Since many predictions of this concept have already been verified by experimental testing (e.g., cell volume regulation,[88-91] ion binding,[9,92] diffusion of cell water[93]) it appears worthwhile to look at the problem of intracellular water movements during dynamic cell processes from the point of view of this theory.

Let us consider the living contracting muscle. It is an established fact that there is no change in volume during the shortening of a contracting muscle, or in other words, that the water is redistributed during contraction inside the muscle cell. In a tentative model based on principles of the association-induction hypothesis, it was proposed by Ling[86] that local liberation of bound K^+ and local differences in water activities may be the cause of water-shifts within sarcomeres of the living contracting muscle.

The experimental proof that a considerable amount of formerly bound K^+ is liberated during contraction must include the demonstration of K^+ binding to certain proteins of the resting muscle, and of a K^+ redistribution during contraction. Both problems have been tackled by using cryomethods including freeze substitution. Some findings will be described next.

2. Experiments with Resting and Contracting Muscles

Experiments were aimed at answering the question whether the main cellular cation K^+ is freely dissolved in cellular water or bound to macromolecules; they were carried out with frog sartorius muscles.[92,94,95] It was postulated that K^+ is weakly bound to β- and γ-carboxyl groups of cellular proteins.[85] Since the highest concentration of these fixed anions is found on myosin in the A bands of the striated muscle, Ling expected a preferential accumulation of K^+ at these sites.[96] In order to visualize the subcellular K^+ distribution with the electron microscope, the method of K^+ replacement by the electron dense ions Rb^+, Cs^+ or Tl^+ was used. In a living frog muscle, most of the cellular K^+ can be replaced reversibly by these ions;[96,97] one may conclude that — if either Rb^+ or Cs^+ or Tl^+ constitutes the main cellular cation of a living muscle — the subcellular distribution of the electron dense ions must be the same as that of K^+ in normal K^+-containing muscle. After freeze-drying and embedding of rapidly frozen Cs^+- and Tl^+-loaded muscles, the subcellular distribution of the electron dense ions could be visualized. The ions do not follow the distribution of water in the cell, but are localized, preferably, in the A bands, especially at the outer regions and at Z lines (Figure 8). These results were confirmed for all four different ions (K^+, Rb^+, Cs^+, Tl^+), and by using different cryomethods combined with autoradiography of frozen hydrated preparations and X-ray microanalysis of freeze-dried cryosections (for review see Reference 92). In order to rule out the possibility that the detected uneven ion distribution is due to freeze-drying or embedding artifacts, frozen-hydrated cryosections of Tl^+-loaded muscle were also investigated.[7,8,97] It was found that the subcellular distribution of Tl^+ in frozen-hydrated muscle is similar to that of K^+, Rb^+ or Cs^+ in dehydrated muscle preparations obtained with the above mentioned independent techniques. In addition, a more precise localization of binding sites of the ions was possible. In the resting muscle, the ions are bound to or near the same sites where uranyl and lead ions (as employed in conventional electron microscopic staining techniques) are bound, i.e., preferably to myosin filaments in the A band and to Z line proteins (Figure 9).

The question whether a considerable amount of ions is liberated during contraction was investigated with the freeze substitution technique.[9] A series of experiments showed that freeze substitution in acetone for 7 d at 193 K and embedding in Lowicryl K11M at 213 K yields muscle preparations in which most of the cellular K^+ is retained (unpublished). By using this method, it was then tested whether the subcellular ion distribution in Cs^+- or Tl^+-loaded muscles is the same as that found in frozen hydrated cryosections. In addition,

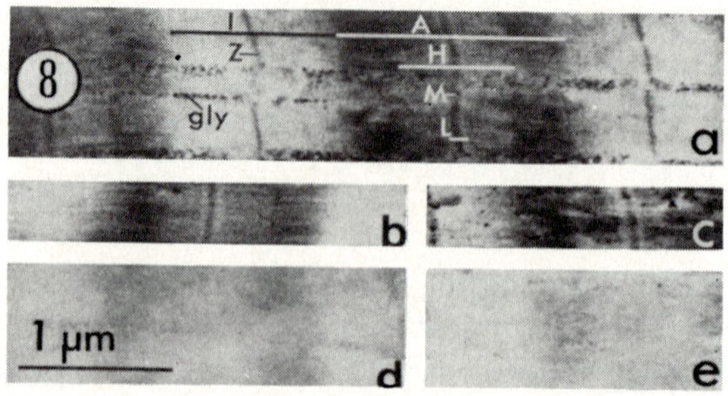

FIGURE 8. 0.3 μm thick dry-cut sections of freeze dried and embedded frog sartorius muscles. (a) Cs^+-loaded muscle, (b, c) Tl^+-loaded muscle; (b) was obtained immediately after sectioning, (c) after exposure of a section to humid atmosphere. (d) Central part of (a) after storage of the section in distilled water (loss of the electron dense Cs^+ ions). (e) Control: normal K^+-containing muscle, *A*, A band; *I*, I Band; *Z*, Z line; *H*, H zone, *M*, M line; *L*, L zone; *gly*, glycogen. The intense staining shown in (a), (b), and (c) is due to the uneven distribution of the electron dense Cs^+ and Tl^+ ions in the muscle preparation. The ions accumulate preferentially in the A bands and at Z lines. (From Edelmann, L., *Physiol. Chem. Phys.*, 9, 313, 1977. With permission.)

FIGURE 9. Frozen hydrated cryosections of frog sartorius muscle. (a) Normal K^+-containing muscle. Only very faint ultrastructural details can be seen. *A*, A band; *Z*, Z line. (b) Tl^+-containing muscle. Dark myosin filaments (arrows) in the A bands (A) and dark Z lines (Z) indicate sites of preferential Tl^+-accumulation in the living cell. The Tl^+-accumulation pattern is similar to the Cs^+ and Tl^+ patterns shown in Figure 8 (From Edelmann,L., *Scanning Microsc.*, 2, 851, 1988. With permission.)

Cs^+-loaded muscles, rapidly frozen during contraction, were investigated. The findings are summarized in Figures 10 and 11 and can be described as follows:

1. The contracting muscle shows A band shortening (Figure 10).
2. Resting Cs^+-loaded or Tl^+-loaded muscles reveal uneven distributions of the electron dense ions similar to those obtained with other cryomethods, namely preferential accumulation of the ions at the A bands and at Z lines (Figure 11c).
3. Contracted Cs^+-loaded muscle reveals a rather poor contrast between I bands and slightly shortened A bands. (Figure 11d). By comparing Figure 11c with Figure 11d, it appears that the electron dense Cs^+ ions have been redistributed during contraction.

Some implications can be derived from these results. By using long freeze substitution times at low temperatures, the freeze substitution technique may yield preparations in which even mobile ions can be localized at those places which they occupy in frozen-hydrated preparations. The results obtained with contracting muscles are consistent with the above

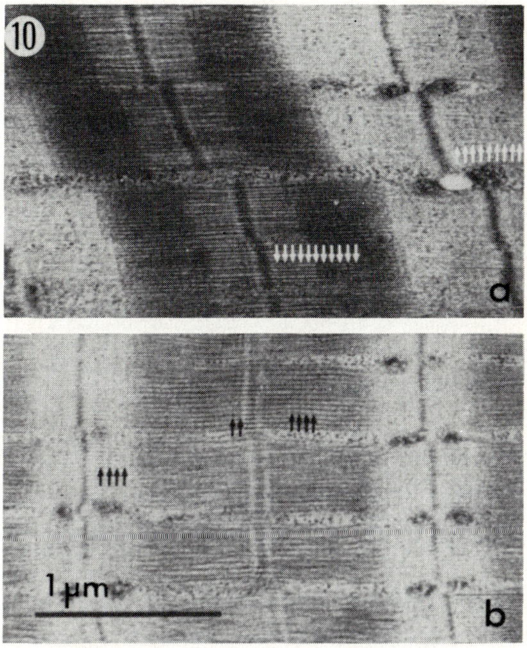

FIGURE 10. Freeze substituted frog sartorius muscles, uranyl acetate, and lead citrate staining. The muscles have been cryofixed by contact with a LN_2 cooled copper block, freeze substituted for 7 d in pure acetone at 193 K, embedded at 213 K in Lowicryl K11M, and polymerized at the same temperature. Freeze substitution, embedding, and polymerization have been carried out in a Reichert-Jung CS auto. (a) Slightly stretched muscle; a spring kept the muscle at a tension of 1 g. (b) Contracting muscle. The muscle, kept at a tension of 1g, has been frozen after five stimuli at a rate of $50s^{-1}$ (100 ms). The muscle shows A band shortening. Arrows in (a) and (b) point to faintly visible axial periodicities of approximately 40 nm (From Edelmann, L., *Scanning Microsc.*, 2, 851, 1988. With permission.)

mentioned hypothesis of water movement during muscle contraction. It appears that dynamic cell processes, particularly those which are associated with a detectable redistribution of cellular water, can only be understood by taking into account rapid release or binding of ions, and probably also other events which lead to local changes in cellular water activity. Therefore, future examination of dynamic cell processes by electron microscopy should include not only the investigation of morphological changes, but also the localization of the main cellular ions at different times of the dynamic events. It is generally accepted that freeze substitution is particularly well suited for morphological studies. Whether a freeze substitution procedure can be found which always yields the ''true'' subcellular distribution of mobile ions remains to be determined in future studies by comparing microanalytical results obtained from freeze-dried or frozen-hydrated cryosections.

III. CONCLUSIONS

Rapid freezing by means of the impact method (LN_2 cooled copper block) is capable of capturing very rapid dynamic cell processes, like membrane fusion or redistribution of water and ions. This technique allows the recognition of ultrastructural changes (e.g., the formation of a fusion pore between membranes during exocytosis in blood platelets followed by swelling of the secretory contents) up to a distance of approximately 3 μm from the cooled plane, as seen after freeze substitution.

Freeze substitution after rapid freezing is a simple procedure to obtain reproducible and durable specimens for investigation in the transmission electron microscope. The use of

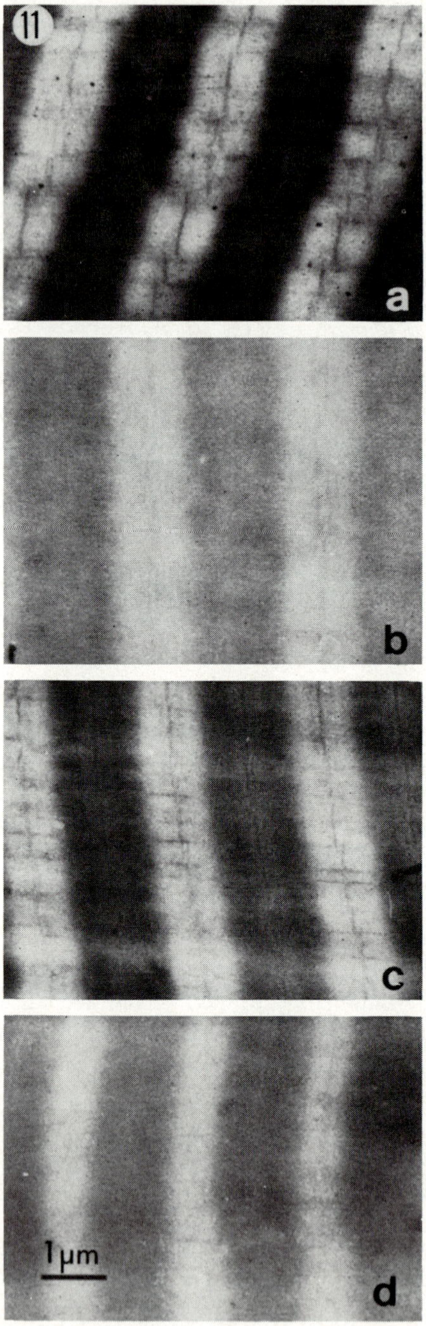

FIGURE 11. 0.3 μm thick sections of freeze substituted and K11M embedded frog muscle (procedure as described in the legend of Figure 10: (a) Dry cut section of a Tl^+ loaded muscle. (b) Wet cut section of a Cs^+-loaded muscle. (c) Dry cut section of a Cs^+-loaded muscle. (d) Dry cut section of a Cs^+-loaded contracting muscle. The muscle, kept at a tension of 1 g, has been frozen after stimulation for 1 s at a rate of $50s^{-1}$. The poor contrast of (b) indicates loss of electron dense Cs^+ ions during wet cutting. The rather poor contrast of the contracting muscle (compare d with c) suggests liberation and redistribution of the electron dense Cs^+-ions during contraction. (From Edelmann, L., *Scanning Microsc.*, 2, 851, 1988. With permission.)

automatic freeze substitution units (commercially available) permits a precise control of the temperature of the specimen during dehydration (freeze substitution). Acetone containing OsO_4 as substitution medium appears to be particularly well suited for the preservation of transient states of cell structures captured by rapid freezing. Ultrathin sections of freeze substituted material permit an optimum control of structural preservation after rapid freezing; an optimum resolution of the cellular ultrastructure in the transmission electron microscope is possible. Hence, freeze substitution is well suited for the investigation of rapid dynamic morphological alterations, such as changes in the structure of membranes and changes in the volume of cell compartments, indicating physicochemical changes during cell function (redistribution of ions, water, or macromolecules). The question, to what extent freeze substitution can be used to prepare a large variety of biological specimens in such a way that the ultrastructure as well as the subcellular distribution of mobile ions is preserved, remains to be determined. The successful experiments carried out with frog muscle indicate that freeze substitution exclusively at low temperatures and low temperature embedding are prerequisites to achieve this goal.

ACKNOWLEDGMENTS

This work is dedicated to Professor Dr. Hellmuth Sitte on the occasion of his 60th birthday. We are particularly grateful for his advice and support in cryowork. We thank our colleague Dr. Klaus Neumann for his helpful criticisms during the preparation of the manuscript. We appreciate the help of Sabine Lipfert and Andreas Ruf in translation and reading, and the excellent technical assistance of Anne Vecerdea. Finally, we would like to thank Professor Dr. Heinrich Patscheke for his successful cooperation. This work was supported by Deutsche Forschungsgemeinschaft (Grant Mo 124/4-2/3) and by a grant from the city of Homburg/Saar (F.R.G.).

REFERENCES

1. **Verkleij, A. J. and Ververgaert, P. H. J. T.,** Architecture of biological and artificial membranes, as visualized by freeze-etching, *Ann. Rev. Phys. Chem.*, 26, 101, 1975.
2. **Heuser, J. E.,** Synaptic vesicle exocytosis and recycling during transmitter discharge from neuromuscular junction, in *Transport of Macromolecules in Cellular System,* Silverstein, S.C., Ed., *(Life Sci. Res. Rep.),* 11, 1978, 445.
3. **Van Harreveld, A., Crowell, J., and Malhotra, S. K.,** A study of extracellular space in central nervus tissue by freeze-substitution, *J. Cell Biol.*, 25, 117, 1965.
4. **Läuchli, A. R., Spurr, A. R., and Wittkopp, R. W.,** Electron analysis of freeze-substituted, epoxy resin embedded tissue for ion transport studies in plants, *Planta,* 95, 341, 1970.
5. **Harvey, D. M. R.,** The preparation of botanical samples for ion localization studies at the subcellular level, *Scanning Electron Microsc.*, II, 409, 1980.
6. **Marshall, A. T.,** Freeze-substitution as a preparation technique for biological x-ray microanalysis, *Scanning Electron Microsc.*, 2, 395, 1980.
7. **Edelmann, L.,** Freeze-dried embedded specimens for biological microanalysis, *Scanning Electron Microsc.*, 4, 1337, 1986.
8. **Steinbrecht, R. A. and Müller, M.,** Freeze-substitution and freeze-drying, in *Cryotechniques in Biological Electron Microscopy,* Steinbrecht, R. A. and Zierold, K., Eds., Springer-Verlag, New York, 1987, 149.
9. **Edelmann, L.,** The cell water problem posed by electron microscopy studies of ion binding in muscle, *Scanning Microsc.*, 2, 851, 1988.
10. **Sitte, H., Neumann, K., and Edelmann, L.,** Cryosectioning according to Tokuyasu versus rapid freezing, freeze-substitution and resin embedding, in *Immuno-Gold Probes in Cell Biology,* Verkleij, A. J., Ed., CRC Press, Boca Raton, FL, 1988, 2.
11. **Ornberg, R. L. and Reese, T. S.,** Beginning of exocytosis captured by rapid-freezing of Limulus amebocytes, *J. Cell Biol.*, 90, 40, 1981.

12. **Ichikawa, M. and Ichikawa, A.,** The fine structure of sublingual gland acinar cells of the mongolian gerbil, *Meriones unguiculatus,* processed by rapid freezing followed by freeze-substitution fixation, *Cell Tissue Res.,* 250, 305, 1987.
13. **Morgenstern, E., Neumann, K., and Patscheke, H.,** The exocytosis of human platelets. A fast freezing and freeze-substitution analysis, *Eur. J. Cell Biol.,* 43, 273, 1987.
14. **Plattner, H.,** Membrane behaviour during exocytosis, *Cell Biol. Int. Rep.,* 5, 435, 1981.
15. **Plattner, H. and Zingsheim, H. P.,** Electron microscopic methods in cellular and molecular biology, in *Subcellular Biochemistry,* Vol. 9, Roodyn, D. B., Ed., Plenum Press, New York, 1983, 1.
16. **Kellenberger, E.,** The response of biological macromolecules and supramolecular structures to the physics of specimen cryopreparation, in *Cryotechniques in Biological Electron Microscopy,* Steinbrecht, R. A. and Zierold, K., Eds., Springer-Verlag, New York, 1987, 36.
17. **Hols, H., Sixma, J. J., Leunissen-Bijvelt, J., and Verkleij, A.,** Freeze fracture studies of human blood platelets activated by thrombin using rapid freezing, *Thromb. Haemost.,* 54, 574, 1985.
18. **Cheng, T. P. O. and Reese, T. S.,** Polarized compartmentalization of organelles in growth cones from developing optic tectum, *J. Cell Biol.,* 101, 1473, 1985.
19. **Fernandez, D. E. and Staehelin, L. A.,** Structural organization of ultrarapidly frozen barley aleuron cells actively involved in protein secretion, *Planta,* 165, 455, 1985.
20. **Ichikawa, M., Ichikawa, A., and Kidokoro, S.,** Secretory process of mucus-secreting cells in mouse colonic mucosa studied by rapid freezing and freeze-substitution, *J. Electron Microsc.,* 36, 117, 1987.
21. **Plattner, H. and Knoll, G.,** Ultrastructural analysis of dynamic cellular processes: A survey of current problems, pitfalls and perspectives, *Scanning Microsc.,* 1, 1199, 1987.
22. **Chandler, D. E.,** Comparison of quick-frozen and chemically fixed sea-urchin eggs: Structural evidence that cortical granule exocytosis is preceded by a local increase in membrane mobility, *J. Cell Sci.,* 72, 23, 1984.
23. **Casley-Smith, J. R., Freeze-substitution of capillary endothelium: The artefactual nature of transendothelial channels and the forms of attached vesicles *Micron,* 11, 461, 1980.**
24. **Sugai, N., Ito, S., and Ichikawa, A.,** The fine structure of the tubulovesicular system in mouse gastric parietal cell processed by cryofixation method, *J. Electron Microsc.,* 34, 113, 1985.
25. **Robards, A. W. and Sleytr, U. B.,** Freeze-substitution and low temperature embedding, in *Low Temperature Methods in Biological Electron Microscopy, Practical Methods in Electron Microscopy,* Vol. 10, Glauert, A. M., Ed., Elsevier, New York, 1985, 461.
26. **Knoll, G., Verkleij, A. J., and Plattner, H.,** Cryofixation of dynamic processes in cells and organelles, in *Cryotechniques in Biological Electron Microscopy,* Steinbrecht, R. A. and Zierold, K., Eds., Springer-Verlag, New York, 1987, 259.
27. **Sitte, H., Edelmann, L., and Neumann, K.,** Cryofixation without pretreatment at ambient pressure, in *Cryotechniques in Biological Electron Microscopy,* Steinbrecht, R. A. and Zierold, K., Eds., Springer-Verlag, New York, 1987, 87.
28. **Pinto-DaSilva, P. and Kachar, B.,** Quick freezing vs. chemical fixation: Capture and identification of membrane fusion intermediates, *Cell Biol. Int. Rep.,* 4, 625, 1980.
29. **Pscheid, P., Schudt, C., and Plattner, H.,** Cryofixation of monolayer cell cultures for freeze-fracturing without chemical pre-treatments, *J. Microsc.,* 121, 149, 1981.
30. **Carlemalm, E., Villinger, W., Acetarin, J. D., and Kellenberger, E.,** Low temperature embedding, in *Science of Biological Specimen Preparation 1985,* Müller, M., Becker, R. P., Boyde, A., and Wolosewick, J. J., Eds., SEM Inc., AMF O'Hare, IL, 1986, 147.
31. **Humbel, B. and Müller, M.,** Freeze substitution and low temperature embedding, in *Science of Biological Specimen Preparation 1985,* Müller, M., Becker, R. P., Boyde, A., and Wolosewick, J. J., Eds., SEM Inc. AMF O'Hare IL, 1986, 175.
32. **Müller, M., Marti, T., and Kriz, S.,** Improved structural preservation by freeze substitution, in *Proc. 7th Eur. Congr. Microsc.,* (Leiden), Vol. 2, Brederoo, P. and De Priester, W., Eds., 1980, 720.
33. **Sitte, H., Neumann, K., and Edelmann, L.,** Cryofixation and cryosubstitution for routine work in transmission electron microscopy, in *Science of Biological Specimen Preparation 1985,* Müller, M., Becker, R. P., Boyde, A., and Wolosewick, J. J., Eds., SEM Inc., AMF O'Hare, IL, 1986, 103.
34. **Torri-Tarelli, F., Grohovaz, F., Fesce, R., and Ceccarelli, B.,** Temporal coincidence between synaptic vesicle and quantal secretion of acetylcholine, *J. Cell Biol.,* 101, 1386, 1985.
35. **Ikeda, H., Ichikawa, A., and Ichikawa, M.,** The effects of freeze-substitution media on the ultrastructure of inclusion bodies in type II pneumocytes of mouse lung processed by the cryofixation method, *J. Electron Microsc.,* 33, 242, 1984.
36. **Nassar, R., Wallace, N. R., Taylor, I., and Sommer, J. R.,** The quick-freezing of single intact skeletal muscle fibers at known time intervals following electrical stimulation, *Scanning Electron Microsc.,* 1, 309, 1986.
37. **Van Harreveld, A. and Crowell, J.,** Electron microscopy after rapid freezing on a metal surface and substitution fixation, *Anat. Rec.,* 149, 381, 1964.

38. **Wagner, R. C. and Andrews, S. B.,** Ultrastructure of the versicular system in rapidly frozen capillary endothelium of the rete mirabile, *J. Ultrastruct. Res.*, 90, 172, 1985.
39. **Steinbrecht, R. A.,** Recrystallization and ice-crystal growth in a biological specimen, as shown by a simple freeze substitution method, *J. Microsc.*, 140, 41, 1985.
40. **Elder, H. Y., Gray, C. C., Jardine, A. G., and Chapman, J. N.,** Optimum conditions for cryoquenching of small tissue blocks in liquid coolants, *J. Microsc.*, 126, 45, 1981.
41. **Morgenstern, E., Edelmann, L., Reimers, H. J., Miyashita, C., and Haurand, M.,** Fibrinogen distribution on surfaces and in organelles of ADP stimulated human blood platelets, *Eur. J. Cell Biol.*, 38, 292, 1985.
42. **Ruf, A., Morgenstern, E., Janzarik, H., and Lüscher E. F.,** Morphology of the interaction of collagen fibrils with normal human platelets and thrombasthenic platelets, *Thromb. Res.*, 44, 477, 1986.
43. **Morgenstern, E. and Reimers, H.-J.,** The platelet contacts during aggregation, *Blut,* 48, 81, 1984.
44. **Ling, G. N. and Bohr, G.,** Studies on ionic distribution in living cells. IV. Effect of ouabain on the equilibrium concentrations of Cs^+, Rb^+, K^+, Na^+ and Li^+ ions in frog muscle cells, *Physiol. Chem. Phys.*, 3, 573, 1971.
45. **Ling, G. N.,** Thallium and cesium in the muscle cells compete for the adsorption sites normally occupied by K^+, *Physiol. Chem. Phys.*, 9, 217, 1977.
46. **Plattner, H.,** Synchronous exocytosis in *Paramecium* cells, in *Cell fusion,* Sowers, A. E., Ed., Plenum Press, New York, 1987, 69.
47. **Akabas, M. H., Cohen, F. S., and Finkelstein, A.,** Separation of the osmotically driven fusion event from vesicle-planar membrane attachment in a model system for exocytosis, *J. Cell Biol.*, 98, 1063, 1984.
48. **Zimmerberg, J.,** Molecular mechanisms of membrane fusion: Steps during phospholipid and exocytotic membrane fusion, *Biosci. Rep.*, 7, 1987, 251.
49. **Morgenstern, E.,** Ultracytochemistry of blood platelets, in *Progress in Histochemistry and Cytochemistry,* Vol.12, Graumann, W., Lojda, Z., Pease, A. G. E., and Schiebler, T. H., Eds., 1980, 4.
50. **Holmsen, H.,** Platelet secretion, in *Hemostasis and Thrombosis,* Colmann, R. V., Hirsh, J., Marder, V. J., and Salzman, E. W., Eds., Lippincott, Philadelphia, 1982, 390.
51. **Stenberg, P. E. and Bainton, D. F.,** Storage organelles in platelets and megakaryocytes, in *Biochemistry of Platelets,* Phillips, D. R. and Shuman, M. A., Eds., Academic Press, Orlando, 1986, 257.
52. **Bendayan, M., Becich, M. J., and Reddy, J. K.,** Immunocytochemical localization of exocrine enzymes in rat pancreatic acinar carcinoma, *Eur. J. Cell Biol.*, 34, 151, 1984.
53. **Kartenbeck, J.,** Coated secretory vesicles, in *Coated Vesicles,* Ockleford, C. D. and White, A., Eds., Cambridge University Press, Cambridge, U.K., 1980, 244.
54. **Fine, R. E. and Ockleford, C. D.,** Supramolecular cytology of coated vesicles, *Int. Rev. Cytol.*, 91, 1, 1984.
55. **Tooze, J. and Tooze, S. A.,** Clathrin-coated vesicular transport of secretory proteins during the formation of ACTH-containing secretory granules in AtT20 cells, *J. Cell Biol.*, 103, 839, 1986.
56. **Siegel, D. P.,** Inverted micellar structures in bilayer membranes, *Biophys. J.*, 45, 399, 1984.
57. **Verkleij, A. J., Mombers, C., Gerritsen, W. J., Leunissen-Bijvelt, J., and Cullis, P. R.,** Fusion of phospholipid vesicles in association with the appearance of lipidic as visualization by freeze-fracturing, *Biochim. Biophys. Acta,* 555, 358, 1979.
58. **Verkleij, A. J.,** Lipidic intermembranous particles, *Biochim. Biophys. Acta,* 779, 43, 1984.
59. **Schmidt, W., Patzak, A., Lingg, G., Winkler, H., and Plattner, H.,** Membrane events in adrenal chromaffin cells during exocytosis: A freeze-etching analysis after rapid cryofixation, *Eur. J. Cell Biol.*, 32, 31, 1983.
60. **Düzgünes, N.,** Membrane fusion, in *Subcellular Biochemistry,* Vol. 11, Roodyn, D. B., Ed., 1985, 195.
61. **Chandler, D. E. and Heuser, J. E.,** Arrest of membrane fusion events in mast cells by quick-freezing, *J. Cell Biol.*, 86, 666, 1980.
62. **Rand, R. P., Reese, T. S., and Miller, R. G.,** Phospholipid bilayer deformations associated with interbilayer contact and fusion, *Nature,* 293, 237, 1981.
63. **Holz, R. W.,** The role of osmotic forces in exocytosis from adrenal chromaffin cells, *Ann. Rev. Physiol.*, 48, 175, 1986.
64. **Finkelstein, A.,** Osmotic swelling of vesicles: Its role in the fusion of vesicles with planar phospholipid bilayer membranes and its possible role in exocytosis, *Ann. Rev. Physiol.*, 48, 163, 1986.
65. **Zimmerberg, J., Curran, M., Cohen, F. S., and Brodwick, M.,** Simultaneous electrical and optical measurements show that membrane fusion precedes secretory granule swelling during exocytosis of beige mouse mast cells, *Proc. Natl. Acad. Sci. U.S.A.*, 7, 251, 1987.
66. **Breckenridge, L. J. and Almers, W.,** Final steps in exocytosis observed in a cell with giant secretory granules, *Proc. Natl. Acad. Sci. U.S.A.*, 84, 1945, 1987.
67. **Morgenstern, E. and Kho, A.,** Morphometrische Untersuchungen an Blutplättchen, Veränderungen der Plättchenstruktur bei Pseudopodienbildung und Aggregation, *Eur. J. Cell Biol.*, (formerly *Cytobiology)* 15, 233, 1977.

68. **White, J. G.,** The secretory pathway bovine platelets, *Blood,* 69, 878, 1987.
69. **White, J. . and Krumwiede, M.,** Further studies of the secretory pathway in thrombin stimulated human platelets, *Blood,* 69, 1196, 1987.
70. **Santoso, S., Zimmerman, U., Neppert, J., and Müller-Eckardt, C.,** Receptor patching and capping of platelet membranes induced by monoclonal antibodies, *Blood,* 67, 343, 1986.
71. **Chandler, D. E. and Heuser, J.,** Membrane fusion during secretion, *J. Cell Biol.,* 83, 91, 1979.
72. **Martin, T. W. and Lagunoff, D.,** Mast cell secretion, in *Cell Biology of the Secretory Process,* Cantin, M., Ed., S. Karger, Basel, 1984, 481.
73. **Mosher, D. F., Pesciotta, D. M., Loftus, J. C., and Albrecht, R. M.,** Secreted alpha granule proteins, the race for receptors, in *Platelet Membrane Glycoproteins,* George, J. N., Nurden, A. T., and Phillips, D. R., Eds., 1985, 171.
74. **White, J. G. and Estenson, R. D.,** Degranulation of discoid platelets, *Am. J. Path.,* 68, 289, 1972.
75. **Patscheke, H. and Wörner, P.,** Platelet activation detected by turbidometric shape-change analysis. Differential influence of cytochalasin B and prostaglandin E_1, *Thromb. Res.,* 12, 485, 1978.
76. **Morgenstern, E. and Patscheke, H.,** unpublished.
77. **Rubin, R. P.,** Stimulus-secretion coupling, in *Cell Biology of Secretory Process,* Cantin, M., Ed., S. Karger, Basel, 1984, 52.
78. **De Lisle, R. C. and Williams, J. A.,** Regulation of membrane fusion in secretory exocytosis, *Ann. Rev. Physiol.,* 48, 225, 1986.
79. **Röhlich, P., Andersson, P., and Uvnäs, B.,** Electron microscope observation on compound 48/80-induced degranulation in rat mast cells, *J. Cell Biol.,* 51, 465, 1971.
80. **Chandler, D. E. and Kazilek, C. J.,** Chemotactic peptide-induced exocytosis in neutrophils: Granule fusion patterns depend on the source of messinger calcium, *J. Cell Sci.,* 83, 293, 1986.
81. **White, J. G.,** The secretory pathway in bovine platelets, *Blood,* 69, 878, 1987.
82. **Zucker-Franklin, D., Benson, K. A., and Myers, K. M.,** Absence of a surface-connected canalicular system in bovine platelets, *Blood,* 65, 241, 1985.
83. **Dierichs, R. and Göhler, S.,** Interactions of the cytoskeleton during platelet secretion, *Haemostasis,* 15, 53, 1985.
84. **Tosteson, D. C. and Hoffman, J. F.,** Regulation of cell volume by active cation transport in high and low potassium sheep red cells, *J. Gen. Physiol.,* 44, 169, 1960.
85. **Ling, G. N.,** *A Physical Theory of the Living State. The Association-Induction Hypothesis,* Blaisdell, Waltham, MA, New York, 1962.
86. **Ling, G. N.,** *In Search of the Physical Basis of Life,* Plenum Press, New York, 1984.
87. **Ling, G. N.,** A physical theory of the living state: Application to water and solute distribution, *Scanning Microsc.,* 2, 899, 1988.
88. **Ling, G. N. and Negendank, W. G.,** The physical state of water in frog muscle, *Physiol. Chem. Phys.,* 2, 15, 1970.
89. **Ling, G. N. and Walton, C. L.,** What retains water in living cells? *Science,* 191, 293, 1976.
90. **Negendank, W.,** Ionic distributions and volume maintenance in human lymphocytes, *Physiol. Chem. Phys. Med. NMR,* 16, 3, 1984.
91. **Negendank, W.,** The state of water in the cell, in *Science of Biological Specimen Preparation,* Müller, M., Becker, R. P., Boyde, A., and Wolosewick, J. J., Eds., SEM Inc., AMF O'Hare Chicago, IL, 1986, 21.
92. **Edelmann, L.,** Subcellular distribution of potassium in striated muscles, *Scanning Electron Microsc.,* 2, 875, 1984.
93. **Rorschach, H. E., Bearden, D. W., Hazlewood, C. F., Heidorn, D. B., and Nicklow, R. M.,** Quasi-elastic scattering studies of water diffusion, *Scanning Microsc.,* 1, 2043, 1987.
94. **Edelmann, L.,** Potassium adsorption sites in frog muscle visualized by cesium and thallium under the transmission electron microscope, *Physiol. Chem. Phys.,* 9, 313, 1977.
95. **Edelmann, L.,** Potassium binding sites in muscle: Electron microscopic visualization of K, Rb, and Cs in freeze-dried preparations and autoradiography at liquid nitrogen temperature using ^{86}Rb and ^{134}Cs, *Histochemistry,* 67, 233, 1980.
96. **Ling, G. N.,** K^+ localization in muscle cells by autoradiography, and identification of K^+ adsorbing sites in living muscle with uranium binding sites in electron micrographs of fixed preparations, *Physiol. Chem. Phys.,* 9, 319, 1977.
97. **Edelmann, L.,** Frozen hydrated cryosections of thallium loaded muscle reveal subcellular potassium binding sites, *Physiol. Chem. Phys. Med. NMR,* 16, 499, 1984.

Chapter 8

TOPOGRAPHY AND INTERNALIZATION OF CELL SURFACE RECEPTORS AS ANALYZED BY AFFINITY- AND IMMUNO-LABELING COMBINED WITH SURFACE REPLICATION AND ULTRATHIN SECTIONING TECHNIQUES

Horst Robenek

TABLE OF CONTENTS

I. INTRODUCTION

Receptor-mediated endocytosis is the prinicpal means by which mammalian cells selectively internalize informational, transport and nutritional macromolecules, such as hormones and growth factors, transferrin and lipoproteins, across the plasma membrane.[1] Over the past several years the continuing refinement of morphological procedures and the concomitant development of probes that can be visualized at the ultrastructural level have sparked major advances in our understanding of the sequence of the internalization pathway. Such sophisticated morphological techniques have made possible the recognition of well defined steps in receptor-mediated endocytosis.

Cells internalize diverse receptor-ligand complexes in astonishingly similar fashion, but nevertheless there are striking differences in the mechanisms that trigger receptor movement to coated pits, and in the intracellular fate of ligands and receptors. Thus, endocytotic systems may be classified into several groups. The receptor-mediated endocytosis of low density lipoproteins (LDL) by fibroblasts,[1] and of asialoglycoproteines by hepatocytes,[2] are representatives of systems in which ligands are catabolized in lysosomes and receptors reutilized. Receptor-mediated endocytosis of insulin[3] or epidermal growth factor[4] leads to degradation of both the ligand and the receptor in lysosomes. In contrast, when diferric transferrin is internalized, neither transferrin nor the transferrin receptor are destroyed; iron is unloaded intracellularly from the transferrin-receptor complex, and both the apolipoprotein and the receptor are returned intact to the cell exterior.

Lipoprotein receptors play a major role in regulating the concentration of plasma lipoproteins.[6] The ultimate metabolism of lipoproteins is dependent upon how these are taken up and catabolized by cells, processes that are, in turn, mediated by several cell surface receptors. Therefore, we have examined lipoprotein receptors as a model system for cell surface receptors in general.

In this short review I will describe some of our work, which has employed morphological techniques tailored to elucidate the topography and internalization of different lipoprotein species and their receptors in cultured cells, under a variety of experimental conditions. Some specific goals of our studies have been to

1. Analyze the native distribution of lipoprotein receptors on the cell surface.
2. Characterize the different mechanisms that trigger the movement of receptors to coated pits.
3. Trace the intracellular pathways taken by the different lipoprotein species and their receptors and
4. Identify the intracellular sites and mechanisms of lipoprotein assembly.

II. LIPOPROTEINS AND LIPOPROTEIN RECEPTORS

A. LIPOPROTEIN COMPOSITION

The plasma lipoproteins are water-soluble spherical macromolecules representing complexes of lipids (triglycerides, cholesterol and phospholipids) and one or more proteins, called apolipoproteins or apoproteins. There are five major classes of human lipoproteins:

1. Chylomicrons
2. Very low-density lipoproteins (VLDL)
3. Intermediate density lipoproteins (IDL)
4. Low Density lipoproteins (LDL)
5. High density lipoproteins (HDL). (For review see Reference 7).

The known functions of the four major apolipoproteins A, B, C, and E may be broadly divided into three categories: enzyme activation, receptor binding, and *de novo* synthesis of lipoprotein particles. Each class of lipoprotein, except LDL, contains a variety of apolipoproteins in differing proportions. LDL contains only apolipoprotein B. Apolipoprotein A is the major protein in HDL. Two forms of apolipoprotein B exist: apolipoprotein B-100 and apolipoprotein B-48. Apolipoprotein B-100 is an obligate constituent of VLDL, IDL and LDL. In man, apolipoprotein B-48 is found in chylomicrons and chylomicron remnants. Apolipoprotein C and E are present in various proportions in all lipoproteins except LDL.[8]

B. LIPOPROTEIN RECEPTORS

The delivery of lipids to target cells involves the faithful recognition of specific apolipoproteins by cell surface receptors. Binding of the apolipoprotein to the receptor is a prerequisite for the uptake and degradation of the lipoprotein. There exist at least four different receptors that bind apolipoprotein B and/or apolipoprotein E-containing lipoproteins. These are the LDL receptor, the scavenger or acetyl-LDL receptor, the hepatic chylomicron remnant (apolipoprotein E) receptor, and the β-VLDL receptor. These genetically distinct receptors show great selectivity, not only with respect to their apolipoprotein ligands, but also to their distribution among different types of cells in the body.[6] The LDL receptor is the most extensively studied lipoprotein receptor, and its role in receptor-mediated transport of LDL is firmly established.[3] Much less is known about the other types of lipoprotein receptors.

Recent studies have provided evidence for the existence of a specific receptor for apolipoprotein A-I-containing HDL on various cell types.[9-14] The present review will focus on the LDL receptor, the acetyl-LDL receptor, the β-VLDL receptor, and the HDL receptor.

C. ULTRASTRUCTURAL DETECTION OF LIPOPROTEIN RECEPTORS

1. Methodical Limitations

Almost all studies on high resolution localization of lipoprotein receptors in the electron microscope exploit the affinity of a particular lipoprotein for its specific receptor. The receptors are localized indirectly when they form complexes with the lipoprotein, which has been coupled before to an electron dense marker (e.g., ferritin or colloidal gold particles). Alternatively, the receptors may be detected using a primary antibody directed towards the lipoprotein, followed by a labeled secondary antibody. These immunostaining techniques are limited to demonstrating lipoprotein receptors on cell surfaces, which are directly accessible to solutions containing the lipoprotein-marker particle conjugates, or to the antibodies. Information obtained so far by means of such procedures is confined to the luminal membranes of surface epithelia[15] and surfaces of cell monolayers.[16-18]

The direct localization of receptor molecules themselves within tissues or cells can only be achieved by immunocytochemical means (see also Chapters 10-12, 16 in this volume). Whereas an impressive array of polyclonal and monoclonal antibodies has become available for visualizing all of the lipoproteins and their apolipoprotein components, suitably efficient antibodies could only be raised against one type of lipoprotein receptor, the LDL receptor.[19] Only two papers have reported the use of monoclonal anti-LDL receptor antibodies for the direct ultrastructural localization of LDL receptors in the plasma membrane of cultured fibroblasts,[18,19] and up to now, no ultrastructural data exist on the intracellular distribution of lipoprotein receptors.

This probably reflects deficiencies in the antibodies and/or procedures for immunoelectron microscopic labeling. Despite intensive effort in our laboratory, we have not been successful in localizing intracellular LDL receptors directly.

2. Probes for the Visualization of Lipoprotein Receptors

Since means for the direct localization of lipoprotein receptors are rather limited, we still depend on indirect procedures for studying lipoprotein receptor distribution and dynamics. Studies on lipoprotein receptors in our laboratory have been carried out exclusively on isolated and cultured cells (i.e., macrophages, fibroblasts and hepatocytes).

We have employed three principally different approaches for visualizing lipoprotein receptors. The first method was based on the selective binding of gold-labeled lipoproteins (i.e., LDL, acetyl-LDL, malondialdehyde-modified LDL, HDL, and β-VLDL) to localize directly the corresponding receptors at the plasma membrane surface. Lipoproteins conjugated to colloidal gold particles (Figure 1) were added to the culture medium of living cells, and the cells were incubated at 4°C and/or 37°C for various periods of time. The experiments were terminated by thorough washing with ice-cold buffer. The cells were then fixed and processed for electron microscopy. Control experiments were carried out to confirm the specificity of the method by exposing the cells first to unlabeled ligands, and then to the same ligands conjugated to colloidal gold. In these experiments the cell surface was always completely devoid of gold label; the unlabeled ligand had competed with the gold conjugate, thus preventing subsequent binding of the labeled ligand. This indicates that the ligand-gold conjugates interacted specifically with their receptors, and that there were no binding sites for the gold itself under the experimental conditions used. Why the lipoprotein-gold conjugates displayed binding properties similar to those of their unlabeled counterparts became clear when negatively stained preparations are examined. The examples in Figure 1 depict gold particles, LDL particles, and gold conjugates with LDL, HDL and β-VLDL. In the HDL and LDL-gold conjugates each gold particle was completely buried within a sheath of lipoprotein molecules. Thus, only the ligand, and not the gold itself, was exposed and available for binding to any potential receptor at the cell surface. Optimally formed LDL and HDL-gold conjugates were produced when the gold particles were not less than 12 nm in diameter. β-VLDL-gold conjugates were less favorably formed than HDL and LDL-gold conjugates, because some of the gold particles remained exposed due to the large size of the β-VLDL molecules. Most of the gold particles, however, were completely covered with the lipoprotein. Therefore, in order to reduce possible adhesion of the conjugates to the cell surface due to extraneous interaction with exposed areas of the gold particles, the amount of β-VLDL relative to colloidal gold was always maintained far above saturation in the β-VLDL-gold conjugates.

The second method involved indirect immunocytochemical labeling of receptor-bound lipoproteins, or their receptors prior to embedding. For this purpose, living cells were exposed first to specific antibodies directed against either the lipoproteins, or their receptors followed by gold-labeled protein A or peroxidase-labeled secondary antibodies. The cells were incubated with the marker at 4°C and/or 37°C for varying periods of time. Incubations were stopped by washing with ice-cold buffer. The cells were then fixed and processed for electron

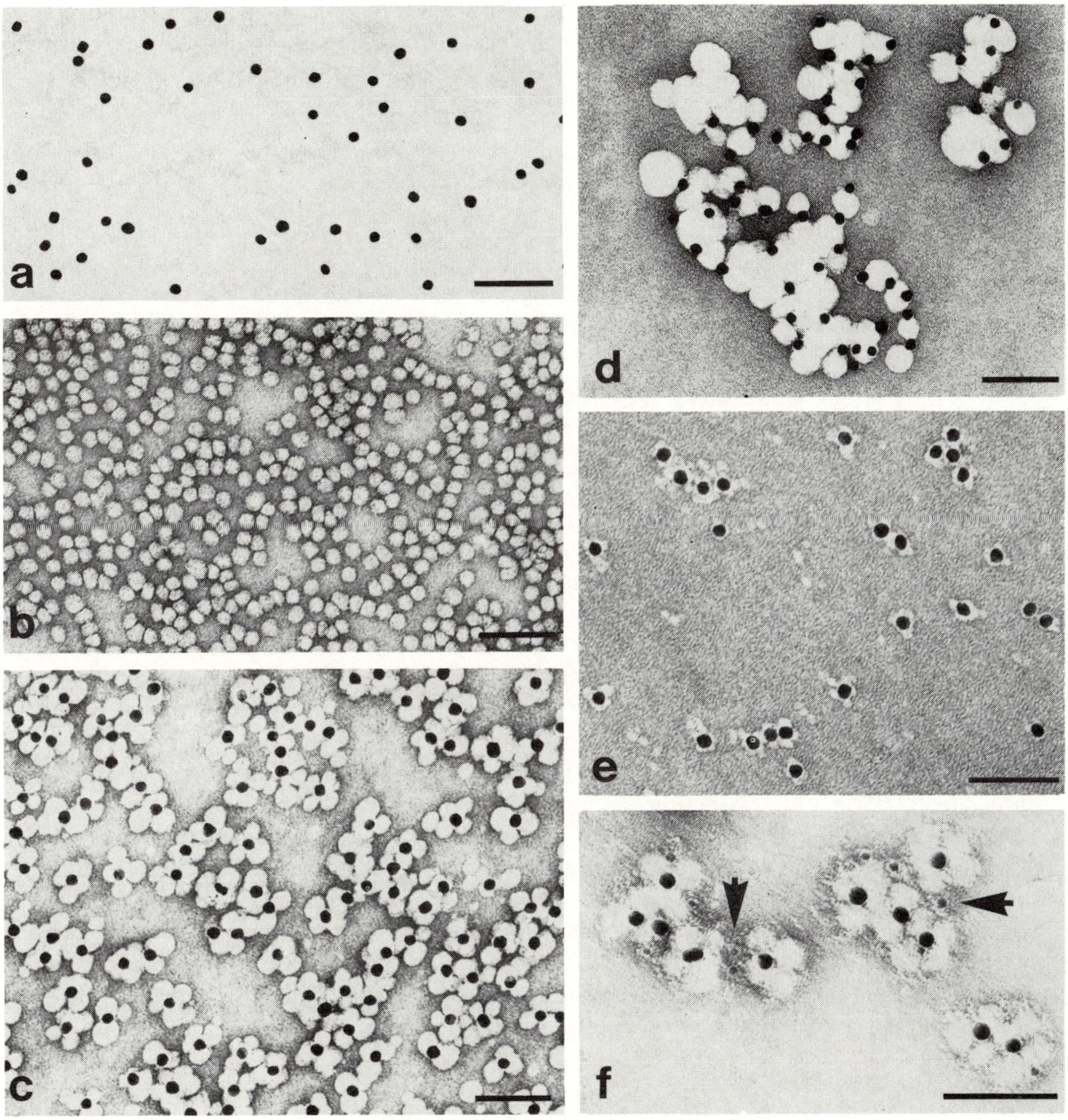

FIGURE 1. Colloidal gold particles, lipoprotein molecules and negatively stained lipoprotein-gold conjugates. a) Colloidal gold particles, average diameter 15 nm; b) unconjugated LDL molecules, c) LDL-gold conjugate; d) β-VLDL-gold conjugate; e) HDL-gold conjugate; f) LDL-gold conjugate (15 nm diameter gold particles labeled with polyclonal anti-apolipoprotein B-100 antibody which has been detected in turn with protein A-gold (5 nm diameter particles, arrows). Bars: 0.1 μm.

microscopy. The efficacy of such immune reactions could be tested *in vitro*. For example, Figure 1(f) shows that an anti-apolipoprotein B-100 antibody, detected here with a 5 nm protein A-gold complex, reacts strongly with a LDL-gold conjugate.

The third method involved postembedding immunocytochemical labeling of lipoproteins. In a first set of experiments, cells were incubated with unlabeled lipoproteins at 4°C and/or 37°C for various periods of time, and after fixation they were processed for cryoultrathin sectioning, or embedded in LR White or Lowicryl K4M resins. Thin sections were treated sequentially with antibodies directed against specific apolipoproteins and with gold-labeled protein A.

The specificity of immunostaining in methods two and three was controlled by omitting the primary antibody. As a second control, primary antibody-labeled lipoproteins were incubated with unlabeled protein-A, or unlabeled secondary antibodies followed by exposure to protein A-gold, or peroxidase-labeled secondary antibodies. Another control involved immunostaining of cells not incubated with lipoproteins.

3. Application of Electron Microscopic Techniques

a. Surface Replication

We have used platinum-carbon (Pt/C) surface replicas of gold-labeled cells to analyze the distribution of lipoprotein receptors on the cell surface.(For this method,see also Chapter 14 by Bohn). The advantage of this technique is that extensive areas of the cell surface can be surveyed. Combined with colloidal gold markers, this technique permits high resolution visualization and reliable identification of cell surface receptors in cultured cells. It also yields more detailed information about the distribution and dynamics of receptors than any other technique (for review see References 18, 20, and literature cited therein). Briefly, surface replicas of gold-labeled fixed cells in monolayers are prepared directly on 1 × 1 cm pieces of plastic cut from culture vessels using a soldering iron. The cells on the plastic pieces are dehydrated and dried in a critical-point drying device and shadowed with Pt/C at an angle of 38°, followed by carbon at 90°, in a vacuum evaporating unit. The preparations are left overnight in household bleach, which separates the metal replicas from the substrate and removes cellular material. The replicas are washed in distilled water and picked up on copper grids prior to examination in the electron microscope.

Surface replicas can also be prepared from the protoplasmic aspect of both apical and ventral plasma membranes of cultured cells. The apical protoplasmic surface is exposed by tearing away the apical portions of the cell for shadowing, using poly-lysine coated coverslips. The lysis-squirting technique has proven most successful for exposing the protoplasmic surface of the ''ventral'' plasma membrane. In this case, the cells are lysed in hypotonic buffers to expose the plasma membrane, and the cytoplasmic components are removed with a stream of buffer from a syringe or Pasteur pipette. The combination of lysis-squirting, critical-point drying, and subsequent replication yields good structural preservation of the plasma membrane comparable to that obtained by deep-etching techniques. In addition, this method has the obvious advantage that it is amenable to immunolabeling procedures. As such, it may contribute substantially to our understanding of the dynamis of receptors and their relationship to cytoskeletal elements.

b. Thin Sectioning

Thin sectioning, though indispensable for analyzing the intracellular pathway of ligands after endocytosis, does not provide adequate information on the surface distribution of ligand-receptor complexes. In the present investigation, however, thin sectioning was particularly useful for correlating the distribution of lipoprotein-receptor complexes on the cell surface with the pathway, and the fate of the complexes after internalization.

III. RECEPTOR-MEDIATED ENDOCYTOSIS OF LDL IN FIBROBLASTS

A. DISTRIBUTION, RECYCLING AND CLUSTERING OF LDL RECEPTORS AS REVEALED BY SURFACE REPLICATION TECHNIQUES

1. Structure and Role of Coated Pits

Coated pits and vesicles have been implicated in a variety of cellular processes, such as membrane transfer, membrane recycling, and the intracellular shuttling of macromolecules. However, their main role appears to be in receptor-mediated endocytosis.[21]

Visualized by thin-section electron microscopy, coated regions of the plasma membrane can be distinguished from other membrane areas by the fuzzy, bristle-like appearance of their cytoplasmic surface (Figure 2a). In Figure 2b, the binding and internalization of LDL-gold particles in this specialized membrane region are demonstrated.

Freeze fracture replicas of cell membranes from normal cultures of fibroblasts render coated pits as regions of the membrane, which differ from the remaining membrane in terms of particle density and size. If fibroblasts are treated with filipin, which forms complexes with cholesterol, freeze fracture replicas display protuberances 20 to 25 nm in size on the fracture faces of the membrane. Coated pits can be clearly recognized by the fact that they are free of filipin-cholesterol complexes (Figure 2d). There is convincing experimental evidence that the membrane-associated clathrin coat prevents the formation of the characteristic filipin protuberances[22] (for further details see Reference 20).

In surface replicas, coated pits were usually slightly indented, so that they could be identified under favorable conditions of metal shadowing. When fibroblasts were exposed to LDL-gold complexes, the complexes bound only to these indented regions of the membrane (Figure 2c). Thus, LDL-gold complexes may serve as markers for localizing coated pits in surface replicas.[23]

Using a combination of quick freeze, freeze fracture, deep etching, and Pt/C replication, Heuser has provided striking views of coated pits on the protoplasmic plasma membrane surface of fibroblasts.[24] Unfortunately, this technique cannot be combined with immunolabeling. Recently, we have combined the lysis-squirting technique with immunogold labeling of selected structural components associated with the protoplasmic surface of the plasma membrane.[25] The lysis-squirting replication method preserves fine structural details comparable to those observed after deep etching. In addition, this method has the obvious advantage that is amenable to immunolabeling procedures.

When we incubated the exposed protoplasmic surfaces of lysis-squirted fibroblasts with antibodies against a minor component of coated vesicles followed by protein A-gold, the gold label appeared specifically bound to coated pits and vesicles (Figure 2e, f). Thus, this procedure allows unambiguous immunocytochemical identification of structural elements originally associated with the plasma membrane. It may help to extend our understanding of the structure of coated vesicles in particular and of membrane dynamics in general.

2. Distribution of LDL Receptors

When fibroblasts were incubated with LDL-gold complexes (Figure 1c) at 4°C for 1 h, the gold particles were observed almost exclusively in clusters (Figure 3a). The number of solitary gold particles outside these clusters was negligible. The same labeling pattern was seen when glutaraldehyde-fixed cells were incubated with LDL-gold conjugates. These clusters generally consisted of densely packed gold particles, and their distribution coincided largely with that of coated pits.

It may be objected that the observed distribution of LDL-gold complexes does not reflect the true distribution of LDL receptors in the membrane, for the complexes represent multivalent ligands that might be expected to bind more efficiently to aggregated, rather than to solitary receptor molecules. This was disproved by the following experiments. When we incubated fibroblasts with LDL-gold conjugates in the presence of an excess of unlabeled LDL at 4°C for 1 h, both LDL-gold complexes and free LDL molecules were visualized in favorable surface replicas in clusters only; neither gold particles nor unlabeled LDL molecules were seen outside the clusters (Figure 3b). Furthermore, gold clusters equivalent to those already described were also obtained

1. When free LDL was allowed to bind to the LDL receptors at 4°C for 1 h before the LDL-receptor complexes were visualized by protein A-gold labeling of antibodies directed against the apolipoprotein B-100 moiety of the LDL molecule (Figure 3c)

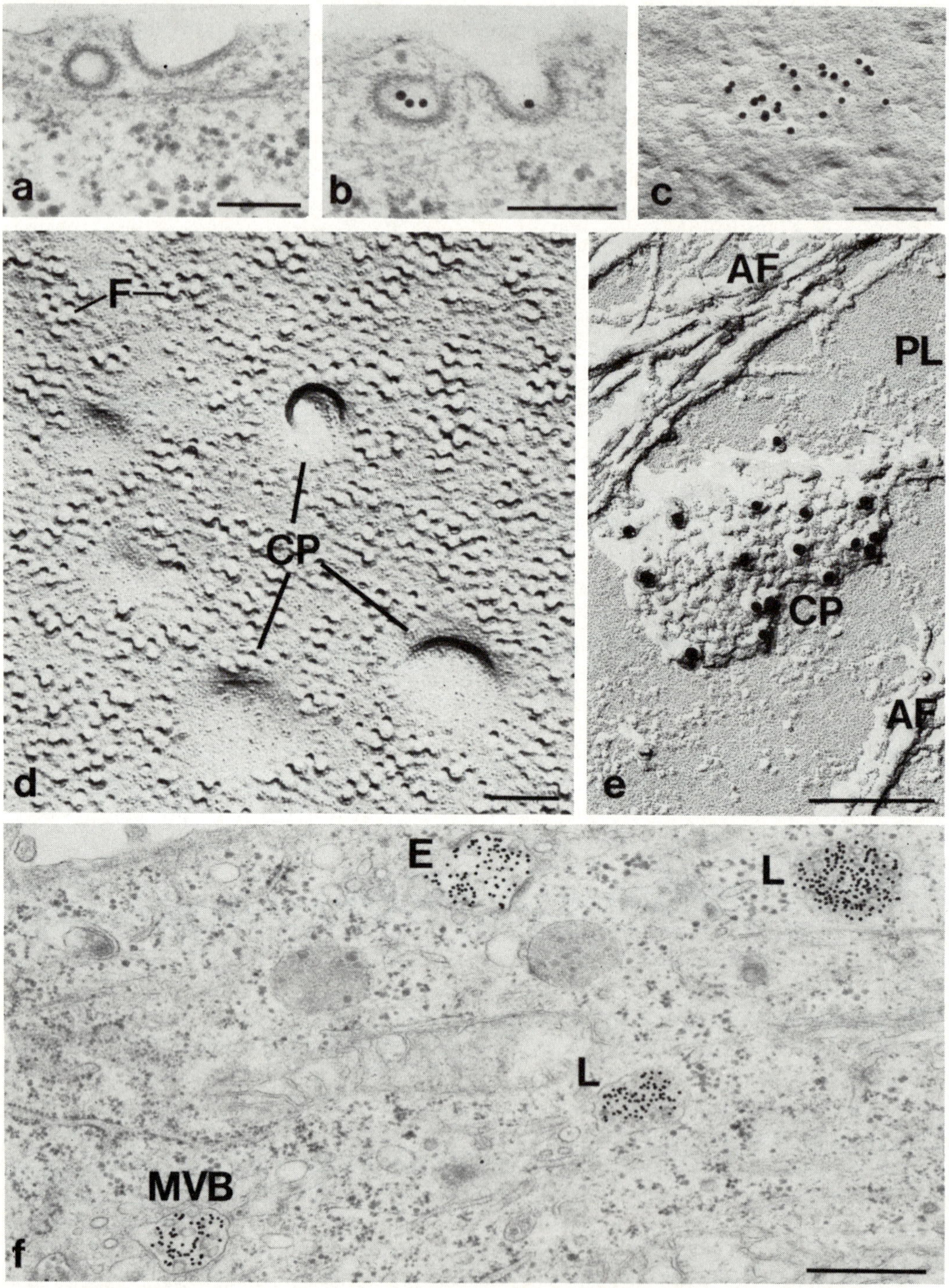

FIGURE 2. Coated pits in cultured human fibroblasts visualized by various electron microscopic techniques. a,b) Conventional thin sectioning; b) coated pit and coated vesicle labeled with LDL-gold conjugate; c) coated pit labeled with LDL-gold conjugate, surface replica; d) coated pits (CP) and filipin-cholesterol complexes (F) on the protoplasmic fracture face after filipin treatment, freeze fracture. e) Surface replica of the "ventral" protoplasmic aspect of the plasmalemma (PL) after labeling with antibodies directed against a coated pit component followed by protein A-gold; coated pits (CP) and actin filaments (AF) are seen. f) Fibroblast incubated with LDL-gold conjugates at 37°C for 1 h, showing gold label internalized in endosomes (E), multivesicular bodies (MVB) and lysosomes (L). Bars: (a to e) 0,2 μm; (f) 0,5 μm.

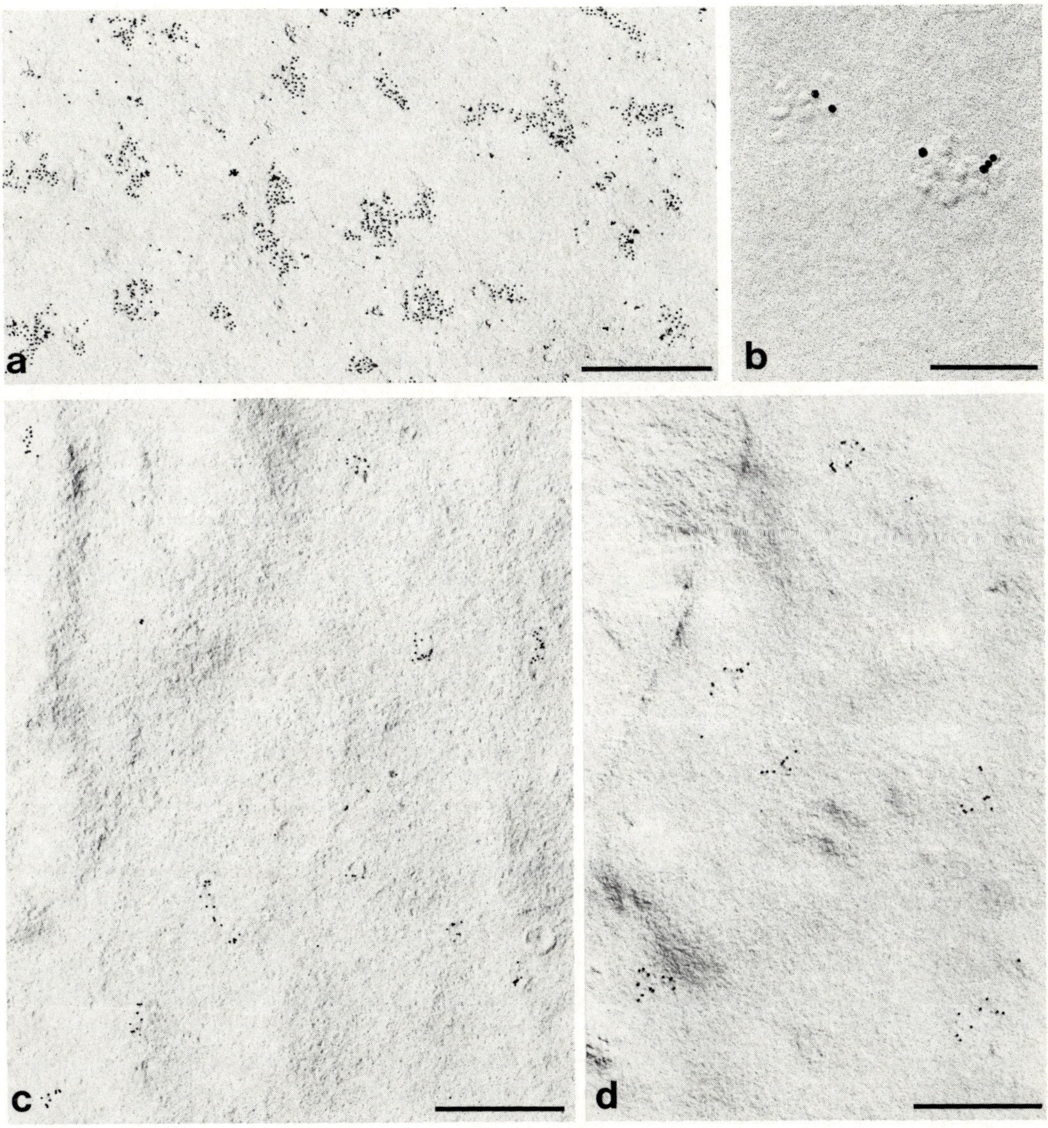

FIGURE 3. Distribution of LDL receptors in the plasma membrane of cultured human fibroblasts. a) Fibroblast incubated with LDL-gold conjugate at 4°C for 1 h. Localization of LDL receptors b) with the aid of a mixture of gold-labeled and unlabeled LDL molecules, c) by immunolabeling of surface-bound LDL molecules using polyclonal antibodies directed against apolipoprotein B-100 and protein A-gold, and d) using monoclonal antibodies directed against the LDL receptor and immunogold antibody labeling. Bars: (a,c,d) 1 μm; (b) 0.2 μm.

2. When immunolabeling of LDL receptors themselves was performed using monoclonal antibodies directed against the receptors followed by protein A-gold labeling (Figure 3d).

The binding of inherently multivalent LDL-gold complexes might also be expected to cause a redistribution of receptors and/or coated pits in the plane of the membrane as does antibody binding in other systems.[26] Therefore, in a further set of experiments we prefixed the cells to ensure that the original topography of the receptors was preserved prior to immunogold labeling of LDL-receptor complexes or LDL receptors. Under all incubation protocols the results were comparable to those obtained with LDL-gold complexes and unfixed cells. Thus, it can be concluded that unoccupied LDL receptors are arranged in the form of clusters in the native cell membrane.

3. Recycling and Clustering of LDL Receptors

Receptor-bound molecules are internalized at the cell surface, when they cluster into coated pit regions which subsequently invaginate and pinch off to form coated intracellular vesicles. The sequence of events and the mechanisms by which receptor-ligand complexes are triggered to cluster in coated pits are still poorly understood.[27] Conflicting views have been proffered concerning how LDL receptors are inserted into the plasma membrane following recycling, and how they move to coated pits. Receptors are believed to be inserted into the cell membrane at random locations and to migrate to coated pits either by diffusion[28] or by active flow.[29]

To probe these assumptions, we performed an experiment designed to detect only newly inserted receptors. Fibroblasts were first loaded with an excess of LDL by incubation with unlabeled LDL at 4°C. The unbound LDL was then removed by exchanging the medium, and the cells were incubated with LDL-gold complexes and warmed to 37°C. Samples were fixed at 2 min intervals. Gold particles began to adhere at the cell surface 4 min after warming to 37°C. The marker particles and, hence, newly inserted LDL receptors, were randomly and uniformly distributed within clusters in varying size and shape and in morphologically undifferentiated regions of the cell membrane. The gold particles became increasingly concentrated in the clusters with time. After 6 to 8 min, almost all the gold particles were observed in clearly delineated membrane regions representing coated pits about to undergo endocytosis.[30] The changing marker particle patterns obviously represented intermediate stages in receptor clustering into coated pits and LDL-receptor endocytosis.

These findings are inconsistent with the model that LDL receptors are incorporated at random into the membrane, and that they migrate appreciably in the plane of the membrane before being trapped by coated pits. Our studies provide convincing evidence that the site of replacement and incorporation of LDL receptors into the plasma membrane after recycling corresponds to the site at which new coated pits are formed and internalized. From these observations, it can be concluded that after their return to the surface, LDL receptors remain clustered so that they can be incorporated rapidly into newly formed coated pits. In contrast to LDL receptors, other recycling receptors, such as those for IgA[27] or transferrin, seem to undergo a transient phase of dispersion on the cell surface before clustering and internalizing again.

B. ANALYSIS OF THE INTRACELLULAR FATE OF LDL BY THIN SECTIONING

In thin sections of fibroblasts incubated with LDL-gold complexes for 1 h at 4°C, the label often appeared over indented and coated regions of the plasma membrane. Some gold clusters were observed over plane or slightly concave uncoated expanses of the membrane. Such gold aggregates probably mark newly inserted receptor clusters. These features were correlatable with their counterparts in surface replicas. Examinations of thin sections of cells that had been warmed from 4 to 37°C for 10 min confirmed that the LDL-gold complexes were internalized in typical coated vesicles of the endocytosis pathway (Figure 2b). Continuous incubation of cells with LDL-gold complexes at 37°C yielded similar results. However, the uptake of ligands was more pronounced, and after an incubation period of 1 h, much label was found in electron-lucent endosomes, multivesicular bodies and electron-dense lysosomes (Figure 2f).

IV. RECEPTOR DOMAINS IN MOUSE PERITONEAL MACROPHAGES

A. DISTRIBUTION AND DYNAMICS OF SCAVENGER RECEPTORS

1. Interaction of Acetylated LDL with Macrophages

As first reported by Goldstein et al.,[31] incubation of cultured macrophages with acetylated LDL (AcLDL) particles results in massive intracellular accumulation of cholesterol esters. The receptor that mediates this uptake has been called the AcLDL receptor, or the scavenger cell receptor. Our present knowledge of this receptor is based mainly on biochemical studies.

Although a few morphological studies have been undertaken,[32-35] the endocytotic behavior of the scavenger receptor has remained insufficiently characterized. Therefore, we studied the surface distribution of AcLDL receptors and their endocytotic behavior in cultured mouse peritoneal macrophages in detail. Exposure of the macrophages to AcLDL-gold conjugates for 1 h at 4°C yielded preferential binding of the gold label in the intermediate zone of the plasma membrane surface (Figure 4a). Higher magnification of the plasma membrane surface of cells labeled at 4°C with AcLDL-gold revealed that the AcLDL receptors were generally diffusely distributed on the plasma membrane (Figure 4b).

When macrophages that had been exposed to AcLDL-gold conjugates at 4°C were washed with fresh medium and warmed to 37°C, a time-dependent redistribution of gold particles took place (Figures 4c-d). After 4 min at 37°C the proportion of gold complexes present in larger aggregates increased considerably (Figure 4c), and after 8 min at 37°C virtually all the gold particles were clustered in coated pits (Figure 4d). After 12 to 15 min the AcLDL-gold complexes had almost completely disappeared from the cell surface. That coalescence and endocytosis of gold-labeled AcLDL-receptor complexes takes place in this experiment can be confirmed using thin sectioning techniques.

Prolonged exposure of macrophages to unlabeled AcLDL for 12 h at 37°C caused the transformation of macrophages into foam cells which were characterized by large intracellular organelles, so-called "foamy organelles" 0.5 to 1.5 μm in diameter (Figures 5a, 8c). Using an enzyme cytochemical test we ascertained that they contained much acid phosphatase characteristic of lysosomes (Figure 5b). When cells which had been pretreated with unlabeled AcLDL for 12 h at 37°C were incubated with AcLDL-gold conjugates, a rapid accumulation of gold particles within foamy organelles was observed (Figure 5a).

The following conclusions can be drawn from these results:

1. AcLDL receptors are distributed diffusely on the intermediate region of the cell surface of cultured mouse peritoneal macrophages.
2. The binding of AcLDL molecules to the receptors induces long-range clustering of the occupied receptors into coated pits. These two results show that both the distribution and dynamics of AcLDL and LDL receptors differ markedly, because LDL receptors, in contrast to receptors for AcLDL, are always confined to presumptive coated pit regions, even prior to ligand binding, and do not migrate appreciably in the plane of the membrane (see section II. A. 3).
3. AcLDL is removed rapidly from the plasma membrane and is delivered via coated vesicles to lysosomes for degradation.

2. Acetylated LDL/Malondialdehyde-Modified LDL Competition Experiments and Double Labeling

Biochemical studies[36,37] have indicated that AcLDL and malondialdehyde-modified LDL (MDA-LDL) are endocytosed by the same scavenger receptor. However, our morphological findings using colloidal gold particles conjugated with these ligands showed that AcLDL receptors are dispersed, whereas MDA-LDL receptors are clustered on the plasma membrane

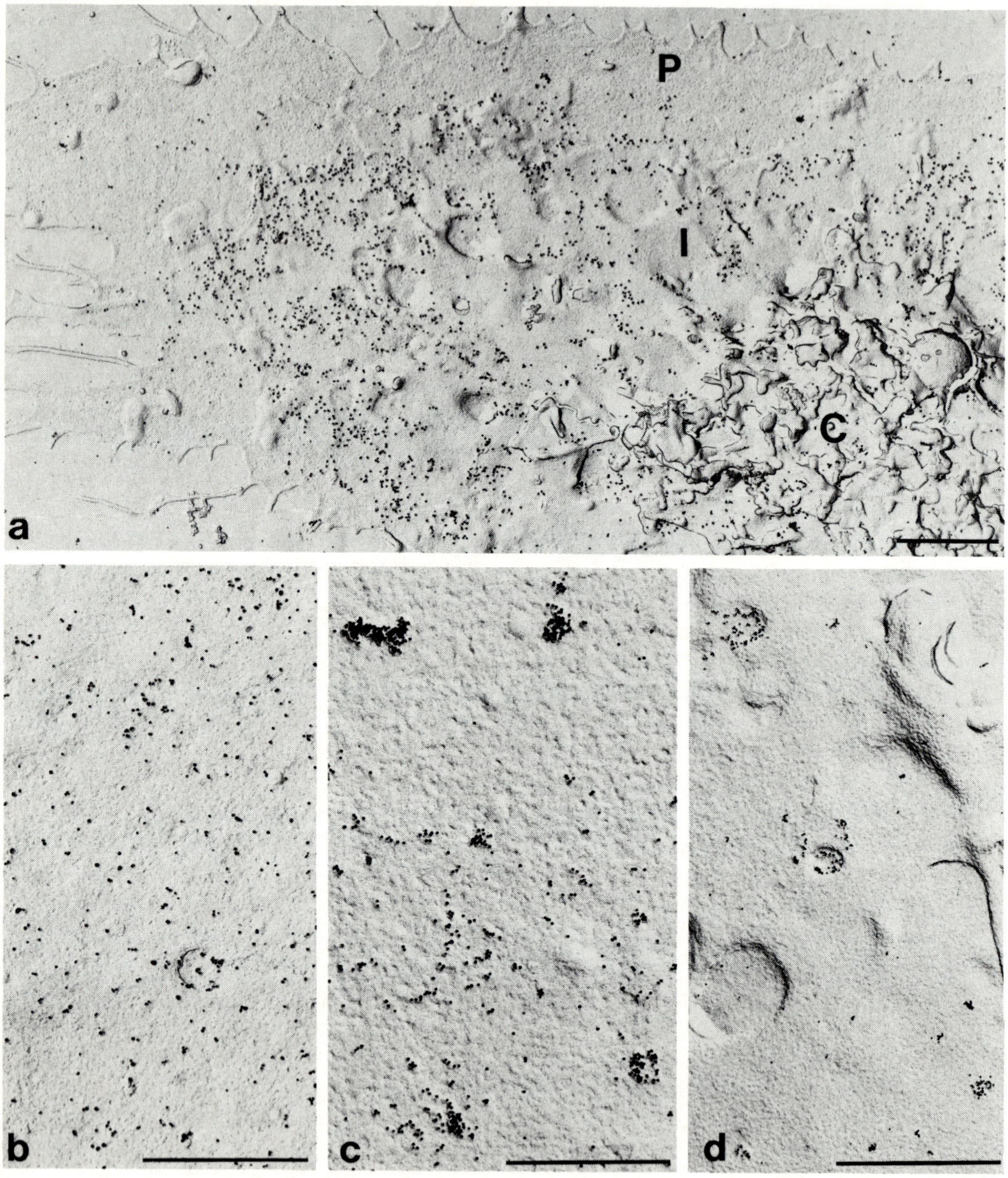

FIGURE 4. a) Low magnification surface replica of the plasma membrane of a cultured mouse peritoneal macrophage showing the distribution of AcLDL receptors after incubation with AcLDL-gold conjugates at 4°C for 1 h. b to d) Time-dependent aggregation of AcLDL receptors in cultured mouse peritoneal macrophages. b) AcLDL receptors in macrophages incubated for 1 h at 4°C with AcLDL-gold conjugate are diffusely distributed on the plasma membrane. c) After warming to 37°C for 4 min most gold marker particles are present in aggregates, and d) after 8 min at 37°C, almost all gold particles are clustered in coated pits. P = cell periphery, I = intermediate zone, C = central region. Bars: (a) 2 μm; (b to d) 1 μm.

of macrophages. This suggests that the two ligands are endocytosed by different receptors. However, it is not possible from these results alone to ascertain whether each type of modified lipoprotein has its own specific receptors, or whether the two lipoproteins are able to interact with the same binding sites. To address this question, we exposed cells first to unlabeled lipoproteins, and then to a second lipoprotein species labeled with gold. We also incubated

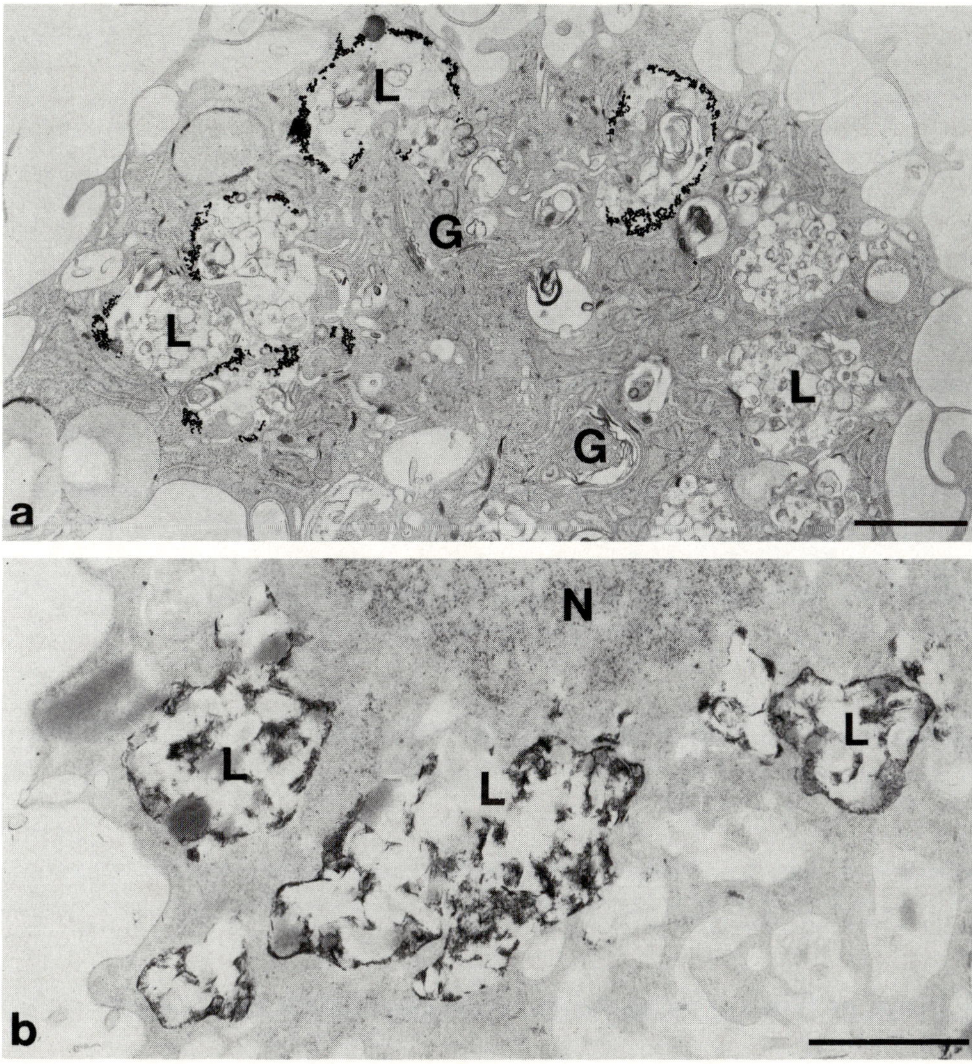

FIGURE 5. "Foamy organelles" (L) of macrophages exposed to unlabeled AcLDL for 12 h at 37°C. a) Cell pretreated with unlabeled AcLDL for 12 h at 37°C before exposure to AcLDL-gold conjugate for 30 min at 37°C, showing accumulated gold particles in foamy organelles. b) Acid - phosphatase reaction products in foamy organelles. G = Golgi apparatus, N = nucleus. Bars: 1 μm.

macrophages simultaneously with AcLDL and MDA-LDL conjugated to gold particles of different sizes. These double-labeling experiments allowed us to identify bound molecules of the two lipoproteins simultaneously, and hence to test the hypotheses of receptor specificity directly. We could also examine the spatial relationships between the receptors for both lipoprotein species. The experiments showed the presence of both types of lipoprotein-receptor complexes on the same membrane area and in common endocytotic vesicles (Figures 6a,b). Taken together, the results are consistent with the supposition that two distinct classes of receptor, one for each of the modified lipoprotein species, exist. Why this conclusion is at variance with the biochemical evidence needs to be clarified.

B. POSTEMBEDDING IMMUNOCYTOCHEMICAL LOCALIZATION OF β-VLDL RECEPTORS IN MACROPHAGES

A naturally occurring lipoprotein capable of causing cholesterol ester-accumulation in

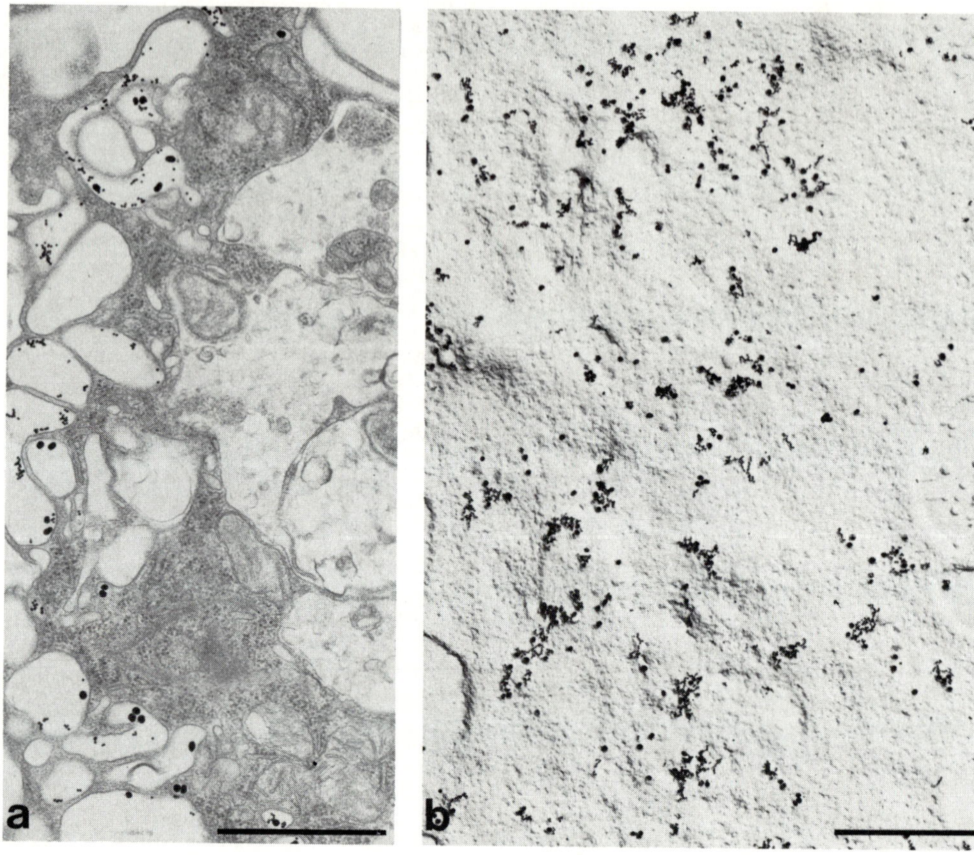

FIGURE 6. a) Thin section and b) surface replica of macrophages incubated with AcLDL-gold (40 nm gold particles) and MDA-LDL-gold (17 nm gold particles) showing simultaneous visualization of both types of lipoproteins on the same membrane surface. Bars: 1 μm.

macrophages is β-VLDL.[38] The two main apolipoprotein constituents of β-VLDL are apolipoprotein B-100 and apolipoprotein E. Studies on murine peritoneal macrophages have shown that neither of these apolipoproteins alone is sufficient to cause binding.[39] Altogether, these studies present a confused picture as to the specificity of the receptor-mediated uptake of β-VLDL. Therefore, we undertook a first study of the distribution of human β-VLDL on the cell surface and of the intracellular fate of human β-VLDL in cultured peritoneal mouse macrophages. For this purpose we used colloidal gold-labeled β-VLDL (Figure 1d) derived from patients with familial dysbetalipoproteinemia. In addition, we developed a postembedding immunoelectron microscopic procedure for optimal labeling of β-VLDL (i.e., apolipoprotein B) based on the protein A-gold technique.

Our results showed that β-VLDL-gold conjugates were specifically recognized by a cell surface receptor on the macrophages. The distribution of β-VLDL receptors and the uptake of β-VLDL are analogous to that for AcLDL via AcLDL receptors.

For immunocytochemical studies, macrophages were incubated with native β-VLDL at 4°C and 37°C for various periods of time. The immunoelectron microscope labeling experiments were carried out on thin sections of LR White, Epon 812, or Lowicryl K4M-embedded cells using affinity purified polyclonal rabbit antibodies against apolipoprotein B-100, in conjunction with the protein A-gold technique. Best labeling results were obtained with fixation at relatively high formaldehyde concentrations (4%), with a low concentration of glutaraldehyde (0.2%) added, and with thin sections of cells embedded in LR White resin.

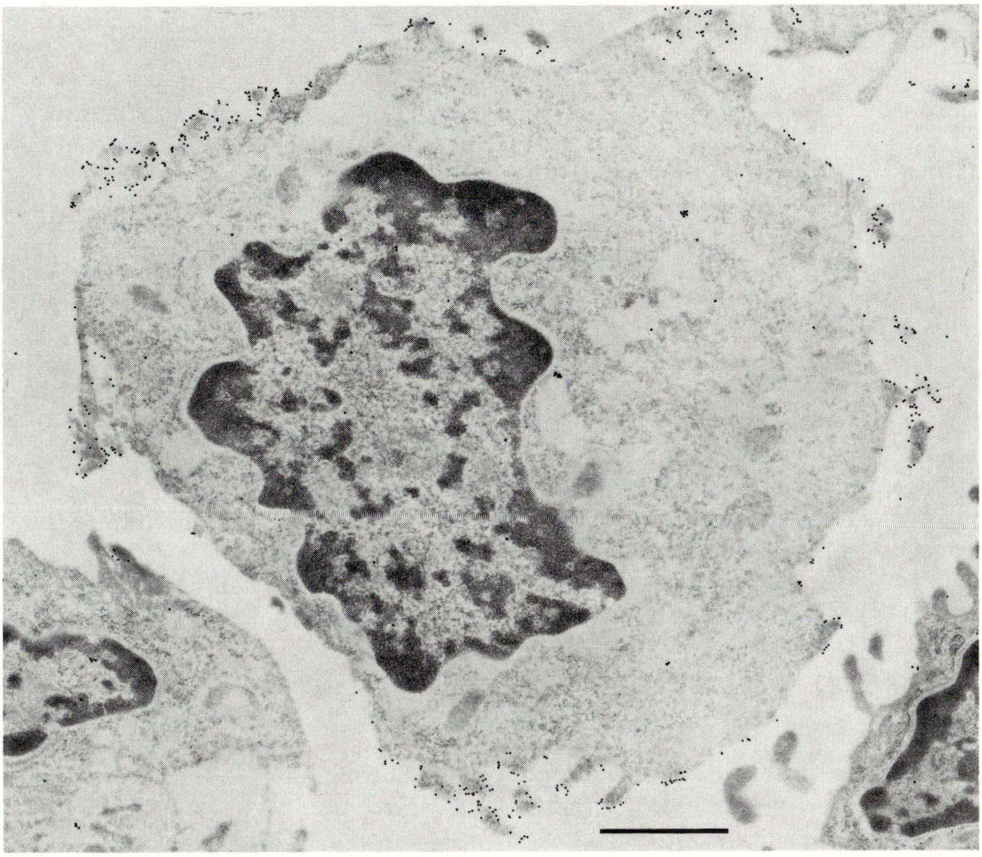

FIGURE 7. Immunocytochemical localization of β-VLDL receptors in the plasma membrane of a cultured mouse peritoneal macrophage following incubation with native β-VLDL for 1 h at 4°C and embeddment in LR White resin. β-VLDL was labeled using affinity purified polyclonal rabbit antibodies against apolipoprotein B-100 followed by protein A-gold. Bar: 1 μm.

In macrophages exposed to β-VLDL for 30 min at 4°C, immunolabel was limited, essentially, to the surface of the plasma membrane (Figure 7). Occasionally, marker particles were found concentrated in coated pits. When macrophages were warmed up to 37°C for 15 min, β-VLDL (i.e., apolipoprotein B) was found in organelles of the peripheral cytoplasm (i.e., endosomes). We conclude that the results obtained by indirectly visualizing β-VLDL were comparable to those obtained by probing with gold-labeled β-VLDL.

V. THE ROLE OF HDL RECEPTORS IN REVERSE CHOLESTEROL TRANSPORT

A. ELECTRON MICROSCOPIC EVIDENCE FOR RECEPTOR-MEDIATED ENDOCYTOSIS OF HDL IN MACROPHAGES

The regulation of cholesterol homeostasis in macrophages is clearly different from that in cells, such as fibroblasts, that bear apolipoprotein B, E (LDL) receptors. Macrophages are highly dependent on effective cholesterol acceptors to prevent their transformation into foam cells during times of increased cholesterol intake.[36,40] Among the most effective cholesterol acceptors are HDL molecules.[40]

Although the presence of HDL receptors has been demonstrated in various cells,[9-14,41-43] a direct link between the specific binding of HDL to its receptor, and HDL-mediated cholesterol transfer to or from cells, has not been firmly established. Furthermore, a consensus has not been reached as to whether HDL undergoes endocytosis, as do ligands known to be endocytosed via the LDL receptor pathway. Therefore, we tried to characterize the interaction of HDL with peritoneal macrophages and human monocytes in order to understand the mechanisms of HDL-macrophage interactions more precisely.

1. Direct Gold-Labeling Techniques

Using biochemical and morphological methods, we were able to show that apolipoprotein E-depleted HDL particles bind to specific surface receptors of mouse peritoneal macrophages.[19,44] Probing with gold-labeled human HDL, we also detected HDL receptors on the surface of autologous monocyte-derived macrophages. The macrophages were loaded with cholesterol ester by incubating with unlabeled AcLDL for 18 h and 24 h prior to chasing with gold-labeled HDL. The gold-HDL particles were found preclustered and randomly distributed on the cell surface, when the experiments were performed at 4°C (Figure 8a). Thus, the native pattern of HDL receptors appeared to be of an intermediate type between exclusive clustering (cf. LDL receptors) and diffuse distribution (cf. AcLDL receptors). Apolipoprotein A-I, the major HDL apolipoprotein is probably the ligand of this receptor.[45]

When the incubation temperature was increased to 37°C, HDL recognized by the receptor entered the macrophages via coated pits and coated vesicles (Figure 8b), indicating that the uptake of HDL in macrophages occurs by receptor-mediated endocytosis. Gold-labeled HDL particles were transported to endosomes which did not fuse with lysosomes (Figure 8c). The intracellular location of the HDL-gold conjugates was confirmed using ruthenium red staining. Ruthenium is a polycationic dye that binds specifically with glycosaminoglycans of the cell surface so that it can be used to distinguish surface from internal membranes.

2. Indirect Immunoperoxidase and Immunogold Techniques

In order to verify the results obtained with the direct gold-labeling technique, we employed indirect labeling with immunoperoxidase and immunogold. Macrophages were incubated with native HDL at 4°C and/or 37°C for various periods of time and subjected to an antibody recognizing apolipoprotein A-I. Then the cells were treated with peroxidase-conjugated goat anti-rabbit IgG (H + L)-F(ab′)$_2$ followed by reacting with diaminobenzidine. In macrophages incubated with native HDL for 1 h at 4°C, the immunoreaction products were diffusely distributed on the plasma membrane, and notably, particularly dense immunoprecipitates were observed in coated pits (Figure 9a). In macrophages warmed to 37°C, the immunolabel was found in endosomes (Figure 9b). For the intracellular localization of HDL the cells were permeabilized with 0.05% saponin. When protein A-gold was used instead of peroxidase-labeled secondary antibodies to visualize receptor-bound native HDL at 4°C, virtually the same results were obtained. After warming the cells to 37°C, the receptor-HDL complexes, which had been labeled with anti-apolipoprotein A-I and protein A-gold, were taken up in coated pits and coated vesicles and transported to endosomes. Addition of ruthenium red clearly showed that these endosomes were without connection to the extracellular space (Figure 9c). To summarize, indirect visualization of native HDL by these immunocytochemical techniques produced results comparable to those which were obtained using HDL coupled directly to colloidal gold.

B. DIFFERENTIATION OF THE MODES OF RECEPTOR-MEDIATED ENDOCYTOSIS OF HDL, AcLDL AND TRANSFERRIN USING DOUBLE LABELING

Cholesterol ester-laden macrophages, when chased with gold-labeled AcLDL, specifically shunt the AcLDL particles to the lysosomal compartment (Figure 5a). In contrast, these cells partition HDL exclusively to nonlysosomal endosomes (Figure 8c). To differentiate further between the AcLDL- and HDL-pathways in macrophages, cholesterol-laden mac-

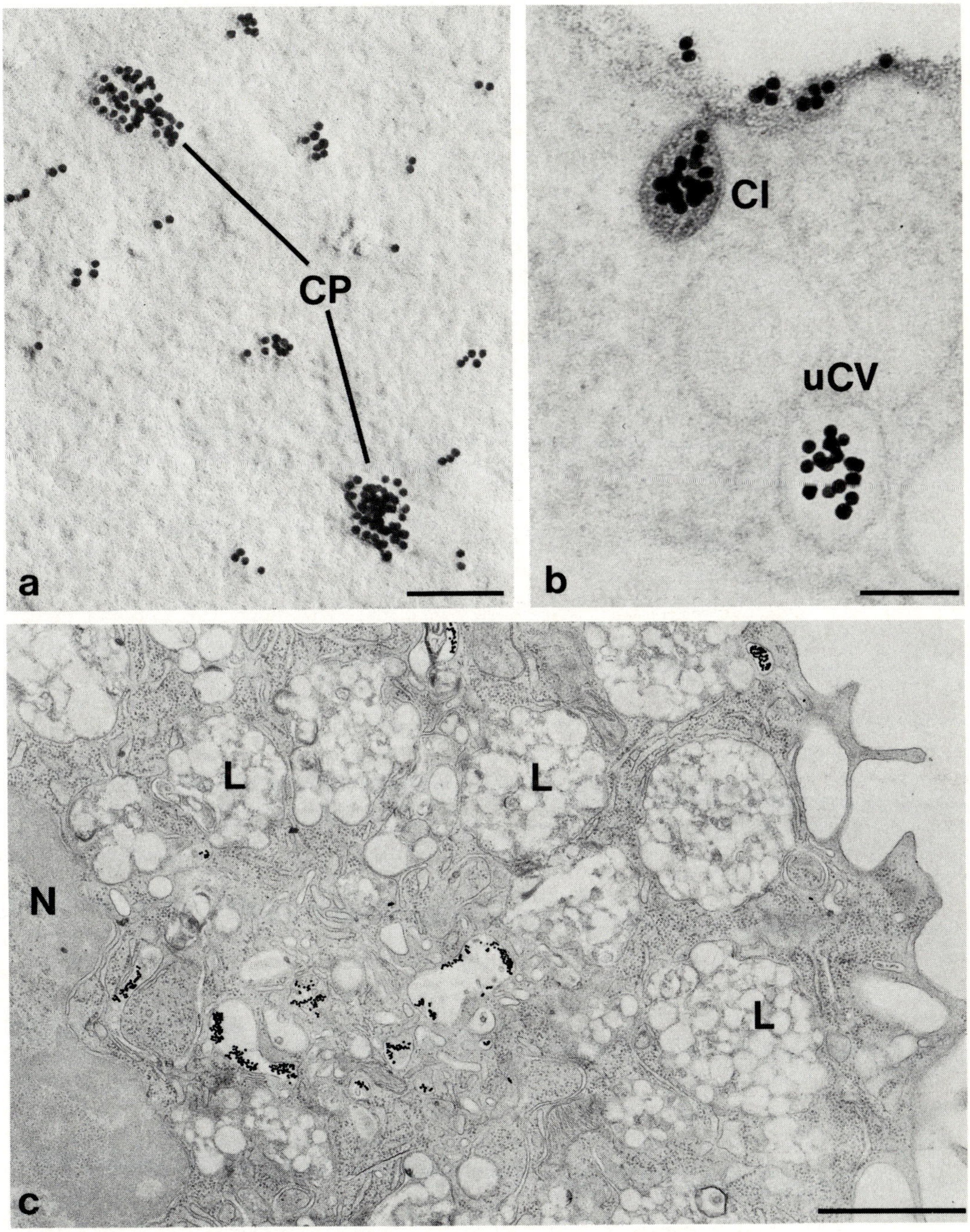

FIGURE 8. a,b) Localization of HDL receptors in cultured human monocyte-derived macrophages as revealed after incubation with HDL conjugated to colloidal gold for 1 h at 4°C. a) Surface replica depicting clustered HDL receptors over coated pits (CP) and randomly distributed solitary HDL receptors. b) Thin section showing HDL-gold complexes in a coated invagination (CI) and an uncoated vesicle (uCV) 10 min after warming the cell from 4 to 37°C. Ruthenium red staining of the glycocalyx indicates that the coated invagination communicates with the exterior, but that the uncoated vesicle does not. c) Mouse peritoneal macrophage after successive incubation with unlabeled AcLDL for 12 h at 37°C and with a HDL-gold conjugate for 30 min at 37°C. Internalized HDL-gold complexes are seen in an endosomal compartment but not in foamy vesicles (L). N = nucleus. Bars: (a) 0.2 μm; (b) 0.1 μm; (c) 1 μm.

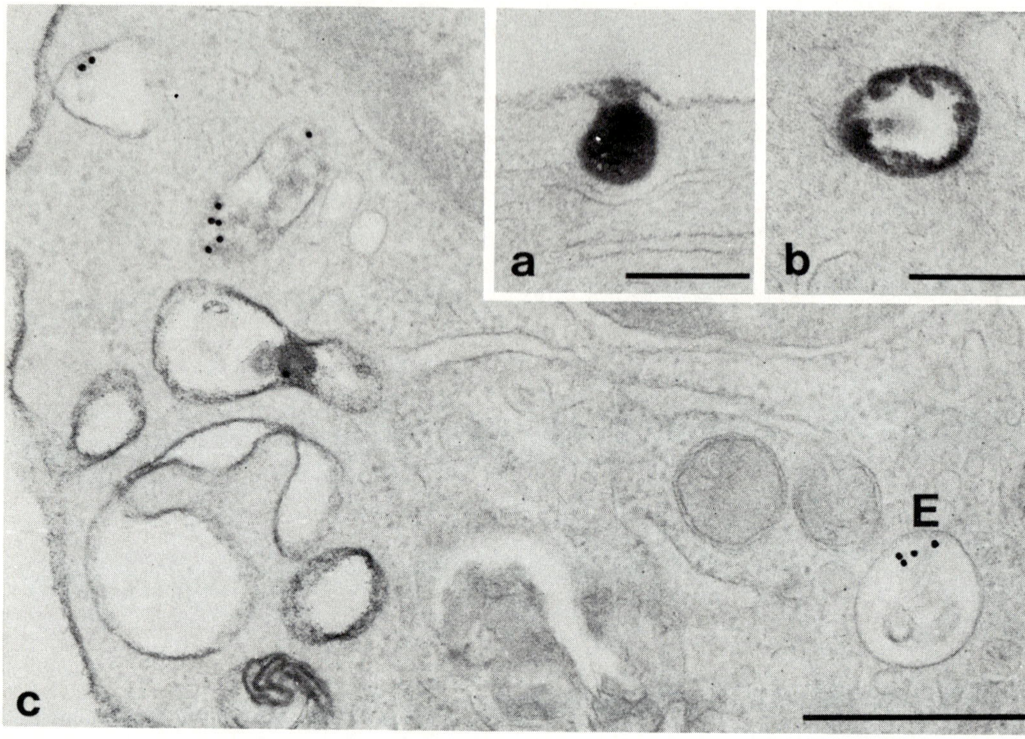

FIGURE 9. (a,b) Human monocyte-derived macrophages incubated with native HDL at 4°C (a) and 37°C (b), and with an antibody to apolipoprotein A-I. The cells were permeabilized with 0.05% saponin and treated with peroxidase-conjugated secondary antibodies, followed by reaction with diaminobenzidine. The peroxidase reaction produce is seen as a thin layer on the plasma membrane and concentrated in coated pits (a) and in endosomes (b). (c) Macrophage incubated with native HDL, anti-apolipoprotein A-I antibody and protein A-gold at 4°C before warming to 37°C. The gold-labeled HDL receptor-complexes are taken up in coated vesicles and transported to endosomes (E). Ruthenium red staining showed that endosomes do not communicate with the extracellular space.

rophages were sequentially incubated for 1 h at 37°C with AcLDL labeled with 16 nm gold particles, followed by HDL coupled to 40 nm gold particles. The AcLDL-gold complexes were found in foamy organelles, whereas the HDL-gold complexes appeared in nonlysosomal compartments (Figure 9d)

In another experiment, iron-saturated transferrin was used instead of AcLDL. Cholesterol-laden macrophages were sequentially incubated for 30 min with HDL conjugated to 40 nm gold particles and iron-saturated transferrin coupled to 16 nm gold particles. We chose transferrin as a probe because the endocytosis of transferrin contrasts with endocytosis of other ligands. In particular, transferrin is not degraded, but rather released intact from cells after endocytosis. Using transferrin instead of AcLDL, we could show that both transferrin and HDL are transported to the same endosomal compartments (Figure 9e). Thus, a similar pathway of transferrin and HDL in macrophages could be substantiated in our studies, a pathway which is clearly different from that of AcLDL.

C. RETROENDOCYTOSIS OF HDL AS REVEALED BY SURFACE REPLICATION

It appeared from both our biochemical and morphological experiments that HDL undergoes retroendocytosis. The time-course of HDL released from the macrophages can be followed using the surface replication technique. For this purpose, cholesterol-laden macrophages were treated with HDL-gold conjugates for 1 h at 37°C. The cells were then held

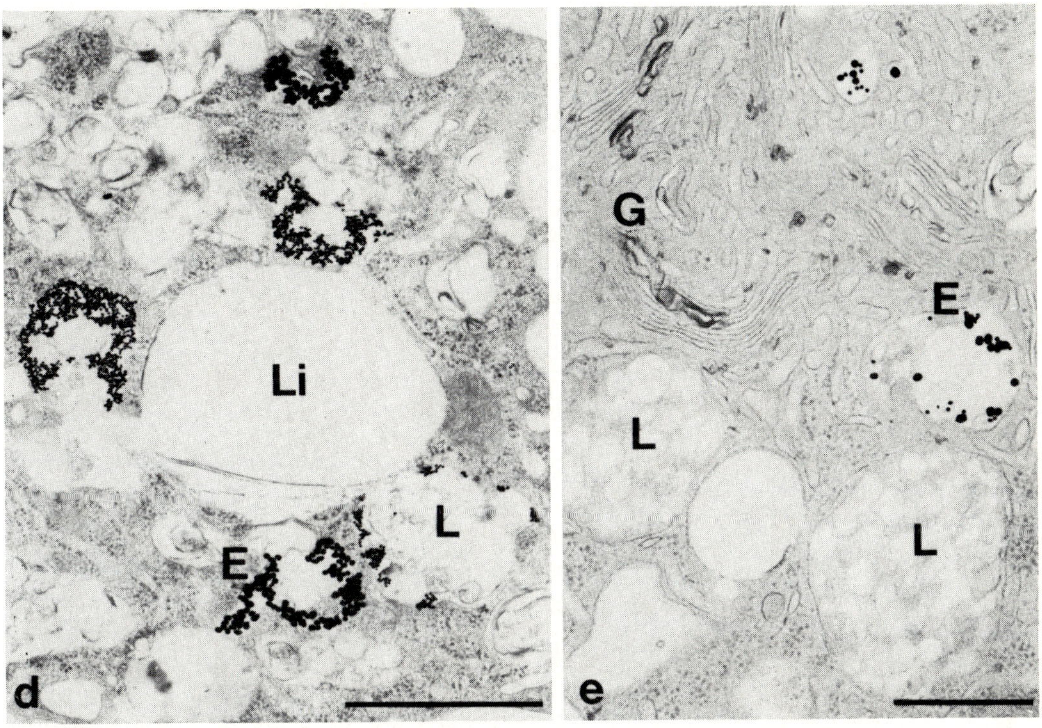

d) Thin section of a cholesterol-laden (AcLDL, 18 h, 37°C) mouse peritoneal macrophage treated with AcLDL coupled to 16 nm gold particles for 1 h at 37°C, chased in normal culture medium for 1 h at 37°C, and further incubated with HDL coupled to 40 nm gold particles. The respective markers denoting the lipoprotein types are segregated into distinct cellular compartments. Note that labeling is almost identical to that obtained when the cells are incubated separately with AcLDL and HDL-gold conjugates. e) Thin section of cholesterol-laden (AcLDL, 18 h, 37°C) mouse peritoneal macrophage after sequential exposure to HDL coupled to 40 nm gold particles and iron-saturated transferrin coupled to 16 nm gold particles (each 30 min, 37°C). Gold particles marking both types of ligand appear in the same endosomal compartment. E = endosome, G = Golgi apparatus, Ly = lysosome, Li = lipid droplet. Bars: (a,e) 0.5 μm; (b,c) 0.2 μm; (d) 1 μm.

in lipoprotein-free culture medium, and the loss of labeled complexes was followed. The receptor bound HDL-gold conjugates disappeared rapidly from the plasma membranes, and after 15 min of incubation in the lipoprotein-free culture medium, the membrane surfaces were almost devoid of gold label. After 1.5 h of incubation, gold particles reappeared on the macrophage surface (Figure 10a). When incubation was continued for up to 10 h, the gold particles increased in number (Figure 10b). We believe that HDL particles take up nonesterified cholesterol from lipid droplets, and that the cholesterol-enriched HDL particles are ultimately resecreted from the cells as native lipoproteins. The detailed mechanism of cholesterol removal and transfer from lipid droplets to the HDL particles during this retroendocytosis pathway remains unclear. In monocyte-derived macrophages of patients suffering from Tangier disease, resecretion of HDL is inoperative, and HDL is erroneously directed into lysosomes. This results in the storage of cholesterol by macrophages and various other cells. Since retroendocytosis is impaired, HDL is absent in Tangier patient plasma.[43] It should be noted that we recently demonstrated that cholesterol efflux from normal macrophages is also facilitated by an HDL-independent secretion of lamellar bodies, which are rich in cholesterol and of lysosomal origin.[14,46]

D. ASSEMBLY OF HDL AND APOLIPOPROTEIN E

The current concept of reverse cholesterol transport is based on the assumption that HDL not only takes up unesterified cholesterol, but also apolipoprotein E recognized by the hepatic

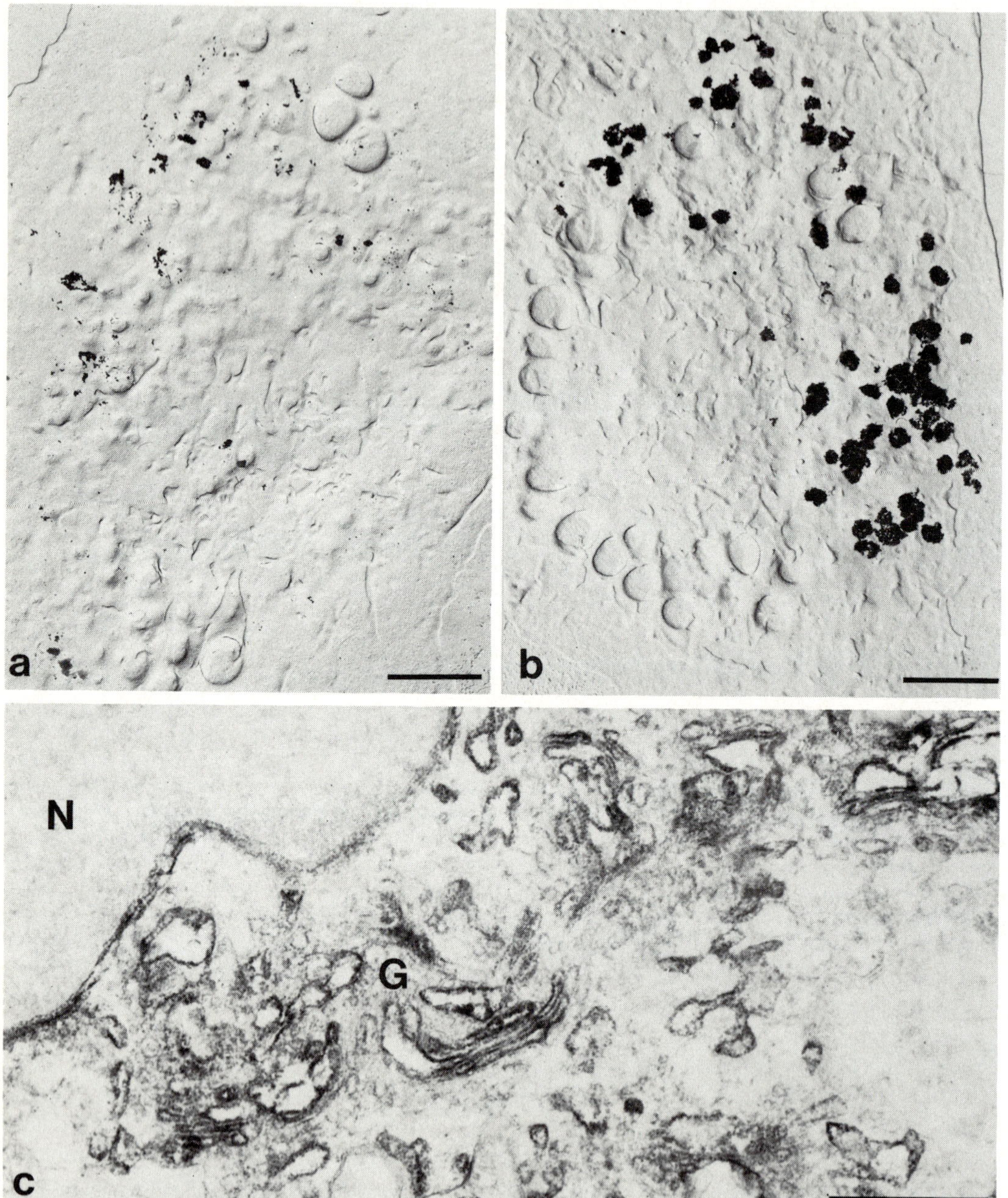

FIGURE 10. Cholesterol-laden (AcLDL, 48 h, 37°C) mouse peritoneal macrophages incubated with HDL-gold conjugate for 1 h at 37°C and chased with normal culture medium at 37°C showing resecreted gold-particles on the plasma membrane surface after a) 1.5 h and b) 5 h of chase. c) Intracellular localization of apolipoprotein E in a cholesterol-laden (AcLDL, 12 h, 37°C) human monocyte-derived macrophage. The cell was permeabilized with 0.05% saponin, incubated with an antibody to apolipoprotein E and treated with a peroxidase-conjugated secondary antibody, followed by reaction with diaminobenzidine. Peroxidase reaction products are found in the nuclear membrane, Golgi apparatus (G) and endoplasmic reticulum. N = nucleus. Bars: (a,b) 3 μm; (c) 0.5 μm.

apolipoprotein E receptor.[46] However, a consensus has not been reached as to whether HDL particles acquire the apolipoprotein E molecules during intracellular processing, or extracellularly after secretion of apolipoprotein E. Simultaneous identification of both apolipoprotein E and HDL using double label experiments is germane to achieving a unified understanding of the assembly of HDL and apolipoprotein E. In our laboratory we addressed this problem by localizing apolipoprotein E in macrophages. Saponin-permeabilized cells were subjected to antibodies recognizing apolipoprotein E, followed by peroxidase-labeled secondary antibodies and visualization with diaminobenzidine. Our preliminary results show that in normal macrophages apolipoprotein E was present in high amounts in the nuclear membrane, rough endoplasmic reticulum, and in the Golgi apparatus (Figure 10c). Experiments to colocalize HDL (using cells pretreated with HDL-gold conjugates at 4°C and 37°C for various periods of time) and apolipoprotein E, with immunoperoxidase techniques, are in progress.

VI. CONCLUSIONS AND PERSPECTIVES

The basic architecture of biological membranes is similar in all cells, but certain specific features of membranes, such as their protein and lipid constituents, vary considerably. These differences may suffice to partition a given membrane into topographically distinct areas with unique local functional properties. The distribution of particular protein species, receptors in particular, on the surface of cultured cells, is far from uniform. Therefore, investigations on the complement and distribution of receptors are of great significance for determining how receptors modulate cellular activities.

The increased interest in studying cell surface receptors in recent years has sparked the improvement of morphological methods, and the development of visual probes for localizing receptor sites in the electron microscope. The studies I have summarized show that the application of affinity and antibody labeling and surface replication can aid our understanding of cell function as related to membrane receptors and their dynamics. We have used various lipoprotein receptors as model systems. I have illustrated how gold probes of different sizes can be employed to detect two receptor species simultaneously, and how experiments can be designed to differentiate between different receptor subtypes and domains on a given cell membrane and in various types of cells.

The surface replication technique is extremely potent and can be employed to study many diverse features of diverse receptor systems. Surface replication, though well suited for analyzing the distribution of receptors and their dynamics at the cell surface, cannot be employed to illuminate the intracellular pathway of ligands and receptors after endocytosis. Although a plethora of data on the endocytosis and intracellular fate of different ligands has been obtained by applying thin sectioning techniques, direct information on the intracellular fate of receptors, notably virtually all lipoprotein receptors, is lacking almost completely, because few suitable antibodies against receptors have been produced to date. It can be hoped that more antibodies against the various lipoprotein receptors will become available, and that their use in connection with immunocytochemical techniques will soon extend our knowledge on the intracellular pathways of lipoprotein receptors.

ACKNOWLEDGMENTS

I wish to thank my collaborators B. Harrach and R. Semich, who generously contributed their data and their ideas to this review. I also gratefully acknowledge the excellent assistance of K. Schlattmann and S. Passmann. The studies on lipoprotein receptors were carried out in collaboration with Dr. G. Schmitz (Münster). Antibodies against apolipoprotein B-100 were kindly provided by Dr. V. Armstrong (Göttingen). Dr. D. Troyer (Münster) critically read the manuscript.

Much of the work presented here was supported by the *Minister für Wissenschaft und Forschung des Landes NRW* and by the *Deutsche Forschungsgemeinschaft* (SFB 223 and 310). The author is a Heisenberg Fellow supported by the *Deutsche Forschungsgemeinschaft.*

REFERENCES

1. **Goldstein, J. L., Brown, M. S., Anderson, R. G. W., Russell, D. W., and Schneider, W. J.,** Receptor-mediated endocytosis: Concepts emerging from the LDL receptor system, *Ann. Rev. Cell Biol.,* 1, 1, 1985.
2. **Geuze, H. J., Slot, J. W., and Strous, G. J. A.,** Intracellular site of asialoglycoprotein-receptor-ligand uncoupling: Double-label immunoelectron microscopy during receptor-mediated endocytosis, *Cell,* 32, 277, 1983.
3. **Krupp, M. N. and Lane, M. D.,** Evidence for different pathways for the degradation of insulin and insulin receptor in the chick liver cell, *J. Biol. Chem.,* 257, 1372, 1982.
4. **Dunn, W. A. and Hubbard, A. L.,** Receptor-mediated endocytosis of epidermal growth factor by hepatocytes in the perfused rat liver: Ligand and receptor dynamics, *J. Cell Biol.,* 98, 2148, 1984.
5. **Dautry-Varsat, A., Ciechanover, A., and Lodish, H. F.,** pH and the recycling of transferrin during receptor-mediated endocytosis, *Proc. Natl. Acad. Sci. U.S.A.,* 80, 2258, 1983.
6. **Gianturco, S. H., Via., D. P., Gotto, A. M., and Bradley, W. A.,** Lipoprotein recpeptors, in *The Role of Receptors in Biology and Medicine,* Gotto, A. M. an O'Malley, B. W., Eds., Raven Press, New York, 1986, 167.
7. **Assmann, G.,** *Lipid Metabolism and Atherosclerosis,* F. K. Schattauer Verlag, Stuttgart, Germany, 1982.
8. **Rifai, N.,** Lipoproteins and apolipoproteins, *Arch. Pathol. Lab. Med.,* 110, 694, 1986.
9. **Robenek, H. and Severs, N. J.,** Double labeling of lipoprotein receptors in fibroblast cell surface replicas, *J. Ultrastruct. Res.,* 87, 149, 1984.
10. **Eisenberg, S.,** High density lipoprotein metabolism, *J. Lipid Res.,* 25, 1017, 1984.
11. **Murakami, M., Horiuchi, S., Takata, K., and Morino, Y.,** Distinction in the mode of receptor-mediated endocytosis between high density lipoprotein and acetylated high density lipoprotein: Evidence for high density lipoprotein receptor-mediated cholesterol transfer, *J. Biochem.,* 101, 729, 1987.
12. **Robenek, H. and Schmitz, G.,** Receptor domains in the plasma membrane of cultured mouse peritoneal macrophages, *Eur. J. Cell Biol.,* 39, 77, 1985.
13. **Schmitz, G., Robenek, H., Lohmann, U., and Assmann, G.,** Interaction of high density lipoproteins with cholesteryl ester-laden macrophages: Biochemical and morphological characterization of cell surface receptor binding, endocytosis and resecretion of high density lipoproteins by macrophages, *EMBO J.,* 4, 613, 1985.
14. **Robenek, H. and Schmitz, G.,** Ca^{++}-antagonists and ACAT-inhibitors promote cholesterol efflux from macrophage by different mechanisms. II. Characterization of intracellular morphological changes, *Arteriosclerosis,* 8, 57, 1988.
15. **Handley, D. A., Arbeeny, C. M., Eder, H. A., and Chien, S.,** Hepatic binding and internalization of low density lipoprotein-gold conjugates in rats treated with 17α-ethinylestradiol, *J. Cell Biol.,* 90, 778, 1981.
16. **Anderson, R. G. W., Goldstein, J. L., and Brown, M. S.,** Localization of low density lipoprotein receptors on plasma membrane of normal human fibroblasts and their absence in cells from familial hypercholesterolemia homozygote, *Proc. Natl. Acad. Sci. U.S.A.,* 73, 2434, 1976.
17. **Handley, D. A., Arbeeny, C. M., Witte, L. D., and Chien, S.,** Colloidal gold-low density lipoprotein conjugates as membrane receptor probes, *Proc. Natl. Acad. Sci. U.S.A.,* 78, 368, 1981.
18. **Robenek, H.,** Distribution and mobility of receptors in the plasma membrane, in *Freeze-Fracture Studies of Membrane Structure,* Hui, S. W., Ed., CRC Press, Boca Raton, FL, in press.
19. **Anderson, R. G. W., Brown, M. S., Beisiegel, U., and Goldstein, J. L.,** Surface distribution and recycling of the low density lipoprotein receptor as visualized with antireceptor antibodies, *J. Cell Biol.,* 93, 523, 1982.
20. **Severs, N. J. and Robenek, H.,** Detection of microdomains in biomembranes. An appraisal of recent developments in freeze-fracture cytochemistry, *Biochim. Biophys. Acta,* 737, 373, 1983.
21. **Goldstein, J. L., Anderson, R. G. W., and Brown, M. S.,** Coated pits, coated vesicles and receptor-mediated endocytosis, *Nature,* 279, 679, 1979.
22. **Steer, C. J., Bisher, M., Blumenthal, R., and Steven, A. C.,** Detection of membrane cholesterol by filipin in isolated rat liver coated vesicles is dependent upon removal of the clathrin coat, *J. Cell Biol.,* 99, 315, 1984.

23. **Robenek, H., Hesz, A., and Rassat, J.,** The variability of the topography of low density lipoprotein (LDL) receptors in the plasma membrane of cultured human skin fibroblasts as revealed by gold-LDL conjugates in conjunction with surface replicas, *J. Ultrastruct. Res.*, 82, 143, 1983.
24. **Heuser, J.,** Three-dimensional visualization of coated vesicle formation in fibroblasts, *J. Cell Biol.*, 84, 560, 1980.
25. **Nicol, A., Nermut, M. V., Doeinck, A., Robenek, H., Wiegand, Ch., and Jockusch, B. M.,** Labeling of structural elements at the ventral plasma membrane of fibroblasts with the immunogold technique, *J. Histochem. Cytochem.*, 35, 499, 1987.
26. **McArdle, H. J. and Morgan, E. H.,** The effect of monoclonal antibodies to the human transferrin receptor on transferrin and iron uptake by rat and rabbit reticulocytes, *J. Biol. Chem.*, 259, 1629, 1984.
27. **Gebhardt, R. and Robenek, H.,** Ligand-dependent redistribution of the IgA receptor on cultured rat hepatocytes and its disturbance by cytochalasin B, *J. Histochem. Cytochem.*, 35, 301, 1987.
28. **Anderson, R. G. W., Brown, M. S., and Golstein, J. L.,** Role of the coated endocytic vesicle in the uptake of receptor-bound low density lipoprotein in human fibroblasts, *Cell,* 10, 351, 1977.
29. **Bretscher, M. S.,** Directed lipid flow in cell membranes, *Nature,* 260, 21, 1976.
30. **Robenek, H. and Hesz, A.,** Dynamics of low density lipoprotein receptors in the plasma membrane of cultured human skin fibroblasts as visualized by colloidal gold in conjunction with surface replicas, *Eur. J. Cell Biol.*, 31, 275, 1983.
31. **Goldstein, J. L., Ho, Y. K., Basu, S. K., and Brown, M. S.,** Binding site on macrophages that mediates uptake and degradation of acetylated low density lipoprotein, producing massive cholesterol deposition, *Proc. Natl. Acad. Sci. U.S.A.*, 76, 333, 1979.
32. **Fukuda, S., Horiuchi, S., Tomita, K., Murakami, M., Morino, Y., and Takahasu, K.,** Acetylated low-density lipoprotein is endocytosed through coated pits by rat peritoneal macrophages, *Virchows Arch. B: (Cell Pathol.),* 52, 1, 1986.
33. **Brown, M. S., Goldstein, J. L., Krieger, M., Ho, Y. K., and Anderson, R. G. W.,** Reversible accumulation of cholesteryl esters in macrophages incubated with acetylated lipoproteins, *J. Cell Biol.*, 82, 597, 1979.
34. **Robenek, H., Schmitz, G., and Assmann, G.,** Topography and dynamics of receptors for acetylated and malondialdehyde-modified low density lipoproteins in the plasma membrane of mouse peritoneal macrophages as visualized by colloidal gold in conjunction with surface replicas, *J. Histochem. Cytochem.*, 32, 1017, 1984.
35. **Mommaas-Kienhuis, A. M., Nagelkerke, J. F., Vermeer, B. J., Daems, W.Th., and Van Berkel, Th. J. C.,** Vizualization of the interaction of native and modified low density lipoproteins with isolated rat liver cells, *Eur. J. Cell Biol.*, 38, 42, 1985.
36. **Brown, M. S. and Goldstein, J. L.,** Lipoprotein metabolism in the macrophage: Implications for the cholesterol deposition in atherosclerosis, *Ann. Rev. Biochem.*, 52, 223, 1983.
37. **Shechter, J., Fogelman, A. M., Haberland, M. E., Seager, J., Hokom, M., and Edwards, P. A.,** The metabolism of native and malondialdehyde-altered low density lipoprotein by human monocyte-macrophages, *J. Lipid Res.*, 22, 63, 1981.
38. **Gianturco, S. H., Brown, S. A., Via, D. P., and Bradly, W. A.,** The β-VLDL receptor pathway of murine P388 D_1 macrophages, *J. Lipid Res.*, 27, 412, 1986.
39. **Innerarity, T. L., Pitas, R. E., and Mahley, R. W.,** Modulating effects of canine high density lipoproteins in cholesteryl ester synthesis induced by β-VLDL in macrophages, *Arteriosclerosis,* 2, 114, 1982.
40. **Ho, Y. K., Brown, M. S., and Goldstein, J. L.,** Hydrolysis and excretion of cytoplasmic cholesteryl esters by macrophages: Stimulation by high density lipoproteins and other agents, *J. Lipid Res.*, 21, 391, 1980.
41. **Rajan, V. P. and Menon, K. M. J.,** Metabolism of high-density lipoproteins in cultured rat luteal cells, *Biochim. Biophys. Acta,* 921, 25, 1987.
42. **Brinton, E. A., Kenagy, Oram, J. F., and Bierman, E. L.,** Regulation of high density lipoprotein binding activity of aortic endothelial cells by treatment with acetylated low density lipoprotein, *Arteriosclerosis,* 5, 329, 1985.
43. **Schmitz, G., Assmann, G., Robenek, H., and Brennhausen, B.,** Tangier disease: A disorder of intracellular membrane traffic, *Proc. Natl. Acad. Sci. U.S.A.*, 82, 6305, 1985.
44. **Schmitz, G., Niemann, R., Brennhausen, B., Krause, R., and Assmann, G.,** Regulation of high density lipoprotein receptors in cultured macrophages: Role of acyl Co-A: cholesterol acyl-transferase, *EMBO J.*, 4, 2773, 1985.
45. **Schmitz, G., Robenek, H., and Assmann, G.,** Role of the high density lipoprotein-receptor cycle in macrophage-cholesterol metabolism, in *Atherosclerosis Reviews,* Gotto, A., Ed., Vol. 16, Raven Press, New York, 1987, 95.
46. **Schmitz, G., Robenek, H., Beuck, M., Krause, R., Schureck, A., and Niemann, R.,** Ca^{++}-antagonists and ACAT-inhibitors promote cholesterol efflux from macrophages by different mechanisms. I. Characterization of cellular lipid metabolism, *Arteriosclerosis,* 8, 46, 1988.

Chapter 9

ENZYMATIC LABELING OF CELL-SURFACE GLYCOPROTEINS AS AN AUTORADIOGRAPHIC PROBE FOR MORPHOMETRIC ANALYSIS OF INTRACELLULAR MEMBRANE TRAFFIC

Lutz Thilo and Chantal de Chastellier

TABLE OF CONTENTS

I. INTRODUCTION

The endomembranes of a cell behave as a homeostatic system. An electron microscope picture of a random sample of cell cross-sections contains no time-related information as far as the membrane structures are concerned. This implies that even a time-series of such pictures cannot provide information on ''membrane dynamics''. In order to study the dynamic behavior of cell membranes, a state of nonequilibrium must be introduced for an observable parameter. A time series of electron micrographs, evaluating the restoration to steady state for this parameter, may then allow an insight into the underlying dynamic process.

A state of nonequilibrium can be introduced into the endomembrane system in two ways. First, the physiological state of the cell can be affected by the addition of drugs, or by changing culture conditions such as temperature, acidity, osmotic pressure, etc. Obviously, any such changes will have to be reversible, so that the subsequent return to steady state can be observed. This approach allows one to look at the behavior of the membrane as a morphological entity. Second, a membrane marker can be introduced into one or more endomembrane compartments. This will allow us to follow the redistribution of such a marker to other compartments until a steady state has been established. The kinetics of this process may then reveal the dynamics of membrane traffic involved. A special case of this approach would be a pulse-chase experiment where the marker is exported from the cell or from the structure under consideration. Depending on the type of marker that is used, the results of such a study may not be valid for the membrane as a whole and at all stages.

The present chapter will describe the use of a radioactive membrane marker for the autoradiographic analysis of membrane traffic during endocytosis. This method has been successful in obtaining quantitative information on the flow rates, pathways, and on the relative compositions of the endomembrane structures involved, in the case of bone marrow-derived macrophages and *Dictyostelium discoideum* amoeboid cells.

II. ENZYMATIC LABELING OF PLASMA MEMBRANE GLYCOPROTEINS

In order to track the entire roundtrip of internalized plasma membrane through the cell, a covalently bound membrane marker is required. Such a marker is provided by the terminal glycoslyation of plasma membrane glycoconjugates, as schematically represented in Figure 1.[1] This label has provided the first biochemical method to quantify membrane internalization and recycling: internalized label is no longer accessible to enzymatic release, and can therefore be distinguished quantitatively from label remaining on the cell surface; previously internalized label again becomes accessible to enzymatic release when it is recycled back to the cell surface.

The binding of radioactive galactose to cell-surface constituents provides a marker that can be used for direct electron microscopic autoradiographic observations of membrane traffic. The efficiency of labeling depends on the availability on the cell surface of terminal N-acetylglucosamine moieties. Additional labeling sites can be made available by a pre-treatment with neuraminidase and β-galactosidase. Such a pretreatment can increase the labeling intensity by 20-fold without modifying the kinetics of subsequent internalization of labeled molecular species.[2] The label provides a suitable membrane marker according to the following criteria:

1. More than 20 different molecular species (M_r) are labeled in various cell types.[3-7]
2. All labeled species are internalized to the same relative extent.[3,6]
3. According to its resistance to release from the membrane by high pH, high salt concentration, or mercaptoethanol, more than 95% of label is bound to integral membrane proteins (unpublished results).

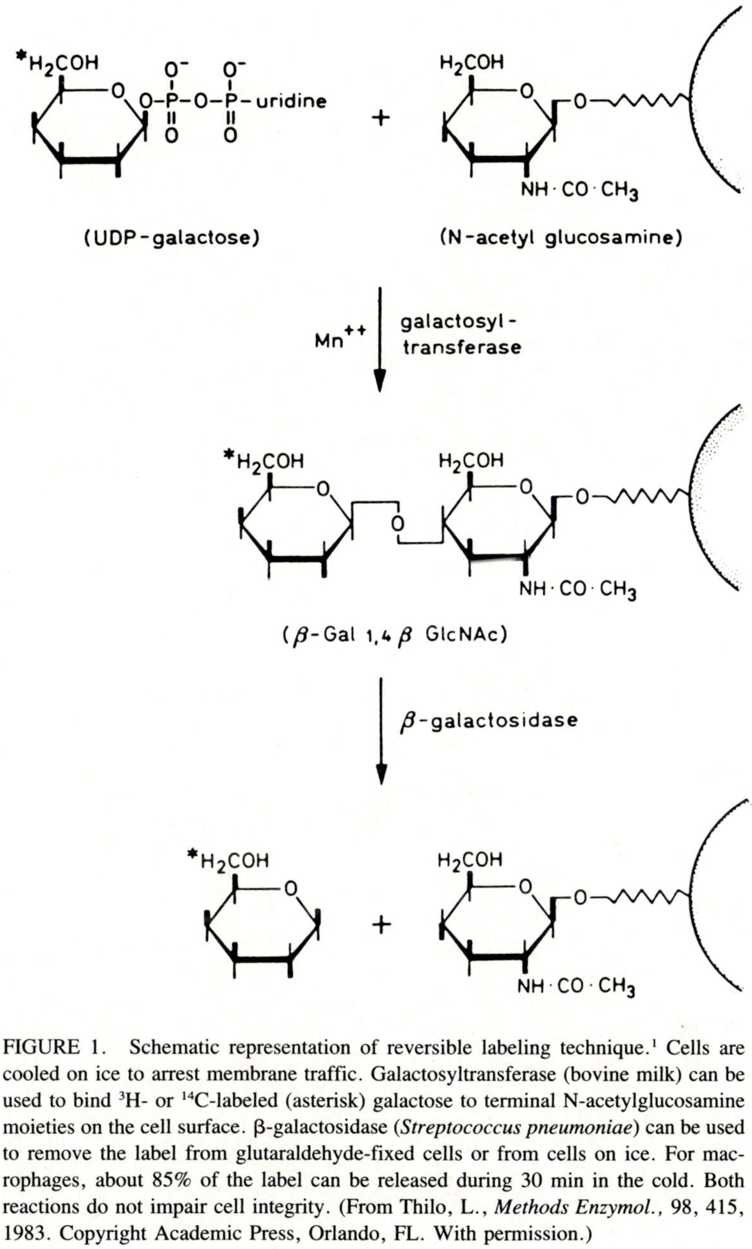

FIGURE 1. Schematic representation of reversible labeling technique.[1] Cells are cooled on ice to arrest membrane traffic. Galactosyltransferase (bovine milk) can be used to bind ^{3}H- or ^{14}C-labeled (asterisk) galactose to terminal N-acetylglucosamine moieties on the cell surface. β-galactosidase (*Streptococcus pneumoniae*) can be used to remove the label from glutaraldehyde-fixed cells or from cells on ice. For macrophages, about 85% of the label can be released during 30 min in the cold. Both reactions do not impair cell integrity. (From Thilo, L., *Methods Enzymol.*, 98, 415, 1983. Copyright Academic Press, Orlando, FL. With permission.)

4. Label is lost into the medium according to turnover times normally found for membrane proteins, and involves all M_r-species to the same extent.[3,4]
5. Cell associated label remains fully membrane bound.[3]
6. Label is not modified as judged by its constant susceptibility to removal by β-galactosidase and neuroaminidase.[1,3]

When cells are fixed immediately after labeling on ice, autoradiographs clearly show that the label is only on the cell surface, Figure 2A. In comparison, when cells are fixed after a period of postlabeling membrane internalization, a significant fraction of grains can be found in association with endomembrane compartments, Figure 2B.

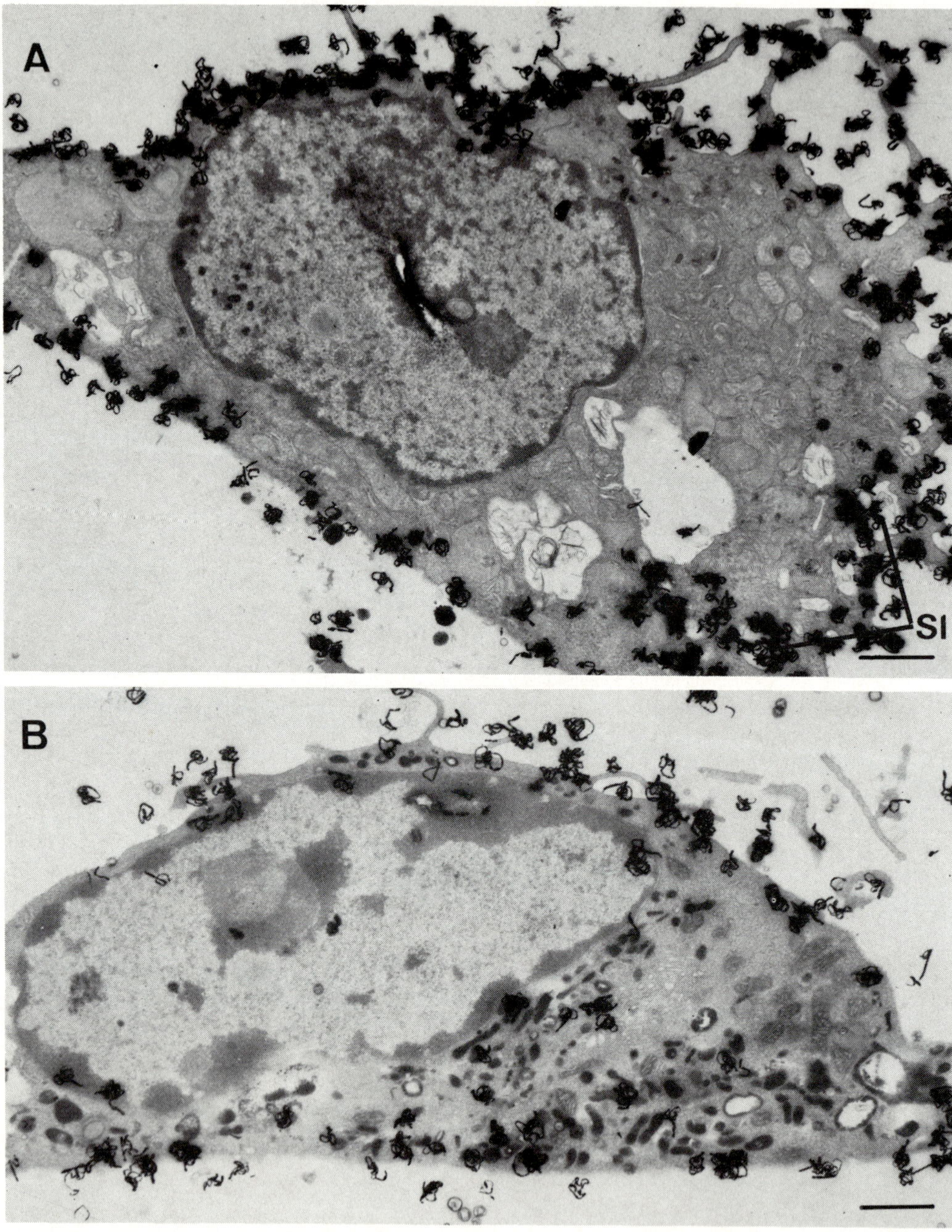

FIGURE 2. Overexposed electron microscope autoradiographs (8 months exposure) of bone marrow derived macrophages labeled with (^{3}H)galactose. A: Distribution of label for cells fixed immediately after labeling on ice, prior to warming; the silver grains are associated with the plasma membrane. B: Distribution of label after 30 min of postlabeling membrane flow; many grains are still associated with the cell surface, but a substantial fraction of the label is associated with intracellular membranes. SI: surface invaginations. Bars = 1 μm. (From de Chastellier, C., *Eur. J. Cell Biol.*, 44, 112, 1987. Copyright Wissenschaftliche Verlagsgesellschaft mbH, Stuttgart. With permission.)

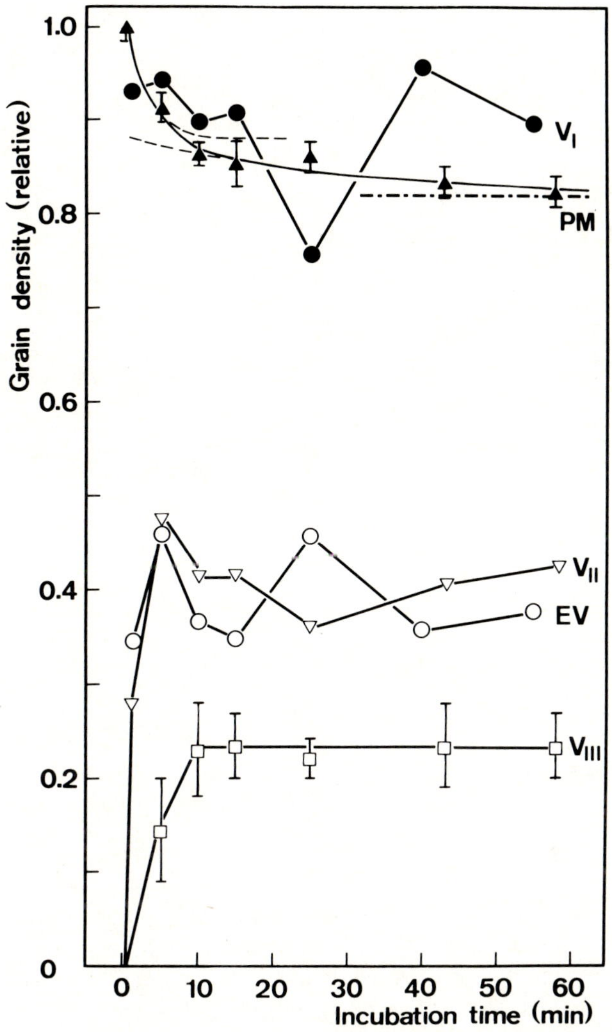

FIGURE 3. Kinetics of relative labeling densities of endocytic membrane compartments during pinocytosis in bone marrow-derived macrophages: plasma membrane (PM), coated vesicles (V_I), endosomes (V_{II}), empty vesicles (EV), and secondary lysosomes (V_{III}). Grain density is given by the ratio of the number of autoradiographic grains per membrane area. Values are the average (error: mean variation) for two independent experiments. For the PM, the grain density at $t = 0$ has been normalized to 1 (and all other values accordingly), reflecting initial cell-surface labeling (cf. Figure 2A.) Correspondingly, the other time poins reflect a loss of label from the PM to intracellular pools (area of PM remaining constant): the curve as indicated is given by Equation 4 (Appendix), with $\lambda_1 = 0.28$ min^{-1}, $\lambda_2 = 0.033$ min^{-1}, $\rho_1 = 0.88$, and $\rho_2 = 0.82$. The dotted lines indicate the two stages of the biphasic behavior. (From de Chastellier, C., *Eur. J. Cell Biol.*, 44, 112, 1987. Copyright Wissenschaftliche Verlagsgesellschaft mbH, Stuttgart. With permission.)

III. QUANTITATIVE ANALYSIS OF MEMBRANE TRAFFIC

A. KINETICS OF MEMBRANE INTERNALIZATION

When macrophages have been labeled on their plasma membrane, subsequent membrane internalization and recycling leads to a redistribution of label from the cell surface to intracellular membranes. This process is observed by measuring the grain density on the plasma membrane as a function of time, Figure 3 (▲). As is indicated by the dotted lines, the redistribution occurs with biphasic kinetics, which can be expressed by the sum of two exponential functions. This behavior has suggested intracellular compartmentation of label.

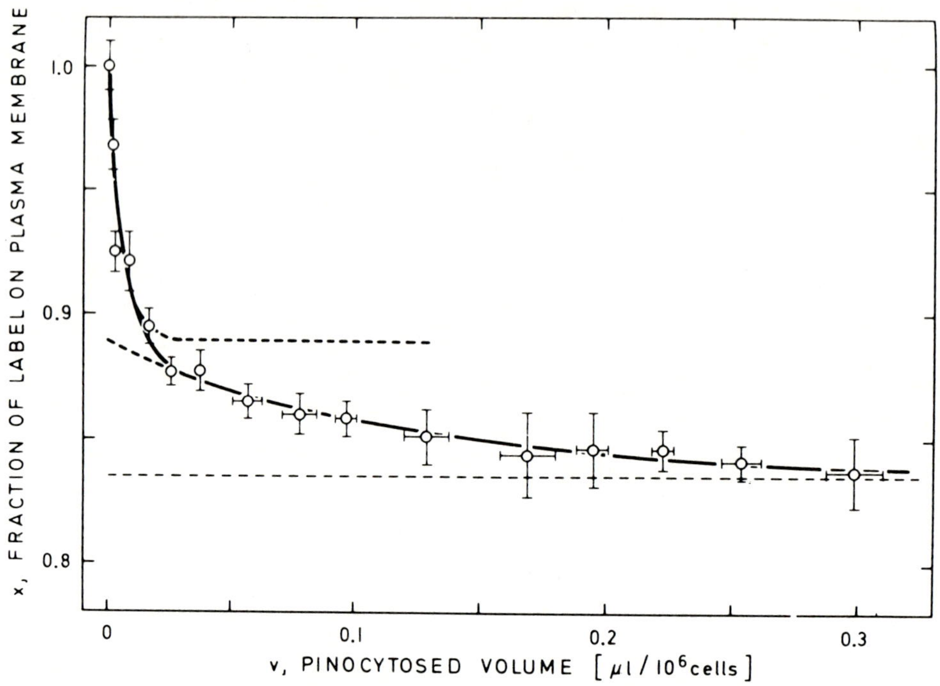

FIGURE 4. Redistribution of labeled plasma membrane glycoconjugates between the cell surface and intracellular membranes, as a function of fluid-phase pinocytosis in cells of the macrophage cell line, $P388D_1$. The curve, indicating biphasic kinetics, is described by the sum of two decaying exponentials. Error bars indicate the mean standard error as measured from five independent experiments. The kinetics can accomodate a model of membrane flow as shown in Figure 9. (From Thilo, L., *Methods Enzymol.*, 98, 415, 1983. Copyright Academic Press, Orlando, FL. With permission.)

The data shown for the grain density on the plasma membrane in Figure 3 (▲) do not allow one to resolve the indicated kinetics. A detailed study of these kinetics can be made by the biochemical method, which quantifies cell-surface label by enzymatic removal (cf. section II). This method is rapid and makes it feasible to obtain a large number of experimental points, Figure 4. These data are in quantitative agreement with the morphometric values observed for the label on the plasma membrane, as in Figure 3 (▲).

From the decay constants of the biphasic kinetics it can be calculated that the rate of membrane internalization amounts to one plasma membrane area equivalent about every 35 min (cf. Appendix. IV. A). This area can be measured in absolute terms as 1300 μm^2 per cell. During 35 min each cell pinocytoses a volume of about 50 μm,[3] as can be measured using fluorescein-labeled dextran as a fluid-phase marker.[3] Therefore, a volume of 50 μm^3 is contained in 1300 μm^2 of membrane, indicating a volume weighted average of $6 \times 50/1300 = 0.23$ μm for the diameter of primary pinosomes.[3] Because primary pinosomes undergo fusion with other endomembranes very rapidly, their average size cannot be determined by direct morphometric observation.

The first detailed morphometric analysis of membrane traffic during pinocytosis in peritoneal macrophages has provided data which allow an estimate of the average lifetime of primary pinosomes.[9] From the increase in volume-to-surface ratios of pinosomes with increasing periods of HRP uptake (horseradish peroxidase served as a fluid-phase marker for cytochemical staining[8]), the increase in average pinosome diameter has been calculated, Figure 5.[9] The size increase resulting from fusion between two vesicles of equal size is 1.26-fold. From the initial slope of the curve in Figure 5, it can be seen that such an increase in average pinosome diameter is reached within the first 15 s, indicating the short lifetime

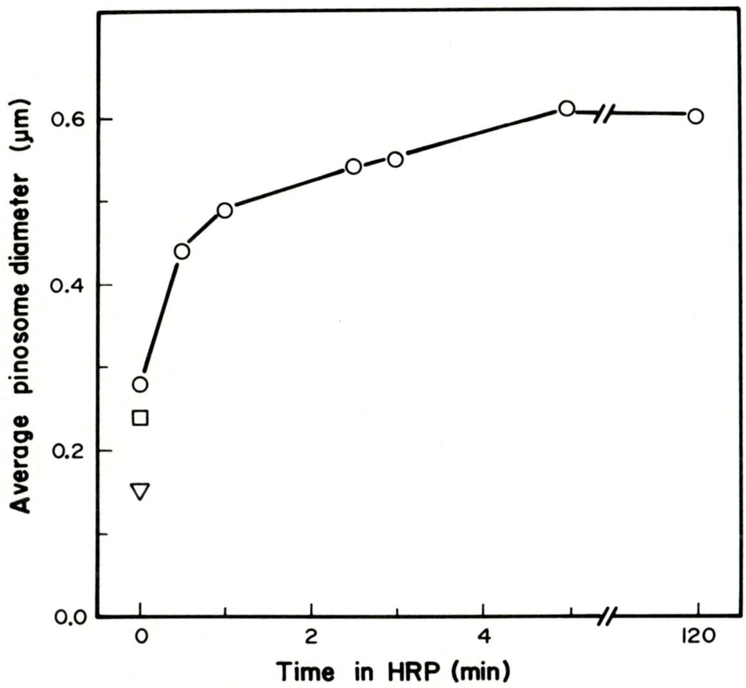

FIGURE 5. Volume weighted average diameter of pinosomes at various times after formation in peritoneal macrophages. Calculated from surface-to-volume ratios of cytochemically stained pinosomes after increasing times of horseradish peroxidase uptake (Table 4 in Steinman et al.[8]) (Courtesy of Elsevier Science Publishers B.V.) (From Thilo, L., *Biochim. Biophys. Acta,* 822, 243, 1985. With permission.)

of primary pinosomes. This estimate can serve as a good example of how a very rapid dynamic process can be observed by morphometric means (using the redistribution of the cytochemical fluid-phase marker, HRP).

When HRP is taken up via mannose receptor-mediated endocytosis, small coated vesicles become stained by the cytochemical reaction product.[2] With their diameter about 0.1 μm, they are significantly smaller than the volume weighted average size of primary pinosomes observed during fluid-phase uptake (cf. above). This suggests that pinocytosis does not occur solely via the relatively small coated vesicles, but that larger structures, although smaller in number than coated vesicles, contribute to the bulk of pinocytic volume uptake. It is still a matter of debate whether or not all pinocytosis occurs via coated membrane structures.

B. MEMBRANE TRAFFIC TO ENDOSOMES AND TO SECONDARY LYSOSOMES

The biphasic pattern of the redistribution kinetics (Figure 3, PM and Figure 4) has been explained in terms of a model for the exchange of membrane between the plasma membrane and two consecutive intracellular membrane compartments (cf. m_1, m_2, and m_3 in Figure 9, and section III E).[3] These compartments have thus been defined purely on a kinetic basis. The only morphometric data available at the time have been based on the consecutive redistribution of horseradish peroxidase as a fluid-phase marker to pinosomes (including endosomes) and secondary lysosomes, respectively.[8] The good correlation with the kinetic parameters have strongly suggested that pinosomes (including endosomes) and secondary lysosomes are the two kinetically resolved intracellular membrane compartments. However, with horseradish peroxidase (and its cytochemical reaction product) not remaining strictly

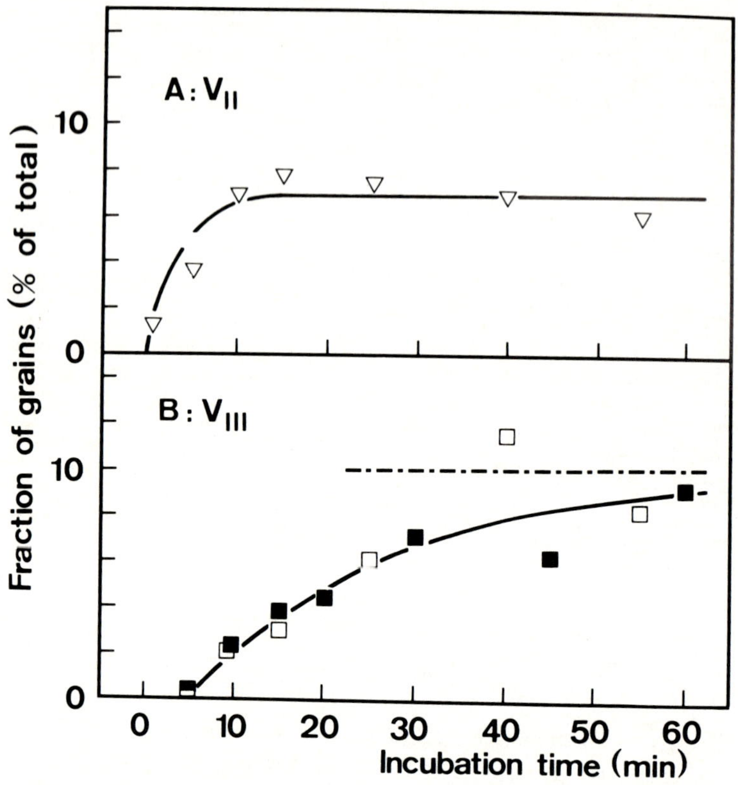

FIGURE 6. Kinetics of redistribution of a covalent cell surface marker to membrane of endocytic compartments during pinocytosis in bone marrow-derived macrophages. The quantitative evaluations were performed on autoradiographs of cell thin sections (exposed for 3 weeks) showing 10 to 15 silver grains per cell profile. The relative amount of radioactive marker is shown as the number of autoradiographic grains observed in association with membrane of endosomes, V_{II} (A), or of secondary lysosomes, V_{III} (B), as characterized via HRP reaction product. The curves indicate first-order kinetics with rate constants, $\lambda = \ln 2/t_{1/2}$, $\lambda_1 = (3.6 \text{ min})^{-1}$ (A) and $\lambda_2 = (22 \text{ min})^{-1}$ (B). Dotted line in (B) indicates the steady-state level given by extrapolation. Different symbols (B) refer to independent experiments. (From de Chastellier, C., *Eur. J. Cell Biol.*, 44, 112, 1987. Copyright Wissenschaftliche Verlagsgesellschaft mbH, Stuttgart. With permission.)

associated with the enclosing membrane, especially at the level of secondary lysosomes, no direct characterization of membrane traffic has been possible. A direct morphometric observation of the kinetically defined compartments has to await the use of a covalently bound membrane marker.

The appearance of label in morphologically defined membranes of bone marrow macrophage endosomes, as a prelysosomal compartment, is represented in Figure 6A.[2] This compartment is filled with label within less than 10 min ($T_{1/2}$ = 2.5 min), confirming that the first intracellular membrane compartment (m_2 in Figure 9) consists of endosomal membrane.

The flow of cell surface-derived label into the membranes of morphologically defined secondary lysosomes is represented in Figure 6B. Label enters this compartment much more slowly ($T_{1/2}$ = 15 min), quantitatively, as predicted by the biochemical kinetic data for the second intracellular compartment (m_3 in Figure 9). A comparison between this rate and the rate of membrane internalization (k_2 vs. k_1, respectively, cf. Figure 9) suggests that 12% of internalized label is recycled via the membrane pool of secondary lysosomes. A value of less than 3% has been observed by cell fractionation and direct detection of label in a high density fraction of secondary lysosomes isolated from a macrophage-like cell line.[4] This discrepancy may be due to different lysosomal membrane populations being observed by the two techniques, although an effect due to cell type cannot be excluded (cf. Table 2 in de Chastellier et al.[2]).

C. RELATIVE MEMBRANE COMPOSITION OF THE PLASMA MEMBRANE, ENDOSOMES, AND SECONDARY LYSOSOMES

The covalent marker will redistribute to all interacting membranes that contain a pool of membrane glycoproteins like those that become labeled at the cell surface. The steady-state distribution of label will reflect the relative pool size of these glycoproteins in each membrane compartment involved. When this pool size (relative grain number) is related to the amount of membrane (morphologically observed membrane area) in the corresponding compartment, the relative membrane composition is indicated in terms of the labeled molecular species.

Coated vesicles display the same grain density as the plasma membrane from which they are derived, Figure 3 (●; the scatter is due to the relatively small number of observations for these structures).[2] They therefore have the same composition. This suggests that for the average labeled M_r-species, no enrichment in coated pits occurs prior to internalization. Such an enrichment is observed for many different cell-surface receptors (about 35-fold for the low density lipoprotein receptor in fibroblasts[10]).

A most interesting observation is that the steady-state grain density on the membranes of endosomes is only about half that on the plasma membrane, Figure 3 (▽ vs. ▲). To our knowledge, this is the first evidence that the composition of endosomal membrane is clearly distinct from that of the plasma membrane. Accordingly, endosomes do not derive their entire membrane from internalized plasma membrane, but rather exist as an independent organelle of composite origin.

According to the relative grain density on secondary lysosomes, about 30% of the morphologically observed lysosomal membrane resembles plasma membrane as far as the labeled M_r-species are concerned, Figure 3 (□). However, in the particular case of high-density lysosomes isolated from a macrophage-like cell line, not all the labeled M_r-species are found to enter lysosomal membrane to the same relative extent. Molecules of M_r = 160 to 190 kDa are excluded, while a group of M_r = 100 to 120 kDa is strongly enriched (such a selectivity may be due to some molecules playing a specific role in interaction between endosomes and lysosomes).[4] If this proves also to be the case for bone marrow macrophages, then the steady-state grain density does not give a quantitative measure for the relative membrane composition of secondary lysosomes.

The use of the covalent membrane marker has allowed us to identify a further membrane compartment as taking part in endocytosis-derived membrane flow. It consists of small empty vesicles (EV), free of pinocytotic marker. These vesicles have often been seen as fusing/budding at the plasma membrane or at endosomes. EV have a grain density similar to that on endosomal membrane, Figure 3 (○), and therefore half that of the PM. Accordingly, this can be interpreted as EV consisting of two vesicle populations of approximately equal numbers: recycling vesicles with the same composition as the plasma membrane to which they are destined, and unlabeled vesicles involved in endosome formation having a composition which is totally different from the cell surface.

D. RAPID MEMBRANE EXCHANGE WITHIN THE ENDOSOME COMPARTMENT

The first-order kinetics observed for the accumulation of label into the respective compartments suggest that each compartment behaves as a rapidly mixed membrane pool: membrane leaving the pool is a representative sample of the entire pool at that moment. The exchange of label within the endocytic membrane pool has been studied in the amoeba

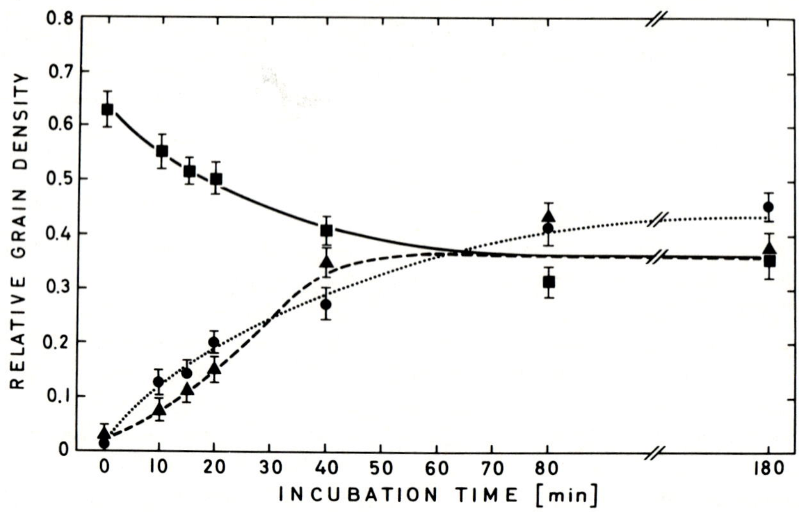

FIGURE 7. Relative surface density of silver grains on the plasma membrane (■), phagosome membrane (●) and pinosome membrane (▲) of *Dictyostelium discoideum* amoebae as determined from EM autoradiographs of cells during pinocytosis after latex phagocytosis. Cells were first incubated with latex beads for 25 min; uningested latex was removed and the cells labeled with (^{3}H)gal and resuspended in axenic medium without latex. For the indicated times during subsequent pinocytosis, the relative grain density on the respective membranes was determined. Error bars represent the standard error of the mean. Curves drawn by eye. (From de Chastellier, C., *Eur. J. Cell Biol.*, 30, 233, 1983. Copyright Wissenschaftliche Verlagsgesellschaft mbH, Stuttgart. With permission.)

Dictyostelium discoideum.[11] In this cell type, there is a single endocytic membrane compartment consisting of all visible vacuolar structures outside the Golgi region. These structures are often termed digestive vacuoles, and include the incoming pinosomes. Morphological evidence suggests that the size of the endocytic membrane compartment is strictly controlled: during phagocytosis the endocytic membrane compartment includes both phagosomal membranes (PhM) and vacuolar membranes (VM) in such a way that an increase in PhM is accompanied by a corresponding decrease in VM.[12] When amoebae phagocytose latex beads part of the endocytic membrane pool can therefore be distinguished as phagosomal membrane, to be compared to the latex-free endocytic membranes. With respect to this differentiation, the exchange of label between endocytic membranes has been studied by morphometric analysis (i.e., newly formed labeled phagosomes and preexisting unlabeled digestive vacuoles, and conversely).[11]

When cells contain unlabeled phagosomal (and pinosomal) membrane, internalization of labeled plasma membrane leads to a rapid increase of grain density on phagosomal membranes, occuring at the same rate and to the same extent as for pinosomes, Figure 7. This shows that preexisting, unlabeled phagosomal membrane exchanges rapidly with newly internalized, labeled pinosomal membrane.

In a complementary way, when phagocytosis of latex beads occurs after labeling of the plasma membrane, the grain density on newly formed phagosomal membrane is much lower than on plasma membrane from which it is derived, Figure 8. Again this shows that (labeled) phagosomal membrane rapidly exchanges with membrane of (unlabeled) preexisting digestive vacuoles.

At steady state both phagosomal and pinosomal membranes have the same grain density as on the plasma membrane. Accordingly, the vacuolar membrane in amoebae seems to have the same composition as the plasma membrane. This result is different from the finding for macrophages (cf. section III C above). It will be interesting to find a reason for this difference between the two cell types.

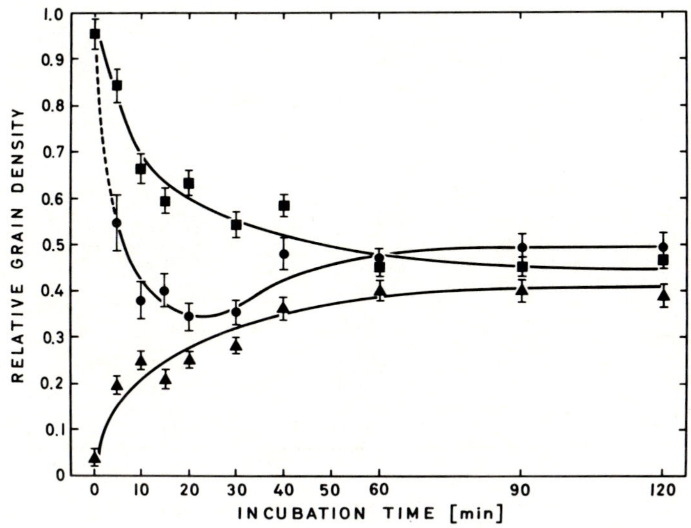

FIGURE 8. Relative surface density of silver grains on the plasma membrane (PM) (■), phagosome membrane (PhM) (●) and Pinosome membrane (PiM) (▲). Cells were labeled with (^{3}H)gal and incubated with latex beads in axenic medium. For the indicated times during phagocytosis the relative grain density on the respective membranes was determined. The relative surface area of the PM and of the sum of PhM and PiM remained constant during an increase of PhM due to phagocytosis. Error bars represent the standard error of the mean. Curves drawn by eye. (From de Chastellier, C., *Eur. J. Cell Biol.*, 30, 233, 1983. Copyright Wissenschaftliche Verlagsgesellschaft mbH, Stuttgart. With permission.)

E. SUMMARY: COMPARTMENTAL MODEL FOR ENDOCYTIC MEMBRANE TRAFFIC

The use of enzymatically labeled cell-surface glycoproteins as an autoradiographic membrane marker for morphometric analysis has provided direct evidence for both quantitative and qualitative aspects of endocytic membrane traffic, Figure 9.

During fluid-phase pinocytosis in macrophages, the plasma membrane (pool size, $m_1 = 100$) is internalized and recycled once every 35 min. Volume uptake occurs by means of primary pinosomes with a volume weighted average diameter of about 0.23 μm and an average lifetime of about 15 s. Due to their high surface to volume ratio, coated vesicles of size of 0.1 μm cannot account for the bulk volume uptake.

Internalized membrane becomes part of a first intracellular membrane compartment (pool size, $m_2 = 11.5$) which includes membranes of coated vesicles and other primary pinosomes, of endosomes, and of recycling vesicles. About 90% of membrane recycling occurs directly from this compartment after an average residence time of about 4 min. Only about 50% of the endosomal membrane area is derived from internalized membrane, with the remainder being contributed by primary endosomal membrane of an independent intracellular origin.

Only about 10% of internalized membrane subsequently enters a second intracellular compartment (pool size, $m_3 = 10.5$). This pool consists of cell surface-derived membrane as part of the membrane of secondary lysosomes. Recycling occurs after an average residence time of about 30 min, and strictly represents "recycling via secondary lysosomes."

IV. APPENDIX

A. KINETIC ANALYSIS OF COMPARTMENTAL EXCHANGE

For the exchange of membrane at the same rate (k as (membrane area/time)) in both directions between 2 membrane compartments (sizes m_1 and m_2 as (membrane area)), the

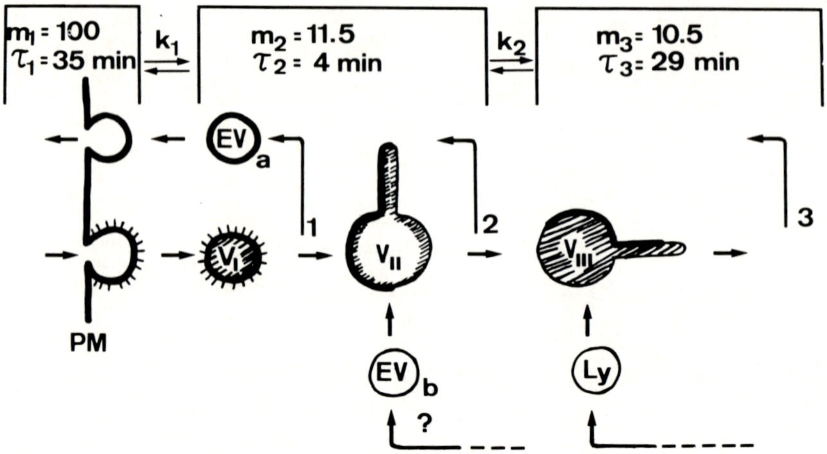

FIGURE 9. Schematic presentation of membrane traffic between endocytic membrane compartments during pinocytosis in bone marrow-derived macrophages: plasma membrane (PM), primary endocytic vesicles (V_I), endosomes (V_{II}), secondary lysosomes (V_{III}), and recycling vesicles (EV_a). Other structures are primary lysosomes (Ly) and membranes (EV_b) destined to endosomes. Shaded areas symbolize HRP reaction product. The partial lengths of thick lines symbolize the relative abundances of cell surface derived constituents (i.e., the used membrane marker) contributing to the total membrane composition of each organelle: V_I and EV_a have the same composition as the PM; V_{II} derives only 50%, V_{III} only 30%, of its membrane constituents from the cell surface; correspondingly, other membrane sources viz EV_b and Ly, respectively, must contribute to V_{II} and V_{III}. The kinetic aspects of membrane flow are indicated by the relative pool sizes (m) and residence time (τ) of cell-surface-derived membrane constituents for each pool. Membrane flow rates (k) are given by (m/τ). Three sites for membrane recycling are indicated: 1) during fusion amongst endocytic vesicles, decreasing their surface to volume ratio, and during endosome processing; 2) during transfer of endosome contents into secondary lysosomes; recycling at sites 1 and 2 involves about 88% of internalized membrane, occurs rapidly within 4 min after internalization, and has not been resolved kinetically or biochemically; 3) during processing of secondary lysosomes (12%, as indicated by $k_2/k_1 = (10.5/29)/(11.5/4)$, after a residence time of about 29 min. Only the third step represents recycling via a lysosome membrane pool. (From de Chastellier, C., *Eur. J. Cell Biol.*, 44, 112, 1987. Copyright Wissenschaftliche Verlagsgesellschaft mbH, Stuttgart. With permission.)

quantities k and m_i can be determined by the kinetics of redistribution of a membrane marker introduced into m_1.[6] The fraction of label in m_1, as given by x, changes with time, as given by t, as

$$x = (1 - \rho) \cdot e^{-\lambda \cdot t} + \rho \tag{1}$$

$$\rho = m_1/(m_1 + m_2) \tag{2}$$

$$\lambda = k/(m_1 \cdot (1 - \rho)) \tag{3}$$

The value of ρ is given by the steady-state distribution of label whereas λ is determined by the half-life with which steady state is approached ($\lambda = \ln 2/T_{1/2}$).

Similarly, for the exchange of membrane between 3 compartments in series (size m_1, m_2, m_3) with bidirectional exchange rates k_1 (m_1 and m_2) and k_2 (m_2 and m_3)[3]

$$x = (1 - \rho_1) \cdot e^{-\lambda_1 \cdot t} + (\rho_1 - \rho_2) \cdot e^{-\lambda_2 \cdot t} + \rho_2 \tag{4}$$

$$\rho_1 = m_1/(m_1 + m_2) \tag{5}$$

$$\rho_2 = m_1/(m_1 + m_2 + m_3) \tag{6}$$

$$\rho_3 = m_2/(m_2 + m_3) \tag{7}$$

$$\lambda_1 = k_1/(m_1 \cdot (1 - \rho_1)) \tag{8}$$

$$\lambda_2 = (k_2/m_2) \cdot ((1 - \rho_3)^{-1} - \rho_1) \tag{9}$$

The values of ρ_i are given by the amplitudes of the two phases (ρ_1: intermediate level; ρ_2: steady-state level) and the values of λ_i by the corresponding half-lives. In general, exchange between n compartments in series can be described by the sum of (n – 1) decaying exponential functions.[13]

B. TECHNICAL ASPECTS OF (^{3} H)GAL LABELED MEMBRANE GLYCOPROTEINS AS AUTORADIOGRAPHIC MARKER

As a rough guide, autoradiography requires a minimum of 1.3 desintegrations per cubic micron of section per 10 d of exposure to get any sensible results with sections 100 nm thick (gold interference color), and a monolayer of Ilford L4 emulsion. In practice, 10^5 to 10^6 dpm of ^3H per 10^6 cells is needed in order to yield 5 to 10 silver grains per cell section (80 nm thick) per week of exposure time, in the case of cells of the size of macrophages or *Dictyostelium discoideum* amoebae.

The quantitative analysis of autoradiographs first requires that the sections be of even thickness. The emulsion thickness, exposure time and conditions, and the development conditions, must also be strictly controlled. Second, as the resolution varies for each isotope and also with autoradiography conditions, it must be determined experimentally with respect to a known source.[14] In the particular case of macrophages and *D. discoideum* amoebae labeled with (^{3}H)gal on their cell surface, the HD (''half distance'') value (distance from a line source within which half of the developed grains fall) was obtained by determining the distribution of silver grains with respect to the plasma membrane on cells fixed immediately after labeling. This was possible because radioactivity was associated exclusively with the plasma membrane at this point. An HD value of 95 nm was obtained. To determine the percentage of grains per source, the grains should be counted in an area of emulsion overlying each source and extending for at least 2 HD on every side, to as much as 5 HD. In the studies discussed in the present chapter, a cutoff distance of 3 HD was chosen because the level of scattered grains then remained constant throughout the experiments.[2,11] All other grains (10 to 20%) resulted from radiation spread and were discarded. At last, the grain counts observed over the specimens will include background grains. For quantitative studies, the grain counts should be corrected by substracting the mean number of background grains found in similar volumes of emulsion away from the radioactive areas in the specimen (In References 2 and 11 it amounted to only 0.3% of the total number of silver grains).

For adequate quantitations, analysis must be performed on a sufficient number of micrographs per time point (30 to 100, depending on the number of grains per micrograph). Care must be taken not to overlap images and to see that all micrographs correspond to different cells.

Morphometric methods are used to determine the relative length of membranes to which silver grains are associated.[15] The test system consists of short lines of equal length (Z), arranged on equidistant and parallel rows, whereby the distance between the end-points of the lines is equal to Z in every direction (in References 2 and 11, 14 mm test lines were used, 14 mm corresponding to 0.5 μm in Reference 2 and 1 μm in Reference 11). The test grid is randomly placed on the micrographs used for silver grain counting, and the total number of intersections between the lines of the test grid and the different membranes is calculated by summing the individual counts for each micrograph. The grain density is obtained from the quotient of the total number of grains and the total number of intersections for each membrane compartment.

REFERENCES

1. **Thilo, L.,** Labeling of plasma membrane glycoconjugates by terminal glycosylation (galactosyltransferase and glycosidase), *Methods Enzymol.,* 98, 415, 1983.
2. **de Chastellier, C., Lang, T., Ryter, A., and Thilo, L.,** Exchange kinetics and composition of endocytic membranes in terms of plasma membrane constituents: a morphometric study in macrophages, *Eur. J. Cell Biol.,* 44, 112, 1987.
3. **Burgert, H. G. and Thilo, L.,** Internalization and recycling of plasma membrane glycoconjugates during pinocytosis in the macrophage cell line, P388D1: Kinetic evidence for compartmentation of internalized membranes, *Exp. Cell Res.,* 144, 127, 1983.
4. **Haylett, T. and Thilo, L.,** Limited and selective transfer of plasma membrane glycoproteins to membrane of secondary lysosomes, *J. Cell Biol.,* 103, 1249, 1986.
5. **Schwarz, H. and Thilo, L.,** Membrane traffic in *Dictyostelium discoideum*: plasma membrane glycoconjugates internalized and recycled during fluid phase pinocytosis enter the Golgi complex, *Eur. J. Cell Biol.,* 31, 212, 1983.
6. **Thilo, L. and Vogel, G.,** Kinetics of membrane internalization and recycling during pinocytosis, in *Dictyostelium discoideum, Proc. Natl. Acad. Sci. U.S.A.,* 77, 1015, 1980.
7. **Thilo, L.,** Selective internalization of granule membrane after secretion in mast cells, *Proc. Natl. Acad. Sci. U.S.A.,* 82, 1711, 1985.
8. **Steinman, R. M., Brodie, S. E., and Cohn, Z. A.,** Membrane flow during pinocytosis. A stereologic analysis, *J. Cell Biol.,* 68, 665, 1976.
9. **Thilo, L.,** Quantification of endocytosis-derived membrane traffic, *Biochim. Biophys. Acta,* 822, 243, 1985.
10. **Anderson, R. G. W., Brown, M. S., and Goldstein, J. L.,** Role of the coated endocytic vesicle in the uptake of receptor-bound low density lipoprotein in human fibroblasts, *Cell,* 10, 351, 1977.
11. **de Chastellier, C., Ryter, A., and Thilo, L.,** Membrane shuttle between plasma membrane, phagosomes, and pinosomes in *Dictyostelium discoideum* amoeboid cells, *Eur. J. Cell Biol.,* 30, 233, 1983.
12. **Ryter, A. and de Chastellier, C.,** Morphometric and cytochemical studies of *Dictyostelium discoideum* in vegetative phase. Digestive system and membrane turnover, *J. Cell Biol.,* 75, 200, 1977.
13. **Rubinow, S. I. and Winzer, A.,** Compartment analysis: an inverse problem, *Math. Biosci.,* 11, 203, 1971.
14. **Salpeter, M. M., Bachmann, L., and Salpeter, E. E.,** Resolution in electron microscope autoradiography, *J. Cell Biol.,* 41, 1, 1969.
15. **Weibel, E. R., Kistler, G. S., and Scherle, W.,** Practical stereological methods for morphometric cytology, *J. Cell Biol.,* 30, 23, 1966.

Chapter 10

LOW-TEMPERATURE EMBEDDING AND CRYOSECTIONING IN THE IMMUNOCYTOCHEMICAL STUDY OF MEMBRANE RECYCLING

Dennis Brown

TABLE OF CONTENTS

I. INTRODUCTION

One of the fundamental problems in cell biology is how specific membrane components are directed towards and inserted into the appropriate membrane domains. This targeting of both proteins and membrane lipids is important not only in the establishment and maintenance of apical, basal, and lateral regions of the plasma membrane of polarized epithelial cells, but is also critical to the specialized function of all intracellular membrane-bounded organelles.[1,2] Over the last few years, a considerable amount of effort has been directed towards understanding how cells are able to sort a multitude of synthesized proteins efficiently, and ensure that they are delivered to their correct "addresses". This work has involved the combined use of different techniques, from the biochemical analysis of separated cell fractions, to the molecular biology of individual membrane proteins. This chapter will consider how the development of new morphological and immunocytochemical techniques has contributed to our investigations of the dynamic interactions that occur among intracellular membranes and the plasma membrane. These techniques have allowed the visualization of endogenous and exogenous (e.g., viral) proteins as they move along the processing railway of the cell, and they have also enabled the detection of a number of membrane components that are specific markers for the various stations along the line. Following an outline of the techniques of low-temperature embedding and ultrathin frozen sectioning, their application to the study of membrane recycling will be illustrated by selected examples rather than by an extensive survey of the field.

II. METHODOLOGY

The two techniques that have been most widely used in studies on the intracellular transport and recycling of membrane proteins, ultracryomicrotomy and low-temperature embedding in Lowicryl K4M, both involve the use of immunogold labeling of thin sectioned material for the detection of antigenic sites (see Chapter 12 by Roth for more details of the immunogold procedure). The advantages of direct labeling of these sections with an electron-dense marker have previously been discussed,[3,4] but include:

1. Direct and unrestricted access of all immunoreagents to antigenic sites on both intracellular compartments and the plasma membrane without the need to permeabilize cells.
2. High resolution visualization of antigenic sites due to the lack of diffusion of the marker, in contrast to immunoperoxidase procedures.
3. Ability to obtain quantitative estimates of labeling intensity by counting gold particles.
4. Possibility of double or triple labeling of cells simultaneously using different sizes of gold particles.

A. LOW-TEMPERATURE EMBEDDING IN LOWICRYL K4M

The technique of low-temperature embedding in hydrophilic resins was introduced by Carlemalm et al.,[5] and has since become widely used for immunocytochemistry at the electron microscope level. The technique involves fixation of a tissue in a way that is expected to retain antigenicity, usually with a mild fixative such as paraformaldehyde lysine-periodate.[6] The tissue is then dehydrated at decreasing temperatures in graded ethanol, so that the 100% ethanol step is performed at -35°C. Subsequent infiltration with the resin, Lowicryl K4M, is also carried out at this low temperature by overnight agitation of small tissue pieces in the resin. Tissue is then transferred into gelatine or Beem capsules (still at low temperature), and polymerization is carried out by indirect irradiation with a UV light mounted in a chest freezer. After an initial 24 h period of polymerization, the capsules can be transferred to room temperature for final hardening under UV light for a further 1 or 2 d.

This method, which is described in detail in previous reports,[5,7] is an excellent technique that has made the postembedding labeling of a multitude of different antigens possible. A simpler, much more rapid alternative, has been described by Altman et al.,[8] who have detected antigenic sites on thin sections of tissue embedded at room temperature in Lowicryl K4M, followed by rapid polymerization under UV for only 45 min. We have used this technique extensively, and have found it ideal for the detection of many antigens. However, the membrane fine structure appears not to be as well retained as with the low-temperature technique, and so it may be less useful for the definition of antigenic sites in some intracellular compartments, such as defined cisternae of the Golgi apparatus and the trans-Golgi network.

Material embedded in either way can be readily cut on glass or diamond knives as usual. However, the hydrophilic nature of the resin can cause some sectioning problems. If water from the knife trough contacts the face of the block, it is necessary to remove the wet surface by cutting several semithin sections before continuing with thin sections. Simply wiping the block face (as with epoxy-resin sectioning) is not usually sufficient. Furthermore, previously cut blocks stored in a damp environment may not be easy to resection immediately; again, semithin sections should be cut to reach a drier part of the resin. It may be necessary to store blocks dessicated in some areas with high humidity.

B. ULTRATHIN FROZEN SECTIONS

Ultrathin sections of frozen tissues can provide excellent results when combined with immunogold labeling for the detection of antigenic sites within cells and on the cell surface. The retention of antigenicity in such sections is generally superior to other techniques because of a reduced number of tissue processing steps. Although for most purposes it is necessary to fix the tissue prior to freezing, subsequent exposure to alcohols and embedding resins is completely eliminated. In order to avoid freezing artifacts, tissue is usually cryoprotected in 2.3M sucrose and then frozen either directly in liquid nitrogen (LN_2), a LN_2 slush, or Freon 22 cooled with LN_2. For some preparations, rapid freezing can be carried out without fixation or cryoprotection (e.g., muscle) and good results obtained. Rapid freezing of unfixed epithelial cells, however, cannot be considered a routine operation at the present time. The disadvantages of ultrathin frozen sectioning are that specialized equipment is required, and the technique is technically more demanding than resin sectioning. However, membranes can be beautifully visualized due to negative staining of the sections (see Figure 7), and high resolution images of immunolabeled membrane compartments can be readily obtained. The precise methodology involved in preparing and staining ultrathin frozen sections is beyond the scope of this review, and has been described in detail elsewhere.[9,10]

As an example of the relative intensities of staining that can be obtained with thin, frozen sections and Lowicryl K4M sections, Figures 1 to 4 show sections of the proximal tubule of the kidney prepared using both methods, and stained with *Helix pomatia* lectin-gold complexes. While both methods give an appreciable degree of labeling, the ultracry-otome sections are more intensely stained, in part because of superior preservation of lectin-binding sites, and in part because multiple layers of gold complexes build up over much of the thickness of the section. This cannot occur with Lowicryl sections, which are stained only at the surface of the resin. In addition, it should also be pointed out that semithin frozen sections can also be cut on the ultracryomicrotome; these sections are also prepared from tissue that has not been dehydrated or embedded, and offer a degree of resolution at the light microscope level equal to that of semithin sections of epoxy-resin embedded material. Combined with the supeior retention of antigenic sites in such frozen sections, they must be considered the best choice for light microscope immunocytochemistry, and are routinely used in our laboratory. Such sections are ideal for double labeling experiments using, for example, lectin-gold complexes and indirect immunofluorescence procedures. As shown in Figures 5 and 6, a combination of these techniques on semithin frozen sections of kidney

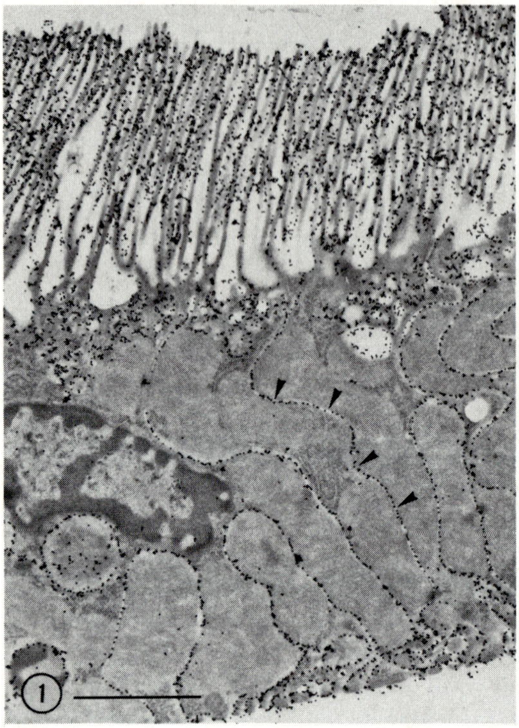

FIGURE 1. Thin section of rat kidney cortex dehydrated and embedded at low temperature using the hydrophilic resin, Lowicryl K4M. The thin section was incubated with *Helix pomatia* lectin-gold complexes for 30 min. Lectin binding sites are revealed by the presence of electron-dense gold particles, and are especially prominent on the apical plasma membrane (including the brush border and apical vesicles and invaginations), as well as on the basolateral plasma membrane (arrowheads). Compare this labeling pattern with that shown in Figure 2 for the medullary portion of the straight proximal tubule. Bar = 2 μm.

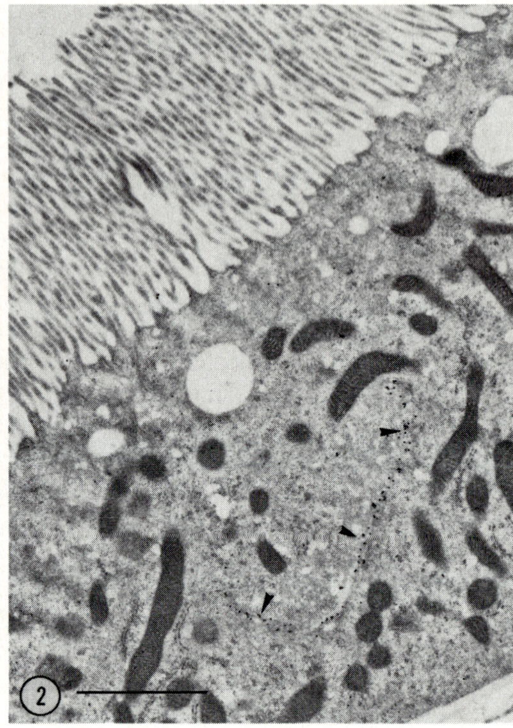

FIGURE 2. Thin section of K4M-embedded kidney, incubated under the same conditions as described for Figure 1, but showing a part of a straight proximal tubule from the outer stripe of the outer medulla. *Helix pomatia* lectin binding sites are virtually absent from all regions of the plasma membrane. However, in this region, some Golgi cisternae (arrowheads) are labeled with this lectin. Bar = 2 μm.

medulla clearly demonstrated that carbonic anhydrase-rich intercalated cells (see below) in this part of the kidney are not labeled with *Helix pomatia* lectin-gold complexes.[11] Such a high level of resolution is not possible with conventional cryostat sections.

III. MEMBRANE RECYCLING

In all eukaryotic cells, integral membrane proteins that are synthesized in the rough endoplasmic reticulum are processed in the Golgi apparatus, where many post-translational modifications can be made to the primary sequence of the protein.[12,13] Different cisternae of the Golgi complex can be identified by their unique complement of processing enzymes; thus, there are now enzymatic markers that can be used to differentiate cis, medial, and trans compartments of the Golgi, as well as a system of tubules and vesicles on the far trans side of the Golgi, known as the trans-reticular network or the trans-Golgi network,[14-17] (see also Chapters 8 and 12). By combining immunocytochemical studies on the distribution of these enzymes with biochemical assays of their functional activity on defined protein sub-

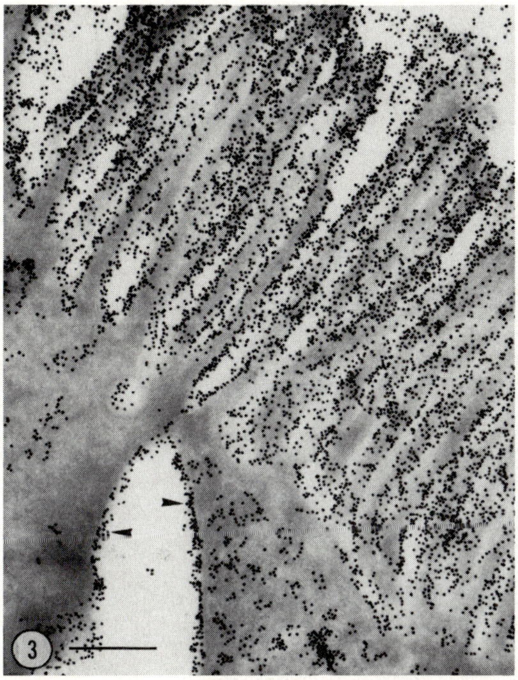

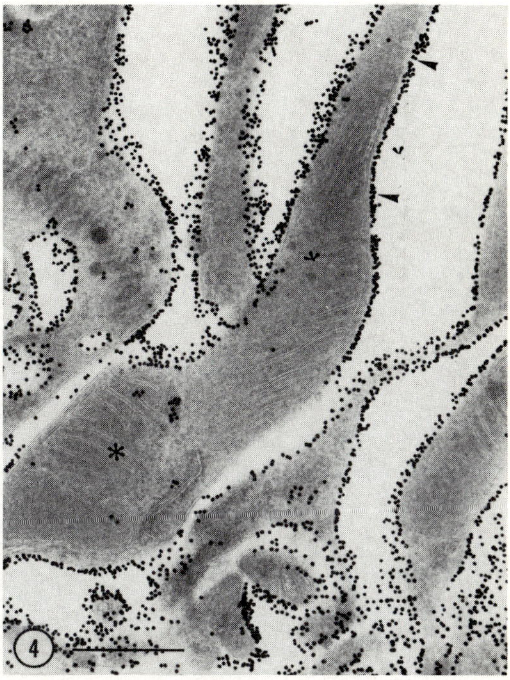

FIGURE 3. Part of an ultrathin frozen section of rat kidney proximal convoluted tubule prepared by ultracryomicrotomy. The tissue was fixed in 1% glutaraldehyde, cryoprotected in 2.3 M sucrose, and frozen directly in liquid nitrogen. Ultrathin frozen sections were prepared by the method of Tokuyasu (see Reference 9) and were incubated for 30 min. with *Helix pomatia* lectin-gold complexes. The sections were stained with uranyl acetate and embedded in methyl cellulose (Tylose) before examination in the electron microscope. The brush border is extremely heavily labeled with gold particles, as are the apical membrane invaginations and apical vesicles. The lateral plasma membrane is also heavily labeled. Bar = 0.5 μm.

FIGURE 4. Ultrathin frozen section of part of a proximal convoluted tubule prepared and incubated as in Figure 3, but showing the basal region of the cell. The plasma membrane (arrowheads) is heavily labeled with gold particles, showing the presence of *Helix pomatia* lectin binding sites. The membranes of mitochondria (asterisks) present in the basal cytoplasm between the infoldings of the plasma membrane are negatively stained, and show no specific labeling. Bar = 0.5 μm.

strates as they pass through the processing system of the cell, it has been possible to identify within the cell a number of major sites at which functionally important modifications are made to newly-synthesized proteins,[15-21] and processing enzymes such as galactosyl transferase and glucosyl transferase have been localized to specific Golgi compartments.[14,15] Furthermore, proteins in transit through the synthetic apparatus of the cell can often be followed and quantified by immunogold labeling of thin sections. The results of such quantification can be directly compared with information obtained by biochemical analysis of synthesized proteins in the processing pathway.[20,21] However, it must be said that the quantitative analysis of immunolabeling by the immunogold procedure is difficult to translate into absolute values for the amount of antigen present in specifically labeled sites. The degree of labeling depends on many factors that are too complex to discuss here, but some of which have been considered elsewhere.[22-25] Immunolabeling should, therefore, be used as a tool for the accurate intracellular localization of antigens, and to give information on the relative amounts of material present in different parts of the cell. By the parallel use of

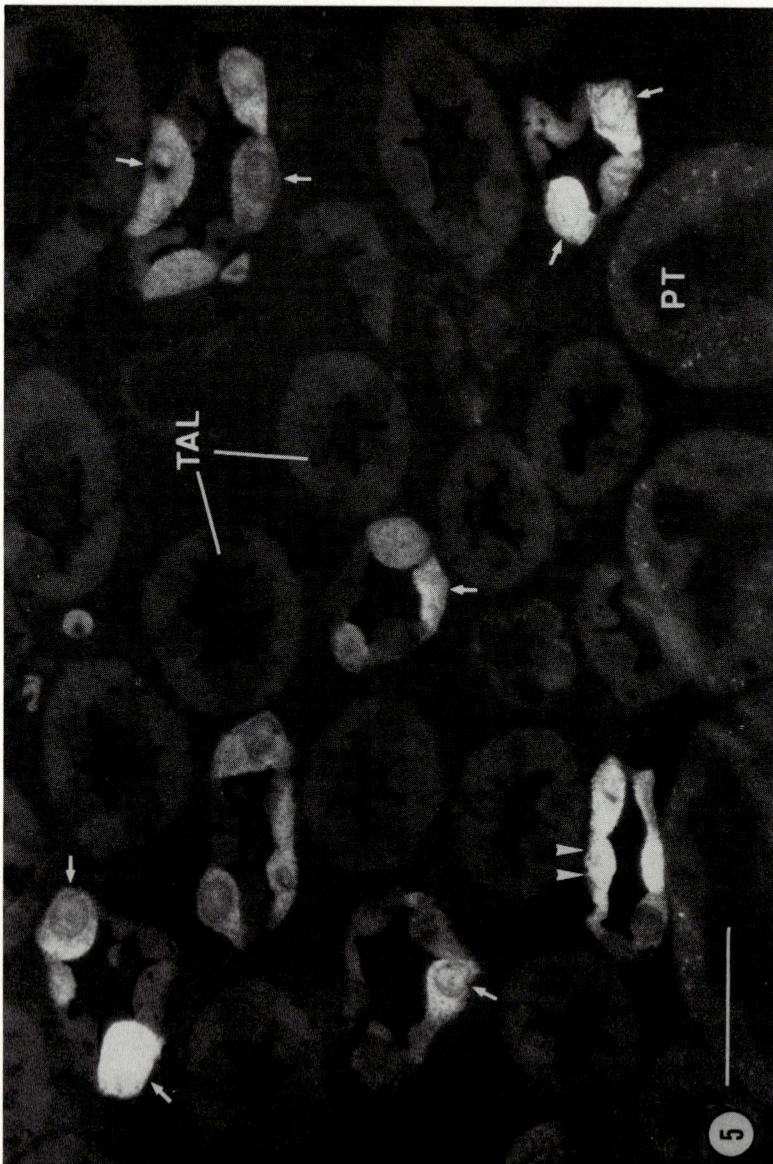
TAL
PT
5

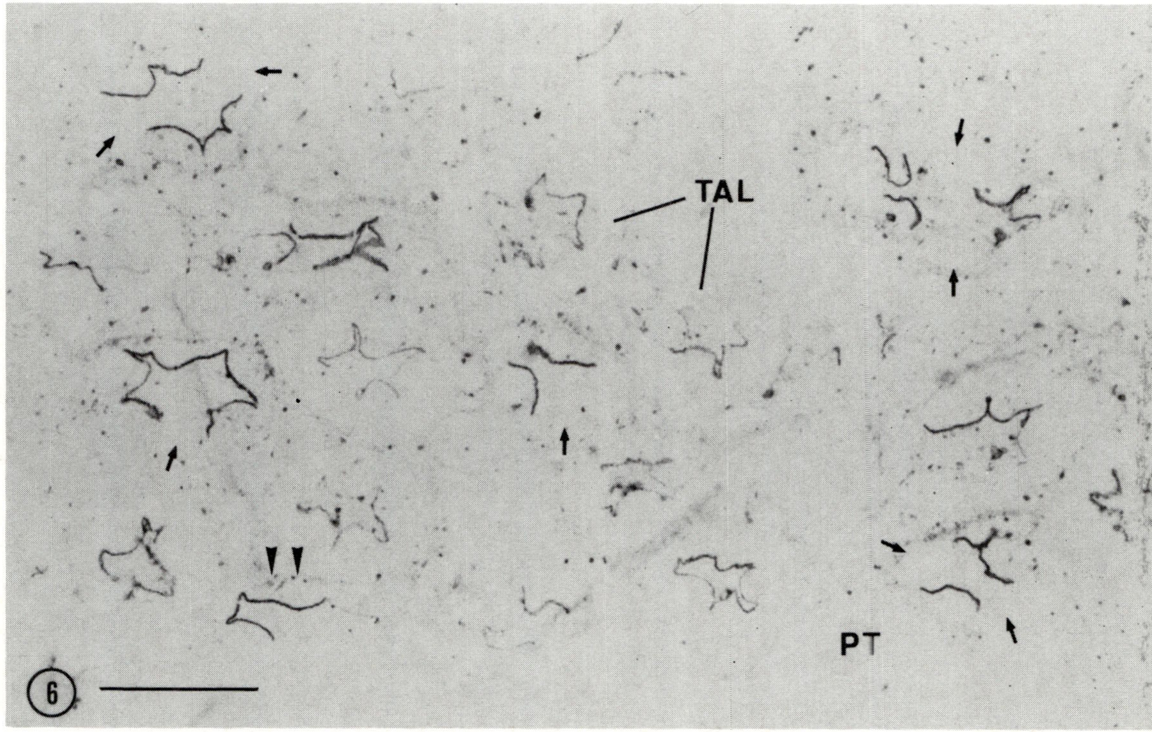

FIGURE 5 and 6. Semithin cryostat section of rat kidney medulla double-stained to reveal carbonic anhydrase by indirect immunofluorescence (Figure 5), and *Helix pomatia* lectin binding sites using lectin-gold complexes (Figure 6). By comparing the same cells in the two figures, it can be seen that the majority of the carbonic anhydrase-rich intercalated cells are unlabeled with lectin-gold complexes (arrows). However, an occasional cell is stained with both reagents (arrowheads). These figures demonstrate the level of resolution that is obtained with the use of semithin frozen sections, and illustrate their suitability for a variety of staining procedures (For details of methodology, see Reference 11). TAL = thick ascending limb of Henle; PT = proximal tubule. Bar = 30 μm. (From Brown, D., Roth, J., and Orci, L., *Am. J. Physiol.*, 248, C348, 1985. With permission.)

cell fractionation and protein chemistry, a more complete picture of the temporal and quantitative aspects of protein processing can be obtained than when these techniques are used in isolation.

The synthesis and processing of membrane and secretory proteins is only one part of the recycling process that occurs in virtually all cells. There is a constant exchange of membrane between the cell surface and intracellular compartments that is brought about by exocytosis and endocytosis of membrane-bounded vesicles. One of the most well known of these processes is receptor-mediated endocytosis, a term originally used for the specific retrieval of LDL receptors and their ligand via clathrin-coated pits.[26] In general, receptor-mediated endocytosis results in the uptake of cell surface molecules (i.e., receptors), often accompanied by their associated ligand (see also Chapter 8 by Robenek). Clathrin-coated vesicles that internalize these receptor-ligand complexes lose their clathrin coat soon after becoming free vesicles in the cytoplasm, and they subsequently deliver their content to the endosomal compartment. Receptors and ligands usually part company in the acidic environment of this compartment,[27] and for this reason it is sometimes referred to as CURL[28] (compartment of uncoupling of receptors and ligand). The fate of the receptors and the ligand following this uncoupling is complex, and has been followed by immunolabeling of a number of receptors and their ligands using ultrathin frozen sections. Using multiple labeling with gold particles of different sizes, Geuze et al.[28] were able to show the segregation of receptors and ligands in the CURL, and to show that recycled receptors (i.e., those returning to the apical plasma membrane) were separated from those involved in transcytosis (i.e., moving to a different part of the plasma membrane) in this endosomal compartment.

IV. THE TRANS-GOLGI NETWORK

The Golgi apparatus plays a central role in the processing of both integral membrane proteins and secreted proteins.[12,13] It has become apparent using fractionation techniques, as well as histochemical and immunocytochemical methods, that the Golgi complex is not composed of a homogeneous population of cisternae. At present, the Golgi can be clearly separated into cis, medial, and trans cisternae,[14,15,17,29] each of which has its characteristic fingerprint of markers. These markers are usually membrane-associated enzymes involved in protein processing (glycosylation, for example), but other features, such as the cholesterol content of the different cisternae, are also variable throughout the Golgi stack.[30] More recently, a post-Golgi compartment, known as the trans-Golgi network, has been described.[17] This compartment can be distinguished from the more proximal Golgi cisternae, both by its content of enzymatic markers, and by functional studies using transported proteins as markers of processing steps. Much work on this compartment has been carried out using virally-infected cells, in which both immunochemical and biochemical studies are facilitated because the entire synthetic machinery of the cell is devoted to producing and processing a small number of well-characterized proteins. When cells from normal rat kidney are infected with vesicular stomatitis virus and kept at 20°C for 2 h, the major viral membrane glycoprotein, G-protein, accumulates in the trans-Golgi network.[31] This G-protein can be readily localized with anti-G protein antibodies and protein A-gold, applied to ultrathin frozen sections of the cells (Figure 7). This accumulation identifies the trans-Golgi network, and double-labeling can be used to colocalize other molecules that are found associated with the network (either material being processed, or resident proteins). In Figure 7, the section has been double-labeled with an antibody against the cation-dependent mannose-6-phosphate receptor and 5 nm protein A-gold. Small but significant amounts of this receptor, which is involved in the sorting of lysosomal enzymes, are detected in this compartment.[16] This, then, is an example of how the technique of ultrathin frozen sectioning, coupled with immunogold labeling, can be used to specifically identify cellular compartments, and to localize molecules

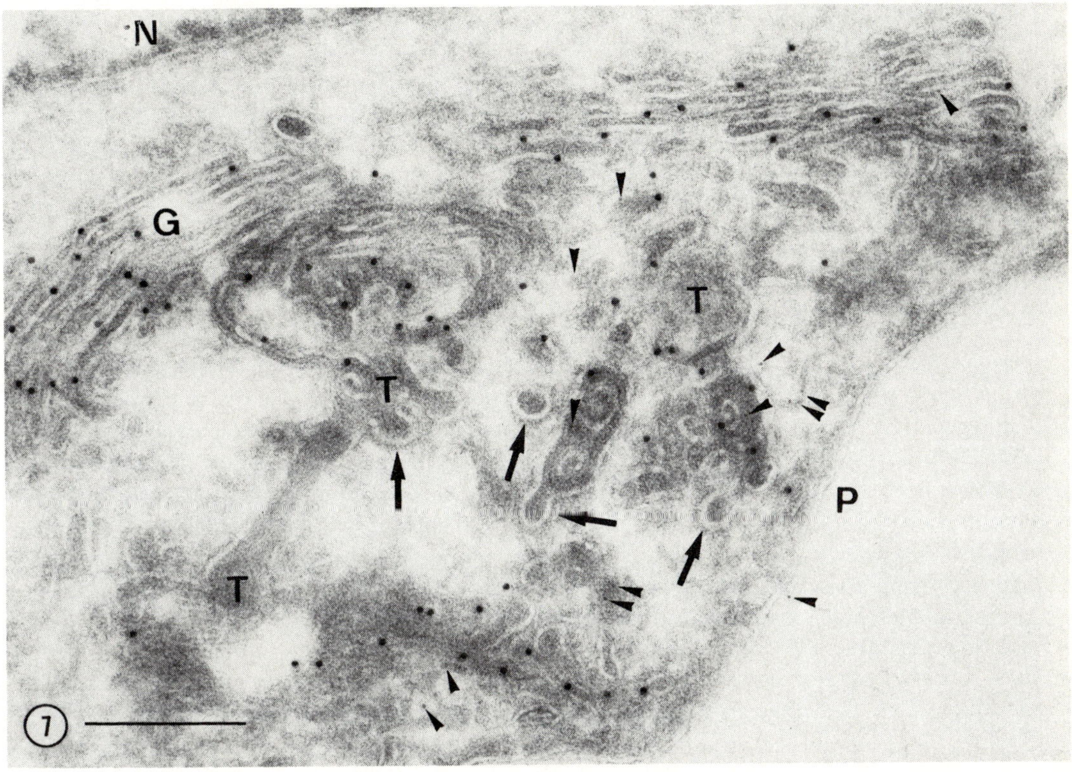

FIGURE 7. Ultrathin cryostat section of a normal rat kidney cell infected with vesicular stomatitis virus and kept at 20°C for 2 h. Under these conditions, the exit of the G protein (the coat glycoprotein) of the virus membrane from the trans-Golgi network (TGN — the last compartment; for details see Griffiths and Simons, 1986) is blocked. Large amounts of G protein accumulate in the now highly-distended TGN, as well as variable amounts in the earlier compartments of the Golgi stack. The section was double-labeled (indirectly) with an affinity-purified antibody against the luminal domain of the G protein (provided by Kai Simons), and 9 nm protein A-gold, followed by incubation with an affinity-purified antibody against the cation-independent mannose-6-phosphate receptor (MPR-provided by Bernard Hoflack) and 5 nm protein A-gold. The G-protein is used here as a marker for the TGN. Small but significant amounts of MPR are detected on the TGN (arrowheads) and on the plasma membrane (P). In this cell, the bulk of the MPR is detected in a late endosomal compartment in the vicinity of the TGN, but functionally distinct from it (not shown in this micrograph, but see Griffiths et al.[16] for details). N = nucleus. G = Golgi, T = trans-Golgi network. Bar = 0.25 μm. (Courtesy of Gareth Griffiths.)

within them that are involved in the processing and sorting of proteins along the synthetic pathway.

V. COATED VESICLES AND MEMBRANE RECYCLING

Recently, we have been examining the role of coated vesicles in cell specific function in the kidney collecting duct. In both the vasopressin-sensitive principal cells and the acid-secreting intercalated cells, the composition, and hence the function, of the plasma membrane is modulated by exocytosis and endocytosis of specific vesicles. In the case of the intercalated cell, these vesicles transport a proton pumping ATPase to and from the cell surface,[32-35] whereas in the principal cells they are believed to shuttle a proteinaceous water channel.[36] This modulation of cell function by membrane recycling is by no means unique to these kidney epithelial cells, but they provide an interesting system which can be used to generate information relevant not only to kidney function, but to the biology of epithelial cells in general. In both cell types, vesicles with a distinct cytoplasmic coat are involved in at least part of the shuttling process, but our studies have shown that the nature of the coating material is quite different in the two systems. In the principal cell, clathrin-coated vesicles are involved, whereas in the intercalated cell, the coat on the transporting vesicles is formed not of clathrin, but of the cytoplasmic domain of proton pumps, the major molecule carried by these vesicles. Much of the data has been obtained using immunocytochemical labeling of tissue embedded in Lowicryl K4M, either at low temperature, or using the rapid-embedding procedure described above.

A. RECYCLING OF WATER CHANNELS IN COLLECTING DUCT PRINCIPAL CELLS

In the kidney collecting duct, the principal cells can respond to antidiuretic hormone (vasopressin) by increasing the water permeability of their apical plasma membrane from a virtually water impermeable state in the absence of hormone, to a normal level of permeability that is generally associated with biological membranes. Over the last two decades, there has been a considerable amount of work devoted to the mechanism by which this dramatic alteration in the permeability of a plasma membrane can be brought about. The prevailing hypothesis is that proteinaceous water channels are located within the membranes of a subpopulation of cytoplasmic vesicles, and these vesicles fuse with the apical plasma membrane of vasopressin-sensitive cells upon stimulation with the hormone; water-permeable patches of membrane are, therefore, incorporated into the previously impermeable apical membrane of these cells.[36,37] This phenomenon is of the utmost physiological importance since it allows osmotic equilibration to occur between the urine within the tubule and the hypertonic interstitial fluid that bathes the tubule on the basolateral surface. In this way, a urine that is hypertonic to plasma can be produced, and water loss from the organism can be finely controlled. Most of the work on which this idea is based has been carried out on the epithelium of the toad urinary bladder, a widely used model of the collecting duct. These specialized membrane patches that are thought to contain water channels can be visualized using the freeze-fracture technique, where they appear as aggregates or clusters of intramembranous particles.[38,39] There is good evidence that such particles represent, at least in part, integral membrane proteins within the lipid bilayer. Since their first discovery in amphibian bladder,[38] a similar phenomenon has also been described in the toad epidermis (another vasopressin-sensitive epithelium), and , most importantly, in principal cells of the mammalian kidney collecting duct.[40] We have examined the recycling of these IMP clusters in rat kidney collecting ducts, and have found that they are internalized by a mechanism that involves clathrin-coated pits.[41]

Upon stimulation of collecting ducts from Brattleboro rats, a strain of Long-Evans rat that has hypothalamic diabetes insipidus due to a lack of vasopressin, IMP clusters appear on apical plasma membranes of principal cells in parallel with the commencement of urinary concentration by these animals.[40] Although direct evidence is lacking in the principal cell, by analogy with the mechanism in the amphibian bladder, the clusters are believed to be inserted into the apical plasma membrane by exocytosis of vesicles carrying preformed clusters on their limiting membranes. As long as hormone is present, these structures appear to continuously recycle rather than staying in place on the membrane. The evidence is that even in the continued presence of hormone, luminal tracers, such as horseradish peroxidase, are avidly endocytosed by the cell.[42] By thin section electron microscopy and the use of anti-clathrin antibodies to label thin sections of kidney embedded at low temperature in Lowicryl K4M, we have shown that clathrin-coated pits are involved in this internalization process (Figures 8, 9).[41,42] Furthermore, by a combination of thin-section (Figure 9) and freeze-fracture electron microscopy (Figures 10, 11) of tissues exposed to the antibiotic, filipin (which disrupts principal cell membranes but leaves clathrin-coated domains intact), we have shown that the IMP clusters are specifically located in clathrin-coated pits. Therefore, the recycling of putative water channels in the collecting duct appears to involve exocytosis of a population of as yet unidentified vesicles, followed by clathrin-mediated endocytosis. It is important to realize that this process is occurring at the luminal (apical) pole of these cells, which is in contact with the urine, and not at the basolateral plasma membrane, where vasopressin receptors are located. Thus, although the recycling of apical membrane molecules is stimulated by vasopressin, this event is quite distinct from the recycling of vasopressin receptors themselves, which presumably occurs at the basolateral pole of the cell. Finally, we have recently made use of this recycling phenomenon to examine the water permeability of recycled vesicles in collecting ducts from Brattleboro rats, in the presence or absence of vasopressin.[43] By allowing the uptake of 6-carboxyfluorescein into endocytotic vesicles, their water permeability can be measured using a fluorescence quenching technique. The rate of vesicle shrinkage upon exposure of isolated vesicles to a hypertonic solution, as well as the activation energy of this process, can indicate the presence of water channels in membranes. This technique demonstrates that endosomes in rat collecting ducts do, indeed, contain a water channel that is recycled in the presence of vasopressin. At present, we are attempting to purify the vesicles sufficiently to isolate the putative channel protein(s). If successful, immunocytochemical localization of the channel using either Lowicryl K4M embedded material, or ultrathin frozen sections, will be an indispensible part of our further studies on the mechanism of recycling of this physiologically important membrane component in the kidney.

B. RECYCLING OF A PROTON-PUMPING ATPase IN INTERCALATED CELLS

Intercalated cells of the kidney are members of a family of cells, known collectively as ''mitochondria-rich'' cells, that are found in a variety of transporting epithelia.[44] They are rich in carbonic anhydrase, they have similar membrane features when examined by freeze-fracture, and as their name suggests, they have large numbers of mitochondria when compared to adjacent cells in the same epithelium. These cells are responsible for proton transport in their respective epithelia; much of our current understanding of the role of these cells has come from studies using the toad, and in particular, the turtle urinary bladder, as model acidifying systems.[45] Several studies have demonstrated morphological alterations in these cells under conditions that are known to stimulate proton secretion,[45] and a large increase in apical membrane surface area is observed. In parallel with this increase in apical surface, there is a reduction in the number of cytoplasmic vesicles in the apical region of these cells.[34] It is now established that when proton secretion is stimulated, specialized cytoplasmic vesicles fuse with the apical membrane, and in doing so, insert proton pumps into the apical plasma membrane.[32-35] We have recently shown, using horseradish peroxidase as a marker of endocytosis,[46] that these vesices are also involved in the endocytotic uptake of plasma membrane components. Our recent work has focused on the localization of proton pumps in these intercalated cells, using immunocytochemical techniques and the rapid-freeze, deep-etch procedure, to examine the structure of the cytoplasmic domain of the pumps. The vesicles that transport pumps to and from the cell surface have a characteristic cytoplasmic coat of stud-like projections that we have shown to be unrelated to clathrin.[47] Thin sections of kidney embedded in Lowicryl K4M were labeled using a polyclonal anti-clathrin antibody (kindly provided by Daniel Louvard, Pasteur Institute, Paris), followed by protein A-gold. Whereas the clathrin-coated vesicles that line the apical plasma membrane of the proximal tubule were extensively labeled (Figure 12), the stud-coated vesicles in the intercalated cells were unlabeled (Figure 13). In the intercalated cells, however, small vesicles with the same diameter as true clathrin-coated vesicles were labeled, providing an internal control for the efficacy of the staining procedure.

To determine the structure of this coating material, the rapid-freeze, deep-etch procedure

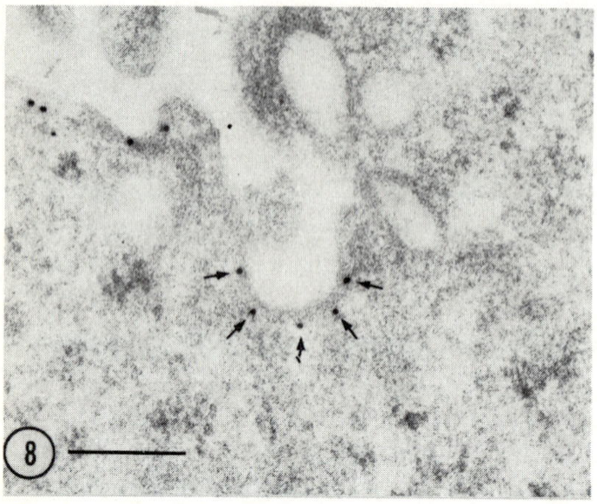

FIGURE 8. Region of apical plasma membrane from a collecting duct principal cell, from kidney embedded at low temperature in Lowicryl K4M. The section was incubated with an anticlathrin antibody (provided by Daniel Louvard, Pasteur Institute, Paris), followed by protein A-gold. This procedure was used to identify the membrane coating material on the apical plasma membrane of principal cells as clathrin. The gold particles (arrows) label a coated pit on the apical membrane, similar to the one shown in Figure 9. Bar = 0.25 μm.

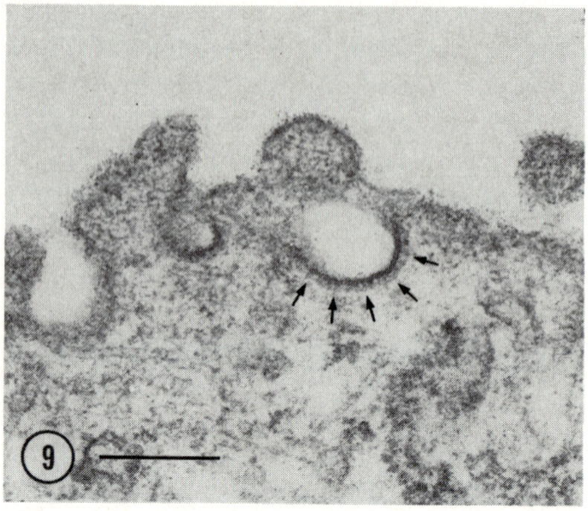

FIGURE 9. Thin section of part of the apical plasma membrane of a homozygous Brattleboro rat that had received exogenous vasopressin. After fixation, kidney tissue was treated with filipin and plasma membranes of collecting duct epithelial cells were disrupted. However, some membrane domains corresponding to clathrin coated pits were spared the action of the antibiotic (arrows). These regions were also visible in freeze-fracture replicas, and contained the vasopressin-induced intramembranous particle aggregates (see Figure 11). Bar = 0.25 μm. (From Brown, D. and Orci, L., *Nature,* 302, 253, 1983. With permission.)

was used, and we were able to obtain images of the cytoplasmic faces of both stud-coated vesicles and portions of the plasma membrane from acid-secreting mitochondria-rich cells of the toad urinary bladder.[32] These views revealed that the material formed paracrystalline, hexagonally-packed arrays of 10 nm studs. Suspecting that this material might represent the cytoplasmic domain of proton pumps that are transported by these vesicles, we examined the structure of bovine kidney proton pumps that had been affinity-purified on a monoclonal antibody column. These purified pumps were incorporated into liposomes, and were examined by rapid-freezing and deep-etching. The purified protein produced arrays in the liposome membrane that were identical to those seen in the epithelial cells.[32]

Using the affinity-purified proton pump as an antigen, polyclonal antibodies were raised in rabbits, and used for the immunocytochemical localization of proton pumps in rat kidney. By light microscopy, all intercalated cells were heavily labeled using the antibodies against different pump subunits, and in the kidney medulla, the label was always concentrated at the apical pole of the cells. Using sections cut from kidney embedded in Lowicryl K4M, immunogold labeling was found on the cytoplasmc side of all stud-coated membranes, including the cytoplasmic vesicles (Figures 14-16).[32] This finding is consistent with physiological data showing that the medullary collecting tubule is involved in hydrogen ion secretion. In the cortex, however, a more complex picture emerged. Some intercalated cells also had uniquely apical labeling, but others showed a clearly-defined basolateral localization of proton pumps.[48] This was confirmed by gold labeling of K4M embedded sections, thus demonstrating directly that cells with opposite polarity with respect to proton pump localization coexist in the cortical collecting duct, as proposed on the basis of earlier studies.[49] It is also consistent with physiological data showing that the cortical collecting duct can secrete either net bicarbonate or acid under different physiological conditions. Our current work is aimed at determining whether, under different conditions of acid-base balance, these cells can reverse their polarity of proton pumping.

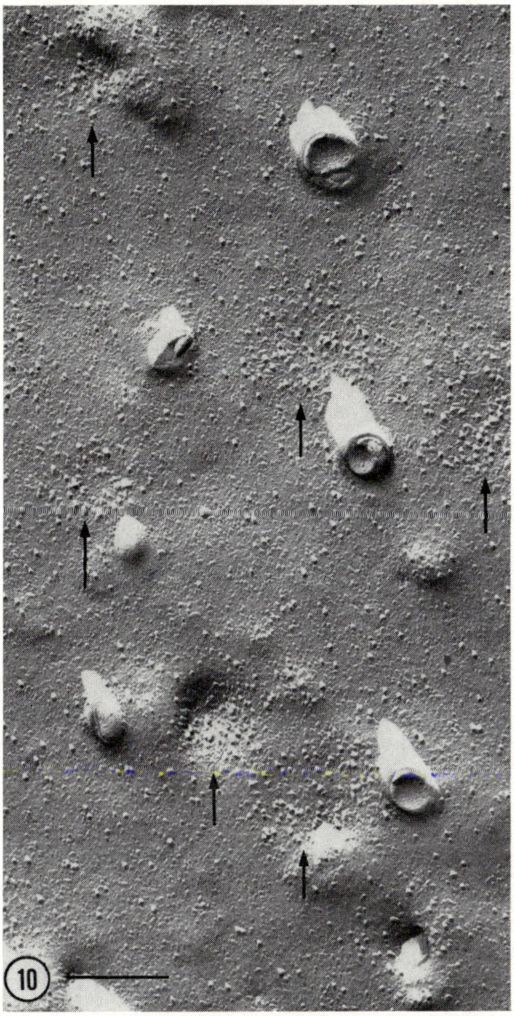

FIGURE 10. Freeze-fracture replica of the apical plasma membrane P-face of a principal cell from a vasopressin-treated Brattleboro rat. Many clusters of intramembranous particle aggregates (arrows) are present on the membrane. These are induced by vasopressin, and are believed to be water permeable patches in the membrane that contain proteinaceous water channels. As demonstrated in Figures 9 and 11, they correspond to clathrin-coated pits on the cell surface. Bar = 0.25 μm. (From Brown, D. and Orci, L., *Nature,* 302, 253, 1983. With permission.)

FIGURE 11. Freeze-fracture replica of a membrane region similar to that in Figure 10, but the tissue was exposed to filipin before freeze-fracture. Filipin deforms the structure of the plasma membrane in all areas, except those occupied by vasopressin-induced clusters. These are visible as round to oval patches of membrane that are devoid of filipin-sterol complexes (pits and protuberances). By comparing this image with that in Figure 9, it can be concluded that the vasopressin-induced clusters are present in clathrin-coated pits, since these two structures are the only membrane domains that are not deformed by the antibiotic. Bar = 0.25 μm. (From Brown, D. and Orci, L., *Nature*, 302, 253, 1983. With permission.)

This observation also poses important questions concerning protein targeting in these cells. As discussed earlier, membrane components must be selectively inserted into specific membrane domains within cells in order for the cell to function. How, then, can a proton pump be inserted into opposite domains in subpopulations of intercalated cells? One possibility is that the subunit composition of this multisubunit enzyme is not identical in the two domains, and that one of the subunits confers specificity on the insertion process.

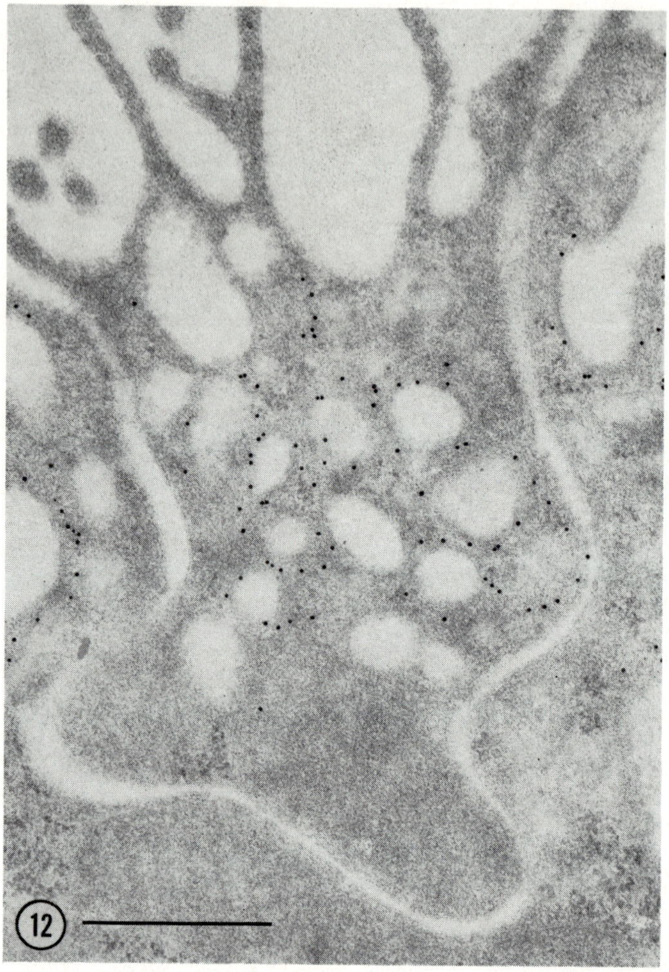

FIGURE 12. Thin section of the apical region of a proximal convoluted tubule cell, embedded in Lowicryl K4M. The section was incubated with an anti-clathrin antibody (provided by Dr. Daniel Louvard, Pasteur Institute, Paris), followed by protein A-gold (15 nm particles). Many apical vesicles and/or invaginations are labeled, indicating the presence of a clathrin coat. Compare this positive labeling with the labeling pattern of the stud-coated vesicles in the intercalated cell shown in Figure 13. Bar = 0.5 μm.

However, we currently have antibodies against four of the pump subunits that together make up almost 80% of the mass of the pump, and all four antibodies colocalize in the same membrane regions.

In summary, we have used a combination of techniques to identify a novel coating material found on intercalated cell vesicles that are involved in a membrane recycling process during which a proton-pumping ATPase is inserted into and removed from the plasma membrane of these cells. Using light and ultrastructural immunocytochemistry, subpopulations of intercalated cells with opposite polarity of the pump were identified. As more antibodies become available, we will search for clues that may indicate how proton pumps are targeted towards different parts of the plasma membrane in these cells. We also hope to use antibody probes on K4M embedded material to determine how and where in the cell the pump is assembled from its component parts, some of which are transmembrane (and may be synthesized in the rough endoplasmic reticulum), while others are peripheral membrane proteins, and are probably synthesized on free ribosomes.

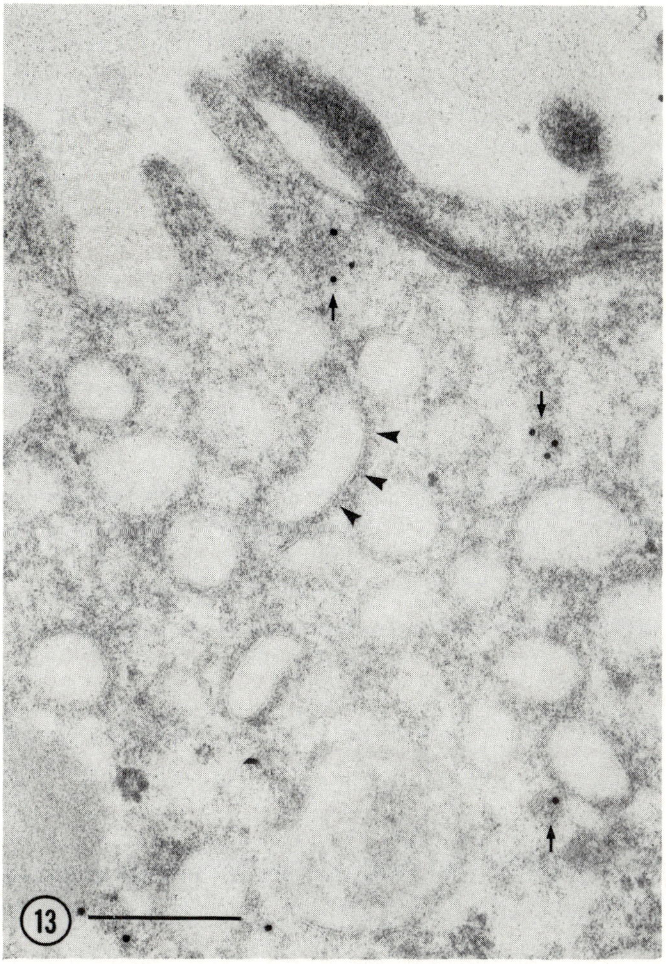

FIGURE 13. Thin section of part of an intercalated cell from a cortical collecting tubule adjacent to the proximal tubule shown in Figure 12. The large vesicles (arrowheads) that characterize intercalated cells are unlabeled with anticlathrin antibodies, whereas smaller, clathrin-coated vesicles (arrows) are labeled with 15 nm gold particles in these cells. In this way, it was shown that the large pool of specialized recycling vesicles in these cells are coated with a material antigenically distinct from clathrin. Bar = 0.25 μm. (From Brown, D. and Orci. L, *Am. J. Physiol.*, 250, C605, 1986. With permission.)

VI. SUMMARY

In order for all eukaryotic cells to function, both newly-synthesized and recycled membrane components must be accurately delivered to precise destinations within the cell. Furthermore, other proteins must be sorted into separate pathways that eventually lead to a variety of fates, such as exocytotic secretion or incorporation into lysosomes. A variety of methods have been used to address the mechanisms by which this complex machinery is regulated, and it is apparent that new techniques that permit the detection of intracellular and cell surface molecules with a high level of resolution have contributed greatly to our emerging picture of these processes. In this review, specific examples of the use of ultrathin frozen sections and Lowicryl K4M-embedded tissue have been described to illustrate their

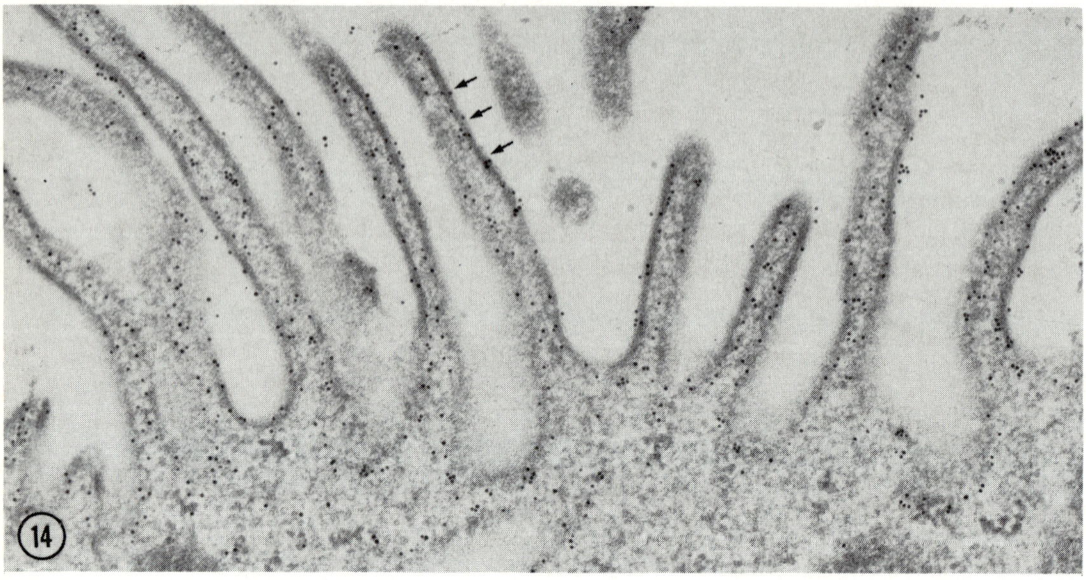

FIGURE 14. Thin section of Lowicryl-embedded kidney showing the apical microvilli and microplicae of an intercalated cell. This section was incubated with a specific, affinity-purified antibody against the 70 kDa subunit of bovine medullary hydrogen ATPase, followed by protein A-gold (8 nm particles). The gold particles are found mostly over a dense band on the cytoplasmic side of the apical membrane (arrows). This band corresponds to a stud-like coating material that we have shown to contain the cytoplasmic domains of proton pumps.[32] Bar = 0.25 μm.

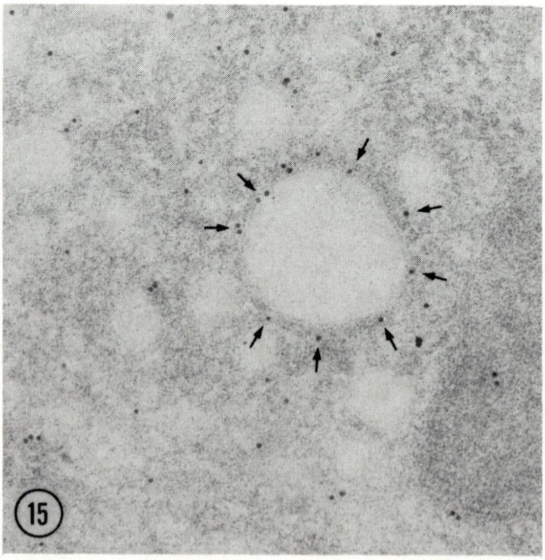

FIGURE 15. Thin section of an intercalated cell, embedded in Lowicryl K4M, showing specific labeling of a large vesicle in the cell, using the same anti-70 kDa antibody used in Figure 14, followed by protein A-gold (8 nm particles). This demonstrates that the specialized vesicles found in intercalated cells are coated with the cytoplasmic domains of proton pumps (arrows). These pumps are inserted into the plasma membrane by vesicle fusion to produce images such as the one in Figure 14. Bar = 0.25 μm.

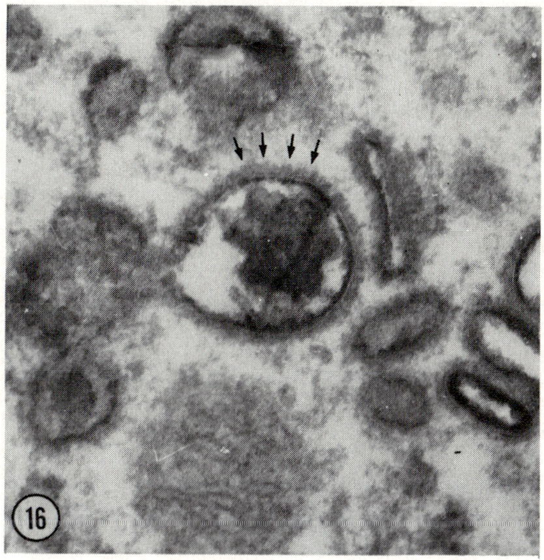

FIGURE 16. Thin section of an intercalated cell from an animal injected with horseradish peroxidase (HRP). The HRP is filtered by the glomerulus and moves into the collecting duct lumen, where it acts as a tracer for endocytosis. It is avidly internalized by a population of intercalated cells, and is mainly located in the specialized stud-coated vesicles (arrows) that carry the proton pumps to and from the cell surface. An immunolabeled vesicle of similar size is shown in Figure 15. Bar = 0.25 μm.

contributions to the study of the recycling of specific membrane components. However, it should be clear that only by combining the powerful tools of molecular biology, protein chemistry, and morphology, will we be able to unravel one of the most fundamental problems of cell biology.

ACKNOWLEDGMENTS

I am grateful to Dr. Gareth Griffiths of the European Molecular Biology Laboratory, Heidelberg, West Germany, for kindly providing Figure 7, and to Dr. Andrea Brown for editorial work on the manuscript. Many colleagues have contributed to various aspects of the work described in this review, and I am indebted to them all. This work has received continued support from the Swiss National Science Foundation, and since 1986, from the National Institutes of Health. D. B. is an Established Investigator of the American Heart Association.

REFERENCES

1. **Rindler, M. J., Ivanov, I. E., Rodriguez-Boulan, E., and Sabatini, D.D.,** Biogenesis of epithelial cell plasma membranes, in *Membrane Recycling, Ciba Foundation Symp.,* Vol. 92, Pitman, London, 1982, 184.
2. **Kelly, R.,** From organelle to organelle, *Nature,* 326, 14, 1987.
3. **Bendayan, M.,** Protein A-gold electron microscopic immunocytochemistry: methods,applicatons and limiations, *J. Electron Microsc. Tech.,* 1, 243, 1984.

4. **Roth, J.**, Post-embedding cytochemistry with gold-labelled reagents: a review, *J. Microsc.*, 143, 125, 1986.
5. **Carlemalm, E., Garavito, R. M., and Villiger, W.**, Resin development for electron microscopy and an analysis of embedding at low temperature, *J. Microsc.*, 126, 123, 1982.
6. **McLean, I. W. and Nakane, P. F.**, Periodate-lysine paraformaldehyde fixative: a new fixative for immunoelectron microscopy, *J. Histochem. Cytochem.*, 22, 1077, 1974.
7. **Armbruster, B. L., Carlemalm, R., Chiovetti, R., Garavito, R. M., Hobot, J. A., Kellenberger, E., and Villiger, W.**, Specimen preparation for electron microscopy using low temperature embedding resins, *J. Microsc.*, 126, 77, 1982.
8. **Altman, L. G., Schneider, B. G., and Papermaster, D. S.**, Rapid embedding of tissues in Lowicryl K4M for immunoelectron microscopy, *J. Histochem. Cytochem.*, 32, 1217, 1984.
9. **Tokuyasu, K. T.**, Immunochemistry on ultrathin frozen sections, *Histochem. J.*, 12, 381, 1980.
10. **Griffiths, G., Simons, K., Warren, G., and Tokuyasu, K. T.**, Immunoelectron microscopy using thin, frozen sections. Application to studies of the intracellular transport of Semliki Forest virus spike glycoproteins, *Methods Enzymol.*, 96, 466, 1983.
11. **Brown, D., Roth, J., and Orci, L.**, Lectin-gold cytochemistry reveals intercalated cell heterogeneity along rat kidney collecting ducts, *Am. J. Physiol.*, 248, C348, 1985.
12. **Farquhar, M. G. and Palade, G. E.**, The Golgi apparatus (complex)-(1954-1981)-from artifact to center stage, *J. Cell Biol.*, 91, 77s, 1981.
13. **Tartakoff, A. M.**, The confined function model of the Golgi complex: center for ordered processing of biosynthetic products of the rough endoplasmic reticulum, *Int. Rev. Cytol.*, 85, 221, 1983.
14. **Roth, J. and Berger, E.**, Immunocytochemical localization of galactosyltransferase in HeLa cells: codistribution with thiamine pyrophosphatase in *trans* Golgi cisternae, *J. Cell Biol.*, 93, 223, 1982.
15. **Dunphy, W. G., Brands, R., and Rothman, J. E.**, Attachment of terminal n-acetylglucosamine to asparagine-linked oligosaccharides occurs in central cisternae of the Golgi stack, *Cell*, 40, 463, 1985.
16. **Griffiths, G., Hoflack, B., Simons, K., Mellman, I., and Kornfeld, S.**, The mannose-6-phosphate receptor and the biogenesis of lysosomes, *Cell*, 52, 329, 1988.
17. **Griffiths, G. and Simons, K.**, The trans Golgi network: sorting at the exit site of the Golgi complex, *Science*, 234, 438, 1986.
18. **Balch, W. E., Dunphy, W. G., Braell, W. A., and Rothman, J. E.**, Reconstitution of the transport of protein between successive compartments of the Golgi measured by the coupled incorporation of N-acetylglucosamine, *Cell*, 39, 405, 1984.
19. **Bergmann, J. E. and Singer, S. J.**, Immunoelectron microscopic studies of the intracellular transport of the membrane glycoprotein (G) of vesicular stomatitis virus in infected Chinese hamster ovary cells, *J. Cell Biol.*, 97, 1777, 1983.
20. **Griffiths, G., Quinn, P., and Warren, G**, Dissection of the Golgi complex. I. Monensin inhibits the transport of viral membrane proteins from *medial* to *trans* Golgi cisternae in baby hamster kidney cells infected with Semliki forest virus, *J. Cell Biol.*, 96, 835, 1983.
21. **Quinn, P., Griffiths, G., and Warren, G.**, Dissection of the Golgi complex. II. Density separation of specific Golgi functions in virally infected cells treated with monensin, *J. Cell Biol.*, 96, 851, 1983.
22. **Kellenberger, E., Durrenberger, M., Villiger, W., Carlemalm, E., and Wurtz, M.**, The efficiency of immunolabel on Lowicryl sections compared to theoretical predictions, *J. Histochem. Cytochem.*, 35, 959, 1987.
23. **Griffiths, G. and Hoppeler, H.**, Quantitation in immunocytochemistry: correlations of immunogold labeling to absolute number of membrane antigens, *J. Histochem, Cytochem.*, 34, 1389, 1986.
24. **Bendayan, M., Nanci, A., and Kan, F. W. K.**, Effect of tissue processing on colloidal gold cytochemistry, *J. Histochem. Cytochem.*, 35, 983, 1987.
25. **Posthuma, G., Slot, J. W., and Geuze, H. J.**, Usefulness of the immunogold technique in quantitation of a soluble protein in ultra-thin sections, *J. Histochem. Cytochem.*, 35, 405, 1987.
26. **Goldstein, J. L., Anderson, R. G. W., and Brown, M. S.**, Coated pits, coated vesicles and receptor-mediated endocytosis, *Nature*, 279, 679, 1979.
27. **Mellman, I., Fuchs, R., and Helenius, A.**, Acidification of the endocytic and exocytic pathways, *Ann. Rev. Biochem.*, 55, 663, 1986.
28. **Geuze, H. J., Slot, J. W., Strous, G. J., Peppard, J., von Figura, K., Hasilik, A., and Schwartz, A. L.**, Intracellular receptor sorting during endocytosis: comparative immunoelectron microscopy of multiple receptors in rat liver, *Cell*, 37, 195, 1984.
29. **Orci, L., Ravazzola, M., Amherdt, M., Brown, D., and Perrelet, A.**, Transport of horseradish peroxidase from the cell surface to the Golgi in insulin-secreting cells: preferential labeling of cisternae located in an intermediate position in the stack, *EMBO J.*, 5, 2097, 1986.
30. **Orci, L., Montesano, R., Meda, P., Malaisse-Lagae, F., Brown, D., Perrelet, A., and Vassalli, P.**, Heterogeneous distribution of filipin-cholesterol complexes across the cisternae of the Golgi apparatus, *Proc. Natl. Acad. Sci.*, U.S.A., 78, 293, 1981.

31. **Griffiths, G., Pfeiffer, S., Simons, K., and Matlin, K.,** Exit of newly synthesized membrane proteins from the *trans* cisternae of the Golgi complex to the plasma membrane, *J. Cell Biol.*, 101, 949, 1985.
32. **Brown, D., Gluck, S., and Hartwig, J.,** Structure of the novel membrane-coating material in proton-secreting epithelial cells and identification as an H^+ATPase, *J. Cell Biol.*, 105, 1637, 1987.
33. **Gluck, S., Cannon, C., and Al-Awqati, Q.,** Exocytosis regulates urinary acidification in turtle bladder by rapid insertion of H^+ATPase into the luminal membrane, *Proc. Natl. Acad. Sci. U.S.A.*, 79, 4327, 1982.
34. **Madsen, K. and Tisher, C. C.,** Structural-functional relationships along the distal nephron, *Am. J. Physiol.*, 250, F1, 1986.
35. **Stetson, D. L. and Steinmetz, P. R.,** Role of membrane fusion in CO_2 stimulation of proton secretion by turtle bladder, *Am. J. Physiol.*, 245, C113, 1983.
36. **Hays, R. M.,** Alteration of luminal membrane structure by antidiuretic hormone, *Am. J. Physiol.*, 245, C289, 1983.
37. **Wade, J. B., Stetson, D. L., and Lewis, S. A.,** ADH action: evidence for a membrane shuttle mechanism, *Ann. N.Y. Acad. Sci.*, 372, 106, 1981.
38. **Chevalier, J., Bourguet, J., and Hugon, J. S.,** Membrane associated particles: distribution in frog urinary bladder epithelium at rest and after oxytocin treatment, *Cell Tissue Res.*, 152, 129, 1974.
39. **Kachadorian, W. A., Levine, S. D., Wade, J. B., DiScala, V. A., and Hays, R. M.,** Relationship of aggregated intramembranous particles to water permeability in vasopressin-treated toad urinary bladder, *J. Clin. Invest.*, 59, 576, 1977.
40. **Harmanci, M. C., Stern, P., Kachadorian, W. A., Valtin, H., and DiScala, V. A.,** Vasopressin and collecting duct intramembranous particle clusters: a dose-response relationship, *Am. J. Physiol.*, 239, F560, 1980.
41. **Brown, D. and Orci, L.,** Vasopressin stimulates formation of coated pits in rat kidney collecting ducts, *Nature,* 302, 253, 1983.
42. **Brown, D. Weyer, P., and Orci, L.,** Vasopressin stimulates endocytosis in kidney collecting duct principal cells, *Eur. J. Cell Biol.*, in press.
43. **Verkman, A. S., Lencer, W., Brown, D., and Ausiello, D. A.,** Endosomes from kidney collecting tubule cells contain the vasopressin-sensitive water channel, *Nature,* in press.
44. **Orci, L., Humbert, F., Brown, D., and Perrelet, A.,** Membrane ultrastructure in urinary tubules, *Int. Rev. Cytol.*, 73, 183, 1981.
45. **Steinmetz, P. R.,** Cellular organization of urinary acidification, *Am. J. Physiol,* 251, F173, 1986.
46. **Brown, D., Weyer, P., and Orci, L.,** Non-clathrin coated vesicles are involved in endocytosis in kidney collecting duct intercalated cells, *Anat. Rec.*, 218, 237, 1987.
47. **Brown, D. and Orci, L.,** The ''coat'' of kidney intercalated cell tubulovesicles does not contain clathrin, *Am. J. Physiol,* 250, C605, 1986.
48. **Brown, D., Hirsch, S., and Gluck, S.,** An H^+ ATPase is present in opposite plasma membrane domains in kidney epithelial cell subpopulations, *Nature,* 331, 622, 1988.
49. **Schwartz, G. L., Barasch, J., and Al-Awqati, Q.,** Plasticity of functional epithelial cell polarity, *Nature,* 318, 368, 1985.

Chapter 11

PRE-EMBEDDING LABELING TECHNIQUES APPLICABLE TO INTRACELLULAR BINDING SITES

Margit Pavelka and Adolf Ellinger

TABLE OF CONTENTS

I. INTRODUCTION

Pre-embedding affinity-cytochemistry is one in a range of approaches to analyze the interior of cells. The electron microscopic techniques considered in this contribution include (a) immunocytochemical approaches that make it possible to specifically label intracellular and cell surface molecules, thus determining their sites in the various cell compartments,[1,2,3] and (b) cytochemical approaches that make use of lectin-sugar interactions, and thus permit the localization of classes of glycoconjugates.[4] It was affinity-cytochemical work, using pre-embedding as well as postembedding techniques and ultrathin frozen sections, that considerably helped to elucidate the organization of cell systems. A great number of findings obtained by means of pre-embedment affinity-cytochemical procedures concern compartments of the major intracellular pathways, i.e., the biosynthetic, endocytic, and recycling pathways. Cell products and internalized molecules have been demonstrated in the organelles involved in the respective routes; secretory molecules, constituents of endomembranes and plasma membrane, as well as lysosomal enzymes, have been localized in the endoplasmic reticulum, transport vesicles, elements of the Golgi apparatus, and various Golgi-derived vesicles, thus labeling biosynthetic routes. Molecules taken up into cells by endocytosis and plasma membrane molecules following internalization have been recovered cytochemically in cell organelles thus helping to unravel endocytic and recycling routes. Furthermore, pre-embedment affinity-cytochemical work has been successful in intracellularly localizing a number of molecules that play key roles in the above-mentioned pathways; these include enzymes of the glycan-processing system, as well as receptor molecules mediating intracellular traffic. (For this aspect, see also Chapters 10 by Brown and 8 by Robenek.)

This contribution concentrates on findings obtained by means of pre-embedment affinity-cytochemical studies that proved to be of great value for elucidating processes of biosynthetic, endocytic, and recycling cellular pathways.

Details of the pre-embedding labeling techniques will be discussed on the basis of lectin-cytochemical preparations performed in the authors' laboratory;[5,6,7] these studies mainly concern the Golgi apparatus and its subcompartments which play a major role in biosynthetic as well as endocytic and recycling routes. (For alternative postembedding techniques, see Chapter 12 by Roth).

II. METHODOLOGY

The pre-embedding affinity-cytochemical procedures used in the authors' laboratory are modifications of a technique originally described by Bernhard and Avrameas.[8] Variations of this technique have been widely employed for the cytochemical demonstration of lectin-sugar interactions in a variety of tissues.[5-10] In the following pages, the procedures used for intracellularly localizing the binding sites of a range of lectins of diverse sugar specificities are described. Various cell types were studied, such as absorptive cells and goblet cells of the rat small intestine at various stages of differentiation, rat hepatocytes, secretory cells of the rat pancreatic acini (adults and during embryonic development), epithelial cells of rat and mouse thyroid follicles, as well as cultured rat fibroblasts.

A. LECTIN-CYTOCHEMICAL PROCEDURES

The tissues and cultured fibroblasts were fixed in a mixture of 0.5% glutaraldehyde (electron microscopy grade, Merck, Darmstadt, F.R.G.) and 4% formaldehyde (freshly prepared from paraformaldehyde) in 0.1 M sodium cacodylate buffer, pH 7.2, for 30 min at 4°C, and subsequently rinsed in the same buffer containing 10% dimethyl sulfoxide (DMSO). For the purpose of blocking free aldehyde groups, the pieces of tissue were treated with 0.05 *M* ammonium chloride for 2 h at 30°C.

TABLE 1
Lectins Used in Pre-Embedding Affinity-Cytochemical Procedures and Their Sugar Specificities

Concanavalin A	Con A	αMan>αGlc>GlcNAc
Lens culinaris lectin	LCA[a]	αMan>αGlc>GlcNAc (+Fuc)
Pisum sativum lectin	PSA	αMan>αGlc = GlcNAc (+Fuc)
Ricinus communis I lectin	RCA I	βGal>αGal≫GalNAc
Erythrina cristagalli lectin	ECA	Galβ1,4GlcNAc>GalNAc>Gal
Helix pomatia lectin	HPA	αGalNAc>αGlcNAc≫Gal
Vicia villosa lectin	VVA	αGalNAc>αGal

Note: Man, mannose, Glc, glucose, GlcNAc, N-acetyl-glucosamine, GalNAc, N-acetyl-galactosamine, Gal, galactose, Fuc, fucose.

[a] "A", agglutinin.

After a 2 to 3 d rinse in DMSO-containing sodium cacodylate buffer, tissues were frozen to −30° in isopentane; 10 μm thick cryosections of the tissues were prepared and thawed in phosphate-buffered saline (PBS). DMSO serves not only as a cryoprotectant, but also as a permeabilizing agent.

Lectin incubations were made as follows. (The lectins used and their sugar-specificities are listed in Table I.) The authors used peroxidase conjugates of the lectins (E-Y Laboratories, San Mateo, CA), with the exception of ConA (Sigma Chem. Co., St. Louis, MO, grade IV); for the purpose of demonstrating ConA binding sites, a 2-step lectin-peroxidase procedure was employed.[8] The cryosections and cultured fibroblasts were incubated in PBS solutions of the lectin peroxidase conjugates.Tissue permeabilization was performed with the aid of saponin added to the incubation media at a concentration of 0.1 mg/ml; the lectin incubation media also contained 1 mg/ml bovine serum albumin. In preliminary experiments, the optimal lectin concentrations were tested; depending on the lectin used and the type of tissue to be examined, the concentrations ranged from 30 μg/ml to 200 μg/ml PBS; incubations were at room temperature for 3 to 6 h. The ConA incubation step was followed by treatment with horseradish peroxidase (Sigma, grade VI, 100 μg/ml PBS; room temperature; 6 h). The cryosections were continuously agitated during the incubations.

The fixation, permeabilization, and incubation procedures listed above have proved optimal for the subcellular demonstration of lectin-sugar interactions in various cells and tissues; this refers to all three parameters, preservation of lectin binding sites, penetration of incubation media throughout the tissues, and preservation of ultrastructural details.

In the field of immunocytochemistry (also taken into consideration in this review), it has been found that, for preservation of antigenicity combined with good tissue stabilization impeding diffusion and relocation of antigen, several other fixatives prove practicable. Fixatives and penetration enhancement procedures valuable in the field of immunocytochemistry, including the McLean and Nakane periodate-lysin-paraformaldehyde recipe[11] and the glutaraldehyde-borohydride-saponin protocol,[12] have recently been summarized.[1,2,3] Furthermore, immunocytochemical amplification procedures for improvement of the sensitivity by increasing the amount of visible reaction product have been found to be of great value, e.g., the PAP (peroxidase-antiperoxidase)[13] and ABC (avidin-biotin-complex) techniques.[14]

Subsequent to the lectin incubations, the specimens were rinsed in PBS and refixed in 1% glutaraldehyde for 1 h at 4°C. For visualization of the binding reactions, cryosections and cultured cells were incubated in diaminobenzidine (DAB, Serva Feinbiochemica, Heidelberg, F.R.G; 0.5 mg/ml 0.05M Tris-HCl buffer, pH 7.6) in the presence of H_2O_2 (1%, 20 μl/ml) for 15 min at room temperature. Following several rinses in bidistilled water, specimens were treated with 1% osmium ferrocyanide for 30 min and subsequently with 1%

veronal acetate-buffered OsO_4 for 8 h; afterwards, they were dehydrated in a graded series of ethanol and embedded in Epon. Ultrathin and semithin sections were examined unstained.

Oxidized DAB reaction products may diffuse through membranes of the reactive compartments and relocate onto cytoplasmic structures; this constitutes a serious problem in DAB affinity-cytochemistry, especially in immunocytochemical preparations performed with formaldehyde-fixed tissues. It has been shown that diffusion of oxidized DAB out of the reactive compartments can be prevented or greatly reduced by treatment of the tissues with hypertonic sucrose solutions following the immunocytochemical binding reactions.[15,16]

B. CONTROLS

Controls were performed by omitting the lectin incubation steps prior to treatment with peroxidase and/or DAB, thus excluding the demonstration of peroxidase binding molecules and of endogenous peroxidase, respectively. The specificity of the binding reactions was tested with the aid of the respective competitive sugars (0.1 to 0.3 M) that were added to the lectin incubation media.

By means of graded series of competitive sugars (0.001 to 0.6 M), it was possible to achieve a differential labeling of various lectin recognizing molecules that coexist in the tissues and interact with the same lectins, but differ as regards their binding affinities.[17]

III. PRE-EMBEDDING AFFINITY-CYTOCHEMICAL APPROACHES TO THE INTERIOR OF CELLS

A. LIGHT MICROSCOPY

Cryosections are monitored in the light microscope subsequent to lectin incubations and DAB reactions, and prior to osmication and embedding (Figure 1 a,b); this makes it possible to judge whether or not the reactions are homogeneous and appear in the entire piece of tissue, (Figure 1a) which is a basic requirement for the efficiency of subsequent electron microscopic studies.

Furthermore, a crude image of the reaction patterns can be obtained. It is possible to attribute reactions to constituents of tissues, e.g., to intercellular structures and to apical and basolateral surface regions of epithelial cells; reactions of intracellular compartments, such as reactions of the Golgi apparatus and of secretory granules, can be differentiated (Figure 1b). However, on the basis of the light microscopical image, it is impossible to judge whether the preparations qualify for subsequent electron microscopic examination. High quality light microscopic images showing intense reaction products and strong intracellular label were insufficient for subsequent ultrastructural exploration in the electron microscope. In these cases, the reactions proved to be too intense; tissues frequently appeared destructed. Diffusion of the DAB reaction products out of the reactive compartments was evident. Those preparations that were optimal for electron microscopic exploration, allowing attribution of the reactions to distinct cell compartments and subcompartments, showed slight reactions in the light microscope which were hardly discernible in the interior of cells.

B. ELECTRON MICROSCOPY

The low magnification micrograph in Figure 2a shows a cross-section throughout a 10 μm thick cryosection of duodenal villus tissue that was treated with lentil lectin. Reaction products are localized in various intracellular compartments, such as endoplasmic reticulum (ER), Golgi apparatus, and various vesicles. The micrograph demonstrates that (1) the reactions appear throughout the entire thickness of the cryosection, and (2) the quality of the reactions is different at the rims and in the central regions of the cryosection.

At the rims, the tissue is injured mechanically; reactive compartments are artificially opened; the cytoplasm is stained by oxidized DAB reaction products. Such regions do not qualify for ultrastructural exploration.

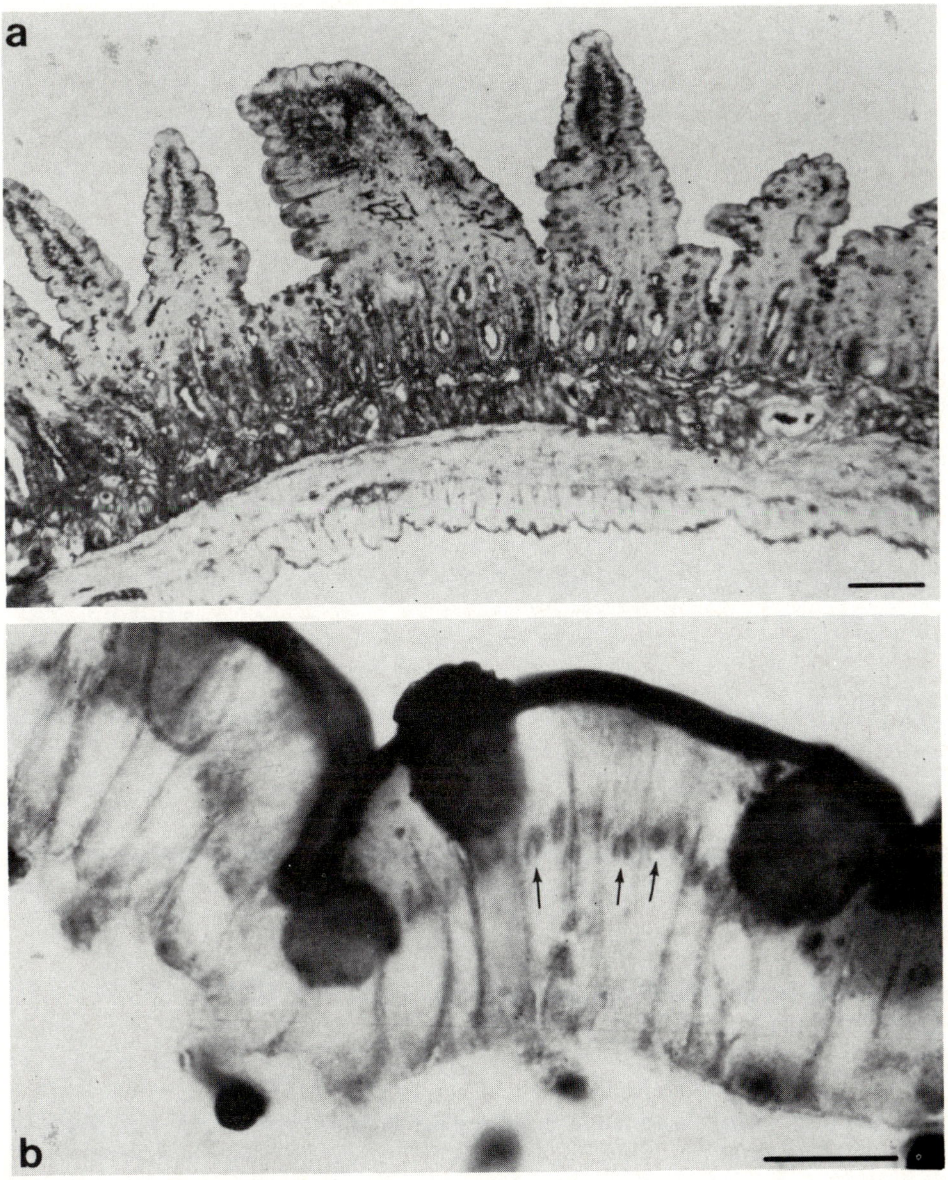

FIGURE 1. Localization of binding sites for the *Erythrina cristagalli* lectin (ECA) in the rat duodenum. (a) Low magnification micrograph showing the distribution of ECA reactions in the duodenal mucous and muscle layer. Bar = 100 μm. (b) At higher magnification, the reactions can be attributed to the brush border, basolateral surface regions, and to the Golgi area of absorptive enterocytes (arrows), as well as to goblet cell secretory granules. Bar = 10 μm.

In the central region of the cryosection, ultrastructural details are well preserved; reactions are confined to particular cell compartments and subcompartments of organelles. Only these central regions of the cryosections are of a quality that permits a clear attribution of the reactions to subcellular structures. Further details are shown in the higher magnification micrograph in Figure 2b: LCA reactions label segments of the endoplasmic reticulum and appear in Golgi apparatus cisternae and various vesicles; the reaction products are enclosed in the different compartments; there is no sign of a destruction of cell organelles and diffusion of DAB reaction products out of reactive compartments.

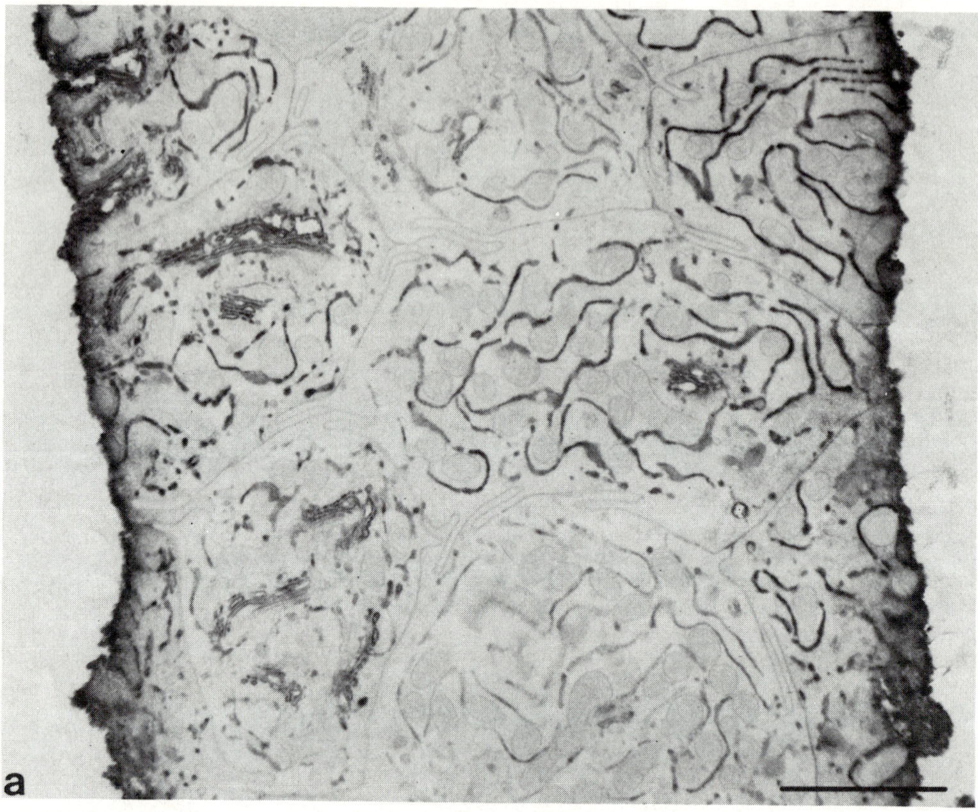

FIGURE 2. Localization of *Lens culinaris* lectin (LCA) binding sites. (a) Low magnification electron micrograph showing a cryosection of villus epithelium of the rat duodenum cut perpendicularly to its surfaces; LCA reactions are throughout the entire cryosection thickness. Reactions of the lentil lectin appear in the ER, Golgi apparatus, and in various vesicles. Bar = 2 μm.

This indicates that, for precise localization of reactions in cell compartments and subcompartments, preparations are required in which the reactions are apparent in the interior of the cryosections, at best throughout the entire cryosection thickness, or, in the cases of cell cultures, throughout the layer of cells. The percentages of cryosections fulfilling this requirement greatly differ, depending on the respective lectin preparations, i.e., on the lectin used and the type of tissue examined. Percentages range from approximately 30% in liver cells treated with ConA, and approximately 80 to 85% in duodenal villi treated with the pea and lentil lectins, as well as in embryonic pancreatic tissue treated with ConA, RCA I, and HPA. By means of varying the lectin concentrations (30 to 200 μg/ml PBS), as well as the times of incubation with these lectins (3 to 6 h), the percentages of cryosections qualifying for further electronic microscopic studies could be increased. Preliminary experiments served to determine the optimal conditions for each type of tissue.

Flat sections throughout the central cryosection regions are ideal objects for the ultrastructural exploration of binding reactions. Such sections make it possible to survey the reactions obtained (Figures 2a, 5a), which is one of the advantages of pre-embedment preparations using peroxidase label. The dense DAB reaction products can be easily attributed to the various cellular compartments and extracellular structures, which allows us to compare different tissue regions and cell types, and to discern variations of the reaction patterns.

Higher magnifications permit the reactions in the diverse cell compartments to be analyzed. Examples are to be found in the illustrations that show binding sites for ConA (Figure

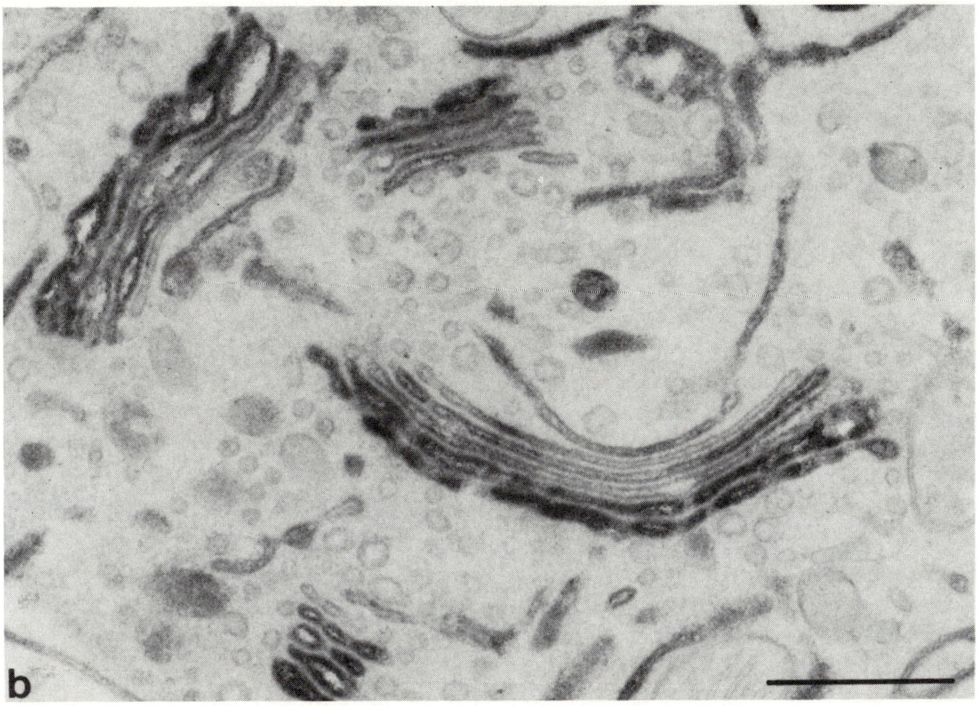

FIGURE 2(b). The higher magnification micrograph of the same preparation shows that the reaction products are confined to particular cell compartments, i.e., cisternae of the ER, Golgi cisternae, and vesicles. The reactions are concentrated at the cis side of the Golgi apparatus stacks. There is no indication for diffusion of reaction products out of reactive compartments; the cytoplasm adjacent to intensely labeled compartments appears unstained; strongly stained Golgi apparatus cisternae are located side by side with unreactive saccules. Within the individual Golgi cisternae a "mosaic pattern" is apparent; intensely stained regions are adjacent to weakly labeled and unreactive areas, thus demonstrating lateral diffusion of DAB reaction products might occur only to a limited extent. Bar = 0.5 μm.

3a), RCA I (Figures 4b, c), the pisum (Figure 5a), lens (Figures 2b, 5b), and *Erythrina* (Figure 3b) lectins, and HPA (Figure 4a) in various cell compartments, such as the endoplasmic reticulum, Golgi apparatus elements, and in lysosomes. Reaction products are either localized in the luminal spaces of the respective compartments or attached to the surfaces of the limiting membranes. The peroxidase reaction products do not permit a quantitative analysis of the binding sites, but the patterns obtained allow a determination of compartment subregions in which the respective binding sites are concentrated. Frequently, a "mosaic-like" image of reactions is apparent in individual compartments, such as in endoplasmic reticulum (Figures 2b, 5b) and Golgi apparatus cisternae (Figures 2b, 4c); reaction products are densely packed in particular regions that are adjacent to weakly stained and unreactive areas. This shows that, in spite of a possible lateral diffusion of oxidized DAB reaction products within a given compartment, a differentiated labeling of compartment subregions can be achieved. Possible limitations are discussed below.

In extracellular spaces, diffusion distances of DAB reaction products have been found to be >100 nm, and readsorption of the reaction products onto negatively charged cell surfaces has been demonstrated;[18] it has been shown that a control of these diffusion artifacts is possible by reducing the amount of DAB generated, thus achieving a resolution of the immunoperoxidase method of <60 nm.[18] At the cell surface of ganglion cells, a lateral spread of DAB reaction products averaging 55 to 80 nm has been observed;[19] another type of increased staining that coincides with the thickness of the plasma membrane has been

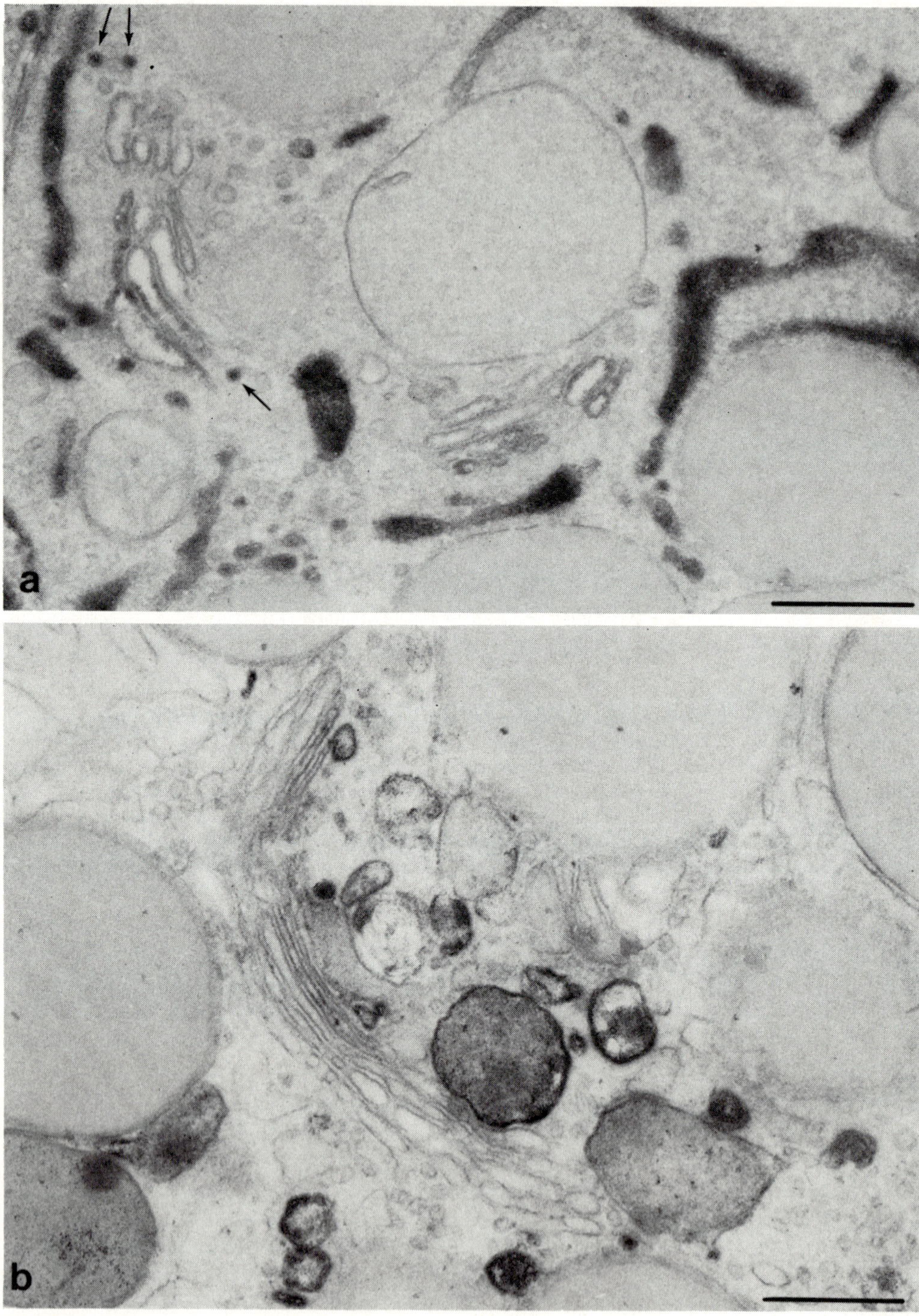

FIGURE 3. Localization of binding sites of Concanavalin A (ConA) and *Erythrina cristagalli* lectin (ECA). (a) ConA — embryonic pancreatic acinar cell of the rat — d 19 of gestation. ConA reactions are intense in cisternae of the endoplasmic reticulum, in transport vesicles (arrows) and cisternae of the cis Golgi side; medial and trans Golgi cisternae are slightly labeled or unreactive. Bar = 0.5 μm. (b) ECA — rat pancreatic acinar cell. Reactions for ECA appear in elements of the trans Golgi side and in various vesicles, whereas endoplasmic reticulum and cis Golgi cisternae are devoid of label. Bar = 0.5 μm.

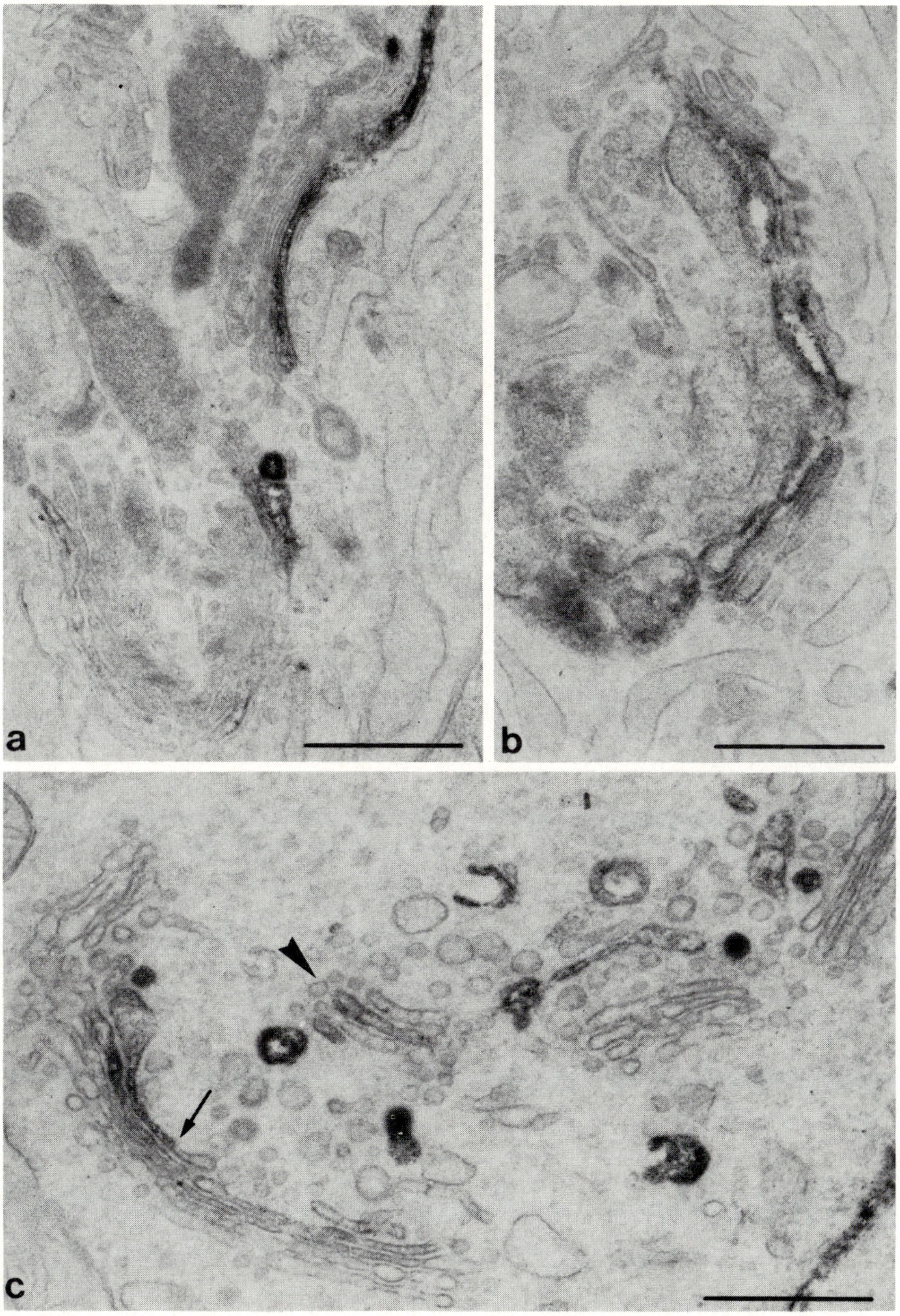

FIGURE 4. Localization of binding sites of *Helix pomatia* lectin (HPA) and *Ricinus communis* I agglutinin (RCA I) (a) HPA, goblet cell. HPA reactions are confined to one cis Golgi apparatus cisterna. Bar = 0.5 μm. (b) RCA I, goblet cell. Medial and trans cisternae are reactive for RCA I. Bar = 0.5 μm. (c) RCA I — embryonic pancreatic acinar cell — d 16 of gestation. RCA I reactions are concentrated in trans and transmost Golgi apparatus cisternae and in several Golgi-associated vesicles; one of the stacks (arrowhead) shows RCA I label in medial cisternae as well. In the individual cisternae, intensely stained regions are adjacent to unreactive areas (arrow). Bar = 0.5 μm.

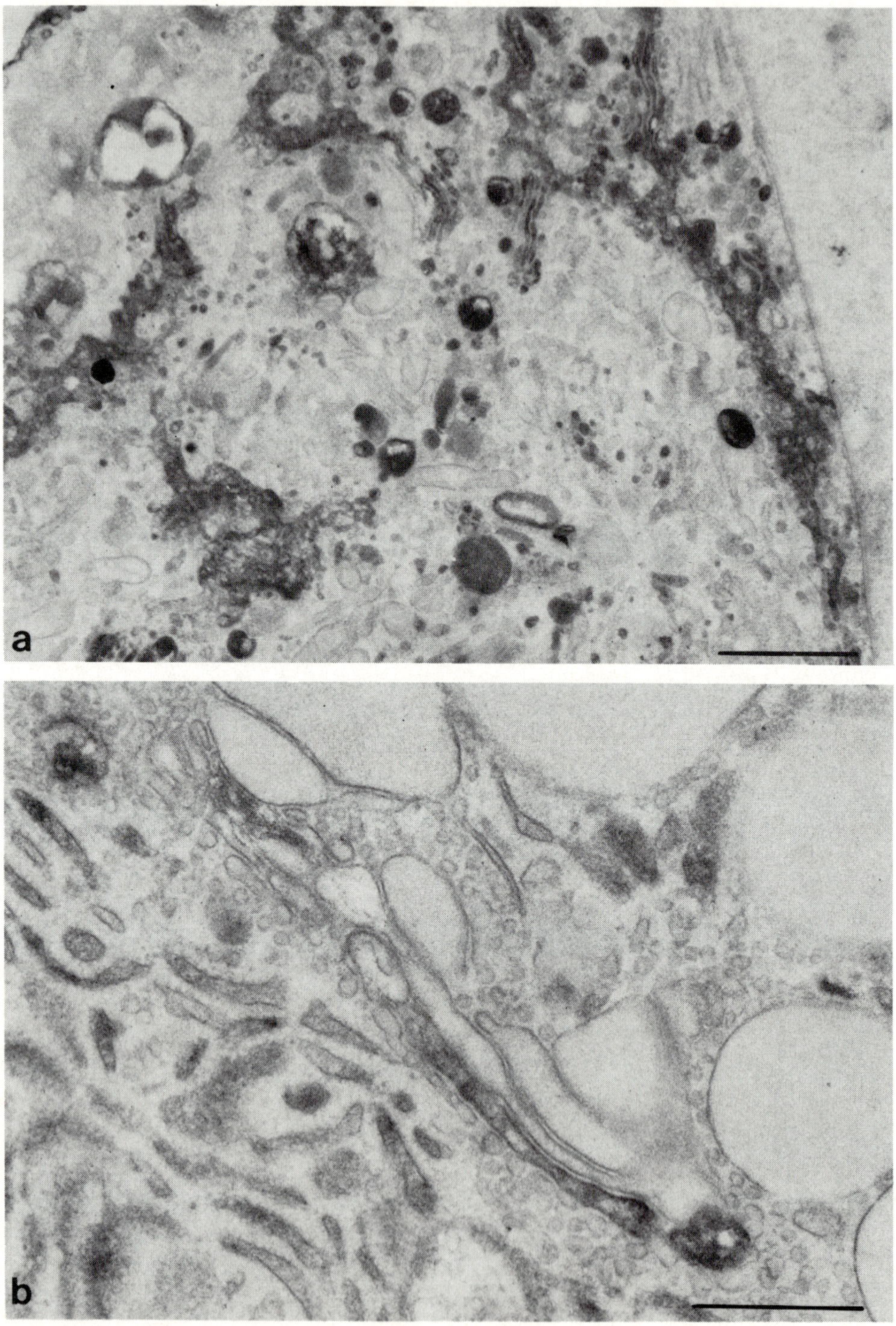

FIGURE 5. Localization of *Pisum sativum* and *Lens culinaris* lectin (PSA; LCA) binding sites. (a) PSA — fibroblast. This low magnification micrograph shows the distribution of PSA reactions in various cell compartments; the reactins are slight in the ER; intense reactions label Golgi apparatus cisternae, various vesicles and lysosomes. Bar = 1 μm. (b) LCA — pancreatic acinar cell. LCA reaction products are enclosed in ER cisternae and one cisterna at the cis side of this Golgi stack; in the ER, the reactions are concentrated in distinct subregions of the cisternae. Bar = 0.5 μm.

shown to extend approximately 160 to 340 nm from the sites of binding.[19] Roughly speaking, in luminal spaces and at the surfaces of endomembranes, the situation as regards diffusion of DAB reaction products may be compared to that in extracellular spaces and at the plasma membrane, respectively. Hence, dense diffusion zones originating from the reaction sites and measuring 100 nm and more have to be expected. On the other hand, in the central cryosection regions of our preparations, there is no sign for a diffusion of DAB reaction product out of reactive compartments (Figures 2b, 3a, b, 4a, b, c, 5b), and "mosaic patterns" (Figures 2b, 4c, 5b) demonstrate that lateral diffusion of reaction products within a given compartment is limited; this shows that, under the conditions used in our preparations, diffusing reaction products do not freely disseminate in a compartmental luminal space or spread all over a compartment surface. "Mosaic patterns" demonstrate that pre-embedment peroxidase labeling is a qualified method for differentiating binding reactions in distinct intracellular (sub-) compartments. Subregions of the endoplasmic reticulum (Figures 2b, 5b) and of Golgi apparatus cisternae (Figures 2b, 4c) were demonstrated in this way.

The fact that reaction products are enclosed in intracellular compartments and are attached to the luminal surfaces of the limiting membranes, makes it possible to study the architecture of intracellular organelles. This is particularly interesting in the case of compartments of complex morphology, such as endoplasmic reticulum and Golgi apparatus. The dense reaction products contained in the luminal spaces and attached to the limiting membranes permit reactive compartments to be followed throughout the entire section thickness; using goniometer equipment for the study of thick sections of e.g., 0.5 μm thickness and more could provide insight into the 3-dimensional structure of reactive compartments and relationships to neighboring organelles.

In the following, possibilities and limitations of the pre-embedment affinity-cytochemical techniques as presented in this contribution are summarized.

1. One has to obtain appropriate labeling reactions in the interior of cryosections and throughout the tissue and cell layers in order to obtain preparations that qualify for electron microscopic exploration.
2. These preparations allow the localization of binding sites in distinct intracellular compartments as well as subcompartments of organelles, such as endoplasmic reticulum and Golgi apparatus subcompartments (Figures 2b, 3a, b, 4a, b, c, 5b).
3. The dense DAB reaction products make it possible to survey the reactions at low electron microscopic magnifications (Figures 2a, 5a) and to compare the patterns in different tissue regions and various cell types.
4. Within individual compartments areas of different reactivity can be recognized; this permits differentiation between compartment subregions, e.g., subregions of endoplasmic reticulum (Figures 2b, 5b) and Golgi apparatus cisternae (Figures 2b, 4c).
5. Owing to the granular appearance of the oxidized DAB reaction product and its possible lateral diffusion, peroxidase affinity-cytochemistry does not allow the exact locations of the reactive molecules to be determined, nor does it permit the binding sites to be quantified.
6. The dense reaction products enclosed in the lumina of intracellular compartments and attached to the limiting membranes provide an approach for studying the 3-dimensional architecture of organelles.

C. IMMUNO- AND LECTIN-CYTOCHEMICAL CHARACTERIZATION OF CELLULAR COMPARTMENTS

Organelles of the major intracellular pathways, the biosynthetic, endocytic and recycling pathways, have been the subject of comprehensive immuno- and lectincytochemical studies with both pre- and postembedding techniques, as well as with ultrathin frozen sections. Here we refer to results obtained by pre-embedding affinity-cytochemistry that concern major compartments of the biosynthetic, and endocytic and recycling pathways: endoplasmic reticulum (ER), Golgi apparatus, vesicular transport systems, endocytic and lysosomal organelles.

1. Biosynthetic Pathways

Cell products destined to be exported out of cells, as well as lysosomal enzymes and newly synthesized molecules of endomembranes and plasma membrane, have been localized in compartments along the biosynthetic pathway.

Secretory molecules, such as albumin and other plasma proteins,[20,21] transferrin,[22] fibronectin and various types of collagen in liver cells,[23] fibronectin in fibroblasts,[24] laminin and type IV collagen in endodermal cells of the yolk sac,[25,26] neurophysin in the hypothalamo-hypophysial system,[27] and plasma cell immunoglobulins,[28] have been demonstrated in the secretory apparatus with the aid of pre-embedment cytochemical methods.

a. Endoplasmic Reticulum — Transport Vesicles

In the case of immunoglobulins, the immunoreactions have been found homogeneously packed in the lumina of nuclear envelope and ER,[28] the sites of their synthesis and initial cotranslational modifications.[29] ER cisternae are also the locations where post-translational alterations of newly synthesized molecules occur and where signal patches are expressed[30] which are crucial for the fate of biosynthetic products, e.g., for either retention in the ER or transport to other cell compartments. Several molecules residing in the ER and involved in the post-translational processing of nascent secretory and membrane proteins have been characterized. One is the immunoglobulin heavy chain binding protein, BiP, that has been localized in the perinuclear cistern and ER of B-cell myelomas, Chinese hamster ovary cells, and hepatocytes;[31] the densely packed BiP-reactions in the nuclear envelope and ER lumina correspond to those obtained for immunoglobulins.[28]

On the other hand, other secretory molecules, such as albumin have been found concentrated in distinct subregions of the ER of liver cells,[21] indicating regions of segregation for this molecule. Subregions of the ER, possibly associated with the segregation of newly synthesized molecules, have also been differentiated with the aid of lectins recognizing high-mannose oligosaccharide chains, such as Concanavalin A (ConA) and the pea and lentil lectins (PSA, LCA).[5,17] In mature absorptive enterocytes (Figure 2b) and in pancreatic acinar cells (Figure 5b), the mannose-binding lectins were found concentrated in distinct ER segments that were frequently situated near the Golgi apparatus. Furthermore, reactions occurred in some of the small (approximately 50 nm in diameter) vesicles (Figure 3a) resembling transport vesicles, and were abundant in the areas surrounding the Golgi apparatus stacks. These vesicles are part of the intracellular vesicular transport system for the transfer of biosynthetic products from the ER to the Golgi cisternae and between the individual Golgi apparatus subcompartments.[32-35]

The mannose-binding lectins mentioned above label the early biosynthetic compartments, such as nuclear envelope, ER, transport vesicles, and Golgi apparatus subcompartments; in combination with other sugar-binding molecules and immunocytochemical approaches, the reaction patterns show how the glycosylation of biosynthetic products can be organized.

b. Endoplasmic Reticulum — Golgi Apparatus: Topology of Glycosylation

ER and Golgi apparatus cooperate in the glycosylation of newly synthesized molecules.[29,32-37] In the case of N-glycosidically-linked glycans, high-mannose oligosaccharide chains terminating in three glucose residues are transferred from dolichol oligosaccharides to the nascent peptide chains in the ER. The glucose residues are cleaved off by two α-glucosidases prior to being exported out of the endoplasmic reticulum and transported to the Golgi apparatus. In the Golgi cisternae, further sugar chain modifications occur, such as mannose-trimming and insertion of N-acetyl-glucosamine, fucose, galactose and sialic acid; several of the N-linked glycans are converted to complex or hybrid-type oligosaccharide species. Furthermore, Golgi apparatus cisternae are the sites where the lysosomal mannose-6-phosphate recognition marker is inserted into the N-linked sugar chains of lysosomal enzymes.[38]

It was pre- (this chapter) and postembedment (Chapter 12 by Roth) immunocytochemical and lectin-cytochemical findings that have provided much information on the sites in the Golgi apparatus subsections where individual glycosylation steps take place. (See also Chapter 10 by Brown).

Glycosidases — Glycosidases have been demonstrated in both ER and Golgi apparatus. With the aid of immunogold labeling, glucosidase II, which is active in cleaving off the two inner α1, 3-linked glucoses, has been localized in cisternae of the ER.[39] Golgi mannosidase II, which is involved in the trimming of α1, 6-mannosyl residues has been demonstrated in the Golgi stack cisternae by means of immunoperoxidase staining.[40]

Glycosyltransferases — By pre-embedment peroxidase labeling, medial cisternae of the Golgi stacks have been recognized as the locations of N-acetyl-glucosaminyl-transferase I in kidney cells and hepatocytes.[41] Galactosyltransferase has been localized by means of both immunogold (postembedding) and immunoperoxidase (pre-embedding) techniques in the trans cisternae of the Golgi apparatus stacks in some cell types, such as HeLa cells, goblet cells and muscle cells.[42,43] Sialyltransferase has been demonstrated by means of immunogold labeling in the trans/transmost Golgi stack subsections in hepatocytes and goblet cells.[44] In the trans Golgi subsections the locations of blood group A α1, 3N-acetyl-galactosaminyltransferase in goblet cells have also been found.[45] By contrast, in the absorptive cells of the large intestine, sialyltransferase and blood group A α1,3N-acetyl-galactosaminyltransferase have not been restricted to the trans/transmost Golgi apparatus subsections, but have been found distributed in all the cisternae across the stacks.[45] These varying locations of sugar chain-processing enzymes agree with findings obtained with lectins demonstrating particular sugars and sugar sequences that have revealed varying reactions in biosynthetic compartments as well (see next paragraph).

The predominance of high-mannose oligosaccharides in the ER is demonstrated by the reactions obtained with mannose-recognizing lectins, such as ConA and the pea and lentil lectins.[5,8,10,17,46-48] ConA preferably binds to tri-antennary high-mannose type glycans; intense ConA reactions have been demonstrated in the perinuclear space and the ER of a number of cell types (Figure 3a). In contrast to the ConA label, the reactions obtained with the pea and lentil lectins were weaker and more frequently confined to ER subregions (Figures 2b, 5b).

c. *Lectin-Cytochemical Patterns of the Golgi Apparatus*

The lectin reactions obtained in the Golgi apparatus cisternae and other Golgi elements of a number of cell types may be related to and mirror the sugar chain modifications taking place in this organelle. The lectin-cytochemical demonstration of sugars and sugar sequences in glycans, just as the immunocytochemical localization of glycan-processing enzymes, provides an approach for differentiating subcompartments of the Golgi apparatus. In a number of cases, the lectin results demonstrating certain sugars and sugar sequences are in agreement with the immunocytochemical localization of the respective sugar-transferring enzymes.

In several cell types, the reactions for the N-acetyl-galactosamine-recognizing *Helix pomatia* lectin (HPA; Figure 4a), just as those for the *Vicia villosa* lectin (VVA), have been found confined to cis Golgi stack elements, thus pointing to compartments where initial steps of the synthesis of O-glycosidically linked sugar chains occur.[49] (In abbreviated lectin designations, A = agglutinin). Reactions for peanut lectin recognizing galactose-β1, 3-N-acetyl-galactosamine-sequences have been demonstrated in medial Golgi cisternae in various cell types of the gastro-intestinal tract.[9,50]

Lectins that preferentially bind with N-glycosidically-linked high-mannose glycans, such as ConA, have been shown to predominate in cis Golgi apparatus cisternae (Figure 3a).[5,10] Similarly, reactions for the mannose/glucose-specific pea and lentil lectins, which for high-affinity binding require a fucose attached to the asparagine-linked N-acetyl-glucosamine, are concentrated in the cis/medial Golgi apparatus subsections (Figures 2b, 5b).[17] In contrast, lectins that recognize terminal N-acetyl glucosamine, galactose, or sialic acid residues, thus indicating complex-type N-linked sugar chains, such as wheat germ agglutinin (WGA), *Ricinus communis* I lectin (RCA I) and *Limax flavus* lectin (LFA), respectively, have been localized in medial, trans and transmost Golgi apparatus elements[5,10,51-53] (Figures 4b, c). Furthermore, *Erythrina cristagalli* lectin (ECA) that recognizes galactose-β1, 4 N-acetyl-glucosamine sequences binds with constituents of trans and transmost Golgi stack elements in absorptive enterocytes, fibroblasts and pancreatic acinar cells (Figure 3b).[54] These patterns are in agreement with a compartmentalized organization of the Golgi stacks with respect to glycosylation,[55-57] and may mirror conversion of high mannose-type to complex- or hybrid-type N-linked glycans. The medial and trans reactions obtained for N-acetyl-glucosamine[10] correspond to the immunocytochemical localization of N-acetyl-glucosaminyl-transferase I in medial Golgi cisternae;[41] the trans/transmost reactions for galactose and sialic acid[5,51-53] coincide with the trans/transmost localizations of galactosyl- and sialyltransferase in some cell types.[42-45] These immunocytochemical and lectin patterns indicate a cis-to-trans orientation of subsequent N-glycosylation steps.

However, major variations of the reaction patterns have been found with sugar-binding molecules[45,48,58,59] such as have been shown for the immunocytochemical reactions of sugar-transferring enzymes. The findings obtained with a variety of cell types have demonstrated that neither ConA, PSA, LCA, HPA and VVA can be seen as cis Golgi markers, nor RCA I and LFA being characteristic of trans/transmost Golgi stack elements. In pancreatic acinar cells of rat embryos, both lectins labeling N- and O- glycosidically linked glycans (such as ConA/RCA I and HPA, respectively) were found in all the cisternae across the Golgi apparatus stacks.[58] Furthermore, the sialic acid-specific LFA has been demonstrated in all the stacked cisternae with the exception of one cisterna at one side of the stacks in fibroblastic cells,[59] and absorptive cells of the large intestine; [45] this agrees with the immunocytochemical sialyltransferase reactions in the Golgi stacks of the latter cell type.[45] These immuno- and lectin-cytochemical results suggest a high flexibility of the Golgi apparatus organization.

d. Endoplasmic Reticulum and Golgi Apparatus Membrane Constituents

With the aid of pre-embedding immunocytochemical incubations a number of membrane proteins have been localized in the ER and subsections of the Golgi stacks.[60-67] A 135 kDa cytes,[60,62] as well as in normal and tumorous prolactin cells,[61] in neurons and astrocytes.[63] These preparations have greatly furthered the definition of various functional states of cells; Golgi apparatus membranes have been discerned in cellular regions where they have not been expected to be, such as in the peripheral regions of astrocyte processes.[63]

Medial Golgi stack cisternae have also been found to be the locations of two cytoplasmically oriented proteins that possibly play a role in membrane-membrane interactions for the maintenance of the stacked Golgi apparatus organization.[64] Furthermore, both membrane proteins of cis[65,66] and trans[65,67] Golgi apparatus subsections, including the trans-tubular network,[65] have recently been characterized and visualized immunocytochemically with the aid of pre-embedding techniques.

Pre-embedment immunoperoxidase methods have also been successful in localizing virus membrane glycoproteins in biosynthetic compartments. Virus protein studies with the aid of both pre- and postembedment procedures have significantly helped to characterize traffic between individual Golgi apparatus subsections from the cis to the trans side of the stacks.[68,69]

In an alternative approach, immunocytochemical localization of virus proteins has been employed for defining sites of molecule sorting for constitutive and regulated secretory routes.[70]

Ultrastructural localization of the regulatory subunit R II of cAMP-dependent (cAMP = 3′, 5′-cyclic adenosinmonophosphate) proteinkinase has been effected with the aid of pre-embedding immunolabeling in cells of the nervous system; the immunoreactions have been found concentrated at one, presumably the trans side, of the Golgi stacks.[71]

e. Lysosomal System

Constituents of the lysosomal system, including lysosomal enzymes as well as lysosomal membrane molecules, have been visualized immunocytochemically in compartments of the biosynthetic pathway.[72-76] Furthermore, it has been pre-embedment labeling and immunocytochemical work with ultrathin frozen sections that have provided approaches for mannose-6-phosphate receptor localization;[15,77,78] reactions for the 215 kDa mannose-6-phosphate receptor have been discerned in cis[15] as well as medial and trans[77,78] subsections of the Golgi apparatus stacks, depending on the cell type. The pre-embedding methods employed for studies of the mannose-6-phosphate receptor include immunoperoxidase labeling,[15] as well as ferritin bridge procedures.[77,79] The mannose-6-phosphate receptor localizations indicate receptor-mediated intracellular routes of newly synthesized lysosomal enzymes from the Golgi apparatus to the lysosomal system involving a prelysosomal CURL system (compartment of uncoupling receptor and ligand).[80] By means of immunocytochemical localization of the mannose-6-phosphate receptor, various influences on Golgi-to-lysosome routes, such as effects of chloroquine[81] and tunicamycin,[81] as well as pathological conditions,[16] have been well characterized.

Moreover, acidic cell compartments have successfully been defined with the aid of immunocytochemical localization of molecules that are concentrated in acidic milieu;[82,83] these studies have proved that elements of the trans Golgi side belong to acidic cell compartments. It has been shown by means of postembedment immunocytochemical localization of fibronectin that acidic trans Golgi subcompartments are not only acceptors of internalized molecules, but compartments of biosynthetic pathways.[82]

2. Endocytic Pathways — Recycling Pathways

The intracellular routes following upon internalization of a number of plasma membrane molecules, as well as molecules taken up into cells by endocytosis, have been followed up immunocytochemically;[84,85] subsequent steps of endocytic and recycling routes have been recorded, the compartments involved having been labeled: endocytic vesicles, early and late endosomes/receptosomes, CURL, multivesiculated bodies, lysosomes, Golgi apparatus elements, recycling tubules and vesicles. (Examples for this are given in Chapter 10 by Brown and Chapter 8 by Robenek.) Antibodies, which recognize components of a proton pump responsible for acidification of endo- and exocytic organelles, have been found to bind to constituents of prelysosomal and lysosomal organelles, as well as to Golgi apparatus cisternae.[74,86]

It has repeatedly been documented that internalized plasma membrane molecules, just as molecules taken up into cells by endocytosis, gain access to Golgi apparatus elements in the course of their endocytic and recycling routes. Several studies have addressed possible common compartments involved in both biosynthetic and endocytic/recycling pathways.[84,85,87-89] Entry into biosynthetic compartments has been documented for asialo-transferrin taken up into cells by endocytosis, as well as for internalized asialo-transferrin receptor that both gain access to sialyltransferase compartments and are resialylated.[90,91] With the aid of the pre-embedding immunoperoxidase technique it has been documented that internalized transferrin receptor is not only taken up into transmost Golgi stack elements, but also visits cis and medial Golgi apparatus cisternae.[92] In accordance with biochemical results that show cellular glycoproteins returning to Golgi mannosidase I compartments,[93] this indicates that not only resialylation, but more extensive modifications may occur with internalized molecules.

IV. SUMMARY

Pre-embedding affinity-cytochemistry is one in a range of approaches that permit looking at the interior of cells. The present contribution refers to two widely different approaches. (1) Immunocytochemical approaches that make it possible to specifically label intracellular and cell surface molecules and to localize them in the various cell compartments. (2) Lectin-cytochemical approaches that permit classes of glycoconjugates to be differentiated. Details of pre-embedding affinity-cytochemical techniques are discussed on the basis of lectin-cytochemical preparations that were performed with 10 μm thick cryosections of various tissues and cultured fibroblasts using peroxidase for visualizing the binding reactions. It is shown that flat sections throughout the central cryosection regions are qualified objects for ultrastructural studies that permit the localization of the lectin-sugar-binding reactions in distinct cellular compartments and subcompartments of organelles.

A great number of findings obtained with the aid of pre-embedding affinity-cytochemical procedures concern compartments of the major intracellular pathways, the biosynthetic, endocytic and recycling pathways. Cell products, including secretory molecules, constituents of endomembranes, and plasma membrane, as well as lysosomal enzymes, have been localized by means of pre-embedment techniques in compartments of the biosynthetic pathway. The cytochemical recovery in cell organelles of molecules taken up into cells by endocytosis and of internalized plasma membrane molecules has been of great value in unraveling endocytic and recycling routes.

ACKNOWLEDGMENTS

The authors gratefully acknowledge the valuable technical assistance of Mrs. Jutta Selbmann, Mrs. Elfriede Scherzer, Mrs. Gerlinde Hartl, Mr. Helmut Oslansky and Mr. Richard Reichhart. This work was supported by the ''Hochschuljubiläumsstiftung der Stadt Wien''.

REFERENCES

1. **Polak, J. M. and Varndell, I. M.,** *Immunolabeling for Electron Microscopy,* Elsevier, Amsterdam, 1984.
2. **Childs, G. V.,** Immunocytochemical Technology, *Am. J. Anat.,* 175, 125, 1986.
3. **Plattner, H. and Zingsheim, H. P.,** *Elektronenmikroskopische Methodik in der Zell- und Molekularbiologie,* Gustav Fischer, Stuttgart, 1987.
4. **Liener, I. E., Sharon, N., and Goldstein, I. J.,** *The Lectins: Properties, Functions, and Applications in Biology and Medicine,* Academic Press, Orlando, FL, 1986.
5. **Pavelka, M. and Ellinger, A.,** Localization of binding sites for Concanavalin A, Ricinus communis I, and Helix pomatia lectin in the Golgi apparatus of rat small intestinal absorptive cells, *J. Histochem. Cytochem.,* 33, 905, 1985.
6. **Pavelka, M. and Ellinger, A.,** Lectin binding patterns of the Golgi apparatus, in *Lectins,* Vol. 5, Bøg-Hansen, T. C. and van Driessche, E., Eds., W. De Gruyter, Berlin, 1986, 485.
7. **Pavelka, M.,** Functional morphology of the Golgi apparatus, *Advances in Anatomy, Embryology and Cell Biology,* Vol. 106, Springer-Verlag, Berlin, 1987.
8. **Bernhard, W. and Avrameas, S.,** Ultrastructural visualization of cellular carbohydrate components by means of Concanavalin A, *Exp. Cell Res.,* 64, 232, 1971.

9. **Sato, A. and Spicer, S. S.,** Ultrastructural visualization of galactose in the glycoprotein of gastric surface cells with a peanut lectin conjugate, *Histochem. J.*, 14, 125, 1982.
10. **Tartakoff, A. M. and Vassalli, P.,** Lectin-binding sites as markers of Golgi subcompartments: Proximal-to-distal maturation of oligosaccharides, *J. Cell Biol.*, 97, 1243, 1983.
11. **McLean, I. W. and Nakane, P. K.,** Periodate-lysine-paraformaldehyde fixative. A new fixative for immunoelectron microscopy, *J. Histochem. Cytochem.*, 22, 1077, 1974.
12. **Willingham, M. C.,** An alternative fixation-processing method for pre-embedding ultrastructural immunocytochemistry of cytoplasmic antigens: The GBS (glutaraldehyde-borohydride-saponin) procedure, *J. Histochem. Cytochem.*, 31, 791, 1983.
13. **Sternberger, L. A., Hardy, P. H., Cuculis, J. J., and Meyer, H. G.,** The unlabeled antibody enzyme method for immunohistochemistry. Preparation and properties of soluble antigen-antibody complex (horseradish peroxidase-antihorseradish peroxidase) and its use in identification of spirochetes, *J. Histochem. Cytochem.*, 18, 315, 1970.
14. **Hsu, S. -M., Raine, L., and Fanger, H.,** Use of avidin-biotin-peroxidase complex (ABC) in immunoperoxidase techniques: A comparison between ABC and unlabeled antibody (PAP) procedures, *J. Histochem. Cytochem.*, 29, 577, 1981.
15. **Brown, W. J. and Farquhar, M. G.,** The mannose-6-phosphate receptor for lysosomal enzymes is concentrated in cis Golgi cisternae, *Cell*, 36, 295, 1984.
16. **Brown, W. J. and Farquhar, M. G.,** Accumulation of coated vesicles bearing mannose-6-phosphate receptors for lysosomal enzymes in the Golgi region of I-cell fibroblasts, *Proc. Natl. Acad. Sci. U.S.A.*, 81, 5135, 1984.
17. **Pavelka, M. and Ellinger, A.,** Affinity cytochemical differentiation of glycoconjugates of small intestinal absorptive cells using *Pisum sativum* and Lens culinaris lectins, *J. Histochem. Cytochem.*, in press.
18. **Courtoy, P. J., Picton, D. H., and Farquhar, M. G.,** Resolution and limitations of the immunoperoxidase procedure in the localization of extracellular matrix antigens, *J. Histochem. Cytochem.*, 31, 945, 1983.
19. **Messing, A., Stieber, A., and Gonatas, N. K.,** Resolution of diaminobenzidine for the detection of horseradish peroxidase on surfaces of cultured cells, *J. Histochem. Cytochem.*, 33, 837, 1985.
20. **Feldman, G., Maurice, M., Bernuau, D., and Rogier, E.,** Morphological aspects of plasma protein synthesis and secretion by the hepatic cells, *Int. Rev. Cytol.*, 96, 157, 1985.
21. **Yokota, S. and Fahimi, H. D.,** Immunocytochemical localization of albumin in the secretory apparatus of rat liver parenchymal cells, *Proc. Natl. Acad. Sci. U.S.A.*, 78, 4970, 1981.
22. **Vassy, J., Rissel, M., Kraemer, M., Foucrier, J., and Guillouzo, A.,** Ultrastructural indirect immunolocalization of transferrin in cultured rat hepatocytes permeabilized with saponin, *J. Histochem. Cytochem.*, 32, 538, 1984.
23. **Clement, B., Rissel, M., Peyrol, S., Mazurier, Y., Grimaud, J. -A., and Guillouzo, A.,** A procedure for light and electron microscopic intracellular immunolocalization of collagen and fibronectin in rat liver, *J. Histochem. Cytochem.*, 33, 407, 1985.
24. **Hedman, K.,** Intracellular localization of fibronectin using immunoperoxidase cytochemistry in light and electron microscopy, *J. Histochem. Cytochem.*, 28, 1233, 1980.
25. **Laurie, G. W., Leblond, C. P., and Martin, G. R.,** Intracellular localization of basement membrane precursors in the endodermal cells of the rat parietal yolk sac. II. Immunostaining for type IV collagen and its precursors, *J. Histochem. Cytochem.*, 30, 983, 1982.
26. **Laurie, G. W., Leblond, C. P., Martin, G. R., and Silver, M. H.,** Intracellular localization of basement membrane precursors in the endodermal cells of the rat parietal yolk sac. III. Immunostaining for laminin and its precursors, *J. Histochem. Cytochem.*, 30, 991, 1982.
27. **Broadwell, R. D., Oliver, C., and Brightman, M. W.,** Localization of neurophysin within organelles associated with protein synthesis and packaging in the hypothalamo-neurohypophysial system: An immunocytochemical study, *Proc. Natl. Acad. Sci. U.S.A.*, 76, 5999, 1979.
28. **Gonatas, N. K., Gonatas, J. O., Stieber, A., Ternynck, T., and Avrameas, S.,** Detection of plasma cell immunoglobulins in tissue sections optimally fixed for ultrastructural immunocytochemistry, *J. Histochem. Cytochem.*, 35, 189, 1987.
29. **Kornfeld, R. and Kornfeld, S.,** Assembly of asparagine-linked oligosaccharides, *Ann. Rev. Biochem.*, 54, 631, 1985.
30. **Pfeffer, S. R. and Rothman, J. E.,** Biosynthetic protein transport and sorting by the endoplasmic reticulum and Golgi, *Ann. Rev. Biochem.*, 56, 829, 1987.
31. **Bole, D. G., Doriaux, M., and Jamieson, J. D.,** Immunocytochemical localization of the immunoglobulin heavy chain binding protein, BIP, *J. Cell Biol.*, 105, 77a, 1987.
32. **Farquhar, M. G. and Palade, G. E.,** The Golgi apparatus (complex) — (1954-1981) — from artifact to center stage, *J. Cell Biol.*, 91, 77s, 1981.
33. **Farquhar, M. G.,** Progress in unraveling pathways of the Golgi traffic, *Ann. Rev. Cell Biol.*, 1, 447, 1985.

34. **Dunphy, W. G. and Rothman, J. E.,** Compartmental organization of the Golgi stack, *Cell,* 42, 13, 1985.
35. **Tartakoff, A. M.,** *The secretory and endocytic paths,* John Wiley & Sons, New York, 1987.
36. **Roth, J.,** Subcellular organization of glycosylation in mammalian cells, *Biochim. Biophys. Acta,* 906, 405, 1987.
37. **Fleischer, B.,** Mechanism of glycosylation in the Golgi apparatus, *J. Histochem. Cytochem,* 31, 1033, 1983.
38. **Sly, W. S. and Fischer, H. D.,** The phosphomannosyl recognition system for intracellular and intercellular transport of lysosomal enzymes, *J. Cell. Biochem.,* 18, 67, 1982.
39. **Lucocq, J. M., Brada, D., and Roth, J.,** Immunolocalization of the oligosaccharide trimming enzyme glucosidase II, *J. Cell Biol.,* 102, 2137, 1986.
40. **Novikoff, P. M., Tulsiani, D. R. P., Touster, O., Yam, A., and Novikoff, A. B.,** Immunocytochemical localization of α-D-mannosidase II in the Golgi apparatus of rat liver, *Proc. Natl. Acad. Sci. U.S.A.,* 80, 4364, 1983.
41. **Dunphy, W. G., Brands, R., and Rothman, J. E.,** Attachment of terminal N-acetyl-glucosamine to asparagine-linked oligosaccharides occurs in central cisternae of the Golgi stack, *Cell,* 40, 463, 1985.
42. **Roth, J. and Berger, E. G.,** Immunocytochemical localization of galactosyltransferase in HeLa cells: Codistribution with thiamine pyrophosphatase in trans Golgi cisternae, *J. Cell Biol.,* 93, 223, 1982.
43. **Tassin, A. M., Paintrand, M., Berger, E. G., and Bornens, M.,** The Golgi apparatus remains associated with microtubule organizing centers during myogenesis, *J. Cell Biol.,* 101, 630, 1985.
44. **Roth, J., Taatjes, D. J., Lucocq, J. M., Weinstein, J., and Paulson, J. C.,** Demonstration of an extensive trans-tubular network continuous with the Golgi apparatus stack that may function in glycosylation, *Cell,* 43, 287, 1985.
45. **Roth, J., Taatjes, D. J., Weinstein, J., Paulson, J. C., Greenwell, P., and Watkins, W. W.,** Differential subcompartmentation of terminal glycosylation in the Golgi apparatus of intestinal absorptive and goblet cells, *J. Biol. Chem.,* 261, 14307, 1986.
46. **Wood, J. G., McLaughlin, B. J., and Barber, R. P.,** The visualization of Concanavalin A binding sites in Purkinje cell somata and dendrites of rat cerebellum, *J. Cell Biol.,* 63, 541, 1974.
47. **Bretton, R. and Bariety, J.,** A comparative ultrastructural localization of Concanavalin A, wheat germ and Ricinus communis on glomeruli of normal rat kidney, *J. Histochem. Cytochem.,* 24, 1093, 1976.
48. **Malchiodi, F., Rambourg, A., Clermont, Y., and Caroff, A.,** Ultrastructural localization of Concanavalin A-binding sites in the Golgi apparatus of various types of neurons in rat dorsal root ganglia: Functional implications, *Am. J. Anat.,* 177, 81, 1986.
49. **Roth, J.,** Cytochemical localization of terminal N-acetyl-D-galactosamine residues in cellular compartments of intestinal goblet cells: Implications for the topology of O-glycosylation, *J. Cell Biol.,* 98, 399, 1984.
50. **Sato, A. and Spicer, S. S.,** Ultrastructural visualization of galactosyl residues in various alimentary epithelial cells with the peanut lectin-horseradish peroxidase procedure, *Histochemistry,* 73, 607, 1982.
51. **Griffiths, G., Brands, R., Burke, B., Louvard, D., and Warren, G.,** Viral membrane proteins acquire galactose in trans Golgi cisternae during intracellular transport, *J. Cell Biol.,* 95, 781, 1982.
52. **Pavelka, M. and Ellinger, A.,** RCA I-binding patterns of the Golgi apparatus, *Eur. J. Cell Biol.,* 41, 270, 1986.
53. **Roth, J., Lucocq, J. M., and Charest, P. M.,** Light and electron microscopic demonstration of sialic acid residues with the lectin from Limax flavus: A cytochemical affinity technique with the use of fetuin-gold complexes, *J. Histochem. Cytochem.,* 32, 1167, 1984.
54. **Pavelka, M. and Ellinger, A.,** Intracellular binding sites for the Erythrina cristagalli lectin, in preparation.
55. **Tartakoff, A. M.,** The confined function model of the Golgi complex: Center for ordered processing of biosynthetic products of the rough endoplasmic reticulum, *Int. Rev. Cytol.,* 85, 221, 1983.
56. **Rothman, J. E.,** The compartmental organization of the Golgi apparatus, *Sci. Am.,* 253, 84, 1985.
57. **Berger, E. G.,** The functional organization of the Golgi apparatus, in Colloque INSERM/CNRS, *Cellular and Pathological Aspects of Glycoconjugate Metabolism,* INSERM, Vol. 126, 111, 1984.
58. **Pavelka, M. and Ellinger, A.,** The Golgi apparatus in the acinar cells of the developing embryonic pancreas: II. Localization of lectin-binding sites, *Am. J. Anat.,* 178, 224, 1987.
59. **Hedman, K., Pastan, I., and Willingham, M. C.,** The organelles of the trans domain of the cell. Ultrastructural localization of sialoglycoconjugates using Limax flavus agglutinin, *J. Histochem. Cytochem.,* 34, 1069, 1986.
60. **Louvard, D., Reggio, H., and Warren, G.,** Antibodies to the Golgi complex and the rough endoplasmic reticulum, *J. Cell Biol.,* 92, 92, 1982.
61. **Tougard, C., Louvard, D., Picart, R., and Tixier-Vidal, A.,** The rough endoplasmic reticulum and the Golgi apparatus visualized using specific antibodies in normal and tumoral prolactin cells in culture, *J. Cell Biol.,* 96, 1197, 1983.
62. **Burke, B., Griffiths, G., Reggio, H., Louvard, D., and Warren, G.,A.,** A monoclonal antibody against a 135 K Golgi membrane protein, *EMBO J.,* 1, 1621, 1982.

63. **Stieber, A., Gonatas, J. O., Gonatas, N. K., and Louvard, D.,** The Golgi apparatus-complex of neurons and astrocytes studied with an anti-organelle antibody, *Brain Res.*, 408, 13, 1987.
64. **Chicheportiche, Y., Vassalli, P., and Tartakoff, A. M.,** Characterization of cytoplasmically oriented Golgi proteins with a monoclonal antibody, *J. Cell Biol.*, 99, 2200, 1984.
65. **Yuan, L., Barriocanal, J. G., Bonifacino, J. S., and Sandoval, I. V.,** Two integral membrane proteins located in the cis-middle and trans-part of the Golgi system acquire sialylated N-linked carbohydrates and display different turnovers and sensitivity to cAMP-dependent phosphorylation, *J. Cell Biol.*, 105, 215, 1987.
66. **Saraste, J., Palade, G. E., and Farquhar, M. G.,** Antibodies to rat pancreas Golgi subfractions: Identification of a 58-kD cis-Golgi protein, *J. Cell Biol.*, 105, 2021, 1987.
67. **Smith, Z. D. J., D'Eugenio-Gumkowski, F., Yanagisawa, K., and Jamieson, J. D.,** Endogenous and monoclonal antibodies to the rat pancreatic acinar cell Golgi complex, *J. Cell Biol.*, 98, 2035, 1984.
68. **Saraste, J. and Kuismanen, E.,** Pre- and post-Golgi vacuoles operate in the transport of Semliki Forest virus membrane glycoproteins to the cell surface, *Cell,* 38, 535, 1984.
69. **Bergmann, J. E. and Singer, S. J.,** Immunoelectron microscopic studies of the intracellular transport of the membrane glycoprotein (G) of vesicular stomatitis virus in infected Chinese hamster ovary cells, *J. Cell Biol.*, 97, 1777, 1983.
70. **Tooze, J., Tooze, S. A., and Fuller, S. D.,** Sorting of progeny coronavirus from condensed secretory proteins at the exit from the trans-Golgi network of AtT20 cells, *J. Cell Biol.*, 105, 1215, 1987.
71. **De Camilli, P., Moretti, M., Donini, S. D., Walter, U., and Lohmann, S. M.,** Heterogeneous distribution of the cAMP receptor protein RII in the nervous system: evidence for its intracellular accumulation on microtubules, microtubule-organizing centers, and in the area of the Golgi complex, *J. Cell Biol.*, 103, 189, 1986.
72. **Novikoff, P. M., La Russo, N. F., Novikoff, A. B., Stockert, R. J., Yam, A., and Le Sage, G. D.,** Immunocytochemical localization of lysosomal β-galactosidase in rat liver, *J. Cell Biol.*, 97, 1559, 1983.
73. **van Dongen, J. M., Barneveld, R. A., Geuze, H. J., and Galjaard, H.,** Immunocytochemistry of lysosomal hydrolase and their precursor forms in normal and mutant human cells, *Histochem. J.*, 16, 941, 1984.
74. **Reggio, H., Bainton, D., Harms, E., Coudrier, E., and Louvard, D.,** Antibodies against lysosomal membranes reveal a 100,000-mol-wt protein that cross-reacts with purified H^+, K^+ ATPase from gastric mucosa, *J. Cell Biol.*, 99, 1511, 1984.
75. **Lewis, V., Green, S. A., Marsh, M., Vihko, P., Helenius, A., and Mellman, I.,** Glycoproteins of the lysosomal membrane, *J. Cell Biol.*, 100, 1839, 1985.
76. **Barriocanal, J. G., Bonifacino, J. S., Yuan, L., and Sandoval, I. V.,** Biosynthesis, glycosylation, movement through the Golgi system, and transport to lysosomes by an N-linked carbohydrate-independent mechanism of three lysosomal integral membrane proteins, *J. Biol. Chem.*, 261, 16755, 1986.
77. **Willingham, M. C., Pastan, I. H., and Sahagian, G. G.,** Ultrastructural immunocytochemical localization of the phosphomannosyl receptor in Chinese hamster ovary (CHO) cells, *J. Histochem. Cytochem.*, 31, 1, 1983.
78. **Geuze, H. J., Slot, J. W., Strous, G. J. A. M., Hasilik, A., and von Figura, K.,** Ultrastructural localization of the mannose-6-phosphate receptor in rat liver, *J. Cell Biol.*, 98, 2047, 1984.
79. **Willingham, M. C.,** Electron microscopic immunocytochemical localization of intracellular antigens in cultured cells: The EGS and ferritin bridge procedures, *Histochem. J.*, 12, 419, 1980.
80. **Geuze, H. J., Slot, J. W., Strous, G. J. A. M., Hasilik, A., and von Figura, K.,** Possible pathways for lysosomal enzyme delivery, *J. Cell Biol.*, 101, 2253, 1985.
81. **Brown, W. J., Constantinescu, E., and Farquhar, M. G.,** Redistribution of mannose-6-phosphate receptors induced by tunicamycin and chloroquine, *J. Cell Biol.*, 99, 320, 1984.
82. **Anderson, R. G. W. and Pathak, R. K.,** Vesicles and cisternae in the trans Golgi apparatus of human fibroblasts are acidic compartments, *Cell,* 40, 635, 1985.
83. **Schwartz, A. L., Strous, G. J. A. M., Slot, J. W., and Geuze, H. J.,** Immunoelectron microscopic localization of acidic intracellular compartments in hepatoma cells, *EMBO J.*, 4, 899, 1985.
84. **Farquhar, M. G.,** Multiple pathways of exocytosis, endocytosis, and membrane recycling: Validity of a Golgi route, *Fed. Proc.*, 42, 2407, 1983.
85. **Pastan, I. and Willingham, M. C.,** The pathways of endocytosis, in *Endocytosis,* Pastan, I. and Willingham, M. C., Eds., Plenum, New York, 1985, 1.
86. **Tougard, C., Louvard, D., Picart, R., and Tixier-Vidal, A.,** Antibodies against a lysosomal membrane antigen recognize a prelysosomal compartment involved in the endocytic pathway in cultured prolactin cells, *J. Cell Biol.*, 100, 786, 1985.
87. **Strous, G. J., Du Maine, A., Zijderhand-Bleekemolen, J. E., Slot, J. W., and Schwartz, A. L.,** Effect of lysosomotropic amines on the secretory pathway and on the recycling of the asialoglycoprotein receptor in human hepatoma cells, *J. Cell Biol.*, 101, 531, 1985.

88. **Robbins, A. R., Oliver, C., Bateman, J. L., Krag, S. S., Galloway, C. J., and Mellman, I.,** A single mutation in Chinese hamster ovary cells impairs both Golgi and endosomal functions, *J. Cell Biol.*, 99, 1296, 1984.
89. **Hedman, K., Goldenthal, K. L., Rutherford, A. V., Pastan, I., and Willingham, M. C.,** Comparison of the intracellular pathways of transferrin recycling and vesicular stomatitis virus membrane glycoprotein exocytosis by ultrastructural double-label cytochemistry, *J. Histochem. Cytochem.*, 35, 233, 1987.
90. **Regoeczi, E., Chindemi, P. A., Debanne, M. T., and Charlwood, P. A.,** Partial resialylation of human asialotransferrin type 3 in the rat, *Proc. Natl. Acad. Sci. U.S.A.*, 79, 2226, 1982.
91. **Snider, M. D. and Rogers, O. C.,** Intracellular movement of cell surface receptors after endocytosis: Resialylation of asialo-transferrin receptor in human erythroleukemia cells, *J. Cell Biol.*, 100, 826, 1985.
92. **Woods, J. W., Doriaux, M., and Farquhar, M. G.,** Transferrin receptors recycle to cis and middle as well as trans Golgi cisternae in Ig-secreting myeloma cells, *J. Cell Biol.*, 103, 277, 1986.
93. **Snider, M. D. and Rogers, O. C.,** Membrane traffic in animal cells: Cellular glycoproteins return to the site of Golgi mannosidase I, *J. Cell Biol.*, 103, 265, 1986.

Chapter 12

ORGANIZATIONAL PRINCIPLES OF THE GOLGI APPARATUS: CONTRIBUTIONS FROM POSTEMBEDDING GOLD LABELING TECHNIQUES

Jürgen Roth and Douglas J. Taatjes

TABLE OF CONTENTS

I. INTRODUCTION

The last ten years have witnessed enormous progress in our understanding of Golgi apparatus structure-function relationship with respect to its involvement in many diverse cellular activities, and have provided a new level of insight into the dynamics of these functions.[1-4] Additional new insight into the complex nature of the architecture of the Golgi apparatus[5,6] has provided much of the structural basis for our current understanding of the Golgi apparatus in functional terms. Our modern functional concept of this organelle would not have evolved without the development of new morphologic and biochemical technologies and their combined use. This has led to the realization that sorting and control of intracellular traffic are two main functions of the Golgi apparatus.[2,7] Obviously, a variety of post-translational protein modifications exerted by the Golgi apparatus have to be considered in context with these basic cellular phenomena as well, and represent a third main function. Over the past years the most actively investigated type of covalent protein modification has been the glycosylation, a multistep process by which the whole variety of oligosaccharide side chains differing in composition and length are added to most proteins. The investigation of this synthetic cascade has provided much of the data with respect to the functional organization of the Golgi apparatus.

In this review we summarize and discuss work on Golgi apparatus-related glycosylation steps obtained during the past ten years. Particular emphasis is placed on studies which have revealed the existence of various degrees of functional subdivision of the Golgi apparatus. For information on earlier *in situ* work about glycosylation at the subcellular level, as well as the biochemical aspects of oligosaccharide biosynthesis and processing, the reader is referred to several excellent reviews.[8-12]

II. METHODICAL BACKGROUND

Biochemcal studies have demonstrated that the process of glycosylation proceeds in distinct steps in each of which specific enzymes are involved which act in a sequential fashion. For example, the biosynthesis of asparagine-linked oligosaccharides can be subdivided into two major steps if one considers only the organelles involved, i.e., the endoplasmic reticulum and the Golgi apparatus. However, in each of these organelles, multiple enzymatic glycosylation reactions take place. Although the order in which the different enzymes apparently act could be resolved *in vitro* and attributed to either the endoplasmic reticulum or the Golgi apparatus, the degree of resolution at the level of subcellular fractions is apparently limited.[13-16]

A. THE MORPHOLOGICAL APPROACH

Electron microscopic gold labeling techniques have provided the most direct means for the high resolution *in situ* localization of enzymes involved in glycosylation, and their products of action.[4] Simultaneously to the biochemical detection of Golgi apparatus glycosylation compartments,[17] Roth and Berger provided in 1982 the first morphological evidence for subcompartmentation of Golgi apparatus-associated glycosylation, when they localized galactosyltransferase by immunoelectron microscopy.[18] Such type of investigation had become possible through the development of two new tools: the protein A-gold technique,[19] and the low temperature Lowicryl K4M embedding procedure.[20] Another important development for cytochemical investigations on glycosylation was the adaption of the lectin-gold technique for the labeling of ultrathin sections.[21,22]

1. The Protein A-Gold Technique

The protein A-gold technique was developed in 1978 by Roth and co-workers[19,23] for the localization of intracellular antigens on thin sections from resin embedded tissues, and

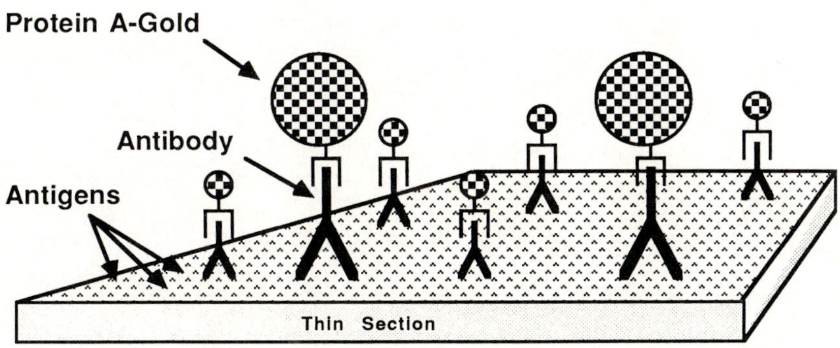

FIGURE 1. Schematic representation of the protein A-gold technique for antigen localization on thin sections. The possibility of double labeling with the use of two different sizes of protein A-coated gold particles is indicated.

subsequently adapted for labeling of ultrathin frozen sections.[24] The distinct advantage of this technique over the immunoperoxidase technique was at least twofold. The use of the particulate and highly electron dense marker colloidal gold, which can be prepared in various sizes down to about 2 nm,[25] provided not only superior resolution, but also unequivocal identification of the immunolabeling over all cellular structures. In addition, a breakthrough in the field of multiple labeling procedures was achieved by the use of monodisperse gold particle preparations of various sizes.[26-29] It is beyond the scope of this review to give a detailed comparative account on pre-embedding (see Chapter 11 in this volume) vs. post-embedding techniques. However, it is quite obvious that at present, postembedding labeling is, in most cases, the technique of choice.[30,31]

The protein A-gold technique is a typical two-step technique (Figure 1). After initial interaction of the antibody with the antigen present on the surface of the thin section, the antigen-antibody complex is visualized by the binding of the protein A-gold complex to the F_c portion of the immunoglobulin molecules. Etching of the thin sections prior to immunolabeling is not required (*vide infra*). Although the spontaneous interaction between resins used for embedding and the protein A-gold complex is minimal under standard conditions,[19] nonspecific binding can be observed if no or inadequate conditioning of the sections with inert proteins, i.e., 0.5% egg albumin, 0.5% bovine serum albumin, or 2% defatted milk powder in PBS was performed prior to labeling. Of particular importance is the use of appropriately diluted protein A-gold.[32] If tannic acid is used for colloidal gold manufacture, 0.05 to 0.1% Triton X-100 and Tween 20 need to be added to the dilution buffer to prevent nonspecific sticking of protein A-gold.[33] For further technical details the reader is referred to recent reviews.[23,32]

In context with the immunolocalization of glycosyltransferases by antibodies, the problem of possible antibody crossreactivity with carbohydrates needs to be mentioned. It was recently demonstrated that affinity-purified antibodies to a human galactosyltransferase recognized blood group specific carbohydrate structures present on the enzyme.[34,35] These results pointed out that antibodies used in the immunolocalization of any glycoprotein should be evaluated for their reaction with carbohydrate epitopes to avoid crossreaction with glycoproteins containing similar carbohydrate groups. Among various possibilities to avoid such crossreactions is the preparation of polypeptide epitope purified antibodies by adsorption to a recombinant glycosyltransferase-β-galactosidase fusion protein produced in bacteria.[36] Since the fusion protein is nonglycosylated, the resulting antibodies will only recognize protein epitopes of the glycosyltransferase.

2. Low Temperature Embedding in Lowicryl K4M

Immunolocalization on thin sections of Epon-embedded tissues or cells of Golgi ap-

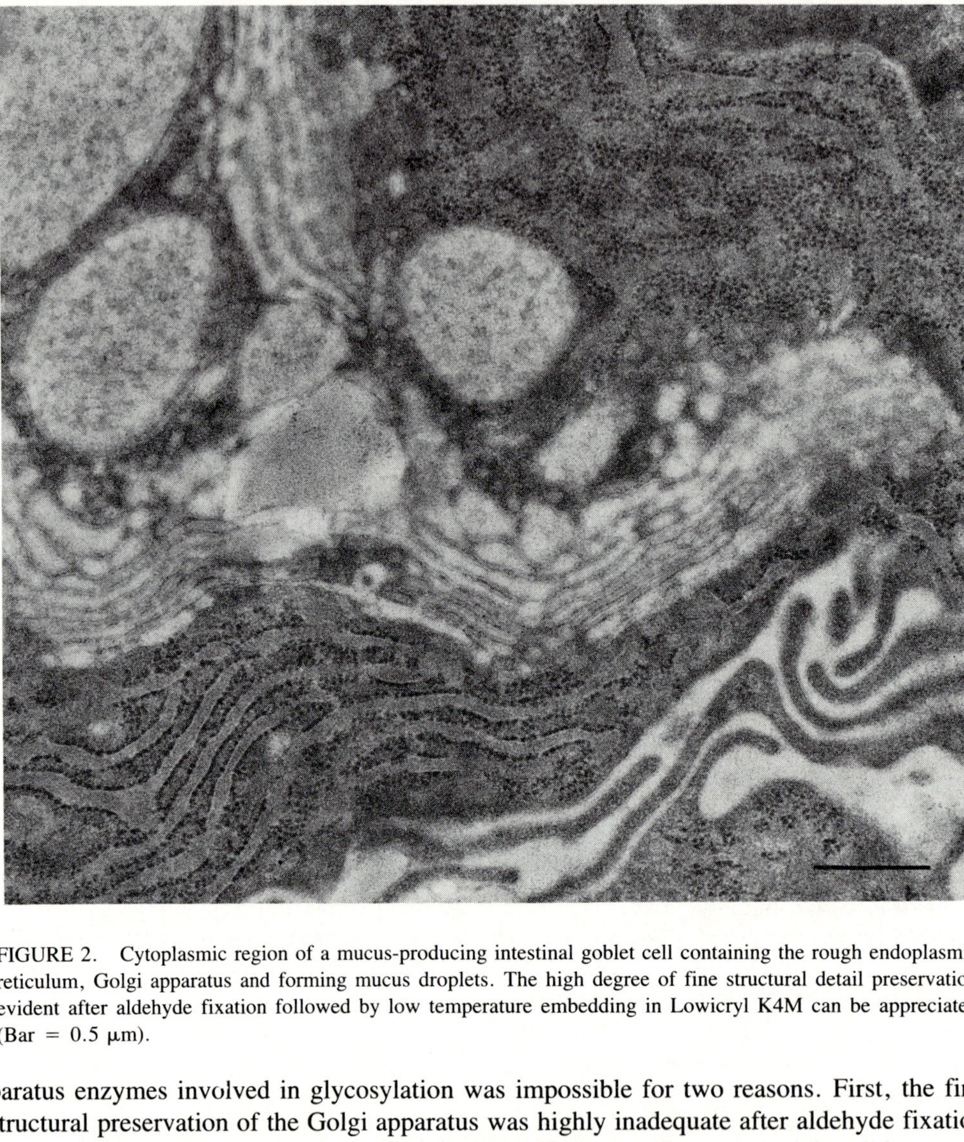

FIGURE 2. Cytoplasmic region of a mucus-producing intestinal goblet cell containing the rough endoplasmic reticulum, Golgi apparatus and forming mucus droplets. The high degree of fine structural detail preservation evident after aldehyde fixation followed by low temperature embedding in Lowicryl K4M can be appreciated (Bar = 0.5 μm).

paratus enzymes involved in glycosylation was impossible for two reasons. First, the fine structural preservation of the Golgi apparatus was highly inadequate after aldehyde fixation without subsequent exposure to osmium tetroxide.[37] Second, Epon embedding did not preserve antigens sufficiently for subsequent immunolocalization with protein A-gold. Fortunately, the degree of preservation of both the fine structure and antigens of the Golgi apparatus (and other cellular structures) was sufficient for subsequent successful immunolocalization when low temperature embedding in Lowicryl K4M was performed[18,19] (Figures 2 and 8). These improvements were due to various interrelated factors of which only a few will be briefly discussed. The low temperature condition significantly lowers the degree of structural (macromolecular) perturbation induced by exposure to the organic solvents used during dehydration,[20] and thereby contributes to the improved preservation of both fine structural detail and reactivity of antigens with antibodies in only mildly aldehyde fixed tissue. The same effect of temperature can be assumed to occur during resin infiltration and UV light induced polymerization. The acrylate-methacrylate based low viscosity resin Lowicryl K4M itself seems to provide advantages of critical importance, which are based on its chemical nature. One of them is its hydrophilic character, which contributes to the better preservation of protein conformation. Very important are aspects related to the cohesion between the

FIGURE 3. Thin section from Lowicryl K4M embedded *E. coli* shadowed with Ta/W at ~16°. Note the strong relief (Bar = 0.5 μm). Photograph by courtesy of W. Villiger (Biocenter, University of Basel, Switzerland).

resin and the biological matter, which have far reaching consequences for the sectioning behavior. In contrast to Epoxy resins, which become bound to biological material by chemical reactions during polycondensation, such cohesion (or copolymerization) with the biological material usually does not occur in the case of radical polymerization of Lowicryl K4M.[38,39] Therefore, during sectioning the cleavage will follow the interface Lowicryl K4M -biological matter, since this is the path of least resitance in the blocks. The result is a specimen-related surface relief of 2 to 6 nm (Figure 3). Since the surface of cleavage follows basically the interface Lowicryl K4M-biological matter, antigenic sites are exposed for immunolabeling.

3. Detection of Carbohydrate Moieties

The immunolocalization of enzymes provides no information about the possible catalytical activity of the protein at this site. However, the intracellular localization of a glycosyltransferase can be compared with the presence and distribution of the corresponding sugar residue, as detected by *in situ* cytochemical means. The combination of these two independent approaches then provides direct evidence for the site at which the glycosyltransferase exerts its *in vivo* function in the glycosylation of proteins. Lectins are the most commonly used reagents for the detection of cellular carbohydrates.[40] Embedding of tissues or cells in Lowicryl K4M, in conjunction with lectin-gold techniques for the intracellular localization of lectin binding sites (Figure 4), has become as powerful a tool as the protein A-gold technique.[21] Moreover, monoclonal antibodies with well characterized specificity for certain carbohydrate epitopes represent another, although less applied, type of reagent. Low temperature embedding of aldehyde-fixed specimens in Lowicryl K4M preserves the conformation of glycoconjugates not only for subsequent interaction with different lectins or carbohydrate specific antibodies, but also for modification by glycosidases or various chemical means.[41-43] This provides the basis for additional important controls of labeling specificity. Very recently, glycosidase-gold complexes were applied to Lowicryl K4M sec-

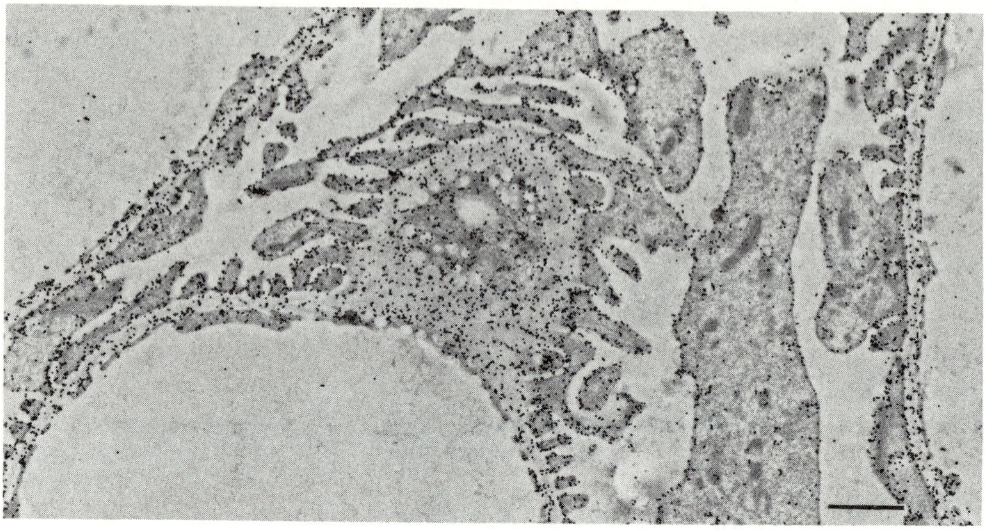

FIGURE 4. Portion from a capillary loop of rat kidney glomerulum; Lowicryl K4M embedding. The thin section was incubated with wheat germ agglutin followed by ovomucoid-gold complex prepared with 15 nm gold particles. Note the intense specific label and the very low degree of background over tissue free regions (Bar = 1 μm).

tions for the detection of the corresponding substrate. A β-glucosidase β-D-glucoside glucohydrolase from almonds was used to detect glucose residues.[44] Its application seems to be somewhat limited, however, since endoplasmic reticulum (ER) intermediates of asparagine-linked oligosaccharides all contain exclusively glucose residues in an α linkage. This may explain the observed lack of labeling over the lumen of the ER which is known to be highly enriched in glucose residues. Similarly, the lack of labeling of the lumen of the ER and other cellular structures with an α-mannosidase (from jack bean)-gold complex[45] is at clear variance with biochemical and lectin cytochemical data.[4,10] Clearly, a more thorough investigation of this technique is required in order to assess its reliability for use in glycoconjugate cytochemistry.

B. THE BIOCHEMICAL APPROACH

In 1981, a most important advance in Golgi apparatus research resulted from the finding by Rothman and his co-workers[17,46] that in vesicular stomatitis virus infected CHO cells many of the glycosylation enzymes could be resolved, and the corresponding Golgi subfractions could be separated on analytical sucrose density gradients (Figure 5). This approach relied on the premise that a vectorial difference in density exists across the stack of Golgi apparatus cisternae. Furthermore, it was assumed that upon cell homogenization the Golgi apparatus cisternae, or their remnants, would be torn apart. Indeed, when CHO membranes from a postnuclear supernatant were placed into a dense sucrose solution at the bottom of a tube, upon which a six-step sucrose gradient was layered, separation could be achieved. Upon ultracentrifugation, membranes accumulated in the sucrose layer corresponding to a density equal to their own. In this way, kinetically early and late enzyme activities were found in heavy and light membrane fractions, respectively. These data were the first biochemical evidence to suggest the existence of distinct Golgi apparatus glycosylation subcompartments.

Other approaches have also been used to determine the location of particular Golgi apparatus functions. One made use of the carboxylic ionophore monensin, introduced by Tartakoff and Vasalli,[47] which had been shown to cause dilation of all Golgi apparatus cisternae and to interrupt intracellular transport of secretory protein at the level of the Golgi

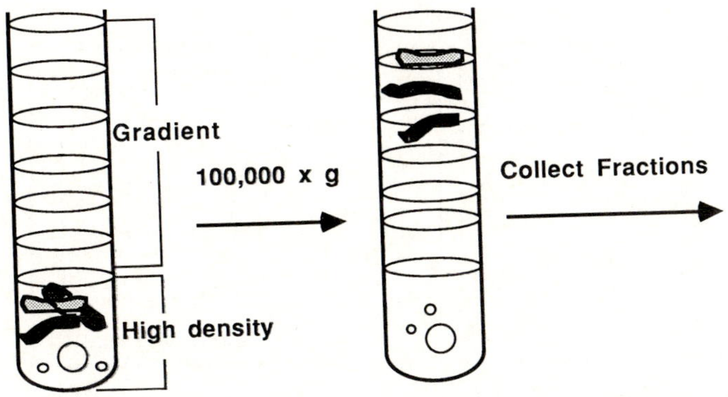

FIGURE 5. Schematic representation of the principle of analytical sucrose density gradient centrifugation for the preparation of Golgi subfractions. For more details, see section B.

apparatus without inhibiting terminal glycosyltransferases.[48] Most of the studies were performed on immunoglobulin M and surface glycoproteins of plasma cells and lymphoid cells, respectively.[49,50] In order to identify the site of monensin action a combination of pulse-chase biosynthetic labeling protocols with amino acids and sugars, subcellular fractionation, and electron microscopic autoradiography was used. Thereby, the extent of post-translational protein processing in the Golgi apparatus, as evidenced by the inhibition of incorporation of certain [^{3}H] sugars, could be assessed and used to identify the site of monensin action. In another experimental protocol involving the use of monensin, advantage was taken of the marked shift in the electrophoretic mobility of proteins due to terminal glycosylation.[50] This shift occurs mainly as a consequence of sialic acid addition, as indicated by its sensitivity to neuraminidase. Essential for the interpretation of the monensin experiments were data about the kinetics of sugar labeling and secretion of the proteins.

A completely different strategy was developed by Rothman and co-workers[51,52] (Figure 6). They prepared mixed monolayer cultures consisting of vesicular stomatitis virus infected CHO cell clones defective in certain glycosylation reactions and of uninfected wild-type CHO cells. Somatic cell hybrids were then obtained by low pH-induced cell fusion. The vesicular stomatitis virus G protein of the mutant cell lines was metabolically labeled prior to cell fusion to permit identification in subsequent lectin binding studies aimed at studying the extent of its further processing. In a typical experiment, the G protein of a mutant cell line capable of adding N-acetylglucosamine, but unable to add galactose, was labeled via [^{3}H] N-acetylglucosamine and then used with another mutant unable to add N-acetylglucosamine, but able to add galactose. Inter-Golgi apparatus transport and continuation of glycosylation by addition of galactose was then monitored in pronase glycopeptides prepared from immunoprecipitated and electrophoretically purified G protein by the binding to the galactose specific *Ricinus communis* lectin I.

C. THE COMBINED MORPHOLOGICAL-BIOCHEMICAL APPROACH

The fortunate combination of biochemical analysis with classical electron microscopy, cytochemistry, and immunolabeling, made high resolution *in situ* investigations possible. An improved resolution in the identification of the monensin-induced block was achieved in studies on virally infected cells.[53-55] The site of accumulation of viral nucleocapsids was easily identifiable by simple electron microscopic inspection.[53] Further characterization was achieved by *in situ* lectin cytochemistry and immunoelectron microscopy, as well as pulse-chase biosynthetic labeling and oligosaccharide analysis.[54] The accumulation of viral nucleocapsids in a particular region of the monensin-perturbed Golgi apparatus allowed its

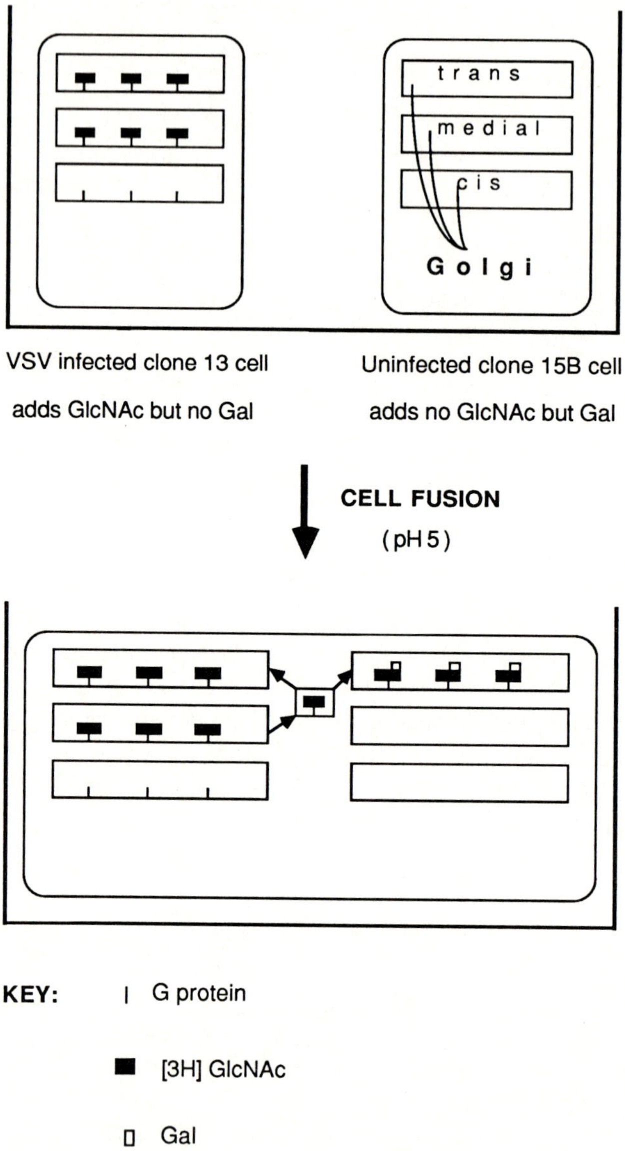

FIGURE 6. Schematic representation of the pH-induced fusion of CHO cell glycosylation mutants to study intercompartmental transportation of protein. For more details, see section B.

separation from the others by density gradient centrifugation, and permitted the assignment of functions to it.

An alternative to the drug monensin was found in temperature-sensitive virus mutants, which could be blocked at nonpermissive temperature at specific intracellular sites,[55] or in solely temperature-induced blocks of intracellular viral transportation.[56-59] In most of these studies, immunoelectron microscopy and cytochemistry were applied for the identification of the arrest site and glycoconjugate analysis to evaluate the degree of glycosylation obtained at these sites.

A very promising approach is the recently reported direct combination between electron microscopic and biochemical methods of *in vitro* glycosylation.[41] Glycosyltransferases exhibit a strict acceptor substrate specificity, and have been shown to be valuable reagents in

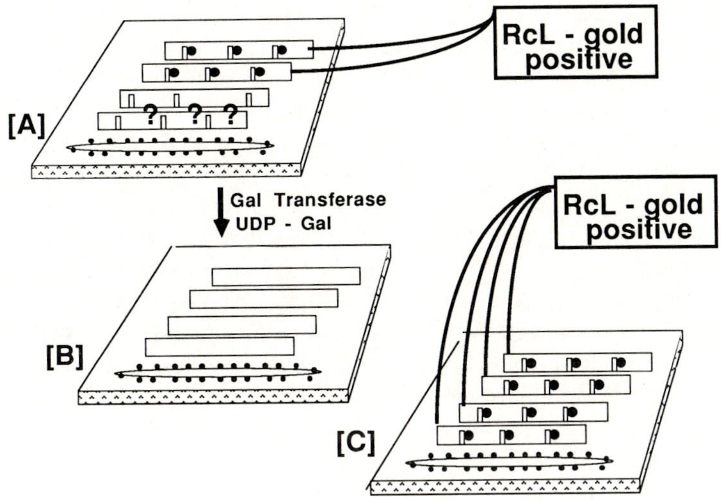

FIGURE 7. Schematic representation of *in vitro* galactosylation of terminal N-acetylglucosamine residues. (A) Section incubated with *Ricinus communis* lectin I-gold complex to reveal the inherent distribution of galactose residues. (B) Section exposed to galactosyltransferase and UDP-galactose to achieve galactosylation of N-acetylglucosamine residues. (C) Subsequent detection of galactosylated N-acetylglucosamine residues with *Ricinus communis* lectin I-gold.

the elucidation of oligosaccharide structures.[60,61] Since sugar residues were found to be sufficiently preserved and accessible for lectin or antibody binding following Lowicryl K4M embedding, it was only logical to incubate such thin sections with a purified glycosyltransferase in the presence of the donor substrate to glycosylate the acceptor substrate present on the section surface. Specifically, purified human milk β-N-acetylglucosaminide β1,4 galactosyltransferase was used to transfer galactose residues from UDP-galactose to terminal nonreducing N-acetylglucosamine residues present on thin sections of Lowicryl K4M embedded liver. Afterwards, the galactosylated transferase product was detected with a *Ricinus communis* lectin I-gold complex. By comparing the distribution of the lectin-gold label for endogenous galactose residues with the distribution of the lectin-gold label after galactosylation, a shift of labeling for galactose residues in the Golgi apparatus could be revealed, indicative of effective galactosylation. The details of the procedure are schematically depicted in Figure 7. Further characterization of the galactosylated product, i.e., its presence on N- or O-linked oligosaccharides, was made possible by endoglycosidase pretreatment of the thin sections.

Collectively, these data demonstrated that sugar residues present on the surface of Lowicryl K4M thin sections are amenable to enzymatic treatment or chemical modification. Such glycosidase treatment of thin sections proved to be a valuable additional tool to characterize the reactivity of an antibody directed against apomucin.[43]

III. ANALYSIS OF THE GOLGI APPARATUS ORGANIZATION OF GLYCOSYLATION

Early electron microscopic studies have provided data that carbohydrates,[62,63] osmiophilic products,[64] and histochemically detectable enzyme activities,[65,66] are not distributed randomly in the Golgi apparatus. They will not be discussed in the following sections. Other developments such as aspects of ER to Golgi apparatus and intra-Golgi apparatus transport,[1,3,51,52] as well as the exciting field of antiport proteins functioning in the sugar nucleotide/nucleotide exchange in the Golgi apparatus,[11] are also beyond the scope of this review primarily due to restrictions in space.

A. DO ER PROTEINS ENTER THE GOLGI APPARATUS, OR THE GOLGI APPARATUS AS TWO ORGANELLES IN TANDEM?

In 1981 Rothman made the proposal "that the Golgi apparatus actually consists of two attached but different organelles (the cis and the trans Golgi apparatus) having related but distinct purposes, and operating sequentially to purify exported proteins from those of the ER on the basis of an underlying principle akin to that of fractional distillation." The trans Golgi apparatus was defined as the last one or two trans cisternae, with the rest of the stack representing the cis Golgi apparatus. This attractive hypothesis was based on two kinds of evidence. First, previous biochemical work had shown the presence of endoplasmic reticulum enzymes, including cytochrome P-450, glucose-6-phosphatase, cytochrome b_5 reductase, and phospholipid biosynthetic enzymes in Golgi apparatus fractions.[67-71] Second, very recent biochemical work from Rothman's group had provided clear evidence for the existence of two functionally separate compartments.[17] On analytical sucrose gradients, activity for mannosidase, a trimming enzyme, banded together with freshly acylated viral G protein, and could be separated from G protein pulsed with [^{3}H] galactose, which banded together with late acting galactosyl- and sialyltransferase activity. Therefore, these data together with certain highly selected histochemical data, were the basis for the working hypothesis of a precisely twofold division of the Golgi apparatus, the so-called distillation hypothesis. The main functional conclusion consisted of the assumption that the cis Golgi apparatus would function in removal of ER proteins from other proteins intended for export, thereby refining the mixture of proteins before entry into the trans Golgi apparatus. Various experimental work was then directed to test the validity of this working hypothesis. These experiments were aimed at high resolution *in situ* electron microscopic localization of typical ER proteins, namely glucosidase II,[72,73] hydrocarbon-induced cytochrome P-450,[74] and phenobarbital-induced cytochrome P-450.[75] All these studies had one finding in common: they provided no evidence for the presence of either of the three ER proteins in any region of the Golgi apparatus. Furthermore, examination of the oligosaccharide chains of glucosidase II and hexose-6-phosphate dehydrogenase, or those of the bulk of glycoproteins of the ER membrane, provided no evidence for the passage of these proteins into the Golgi apparatus.[74] The predominant oligosaccharide chains were those with eight or nine mannose residues, and thus represented typical products of ER localized mannosidase.[4] Subsequent studies on glucosidase II biosynthesis and turnover confirmed and extended these data by showing persistence of Endo-H sensitivity of glucosidase II and retainment of the innermost glucose residue(s) on its oligosaccharides.[76] Thus, two basic predictions made in the distillation hypothesis could not be confirmed. Rather, it could be firmly established that in liver hepatocytes, proteins of the ER were effectively retained within this organelle.

Investigations on further cell types, however, have yielded unexpected results. In kidney tubular cells immunolabel for glucosidase II was not restricted to the nuclear envelope and the ER, but also present in the Golgi apparatus, cytoplasmic post-Golgi apparatus vesicles, and the plasma membrane.[77] In addition, enzymatically active glucosidase II was found to carry sialic acid residues providing direct evidence for Golgi apparatus-associated processing of the enzyme. These findings, with others which will be discussed below, are strong indications for cell-type-related specific differences in the expression of interorganelle boundaries and intraorganelle compartments.

In their publication describing the detection of two functionally distinct Golgi apparatus compartments, Dunphy et al.[17] did not entirely rule out the possibility of the existence of additional Golgi-associated compartments. The following section provides a critical discussion of the experimental evidence obtained in subsequent studies.

B. DIFFERENTIAL COMPARTMENTATION OF GLYCOSYLATION REACTIONS IN THE GOLGI APPARATUS

The first direct demonstration of a functionally defined Golgi apparatus subcompartment became possible through the immunoelectron microscopic localization of a galactosyltransferase in HeLa cells.[18] Immunoreactive sites were present exclusively over some cisternae on one side of the Golgi apparatus cisternal stack. The simultaneous demonstration in a single thin section of thiamine pyrophosphatase enzyme activity, a classical marker for trans cisternae, and of galactosyltransferase immunolabel therein, provided unequivocal evidence for the existence of a trans Golgi apparatus subcompartment. Implications derived from the close topological association of galactosyltransferase and thiamine pyrophosphatase will be disucssed below. Further evidence for the existence of a trans subcompartment involved in late processing of oligosaccharides was obtained through the localization of galactose residues in the Golgi apparatus of Semliki Forest virus infected BHK cells.[78] Altogether, these *in situ* data showed that the late Golgi apparatus function biochemically separated from an early function[17] was indeed located in trans Golgi apparatus cisternae. Another late glycosylation step, attachment of terminal sialic acid residues to galactose residues of an oligosaccharide chain, was investigated in hepatocytes of intact rat liver (Figures 9 and 10)[79] and mucus-producing goblet cells of rat colon (Figures 12 and 13).[6,80] The distribution of immunoreactivity for sialyltransferase and cytochemical label for sialic acid was found to coincide over the trans cisternae and the trans-tubular network of the Golgi apparatus. These data, together with those demonstrating the coincidence of distribution of galactose and sialic acid residues in the Golgi apparatus of both cell types, also provided evidence that the two sequentially occurring terminal glycosylation steps, i.e., galactosylation and sialylation, take place in the same trans Golgi apparatus subcompartment. Moreover, specific activities for galactosyl- and sialyltransferase could not be resolved on sucrose density gradients.[17,81] Another terminal glycosylation step performed by the blood group A α1,3 N-acetylgalactosaminyltransferase on mucin-type (or O-linked) oligosaccharide chains was investigated on intestinal epithelia.[80,82] Immunolabel for both the α1,3 N-acetylgalactosaminyltransferase and its product of action, blood group A substance, were restricted to trans cisternae and the transtubular network of the Golgi apparatus of goblet cells. At present no immunoelectron microscopic data are available on the localization of the other terminal (or late acting) glycosyltransferases.

On analytical sucrose gradients membranes prepared from CHO, mouse lymphoma BW5147.3 and P3388D$_1$ plasma cells equilibrated in such a way that activity for galactosyl- and sialyltranferase could be separated from activities of earlier acting enzymes such as N-acetylglucosaminyltransferase I, II and IV, as well as mannosidase I and II.[81,83] Direct immunoelectron microscopical evidence for the presence of N-acetylglucosaminyltransferase I in a middle region of the Golgi apparatus cisternal stack came from investigations on a rabbit-kidney-derived cell line and rabbit liver hepatocytes.[84] It was noted that a single unstained cisterna existed at one end of the cisternal stack and several negative cisternae at the other end, most probably corresponding to the trans side of the Golgi apparatus. A subsequent cytochemical study established the Golgi apparatus distribution of terminal non-reducing N-acetylglucosamine residues in rat and pig liver hepatocytes.[42] Purified galactosyltransferase was used to galactosylate such residues on thin sections. Subsequent indirect detection by lectin cytochemistry demonstrated their presence in middle as well as cis cisternae, and sometimes in smooth-membraned structures most probably corresponding to the fenestrated first cis cisterna. The latter structures were reported not to contain immunocytochemically detectable N-acetylglucosaminyltransferase I; however, closer inspection of the published micrographs (Figure 6 A-D in Reference 84) gives the impression that they are indeed weakly positive. Further evidence for the existence of a middle Golgi apparatus compartment derived from monensin experiments.[49,54,55] Although the initial data could not be unequivocally interpreted as indicating the existence of such a compartment (they were obtained before the discovery of glycoslyation subcompartments), they clearly showed that the monensin block was proximal to the location of terminal glycosyltransferases.[49] Sub-

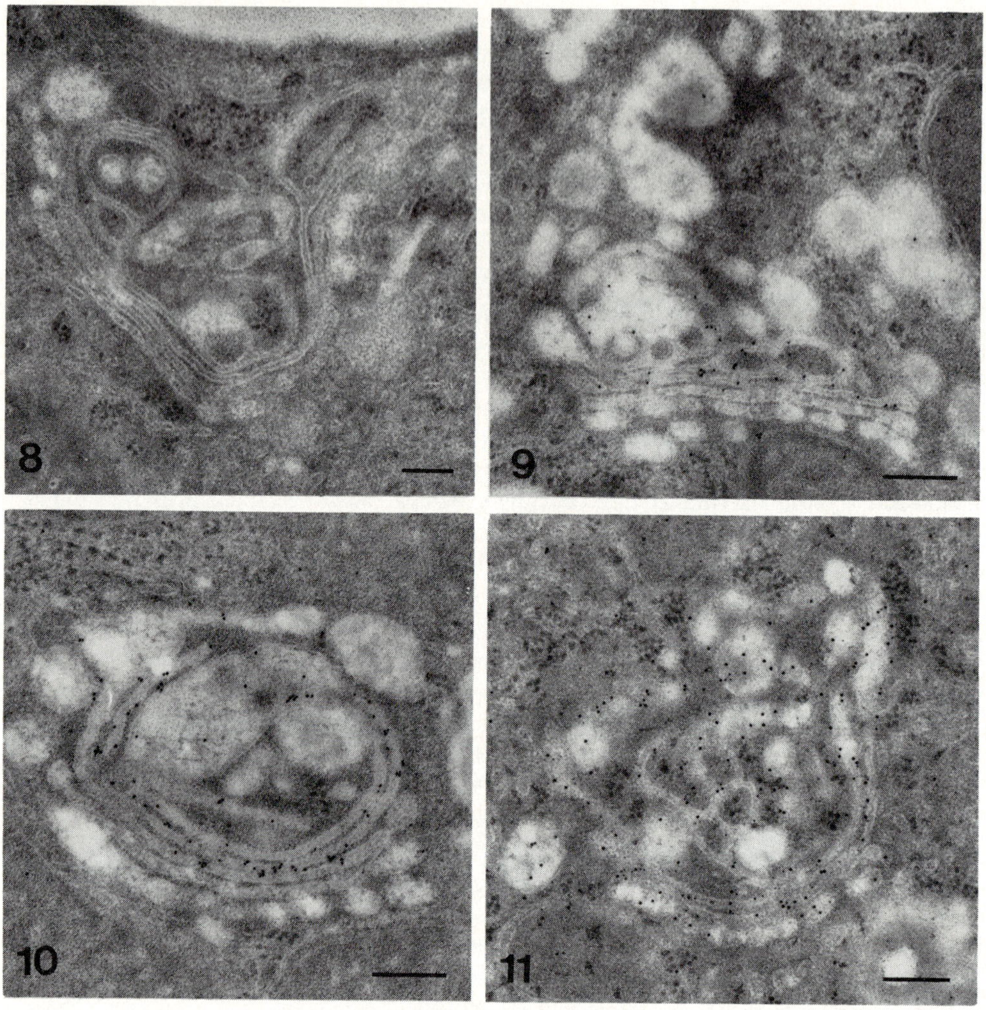

FIGURE 8 to 11. Thin sections from aldehyde fixed and low temperature Lowicryl K4M embedded rat liver. In all photographs the cis side of the Golgi apparatus is to the bottom. Figure 8 shows a stack composed of four cisternae. The two trans cisternae are in continuity with the trans-tubular network. This continuity is also evident in Figure 9 to 11. The presence of immunoreactivity for sialyltransferase in two trans cisternae and the transtubular network is shown in Figures 9 and 10. Immunolocalization of albumin (Figure 11) reveals presence of immunoreactive sites throughout the cisternal stack and the transtubular network. (Bars = 0.2 μm.)

sequent studies on viral membrane transport and glycosylation, as well as fatty acylation, strongly supported the view that the Golgi apparatus cisternal stack can be divided into three distinct compartments.

The existence of a cis Golgi apparatus compartment, although mainly based on negative evidence, is undisputed. The most convincing, although indirect evidence, was obtained in studies on the assembly of O-linked oligosaccharides. In a lectin-gold study, evidence for the onset of O-linked glycosylation in the cis Golgi apparatus was obtained.[85] Subsequent morphological and biochemical studies supported the assumption that the cis compartment can be correlated with the site of onset of O-linked glycosylation. An antibody prepared against enzymatically deglycosylated mucin (apomucin) failed to label Golgi apparatus structures, but gave intense staining of the endoplasmic reticulum.[43] Double labeling with this antibody and the N-acetylgalactosamine specific *Helix pomatia* lectin resulted in very little overlap, supporting the conclusion that core O-glycosylation is a cis Golgi apparatus event.

The enzyme establishing this linkage, UDP-GalNAc: polypeptide N-acetylgalactosaminyltransferase (UDP = uridine diphosphate, GalNAc = N-acetylgalactosamine), has been recovered on a sucrose density gradient in a membrane fraction of intermediate density, between the endoplasmic reticulum markers glucosidase I and II, and the trans Golgi apparatus marker galactosyltransferase.[86] Other candidates and functions for the cis compartment are the UDP-N-acetylglucosamine: lysosomal enzyme N-acetylglucosamine-1-phosphotransferase and the N-acetylglucosamine 1-phospodiester α-N-acetylglucosaminidase[83] involved in the synthesis of the mannose-6-phosphate recognition marker of lysosomal enzymes (see also Chapter 10 by Brown). Fatty acylation of viral membrane proteins could also occur in the cis compartment, but equally as well in the middle compartment.[55]

Collectively, ail the assembled data discussed above indicate a functional subdivision of the Golgi apparatus with respect to post-translationally occurring glycosylation steps. Nevertheless, the body of data suggesting such subcompartmentation is obviously not overwhelming. Only a few of the numerous enzymes involved in glycosylation have been mapped. Indeed, some recent observations indicate that we must reconsider our current concept of three Golgi apparatus compartments. Immunolabel for sialyltransferase and blood group A α1,3 N-acetylgalactosaminyltransferase, both representing terminal transferases, lacked a polarized distribution in absorptive intestinal cells (Figure 14).[80] Similarly, the distribution of sialic acid residues and blood group A substance was not restricted to trans cisternae (Figure 15). In kidney tubular cells, sialyltransferase and sialic acid residues are also present throughout the cisternal stack (unpublished data). These data provided the first evidence of cell-type specific variability in the Golgi apparatus distribution of terminal glycosylation steps. Supporting, although indirect, evidence for cell-type-specific variation in the distribution of glycosylation steps has been obtained by other investigators using lectin cytochemistry.[87,88] A discussion of the significance of these observations was presented in Roth[4] and Roth et al.[80]

A different level of Golgi apparatus organization representative of another extreme situation is worth mentioning. According to the one-linkage one-enzyme hypothesis, several different types of sialyltransferases, for instance, should exist and all the available experimental evidence supports this view. Therefore, one could expect to find cells possessing a particular type of sialyltransferaese and lacking another. Such a situation was recently encountered in studies on rat intestinal mucosal cells.[36] By immunoelectron microscopy, the galactose β1,4 N-acetylglucosamine α2,6 sialyltransferase could be detected in the Golgi apparatus of the epithelia of the colon, but was undetectable in the Golgi apparatus of the same cells in the various regions of the small intestine. This finding was corroborated by specific enzyme assays which, on the other hand, proved the existence of an α2,3 sialyltransferase activity in the mucosa of the small intestine. This differential expression of the galactose β1,4N-acetylglucosamine α2,6 sialyltransferase in the epithelia of the small and large rat intestine seems also to be manifested in the oligosaccharide composition. Colonic mucins contain sialic acid in the α2,6 linkage, whereas mucins of the small intestine contain sialic acid only in the α2,3 linkage.[89]

It was mentioned above that galactosyltransferase in HeLa cells was detected in trans Golgi apparatus cisternae together with thiamine pyrophosphatase activity.[18] This finding was in support of a model proposing a concerted action of both enzymes.[90,91] During the action of galactosyltransferase free UDP is generated, which from certain concentrations upwards would have an inhibitory effect on galactosyltransferase activity. Therefore, there is a requirement for the presence of a nucleoside diphosphatase, together with galactosyltransferase, at the subcellular site of its action to prevent the build-up of such amounts of UDP. The demonstration of thiamine pyrophosphatase activity, a nucleoside diphosphatase, codistributing with galactosyltransferase provided some experimental evidence for the above

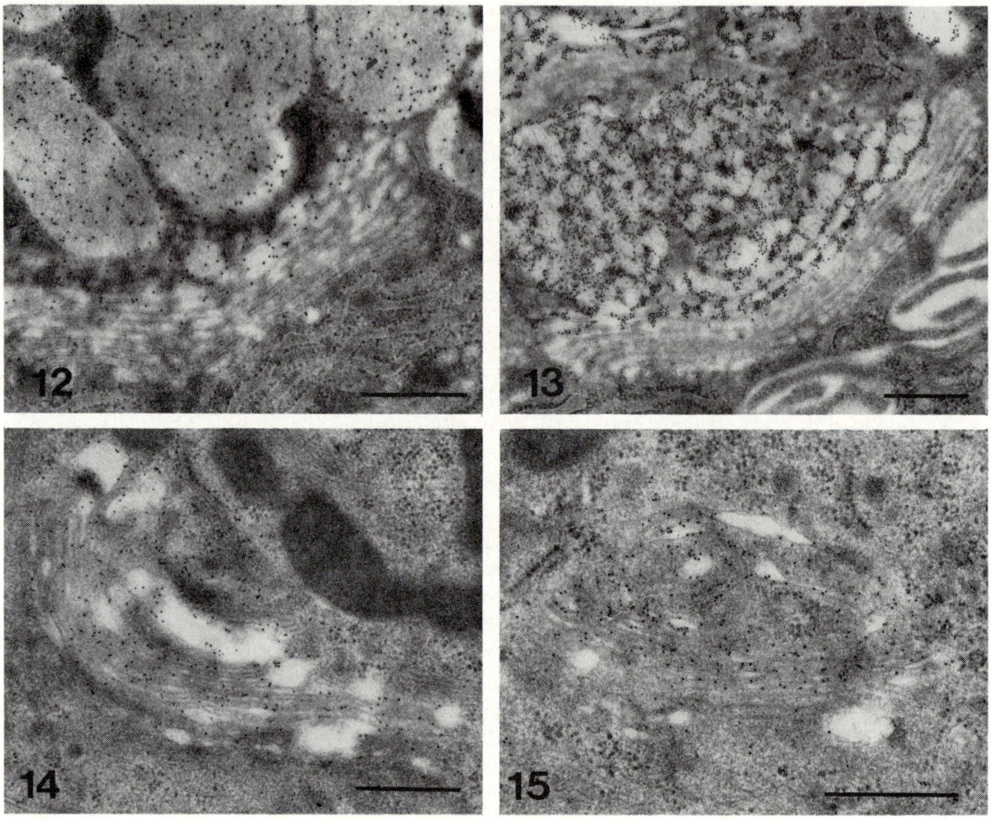

FIGURE 12 to 15. Differential compartmentation of sialylation in rat colonic epithelia. In the goblet cell Golgi apparatus label for both sialyltransferase (Figure 12) and sialic acid residues (Figure 13) is restricted to trans cisternae. The significance of the immunolabel for sialyltransferase as detected with epitope-purified antibodies is discussed in Reference 36. In contrast to the restricted distribution in goblet cells, the entire cisternal stack in absorptive enterocytes, with the exception of the first cis cisterna, is positive for sialyltransferase (Figure 14). Sialic acid residues exhibit the same diffuse distribution in the absorptive enterocyte Golgi apparatus (Figure 15). (Bars = 0.5 μm.)

mentioned proposal. By analogy with the observed codistribution of galactosyltransferase with thiamine pyrophosphatase in trans Golgi apparatus cisternae, one could expect to find a similar situation for sialyltransferase. Since an inhibitory effect of CMP (cytosine monophosphate) on sialyltransferase activity has been observed, codistribution with a CMP splitting enzyme would be logical. Such an enzyme, acid phosphatase (CMP-ase), has been histochemically shown to exist in trans elements of the Golgi apparatus.[6] Intriguingly, most immunolabel for sialyltransferase in rat liver hepatocytes was found in a Golgi apparatus region that displayed mainly thiamine pyrophophatase activity, but little CMP-ase activity. At first this result seemed surprising. However, in the light of recent investigations on the mechanism of sugar nucleotide and nucleoside monophosphate translocation in Golgi apparatus membranes it becomes evident that the above mentioned postulate is no longer valid.[11] These authors presented data for the existence of specific and unique translocation proteins in membranes derived from Golgi apparatus. One such antiport protein has been found to function in a coupled equimolar transmembrane exchange of CMP-sialic acid with CMP. Therefore, these data indicate that sialyltransferase need not to be exclusively associated with a CMP-ase positive Golgi apparatus region, as was shown by Roth et al. for hepatocytes.[79]

C. THE TRANSTUBULAR NETWORK OF THE GOLGI APPARATUS AND GERL

An extensive system consisting of flattened, cisternal elements and of interconnected and blind ending tubules at the trans side of the Golgi apparatus of neurons, which exhibited histochemically demonstrable acid phosphatase activity, was first described by Novikoff[92] and extensively investigated in subsequent studies in this and further cell types.[93,94] This structural complex was named GERL and interpreted to correspond to "a specialized system of smooth endoplasmic reticulum with continuities with rough endoplasmic reticulum" involved in the formation of lysosomes. GERL, as interpreted by Novikoff and his co-workers, was always excluded from the Golgi apparatus cisternal stack, both structurally and functionally. Rambourg and co-workers,[5,95,96] employing various impregnation techniques and acid phosphatase histochemistry on various cell types, sometimes in combination with high voltage electron microscopy (see Chapter 17 by Ris and Pawley), revealed the presence of a broad branching system of anastomosing and blind ending tubules at the trans side of the Golgi apparatus, and emphasized its structural similarity to the one described by Novikoff et al.[94] However, this structure, which was named the transtubular network, was seen in direct continuity with the Golgi apparatus cisternae, which showed a peeling-off configuration and could not be differentiated morphologically from the other cisternae of the stack. In recent studies, we found sialyltransferase and sialic acid in two trans cisternae, and the transtubular network of the Golgi apparatus in rat liver hepatocytes (Figures 9 and 10).[79] In addition, we demonstrated continuity of the transtubular network with these two trans cisternae, and the presence of activity for acid phosphatase and thiamine pyrophosphatase. Therefore, the transtubular network of hepatocytes and other cell types expresses a characteristic feature of GERL. Therefore, GERL has to be considered structurally coincident with the transtubular network, and functionally to be part of the glycosylation machinery of the Golgi apparatus. The recently introduced terms trans reticular (TR) Golgi,[97] trans Golgi network (TGN),[58,98] and trans Golgi reticulum (TGR),[99] are basically synonymous with the transtubular network originally described by Rambourg et al.[96] The transtubular network is not only involved in glycosylation, but also other Golgi apparatus functions, such as secretion (Figure 11),[100] transport of viral glycoprotein[58] or lysosomal enzymes,[99] and endocytosis,[3,101] and represents the *bona fide* exit site of the Golgi apparatus.[58,98] Direct evidence that it is the subcellular site of sorting of proteins was recently reported.[102] The transtubular network was tentatively distinguished from trans Golgi apparatus cisternae based on the nonoverlapping localizations of thiamine pyrophosphatase and acid phosphatase.[98] This distinction, however, discounts the earlier demonstration of a structural continuity between trans cisternae and the transtubular network, as well as the presence of sialyltransferase in both structures.[79] Based on these observations, it seems justified to consider both structures as representing a single entity.

IV. CONCLUDING REMARKS

The various techniques described in this review have helped to reveal certain fine details of Golgi apparatus function in relation to an important cellular activity, namely glycosylation. From initial data indicating a relatively simple principle of a twofold division, the current view of cell-type-specific organization of the Golgi apparatus has emerged. Thus, in many cell types three compartments can be distinguished, whereas in others no such compartmentation is evident. The underlying principle for the establishment of either of these extreme situations is completely unclear at present. The recent successful cloning of genes for galactosyl- and sialyltransferase may open an exciting field for molecular analysis and manipulation.[103-106]

REFERENCES

1. **Farquhar, M. G. and Palade, G. E.,** The Golgi apparatus (complex) — (1954-1981) — from artifact to center stage, *J. Cell Biol.,* 91, 77s, 1981.
2. **Tartakoff, A. M.,** The confined function model of the Golgi complex: center for ordered processing of biosynthetic products of the rough endoplasmic reticulum, *Int. Rev. Cytol.,* 85, 221, 1983.
3. **Farquhar, M. G.,** Progress in unraveling pathways of Golgi traffic, *Ann. Rev. Cell Biol.,* 1, 447, 1985.
4. **Roth, J.,** Subcellular organization of glycosylation in mammalian cells, *Biochim. Biophys. Acta,* 906, 405, 1987.
5. **Rambourg, A., Clermont, Y., and Hermo, L.,** Three-dimensional structure of the Golgi apparatus, in *Methods in Cell Biology,* Vol. 23, Hand, A. R. and Oliver, C., Eds., Academic Press, New York, 1981, 155.
6. **Novikoff, A. B. and Novikoff, P. M.,** Cytochemical contributions to differentiating GERL from the Golgi apparatus, *Histochem. J.,* 9, 525, 1977.
7. **Griffiths, G. and Simons, K.,** The *trans* Golgi network: sorting at the exit of the Golgi complex, *Science,* 234, 438, 1986.
8. **Bennett, G.,** Role of the Golgi complex in the secretory process, in *Cell Biology of the Secretory Process,* Cantin, M., Ed., Karger, Basel, 1984, 102.
9. **Leblond, C. P. and Bennett, G.,** Role of the Golgi apparatus in terminal glycosylation, in *International Cell Biology,* Brinkley, B. R. and Porter, K. R., Eds., Rockefeller University Press, New York, 1976, 326.
10. **Kornfeld, S. and Kornfeld, R.,** Assembly of asparagine-linked oligosaccharides, *Ann. Rev. Biochem,* 54, 631, 1985.
11. **Hirschberg, C. B. and Snider, M.,** Topography of glycosylation reactions in the rough endoplasmic reticulum and Golgi apparatus, *Ann. Rev. Biochem.,* 56, 63, 1987.
12. **Schachter, H.,** Biosynthetic controls that determine the branching and microheterogeneity of protein-bound oligosaccharides, *Biochem. Cell Biol.,* 64, 163, 1986.
13. **Fleischer, B., Fleischer, S., and Ozawa, H.,** Isolation and characterization of Golgi membranes from bovine liver, *J. Cell Biol.,* 43, 59, 1969.
14. **Munro, J. R., Narasimhan, S., Wetmore, S., Riordan, J. R., and Schachter, H.,** Intracellular localization of GDP-fucose glycoprotein and CMP-sialic acid: apolipoprotein glycosyltransferase in rat and pork livers, *Arch. Biochem. Biophys.,* 169, 269, 1975.
15. **Schachter, H., Jabbal, I., Hudgin, R. L., Pinteric, L., McGuire, E. J., and Roseman, S.,** Intracellular localization of liver sugar nucleotide glycoprotein glycosyltransferase in a Golgi-rich fraction, *J. Biol. Chem.,* 245, 1090, 1970.
16. **Bretz, R., Bretz, H., and Palade, G. E.,**, Distribution of terminal glycosyltransferases in hepatic Golgi fractions *J. Cell Biol.,* 84, 87, 1980.
17. **Dunphy, W. G., Fries, E., and Rothman, J. E.,** Early and late functions associated with the Golgi apparatus reside in distinct compartments, *Proc. Natl. Acad. Sci. U.S.A.,* 78, 7453, 1981.
18. **Roth, J. and Berger, E. G.,** Immunocytochemical localization of galactosyltransferase in HeLa cells: codistribution with thiamine pyrophosphatase in *trans*-Golgi-cisternae, *J. Cell Biol.,* 93, 223, 1982.
19. **Roth, J., Bendayan, M., and Orci. L.,** Ultrastructural localization of intracellular antigens by the use of protein A-gold complex, *J. Histochem. Cytochem.,* 26, 1074, 1978.
20. **Carlemalm, E., Garavito, R. M., and Villiger, W.,** Resin development for electron microscopy and an analysis of embedding at low temperature, *J. Microsc.,* 126, 123, 1982.
21. **Roth, J.,** Application of lectin-gold complexes for electron microscopic localization of glycoconjugates on thin sections, *J. Histochem. Cytochem.,* 31, 987, 1983.
22. **Roth, J.,** The colloidal gold marker system for light and electron microscopic cytochemistry, in *Techniques in Immunocytochemistry,* Vol. 2, Bullock, G.R. and Petrusz, P., Eds., Academic Press, London, 1983, 217.
23. **Roth, J.,** The protein A-gold (pAg) technique — a qualitative and quantitative approach for antigen localization on thin sections, in *Techniques in Immunocytochemistry,* Vol. 1, Bullock, G. R., and Petrusz, P., Eds., Academic Press, London, 1982, 107.
24. **Geuze, H. J., Slot, J. W., Scheffer, R. C. T., and van der Ley, P. A.,** Use of colloidal gold particles in double-labeling immunoelectron microscopy of ultrathin frozen tissue sections, *J. Cell Biol.,* 89, 653, 1981.
25. **Baschong, W., Lucocq, J. M., and Roth, J.,** "Thiocyanate gold": small (2-3 nm) colloidal gold for affinity cytochemical labeling in electron microscopy, *Histochemistry,* 83, 409, 1985.
26. **Slot, J. W. and Geuze, H. J.,** Sizing of protein A-colloidal gold probes for immunoelectron microscopy, *J. Cell Biol.,* 90, 533, 1981.
27. **Bendayan, M.,** Double immunocytochemical labeling applying the protein A-gold technique, *J. Histochem. Cytochem.,* 30, 81, 1982.

28. **Roth, J.,** The preparation of protein A-gold complexes with 3 nm and 15 nm gold particles and their use in labeling multiple antigens on ultrathin sections, *Histochem. J.*, 14, 791, 1982.
29. **Slot, J. W. and Geuze, H. J.,** A new method of preparing gold probes for multiple-labeling cytochemistry, *Eur. J. Cell Biol.*, 38, 87, 1985.
30. **Novikoff, A. B.,** DAB cytochemistry: artifact problems in its current use, *J. Histochem. Cytochem.*, 28, 1036, 1980.
31. **Courtoy, P. J., Picton, D. H., and Farquhar, M. G.,** Resolution and limitations of the immunoperoxidase procedure in the localization of extracellular matrix antigens, *J. Histochem. Cytochem.*, 31, 945, 1983.
32. **Roth, J.,** Postembedding cytochemistry with gold-labelled reagents: a review, *J. Microsc.*, 143, 125, 1986.
33. **Roth, J., Warhol, M. J., and Taatjes, D. J.,** Prevention of non-specific interactions of gold-labeled reagents on tissue reactions, *Histochemistry,* in press.
34. **Childs, R. A., Berger, E. G., Thorpe, S. J., Aegerter, E., and Feizi, T.,** Blood-group-related carbohydrate antigens are expressed on human milk galactosyltransferase and are immunogenic in rabbits, *Biochem. J.*, 238, 605, 1986.
35. **Feizi, T., Thorpe, S. J., and Childs, R. A.,** Blood group genetic markers on human milk galactosyltransferase: relevance to the immunohistochemical approach to enzyme localization, *Biochem. Soc. Trans.*, 15, 614, 1987.
36. **Taatjes, D. J., Roth, J., Weinstein, J., and Paulson, J. C.,** Post-Golgi apparatus localization and regional expression of rat intestinal sialyltransferase detected by immunoelectron microscopy with polypeptide epitope specific antibodies, *J. Biol. Chem.*, 263, 6302, 1988.
37. **Bendayan, M., Roth, J., Perrelet, A., and Orci, L.,** Quantitative immunocytochemical localization of pancreatic secretory proteins in subcellular compartments of the rat acinar cell, *J. Histochem. Cytochem.*, 30, 1075, 1980.
38. **Acetarin, J. -D., Carlemalm, E., Kellenberger, E., and Villiger, W.,** Correlation of some mechanical properties of embedding resins with their behaviour in microtomy, *J. Electron. Microsc. Techn.*, 6, 63, 1987.
39. **Kellenberger, E., Dürrenberger, M., Villiger, W., Carlemalm, E., and Wurtz, M.,** The efficiency of immunolabel on Lowicryl sections compared to theoretical predictions, *J. Histochem. Cytochem.*, 35, 959, 1987.
40. **Roth, J.,** *The Lectins. Molecular Probes in Cell Biology and Membrane Research,* (Exp. Pathol., Suppl. 3), Gustav Fischer, Jena, 1978.
41. **Roth, J., Lucocq, J. M., and Charest, P. M.,** Light and electron microscopic visualization of sialic acid residues with the lectin from *Limax flavus*: a cytochemical affinity technique with the use of fetuin-gold complexes, *J. Histochem. Cytochem.*, 32, 1167, 1984.
42. **Lucocq, J. M., Berger, E. G., and Roth, J.,** Detection of terminal *N*-linked *N*-acetylglucosamine residues in the Golgi apparatus using galactosyltransferase and endoglucosaminidase F/peptide *N*-glycosidase F: adaption of a biochemical approach to electron microscopy, *J. Histochem. Cytochem.*, 35, 67, 1987.
43. **Deschuyteneer, M., Eckhardt, A. E., Roth, J., and Hill, R. L.,** The subcellular localization of apomucin and terminal nonreducing *N*-acetylgalactosamine in porcine submandibulary gland, *J. Biol. Chem.*, 263, 2452, 1988.
44. **Bendayan, M. and Benhamou, N.,** Ultrastructural localization of glucoside residues on tissue sections by applying the enzyme-gold approach, *J. Histochem. Cytochem.*, 35, 1149, 1987.
45. **Londono, I. and Bendayan, M.,** Ultrastructural localization of mannoside residues on tissue sections: comparative evaluation of the enzyme-gold and the lectin-gold approaches, *Eur. J. Cell Biol.*, 45, 88, 1987.
46. **Rothman, J. E.,** The Golgi apparatus: two organelles in tandem, *Science,* 213, 1212, 1981.
47. **Tartakoff, A. M. and Vasalli, P.,** Plasma cell immunoglobulin secretion. Arrest is accompanied by alterations of the Golgi complex, *J. Exp. Med.*, 146, 1332, 1977.
48. **Tartakoff, A. M. and Vasalli, P.,** Comparative studies of intracellular transport of secretory proteins, *J. Cell Biol.*, 79, 694, 1978.
49. **Tartakoff, A. M. and Vassalli, P.,** Plasma cell immunoglobulin M molecules. Their biosynthesis, assembly, and intracellular transport, *J. Cell Biol.*, 83, 284, 1979.
50. **Tartakoff, A. M., Hoessli, D., and Vassalli, P.,** Intracellular transport of lymphoid cell surface glycoproteins, Role of the Golgi complex, *J. Mol. Biol.*, 150, 525, 1981.
51. **Rothman, J. E., Urbani, L. J., and Brands, R.,** Transport of protein between cytoplasmic membranes of fused cells: correspondence to processes reconstituted in a cell-free system, *J. Cell Biol.*, 99, 248, 1985.
52. **Rothman, J. E., Miller, R. L., and Urbani, L. J.,** Intercompartmental transport in the Golgi complex is a dissociative process: facile transfer of membrane protein between two Golgi populations, *J. Cell Biol.*, 99, 260, 1985.
53. **Johnson, D. C. and Schlessinger, M. J.,** Vesicular stomatitis virus and Sindbis virus glycoprotein transport to the cell surface is inhibited by ionophores, *Virology,* 103, 407, 1980.

54. **Griffiths, G., Quinn, P., and Warren, G.,** Dissection of the Golgi complex. I. Monensin inhibits the transport of viral membrane proteins from medial to trans Golgi cisternae in baby hamster kidney cells infected with Semliki Forest virus, *J. Cell Biol.*, 96, 835, 1983.
55. **Quinn, P., Griffiths, G., and Warren, G.,** Dissection of the Golgi complex. II. Density separation of specific Golgi functions in virally infected cells treated with monensin, *J. Cell Biol.*, 96, 851, 1983.
56. **Saraste, J. and Kuismanen E.,** Pre- and post-Golgi vacuoles operate in the transport of Semliki Forest virus membrane glycoproteins to the cell surface, *Cell,* 38, 5345, 1984.
57. **Matlin, K. S. and Simons, K.,** Reduced temperature prevents transfer of a membrane glycoprotein to the cell surface but does not prevent terminal glycosylation, *Cell,* 34, 233, 1983.
58. **Griffiths, G., Pfeiffer, S., Simons, K., and Matlin, K.,** Exit of newly synthesized membrane proteins from the trans cisternae of the Golgi complex to the plasma membrane, *J. Cell Biol.*, 101, 949, 1985.
59. **Fuller, S. D., Bravo, R., and Simons, K.,** An enzymatic assay reveals that proteins destined for the apical or basolateral domains of an epithelial cell line share the same late Golgi compartments, *EMBO J.*, 4, 297, 1985.
60. **Sadler, J. E., Rearick, J. I., Paulson, J. C., and Hill, R. L.,** Purification to homogeneity of a β-galactoside 2,3 sialyltransferase and partial purification of an β-N-acetylgalactosamide 2,6 sialyltransferase from porcine submaxillary glands, *J. Biol. Chem.*, 254, 4434, 1979.
61. **Ravindranath, M. H., Higa, H. H., Cooper, L. E., and Paulson, J. C.,** Purification and characterization of an O-acetyl sialic acid specific lectin from a marine crab Cancer antennarius, *J. Biol. Chem.*, 260, 8850, 1985.
62. **Wetzel, M. G., Wetzel, B. K., and Spicer, S. S.,** Ultrastructural localization of acid mucosubstances in the mouse colon with iron-containing stain, *J. Cell Biol.*, 30, 299, 1966.
63. **Rambourg, A., Hernandez, W., and Leblond, C. P.,** Detection of complex carbohydrates in the Golgi apparatus of rat cells, *J. Cell Biol.*, 40, 395, 1969.
64. **Friend, D. S. and Murray, M. J.,** Osmium impregnation of the Golgi apparatus, *Am. J. Anat.*, 117, 135, 1965.
65. **Goldfischer, S.,** The internal reticular apparatus of Camillo Golgi, *J. Histochem. Cytochem.*, 30, 717, 1982.
66. **Hand, A. R. and Oliver, C.,** The role of GERL in the secretory process, in *Cell Biology of the Secretory Process,* Cantin, M., Ed., Karger, Basel, 1984, 148.
67. **Jelsema, C. and Morré, D. J.,** Distribution of phospholipid biosynthetic enzymes among cell compartments of rat liver, *J. Biol. Chem.*, 253, 7960, 1978.
68. **Borgese, N. and Meldolesi, J.,** Localization and biosynthesis of NADH- cytochrome b_5 reductase, an integral membrane protein, in rat liver cells. I. Distribution of the enzyme activity in microsomes, mitochondria, and Golgi complex, *J. Cell Biol.*, 85, 501, 1980.
69. **Hino, Y., Asano, A., Sato, R., and Shimizu, S.,** Biochemical studies of rat liver Golgi apparatus. I. Isolation and preliminary characterization, *J. Biochem.* 83, 909, 1978.
70. **Howell, K., Ito, A., and Palade, G.,** Endoplasmic reticulum marker enzymes in Golgi fractions — what does it mean?, *J. Cell Biol.*, 79, 581, 1978.
71. **Ito, A. and Palade, G.,** Presence of NADPH-cytochrome P-450 reductase in rat liver Golgi membranes. Evidence obtained by immunoadsorption method, *J. Cell Biol.*, 79, 590, 1978.
72. **Lucocq, J. M., Brada, D., and Roth, J.,** Immunolocalization of glucosidase II in pig liver, *J. Cell Biol.*, 99, 354a, 1984.
73. **Lucocq, J. M., Brada, D., and Roth, J.,** Immunolocalization of the oligosaccharide trimming enzyme glucosidase II, *J. Cell Biol.*, 102, 2137, 1986.
74. **Brands, R., Snider, M. D., Hino, Y., Park, S. S., Gelboin, H. V., and Rothman, J. E.,** Retention of membrane proteins by the endoplasmic reticulum, *J. Cell Biol.*, 101, 1724, 1985.
75. **Yamamoto, A., Masaki, R., and Tashiro, Y.,** Is cytochrome P-450 transported from the endoplasmic reticulum to the Golgi apparatus in rat hepatocytes? *J. Cell Biol.*, 101, 1733, 1985.
76. **Strous, G. J., van Kerkhof, P., Brok, R., Roth, J., and Brada, D.,** Glucosidase II, a protein of the endoplasmic reticulum with high mannose oligosaccharide chains and a rapid turnover, *J. Biol. Chem.*, 262, 3620, 1987.
77. **Brada, D., Kerjaschki, D., and Roth, J.,** The glycoprotein processing enzyme glucosidase II is present in post-Golgi apparatus compartments in addition to the endoplasmic reticulum in kidney tubular cells, *J. Cell Biol.*, 105, 80a, 1987.
78. **Griffiths, G., Brands, R., Burke, B., Louvard, S., and Warren, G.,** Viral membrane proteins acquire galactose in *trans* Golgi cisternae during intracellular transport, *J. Cell Biol.*, 95, 781, 1982.
79. **Roth, J., Taatjes, D. J., Lucocq, J. M., Weinstein, J., and Paulson, J. C.,** Demonstration of an extensive *trans*-tubular network continuous with the Golgi apparatus cisternal stack that may function in glycosylation, *Cell,* 43, 287, 1985.
80. **Roth, J., Taatjes, D. J., Weinstein, J., Paulson, J. C., Greenwell, P., and Watkins, W. M.,** Differential subcompartmentation of terminal glycosylation in the Golgi apparatus of intestinal absorptive and goblet cells, *J. Biol. Chem.*, 261, 14307, 1986.

81. **Dunphy, W. G. and Rothman, J. E.,** Compartmentation of asparagine-linked oligosaccharide processing in the Golgi apparatus, *J. Cell Biol.*, 97, 270, 1983.
82. **Roth, J., Greenwell, P., and Watkins, W. M.,** Immunolocalization of blood group A specified α1,3 *N*-acetylgalactosaminyltransferase and blood group A substance in the *trans*-tubular network of the Golgi apparatus and mucus of intestinal goblet cells, *Eur. J. Cell Biol.*, 46, 105, 1988.
83. **Goldberg, D. E. and Kornfeld, S.,** Evidence for extensive subcellular organization of asparagine-linked oligosaccharide processing and lysosomal enzyme phosphorylation, *J. Biol. Chem.*, 258, 3159, 1983.
84. **Dunphy, W. G., Brands, R., and Rothman, J. E.,** Attachment of terminal *N*-acetylglucosamine to asparagine-linked oligosaccharides occurs in central cisternae of the Golgi stack, *Cell*, 40, 463, 1985.
85. **Roth, J.,** Cytochemical localization of terminal *N*-acetyl-*D*-galactosamine residues in cellular compartments of intestinal goblet cells: implications for the topology of O-glycosylation, *J. Cell Biol.*, 98, 399, 1984.
86. **Elhammer, A. and Kornfeld, S.,** Two enzymes involved in the biosynthesis of O-linked oligosaccharides are localized on membranes of different densities in mouse lymphoma BW5147 cells, *J. Cell Biol.*, 99, 327, 1984.
87. **Pavelka, M. and Ellinger, A.,** RCA I-binding patterns of the Golgi apparatus, *Eur. J. Cell Biol.*, 41, 270, 1986.
88. **Hedman, K., Pastan, I., and Willingham, M. C.,** The organelles of the trans domain of the cell. Ultrastructural localization of sialoglycoconjugates using *Limax flavus* agglutinin, *J. Histochem. Cytochem.*, 34, 1069, 1986.
89. **Van Halbeek, H., Gerwig, G. J., Vliegenthart, J. F. G., Smits, H. L., van Kerkof, P. J., and Kramer, J. F.,** Terminal α(β1-4)-linked *N*-acetylglucosamine: a characteristic constituent of duodenal-gland mucous glycoproteins in rat and pig, *Biochim. Biophys. Acta*, 747, 107, 1983.
90. **Kuhn, N. J. and White, A.,** The role of nucleoside diphosphatase in an uridine nucleotide cycle associated with lactose synthesis in rat mammary gland Golgi apparatus, *Biochem. J.*, 168, 423, 1977.
91. **Kuhn, N. J. and White, A.,** Evidence for specific transport of uridine diphosphate galactose across the Golgi membranes of rat mammary gland, *Biochem. J.*, 154, 243, 1977.
92. **Novikoff, A. B.,** Enzyme localizations and ultrastructure of neurons, in *The Neuron*, Heyden, H., Ed., Elsevier, New York, 1967, 255.
93. **Novikoff, A. B. and Novikoff, P. M.,** Cytochemical contributions to differentiating GERL from the Golgi apparatus, *Histochem. J.*, 9, 525, 1977.
94. **Novikoff, P. M., Novikoff, A. B., Quintana, N., and Hauw, J. -J.,** Golgi apparatus, GERL, and lysosomes of neurons in rat dorsal root ganglia, studied by thick section and thin section cytochemistry, *J. Cell Biol.*, 50, 859, 1971.
95. **Rambourg, A.,** L'appareil de Golgi: examen en microscopie électronique de coupes épaisses (0,5-1 μm), colorées par le mélange chlorohydrique phosphotungstique, *C. R. Acad. Sci. Paris, Series D*, 269, 2125, 1969.
96. **Rambourg, A., Clermont, Y., and Hermo, L.,** Three-dimensional architecture of the Golgi apparatus in Sertoli cells of the rat, *Am. J. Anat.*, 154, 455, 1979.
97. **Pastan, I. and Willingham, M. C.,** The pathway of endocytosis, in *Endocytosis*, Pastan, I. and Willingham, M. C., Eds., Plenum Press, New York, 1985, 1.
98. **Griffiths, G. and Simons, K.,** The *trans* Golgi network: sorting at the exit site of the Golgi complex, *Science*, 234, 438, 1986.
99. **Geuze, H. J., Slot, J. W., Strouss, G. J. A. M., Hasilik, A., and von Figura, K.,** Possible pathways for lysosomal enzyme delivery, *J. Cell Biol.*, 101, 2253, 1985.
100. **Taatjes, D. J. and Roth, J.,** The trans-tubular network of the hepatocyte Golgi apparatus is part of the secretory pathway, *Eur. J. Cell Biol.*, 42, 344, 1986.
101. **Van Deurs, B., Sandvig, K., Petersen, O. W., Olsnes, S., Simons, K., and Griffiths, G.,** Estimation of the amount of internalized ricin that reaches the trans-Golgi network, *J. Cell Biol.*, 106, 253, 1988.
102. **Orci, L., Ravazzola, M., Amherdt, M., Perrelet, A., Powell, S. K., Quinn, D. L., and Moore, H. P. H.,** The *trans*-most cisternae of the Golgi complex: a compartment for sorting of secretory and plasma membrane proteins, *Cell*, 51, 1039, 1987.
103. **Shaper, N. L., Shaper, J. H., Menth, J. L., Fox, J. L., Chang, H., Kirsch, I. K., and Hollis, G. F.,** Bovine galactosyltransferase: identification of a clone by direct immunological screening of a cDNA expression library, *Proc. Natl. Acad. Sci. U.S.A.*, 83, 1573, 1986.
104. **Narimatsu, H., Sinha, S., Brew, K., Okayama, H., and Qasba, P. K.,** Cloning and sequencing of cDNA of bovine N-acetylglucosamine (β1-4)galactosyltransferase, *Proc. Natl. Acad. Sci. U.S.A.*, 83, 4720, 1986.
105. **Humphreys-Beher, M. G., Bunnell, B., van Tuinen P., Ledbetter, D. H., and Kidd, V. G.,** Molecular cloning and chromosomal localization of human 4-β-galactosyltransferase, *Proc. Natl. Acad. Sci. U.S.A.*, 83, 8918, 1986.
106. **Weinstein, J., Lee, U. E., McEntee, K., Lai, P.-H., and Paulson, J. P.,** Primary structure of β-galactoside α2,6 sialyltransferase. Conversion of membrane bound enzyme to soluble form by cleavage of the N-terminal signal anchor, *J. Biol. Chem.*, 262, 17735, 1987.

Chapter 13

NEOGLYCOPROTEINS AS TOOLS TO ANALYZE ENDOGENOUS LECTINS AND GLYCOCONJUGATES

Michel Monsigny, Jean Hubert, Angèle Obrénovitch, and Joseph Schrével

TABLE OF CONTENTS

ABBREVIATIONS

BSA Bovine serum albumin
CB di-*N*-acetylchitobiose
CT tri-*N*-acetylchitotriose
DAB 3,3′-diaminobenzidine
EM electron microscope (-py)
Fuc L-fucose
Gal galactose
Gal*N*Ac *N*-acetylgalactosamine
Glc glucose
Glc*N*Ac *N*-acetylglucosamine
HpA *Helix pomatia* agglutinin
Lac lactose
Man mannose
Neu5Ac *N*-acetyl neuraminic acid
PHA *Phaseolus vulgaris* agglutinin
PHA-ELs *Phaseolus vulgaris* agglutinin inducing lymphocyte blast transformation
PHA-E_4 *Phaseolus vulgaris* erythroagglutinin
PNA peanut agglutinin
PTC phenylthiocarbamoyl
PTG porcine thyroglobulin glycopeptides
RcA-120 *Ricinus communis* agglutinin (tetramer)
SBA soybean agglutinin
StA *Solanum tuberosum* agglutinin
UeA *Ulex europeus* agglutinin
WGA wheat germ agglutinin

I. INTRODUCTION

Glycoconjugates (glycolipids, glycoproteins and glycosaminoglycans or proteoglycans) have been known for several decades to be present on cell surfaces and in basal membranes. During the last 20 years it has been discovered that glycoconjugates are also present in intracellular organelles, including endoplasmic reticulum (ER), Golgi apparatus, lysosomes, endosomes, nuclear pores, nuclear envelope, and nuclei. On the other hand, sugar receptors called lectins[1] are also present on the plasma membranes, in endosomes, lysosomes, and even in the nucleus.

Glycoconjugates and endogenous lectins can be evidenced in fluorescence microscopy and in electron microscopy (EM) by using the classical immunochemical approach, i.e., specific polyclonal or monoclonal antibodies raised either against the sugar or the protein moieties. However, another approach has been developed to allow the characterization of the sugar moieties of glycoconjugates based on a two-step method using sequentially soluble lectins and synthetic glycoproteins, such as glycosylated ferritin, peroxidase, or bovine serum albumin. Such synthetic glycoproteins, known as *neoglycoproteins*[2] (NGPs), also allow the direct characterization of endogenous lectins, the binding sites of which selectively bind NGPs bearing the appropriate sugar residues.

In this paper, we present basic data on endogenous glycoconjugates and endogenous lectins, the preparation and properties of NGPs, the main methods for EM analysis using labeled NGPs, and the characterization of endogenous glycoconjugates and lectins with regard to cell dynamics.

II. ENDOGENOUS GLYCOCONJUGATES AND LECTINS

A. GLYCOCONJUGATES

Glycoconjugates are synthetized in the ER and/or in the Golgi apparatus. The structure of the sugar moieties of a large number of glycoconjugates has been established; for reviews on glycolipids, see Hakomori,[3] on glycoproteins, see Montreuil,[4,5] on proteoglycans, see References 6 to 9.

The main biosynthetic pathways of glycoconjugates have been elucidated;[10-14] they involve a number of different glycosyltransferases, glycosylphosphotransferases, glycosidases, epimerases, and sulfotransferases.

Membrane glycoconjugates are carried from the ER through the Golgi complex to the cell surface. Glycoconjugates, such as glycolipids, or membrane glycoproteins which possess an anchorage moiety embedded in the lipid bilayer, are permanently membrane bound. Membrane glycoproteins are arrested (in the ER, in the cis, median, trans-Golgi or in specialized microsomes such as lysosomes) by specific interactions with components present in the lumen, or in the membrane itself, or with cytosolic components associated with the membrane.[15,16] Membrane glycoproteins which lack such a signal are expected to be transported to the plasma membrane. Glycoproteins which are not membrane bound and have no specific recognition signals are transported and processed through the same pathway and are then secreted into the medium.[17]

Membrane glycoproteins present on cell surfaces contain oligosaccharides linked on asparagine residues (*N*-linked-oligosaccharides)[4,5,13] (Figure 1), on serine and/or threonine (*O*-linked oligosaccharides)[4] (Figure 2), or on the C-terminal amino acid (protein-phosphatidylinositol oligosaccharide bridge)[18,19] (Figure 3).

The *O*-linked oligosaccharides are synthesized by the successive action of individual transferases acting directly on the β-hydroxy amino acids (serine or threonine-UDPGal*N*Ac *N*-acetylgalactosamine transferase), and then on the growing peptide-bound oligosaccharides.[10] (UDP = uridine diphosphate, Gal*N*Ac = *N*-acetylgalactosamine). Each glycosyl transferase has a specificity for the sugar nucleotide and for the acceptor; each enzyme will establish only one type of glycosidic linkage with regard to the anomery of the transfered sugar and to the hydroxyl group of the sugar acceptor.[20]

The biosynthesis of *N*-linked oligosaccharides involves still a more complex pathway which can be divided into three phases:

1. The synthesis of a pentamannosyl-di-*N*-acetylglucosaminyl-pyrophosphoryl dolichol on the cytosolic face of the ER, followed by the completion of the tetradecasaccharide on the luminal face of the ER.[57]
2. The en bloc transfer of this oligosaccharide to an asparagine of the protein, usually when the protein is still a nascent polypeptide chain; the asparagine must be a part of a tripeptide, either Asn-X-Ser (Asn = asparagine, Ser = serine) or Asn-X-Thy (Thr = threonine), in which X is any amino acid but proline.[5,22]
3. The processing of protein bound oligosaccharides to generate the final oligosaccharide structure by the sequential actions of specific glycosidases, glycosyltransferases, glycosylphosphoryl transferases, and sulfotransferases.[13,14,23]

The biosynthesis of proteoglycans such as chondroitin, heparin, and heparin sulfate, requires several glycosyltransferases to build the polymeric precursors. Then an epimerase will convert *in situ* glucuronic acid into iduronic acid, and several O-sulfotransferase, *N*-deacetylase and *N*-sulfotransferases[7,9] will further modify the polysaccharide moiety. Altogether, the biosynthesis of the sugar moieties of glycoproteins involves about one hundred specific enzymes leading to a large number of defined oligosaccharides.

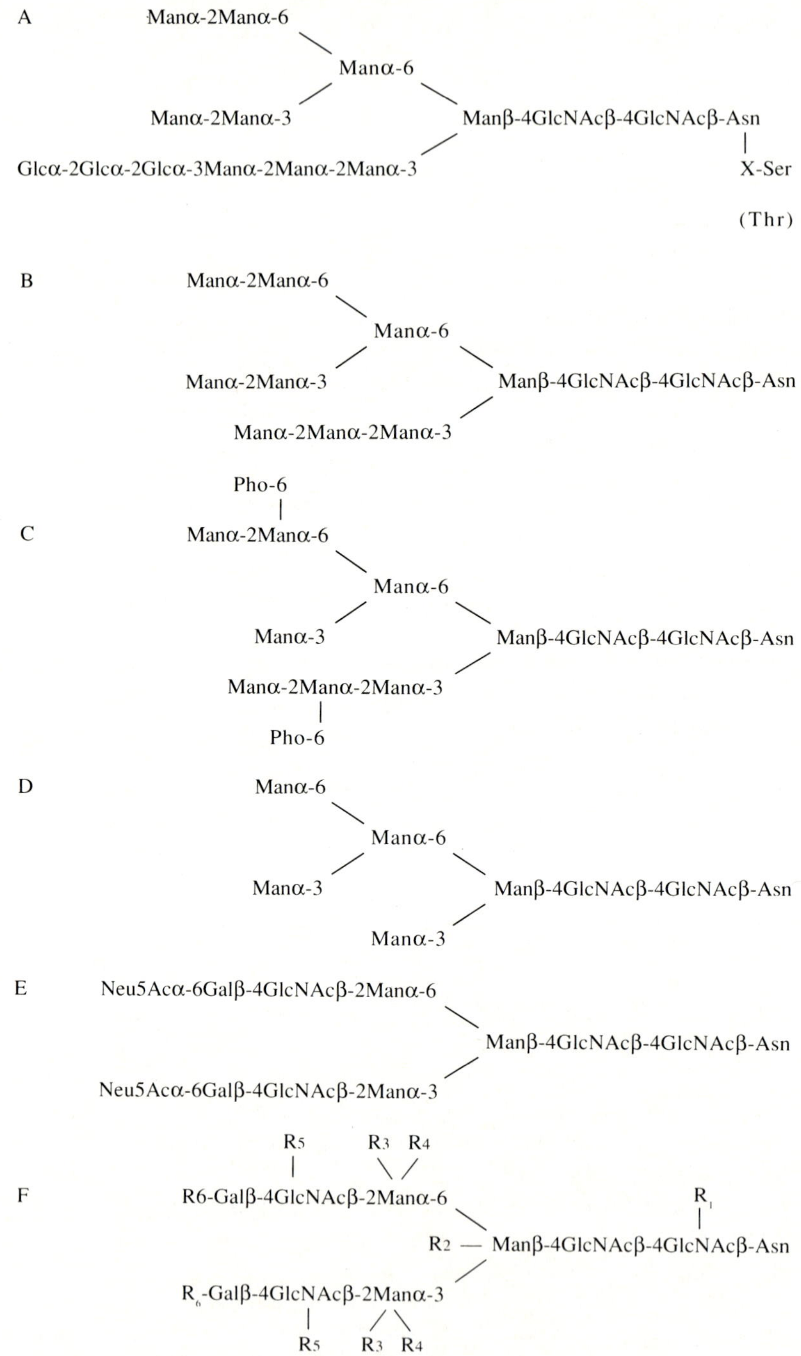

FIGURE 1. Structure of some oligosaccharides linked to *N*-glycosyl-protein (asparagine).[4,5,13] (A) Tetradecasaccharide of glycoproteins in the ER; (B) oligomannoside-type undecasaccharide; (C) phosphorylated oligosaccharide of lysosomal enzymes; (D) heptasaccharide acting as the specific receptor of *N*-acetylglucosamine transferase I; (E) complex oligosaccharide of human transferrin and other glycoproteins; (F) core structure and potential substitutions of complex oligosaccharide types, with X between Asn and Ser (or Thr) can be any amino acid but Pro; Pho: Phosphate residue; R_1: Fucα-6; R_2: Glc*N*Acβ-4; R_3: R_6-Galβ-4Glc*N*Acβ-6; R_4: R_6-Galβ-4Glc*N*Ac-β-4; R_5: Fucα-3; R_6: Neu5Acα-3, or Neu5Acα-6, or Galα-3 or R_6-Galβ-4Glc*N*Ac-β-3. Furthermore, Neu5Ac may be acetylated on position 4, 7, 8 and/or 9, or oxidized on the acetyl group leading to 5-glycoloyl neuraminic acid.

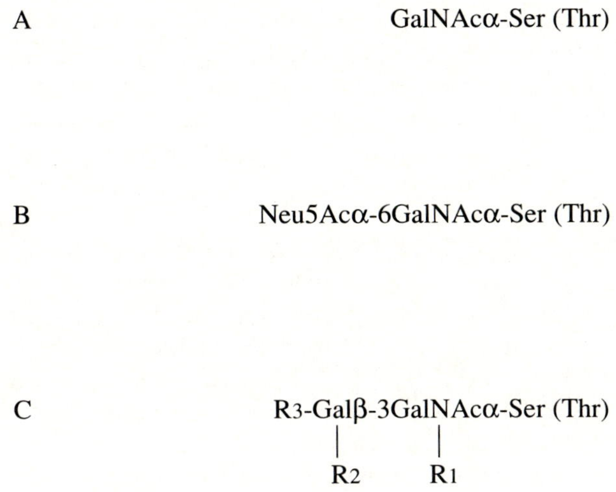

FIGURE 2. Structure of some oligosaccharides linked to *O*-glycosylprotein (serine or threonine). (A) Basic core structure; (B) *O*-linked disaccharide of human submaxillary mucin and other glycoproteins; (C) core structure of complex *O*-linked oligosaccharides and potential substitutions with R_1: Neu5Acα-6 or R_4-Glc*N*Acβ-6; R_2: Fucα-2; R_3: Neu5Acα-3, or GalNAcα-3, or R4-Glc*N*Acβ-3; R_4: R_2, R_3-Galβ-4, etc.

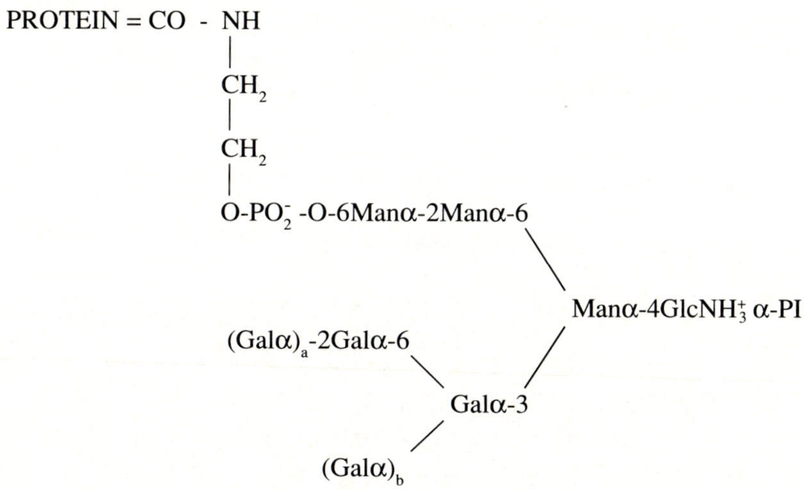

FIGURE 3. Structure of the oligosaccharide bridge between the protein core and the intra membrane phosphatidylinositol of the variant surface glycoprotein of the parasitic protozoan *Trypanosoma brucei*, according to Ferguson et al.[140] Letters a and b mean 0 or 1; PI-stands for phosphatidylinositol; $GlcNH_3^+$ is glycosidically linked to the hydroxyl 6 of PI (dimyristoylphosphatidylinositol).

B. LECTINS

During the last decade, various cell surface receptors were evidenced, including some carbohydrate binding proteins. Cell surface carbohydrate specific receptors may be lectins.[1] ectoglycosyl transferases, or glycosidases. Glycolipids, glycoproteins and proteoglycans have been shown to interact with lectins on the surface of a large number of animal cells.[24-33] The first membrane lectin of animal cells was characterized about 20 years ago by Ashwell and Morell.[24,34] Working on the metabolism of ceruloplasmin, these authors treated this glycoprotein with neuraminidase, labeled the nonreducing terminal galactose residue, and found that this asialo-glycoprotein had a clearance drastically shorter than the native glycoprotein. Subsequently, they showed that asialo-glycoproteins were very efficiently taken up by hepatocytes. Rogers and Kornfeld[35] showed that radio-labeled lysozyme and serum albumin substituted with fetuin asialo-glycopeptides were taken up into rat liver *in vivo*, while the corresponding sugar-free proteins and the intact (sialylated) glycopeptide-protein conjugates were not, confirming the primary role of the nonreducing terminal galactose in the clearance of asialoglycoproteins. Membrane lectins have been evidenced in many other animal cells, especially in tumor cells,[36-39] macrophages,[40,41] and monocytes.[40]

C. GLYCOCONJUGATE-LECTIN INTERACTIONS AND CELL DYNAMICS

Because lectins are able to selectively and reversibly bind the sugar moieties of glycoconjugates,[29,30,42,43] glycolipids and glycoconjugates can conversely act as specific receptors. At the present time, it is well established that glycoconjugate-lectin interactions are involved in essential biological functions.[44] The role of such an interaction in specific endocytosis (receptor-mediated endocytosis), which is one of the most characteristic aspect of the plasma membrane dynamic, is particularly well documented.[45-47] Various kinds of extracellular molecules (referred to as ligands), such as hormones, growth factors, antibodies, viruses, and toxins, are taken up into the cells of multicellular organisms in this way. It is generally admitted that the overall common pathway of internalized ligand begins at the cell surface and ends in lysosomes, where the ligands are degraded, while the membrane receptors are recycled back to the cell surface (Figure 4). Specific organelles (coated pits, coated vesicles, endosomes, receptosomes, Golgi vesicles, lysosomes) are involved in the receptor-mediated endocytosis. However, the modalities of the pathway of ligands via these organelles still pose many unanswered questions. Especially with regard to the organelles where the uncoupling of the ligand-receptor complexes occurs, the factors involved in this dissociation, as well as the recycling of receptors, remain poorly understood. In addition, the presence of glycoconjugates and lectin-like proteins in the nucleus[48,49] suggests that glycoconjugate-lectin interactions may be involved in the dynamics of the nucleus. However, this idea remains to be demonstrated.

In the following parts, we attempt to show that EM studies using labeled NGPs are suitable for a better understanding of cellular dynamics.

III. NEOGLYCOPROTEINS AND GLYCOSYLATED CYTOMARKERS

A. SYNTHESIS OF NEOGLYCOPROTEINS

Endogenous lectins may be directly evidenced by measuring the binding of labeled glycoconjugates. The glycoconjugates used are either native glycoproteins, glycosidase-treated glycoproteins or synthetic glycoproteins called *neoglycoproteins* (NGPs).[2,30,31,50,51.] NGPs are obtained by adding an activated sugar, oligosaccharide or glycopeptide to a protein such as serum albumin. The activated carbohydrate is usually either a glycosyl-alkylimidate, a glycosylphenyl-diazonium, or a glycosylphenyl-isothiocyanate.

The first NGP was synthetized by Goebel and Avery,[52] who prepared diazophenylglycosides, coupled them to serum albumin, and showed that these carbohydrate-protein conjugates were specific antigens.[53] Later, the same authors prepared new NGPs starting with oligosaccharides from type III *Pneumococcus* which were derivatized by using p-nitrobenzylbromide.[54]

Such glycosylphenylazo-protein conjugates were first used by Iyer and Goldstein,[55] and by Privat et al.[56] to study the properties of plant lectins: concanavalin A (ConA) and wheat germ agglutinin (WGA) lectin, respectively. Subsequently, glycosylated cytochemical markers, i.e., glycosylated ferritin or peroxidase, were prepared[57] and shown to be suitable to visualize lectins bound to cell surface associated glycoconjugates.[58-60]

NGPs are also easily obtained by using glycosylphenylisothiocyanate (Figure 5).[31,39,61,62]

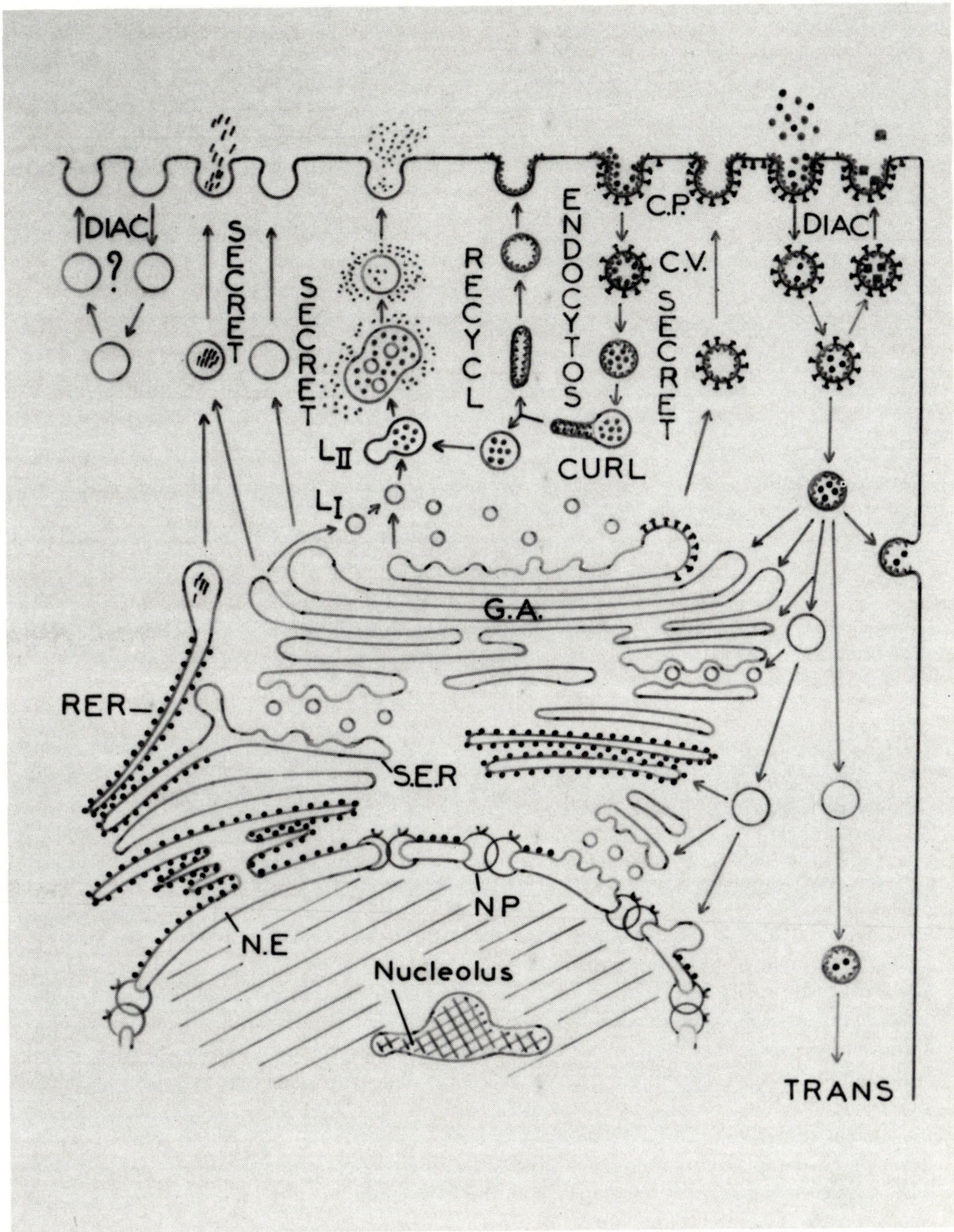

FIGURE 4. Hypothetic scheme of the dynamics of endogenous membrane glycoconjugates and lectins. CP, coated pit; CV, coated vesicle; CURL, compartment of uncoupling receptor ligand; DIAC, diacytosis; ENDOCYTOS, endocytosis; L, primary lysosome; L, secondary lysosome; NE, Nuclear envelope; NP, nuclear pore; RECYCL, recycling; RER, rough ER; SECRET, secretion; SER, smooth ER; TRANS, transcytosis.

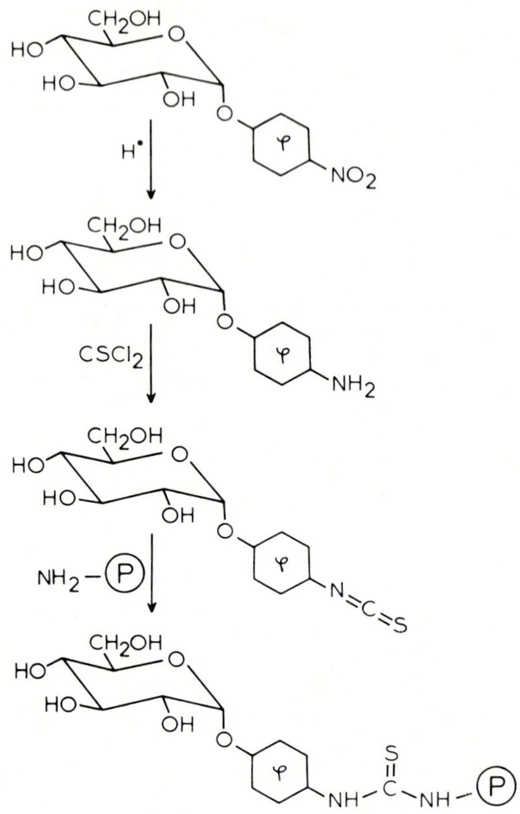

FIGURE 5. Scheme of synthesis of neoglycoproteins. A p-nitrophenyl-ol-D-gluco-pyrannoside is reduced in the presence of palladium on charcoal, the aromatic amine is transformed into phenylisothiocyanate upon treatment with thiophosgene, and finally, the activated sugar is added to a solution of protein at pH 8.5. Under optimal conditions, each step is roughly quantitative. According to the glycosylphenylisothiocyanate/protein ratio used, it is possible to obtain between few and 50 sugar units per serum albumin molecule. P, serum albumin.[31,39]

This type of NGP is quite adapted to prepare fluoresceinylated derivatives because the phenylthiocarbamyl group does not impair the emission of fluorescence.[39] Complex oligosaccharides are easily derivatized by reductive amination of aminophenylethylamine[63] in the presence of cyanoborohydride anion at neutral pH,[64] and then transformed into phenylisothiocyanate derivatives.[65] Alternatively, a reducing oligosaccharide can be directly linked to the amino groups of a protein by reductive amination with cyanoborohydride anion,[66] or can be oxidized and coupled to a protein.[67]

Lee et al.[68] developed NGPs based on the reaction of 2-imino-2-methoxyethyl-1-thioglycosides with the amino groups of a protein, leading to glycosylamidinated proteins. These NGPs were shown to react selectively with the liver lectin when the relevant specific sugar (galactose) was borne by the NGP.[69]

NGPs, such as glycosylated serum albumin containing about 20 sugar residues, are quite useful to study the specificity and the properties of endogenous membrane lectins because their apparent binding constant is much higher than that of the related free sugars or glycosides. Indeed, the concentrations required to inhibit the binding of a lectin to red blood cells are usually two or three orders of magnitude lower when NGPs, instead of free glycosides, are used as agglutination inhibitors.[70] (Table 1).

The binding of NGPs to lectins increased relatively to the number of sugars bound on the NGP; however the binding is more specific when the number of sugars bound to a protein molecule is close to 20.[39]

TABLE 1
NGPs Are More Efficient Than Free Sugars in Interacting with Lectins. Concentrations of Free Sugars and NGPs Required to Inhibit Rabbit Red Blood Cell Agglutination Induced by Concanavalin A (Con A).

Sugar or NGP	Lectin	Concentration A	Concentration B
Met-α-D-Man	ConA	2600	2600
Man PTC_{20} BSA	ConA	0.8	16

Note: NGPs are more efficient than free sugars in interacting with lectins.

A, Concentration expressed as μmol/1
B, Concentration expressed as sugar units in μmol/1

From Roche et al. unpublished data.

Usually, glycoconjugates or NGPs are recognized by cell surface endogenous lectins at an optimal pH close to 7.4. However, the glycoconjugates' receptors could be enzymes that are devoid of enzymatic activity because of the high pH of the medium. Indeed, Rauvala and Hakomori[71] showed that an α-mannosidase at the fibroblast cell surface, which is almost inactive at pH 7.4 but is active at pH values lower than 6, could be involved in cell recognition. These receptors could also by glycosyltransferases working as binders, but unable to transfer any sugar on their ligand because of the absence of nucleotide sugar in the surrounding medium. Along this line, Shur[72] showed that migrating embryonic cells have high levels of cell surface galactosyltransferases which bind to terminal *N*-acetyl-glucosamine within the extracellular matrix. However, in many cases, the membrane lectins bind their ligands at neutral pH[24] and lose their binding capacity upon acidification. This mechanism is the basis of the rapid turnover and recycling of membrane lectins. In other cases, the reverse may also be found, e.g., thyrocyte lectins bind *N*-acetyl-glucosamine terminated glycoconjugate at acidic but not at neutral pH.[73]

B. LABELED NEOGLYCOPROTEINS

1. Neoglycoproteins as Probes to Detect Endogenous Lectins

Synthetic glycoproteins selectively bind endogenous lectins. Consequently, they can be used to detect and localize cellular lectins (Figure 6A). Glycosylated ferritin and glycosylated peroxidase are visible in the EM, either directly or upon an enzyme driven reaction, respectively. Other NGPs are not visible and, therefore, have to be labeled with gold beads or detected by using the biotin-avidin method.

a. Glycosylated Horseradish Peroxidase

Horseradish peroxidase (HRP), a glycoprotein containing an oligomannose-type sugar moiety, has a molecular weight (M_r) of 40,000, can be visualized by the cytochemical method of Graham and Karnovsky[74] with a 3,3′-diaminobenzidine (DAB)-(H_2O_2) treatment for LM and a DAB-H_2O_2-OsO_4 treatment for EM.

According to Seligman et al.,[75] the oxidation of DAB occurs on the amino groups and after polymerization DAB polymers are seen in the LM as brown deposits. In the EM the amino reactive groups are used to reduce OsO_4 to osmium black. HRP has been substituted

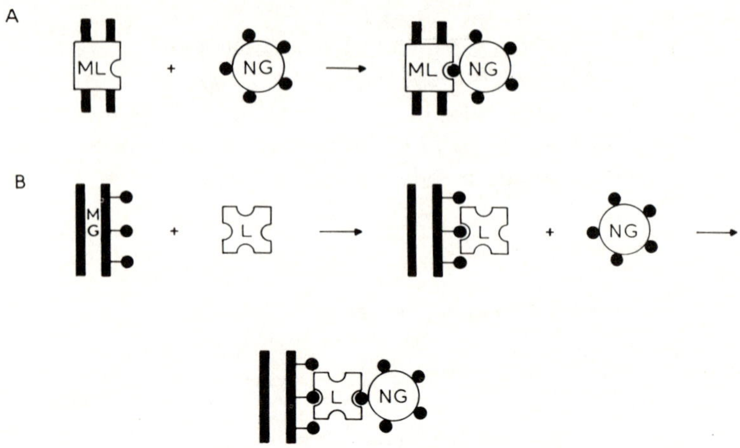

FIGURE 6. (A) Detection of a membrane lectin (ML) using a neoglycoprotein (NG). (B) Mechanism of the two-step procedure to detect a membrane glycoconjugate (MG) using a lectin (L) and a glycosylated cytochemical marker.[58]

TABLE 2
Glycosylated Cytochemical Markers (GCM) Suitable to Visualize Lectins Bound to Cell Surface Carbohydrates[98]

		Ferritin or HRP substituted by					
Lectins	**Simplest inhibitors**	**Man**	**Fuc**	**GalNAc**	**Lac[a]**	**CB[b]**	**PTG[c]**
Concanavalin A[d]	Glc, Man	+	−	−	−	−	+
UeA-Fuc	Fuc	−	+	−	−	−	
RcA-120	Gal	−	−	−	+	−	
PNA	Gal, Gal-Gal*N*Ac	−	−	−	+	−	−(+)[e]
SBA	Gal, Gal*N*Ac	−	−	+	+	−	−(+)[e]
HpA	Gal*N*Ac	−	−	+	−	−	−
PHA-Els	Gal-Glc*N*Ac-Man	−	−	−	−	−	+
PHA-E_4	Gal-Glc*N*Ac-Man	−	−	−	−	−	+
WGA	Glc*N*Ac, CB	−	−	−	−	+	−
StA	CB, CT	−	−	−	−	+	−
Limulin	Neu5Ac-α-Gal*N*Ac	−	−	−	−	−	−

[a] Lac, Galβ-4Glc.
[b] CB, di-*N*-acetylchitobiose; CT, tri-*N*-acetylchitobiose.
[c] PTG, porcine thyroglobulin glycopeptides.
[d] All lectins which bind Man will react with Man-GCM; these lectins also react with unmodified horseradish peroxidase.
[e] This lectin does not react with sialoglycopeptides but reacts with desialylated glycopeptides. (See Abbreviations).

by various glycosides[57] in order to extend its use to visualize other lectins in addition to Concanavalin A or related lectins (Table 2).

b. Gold Bead Labeled Neoglycoproteins

The adsorption of the NGP molecules on gold particles is a current method used to label NGPs.[76-79] The first application of colloidal gold as a transmission EM cytochemical marker was described in 1971 by Faulk and Taylor.[80] Since then this method has been extensively used.[81] The preparation of monodisperse gold colloids have been often described (for details, see e.g., References 82, 83).

Briefly, gold beads are obtained by controlled reduction of tetrachloroauric acid. Interestingly, gold beads of different diameters can be prepared according to the reducing agent used.[82] Because gold particles are negatively charged, they bind to macromolecules by noncovalent electrostatic adsorption.[81] The adsorption of NGPs as other proteins onto colloidal gold particles is pH-dependent, and the colloid should be brought to a value close to the isoelectric point of the proteins. NGPs adsorbed onto gold beads keep their affinities for lectins (see e.g., Reference 76.) and the complexes are stable for months.

c. *Glycosylated Ferritin*

This NGP is directly visible in the EM thanks to the ferric micellar core of the ferritin. Ferritin consists partly of a protein shell of apoferritin, with an outer diameter of 12 nm and an inner diameter of 7 to 7.5 nm. This protein has a M_r of 440,000 and is made of 24 subunits of M_r 18,500.[84] The presence of carbohydrates has been reported in horse spleen ferritin.[85]

The varying iron content, up to 4,500 atoms per ferritin molecule,[86] gives M_r values ranging from 750,000 to 900,000.[87] At high magnifications, the ferric micellar core of ferritin molecules appears as an electron opaque particle with an approximate diameter of 5.5 nm. At low magnification, ferritin is often difficult to detect, and an increase in the size and electron opacity is obtainable using lead ferrocyanide,[88] or by staining ultrathin sections with bismuth subnitrate.[89] In the later case, the twofold increase of the diameter of electron opaque particles results from an addition of metallic Bi to the normally undetectable protein shell.

In glycoconjugate cytochemistry ferritin is coupled with lectins.[90] or covalently substituted with sugars, to give affinity markers.[57,58] The concentration of ferritin may be determined either from the *N* content determined by a micro Kjeldahl method, or from the Fe content measured using the 2-2′ bipyridine method,[91] or from the absorbance of the apoferritin at 280 nm, $A_{1\,cm}^{1\%} = 9.8$.[92] The apoferritin is obtained by reduction of ferritin with sodium dithionite[93] or with thioglycollic acid.[84]

d. *Biotinylated Neoglycoproteins*

NGPs can also be easily substituted with biotin,[94] 2-imino-biotin or their derivatives, such as biotinyl amido caproic acid[95] and biotinyl amido-3,4-dithiapentanoic acid.[96] Biotinylated NGPs bound to endogenous lectins can be evidenced by using avidin, a basic glycoprotein from the egg white, or streptavidin, a slightly acidic non-glycosylated protein from *Streptomyces avidinii*, and a biotinylated cytochemical marker. Biotin-spacer conjugates are recommended in order to limit steric hindrance. Streptavidin gives better results than avidin, which may lead to false positive results due to charges mediated adsorption or to direct binding of its glycan moiety to endogenous lectins.

2. NGPs as Glycosylated Markers for Lectins to Detect Lectins Bound to Endogenous Glycoconjugates

Because lectins have at least two binding sites,[1] they are able to bind both a cellular sugar and a glycosylated marker such as labeled NGPs.[60,97] Thus, NGPs can help to localize endogenous carbohydrates according to the two steps method[59,60,97,98] summarized in Figure 6B.

IV. METHODS FOR ELECTRON MICROSCOPY ANALYSIS USING LABELED NEOGLYCOPROTEINS

The EM detection of membrane glycoconjugates using lectins labeled with NGPs as glycosylated cytomarkers (GCM) can be performed with free cells before or after glutaraldehyde fixation, with slices,[60] small tissue "blocks,"[99] or thin sections of material embedded in glycol methacrylate.[100] The incubations with living cells must be performed at low temperature in order to prevent lectin-ligand complex redistribution during incubation, and to minimize endocytosis.[60] The release of the cell surface lectin bound during GCM incubation was tested with fluoresceinthiocyanate-wheat germ agglutinin (FTC-WGA) on mouse thymocytes, and was found to be low because the concentration of GCM is lower than 5.10^{-6} M.[60]

Comparative observations demonstrated that glycosylated ferritin reaches the cell surface-bound lectins in the regions where the glycoconjugate-lectin network is loose enough to allow the diffusion of these markers. In contrast, glycosylated peroxidase is expected to reach any cell surface-bound receptors. However, glycosylated peroxidase, such as in the method with peroxidase and Con A[101] gives diffuse staining, whereas glycosylated ferritin is able to provide semiquantitative information.[60] Comparison between the sequential method using Con A and mannosylated ferritin, and the direct method using Con A ferritin conjugate, was made with different *Trypanosoma equiperdum* variants; it appeared (Figure 7) that the sequential method gives the best results.[60]

The sequential method with lectin and GCM is quite useful in localizing cell surface glycoconjugates because the sensitivity of the method is very high. For example, living mouse thymocytes were labeled by incubation in the presence of 20 μg WGA per ml phosphate buffered saline (PBS), and then of 100 μg CB ferritin/ml PBS and visualized in the transmission EM.[60]

The analysis of membrane dynamics involving cell surface glycoconjugates and lectins is generally performed on living cells. Either cells are incubated in the presence of labeled exogenous proteins, or animals are perfused with these proteins; the cells or tissues to be studied are subsequently fixed, embedded in Epoxy resins, sectioned, and contrasted. Because the observation of two dimensional images of random sections do not allow the visualization of the spatial relationships between the cellular organelles, a precise ultrastructural analysis of the ligand pathway via different cytoplasmic organelles, for instance, requires serial sectioning.[102]

As far as the nucleus is concerned, the NGPs or the lectins available until now are not able to pass from the cytoplasm to the nucleus.[49,103] This fact is related to the nuclear envelope which acts as a selective barrier, so that only certain proteins (called karyophilic proteins)[104] can enter the nucleus. The detection of intranuclear sugars and lectin binding sites in fluorescence microscopy, or their quantitative analysis by flow cytofluorometry, therefore must be performed using isolated and membrane depleted nuclei.[48,49] The EM analysis allowed one to detect these binding sites in nuclei *in situ*. In this case, ultrathin sections of cells passing through the nucleus are floated on the solution containing the labeled proteins, which are, therefore, in contact with the intranuclear structures. However, in order to conserve the binding activity of endogenous lectins, the cells must be fixed by aldelydes alone and embedded in hydrophilic resins such as glycol methacrylate[105] or Lowicryl.[76] The same method is recommended for the ultrastructural detection of nuclear glycoconjugates.[103] When glycosylated ferritin is used, it is preferable to observe noncontrasted sections, on account of the high contrast of some nuclear and nucleolar structures after staining that makes it difficult to visualize the small ferric micellar core of ferritin.

V. MEMBRANE DYNAMICS

A. MEMBRANE DYNAMICS OF ENDOGENOUS GLYCOCONJUGATES

Wheat germ agglutinin (WGA) binding sites were visualized with chitobiosyl-ferritin or chitobiosyl-peroxidase on murine plasmatocytoma cells,[58,60,98,106] on ciliary membrane of quail oviduct,[99] on *Dictysotelium discoideum* plasma membrane,[107,108] and on mouse thy-

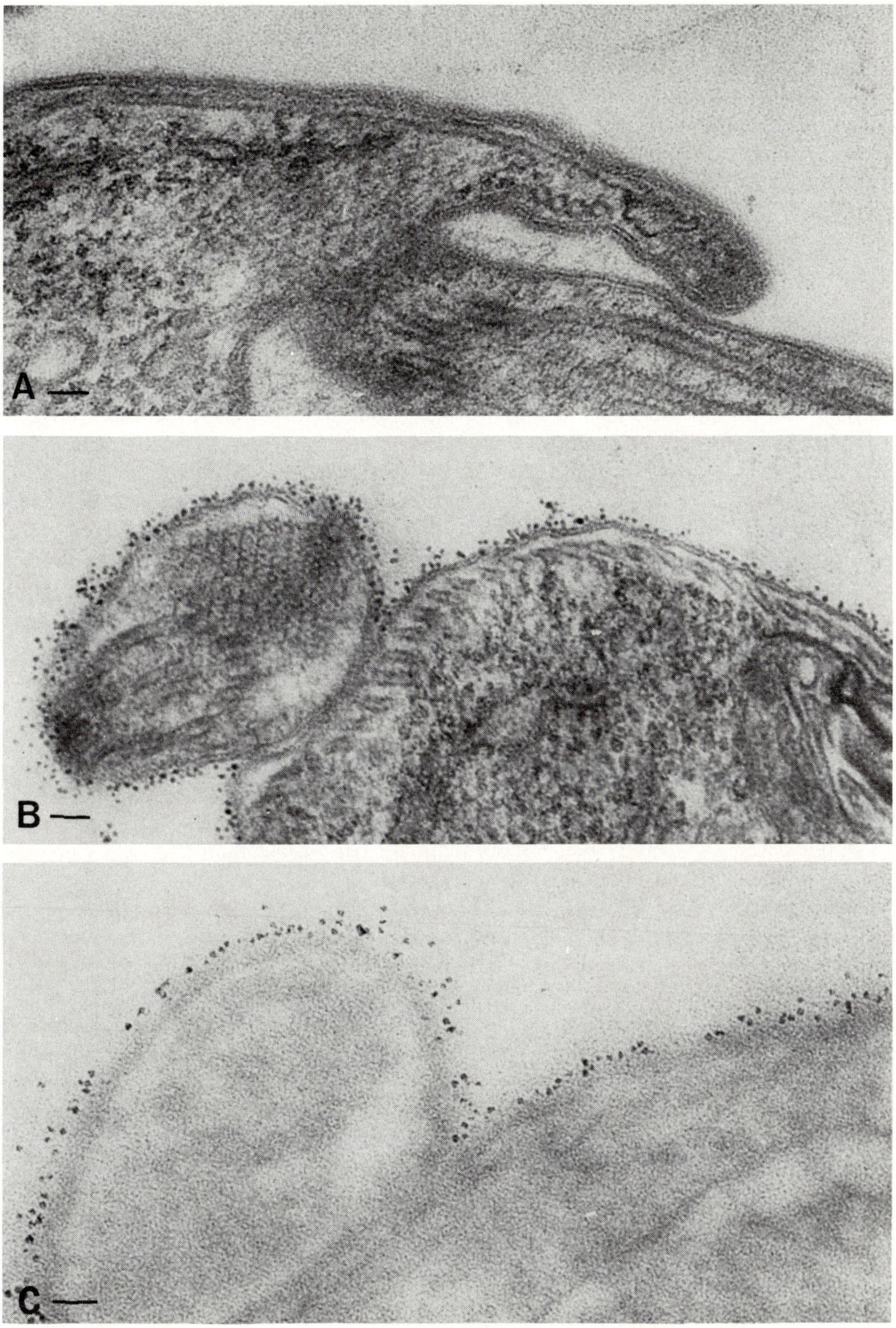

FIGURE 7. Detection of Con A binding sites at the cell coat of BoTat-78 variant of *Trypanosoma equiperdum*. (A) and (B) one-step procedure using Con A-ferritin conjugate (200 μg/ml). (A) no binding site is visualized. (B) when the same sample of Con A-ferritin conjugate is used after a brief trypsin treatment, numerous binding sites are unmasked. (C) when the two-step procedure with Con A and mannosyl-ferritin (200 μg/ml) is used, numerous binding sites are visualized without trypsin treatment. (A) and (B) contrasted with lead; (C) uncontrasted. (Micrographs courtesy of J. Schrével). Bars = 100 nm.

mocytes and myocardium cells.[60] Mauchamp and Hubert[109] have used WGA and chitobiosyl-ferritin to analyze membrane carbohydrates involved in the uptake of molting fluid components during the molt of a butterfly. Con A biniding sites were localized with mannosyl-ferritin on *Amoeba, Gregarina, Trypanosoma equiperdum,*[60] on ciliary membranes of quail oviduct (Figure 8),[99] and on *Dictyostelium discoideum.*[108] Other lectins have been used to detect glycoconjugates: *Ricinus communis* agglutinin I, in the quail oviduct,[99] soybean agglutinin in *Trypanosoma lewisi,*[60] and peanut agglutinin on mouse thymocytes;[60] these lectins were detected using lactosyl-ferritin or lactosyl-peroxidase. In plants, the ultrastructural detection of fucosyl residues at the surface of axenically grown roots of *Zea maïs* has been performed using *Ulex europaeus* lectin and fucosyl-ferritin.[110]

B. MEMBRANE DYNAMICS OF ENDOGENOUS LECTINS

Fluorescein-labeled NGPs have been extensively used for the study of membrane dynamics of lectins by fluorescence microscopy,[111] quantitative spectrofluorometry,[39] or quantitative flow cytofluorometry.[31,112,113] However, the development of methods allowing us to link proteins to markers visible in EM had led a number of investigators to utilize labeled NGPs for EM analysis of lectin-mediated endocytosis, and especially the specific endocytosis of desialylated glycoproteins in mammalian hepatocytes. This process has focused the interest of investigators, since it has been established that the internalization of asialoglycoproteins modulates the clearance of these proteins from blood plasma[34] (for reviews, see References 24, 114, 115). This clearance involved a membrane lectin which selectively binds galactose or *N*-acetylgalactosamine terminal residues of asiologlycoproteins, which are endocytosed and delivered in lysosomes where proteolysis occurs[116,117] (for a review, see Reference 118). NGPs substituted by oligosaccharides, which are recognized by the membrane lectin, have been shown to be suitable probes to precise the intracellular pathway of the natural ligands, since they are similarly internalized. Thus, the *in vivo* clearance of galactosylated ferritin (Lac-fer) is identical to the clearance of asialo-orosomucoid linked to horseradish peroxidase.[102] In addition, Wall and Hubbard[119] reported that by using Lac-ferritin (fer) the quantification of the ligand molecules and of their receptors was better than with asialo-orosomucoid-HRP conjugates, and furthermore, Lac-fer led to a clear visualization of coated membrane regions, while asialo-orosomucoid-HRP conjugates did not.

In vivo experiments[102] showed that Lac-fer was localized first in coated pits and coated vesicles at the base of microvilli along the sinusoidal surface of hepatocytes. Subsequently, Lac-fer was found in uncoated larger vesicular and tubular structures of the peripheral cytoplasm, and ultimately in the Golgi/lysosome region. The analysis of serial sectioning indicated that the Golgi structures containing the ligand fused directly with lysosomes. In other experiments performed at 4°C, a temperature which blocks any ligand internalization, Wall and Hubbard[119] observed a density of Lac-fer particles 70-fold greater over coated pits than over uncoated regions. Another interesting result reported by these authors is the lack of any measurable decrease in receptor numbers or change in their distribution in temperature shift experiments, suggesting that receptor reutilization occurs very fast by a system which maintains a constant population of cell surface receptors.

Using also Lac-fer as tracer to delineate the structures involved in the carbohydrate specified endocytosis in rat hepatocytes, Haimes et al.[120] concluded that an extensive lysosomal compartment exists in hepatocytes.

Galactose specific endocytosis in hepatocytes has been further analyzed both by cell fractionation techniques and EM after perfusion of rats with galactosylated bovine serum albumin coupled with HRP (Gal-BSA-HRP).[121] After differential centrifugation of liver homogenates, the galactosylated ligand was concentrated in the light mitochondrial (L) and the microsomal (P) fractions. After isopycnic centrifugation on sucrose gradients of L and P fractions, Gal-BSA-HRP containing structures equilibrated as a narrow band in the low

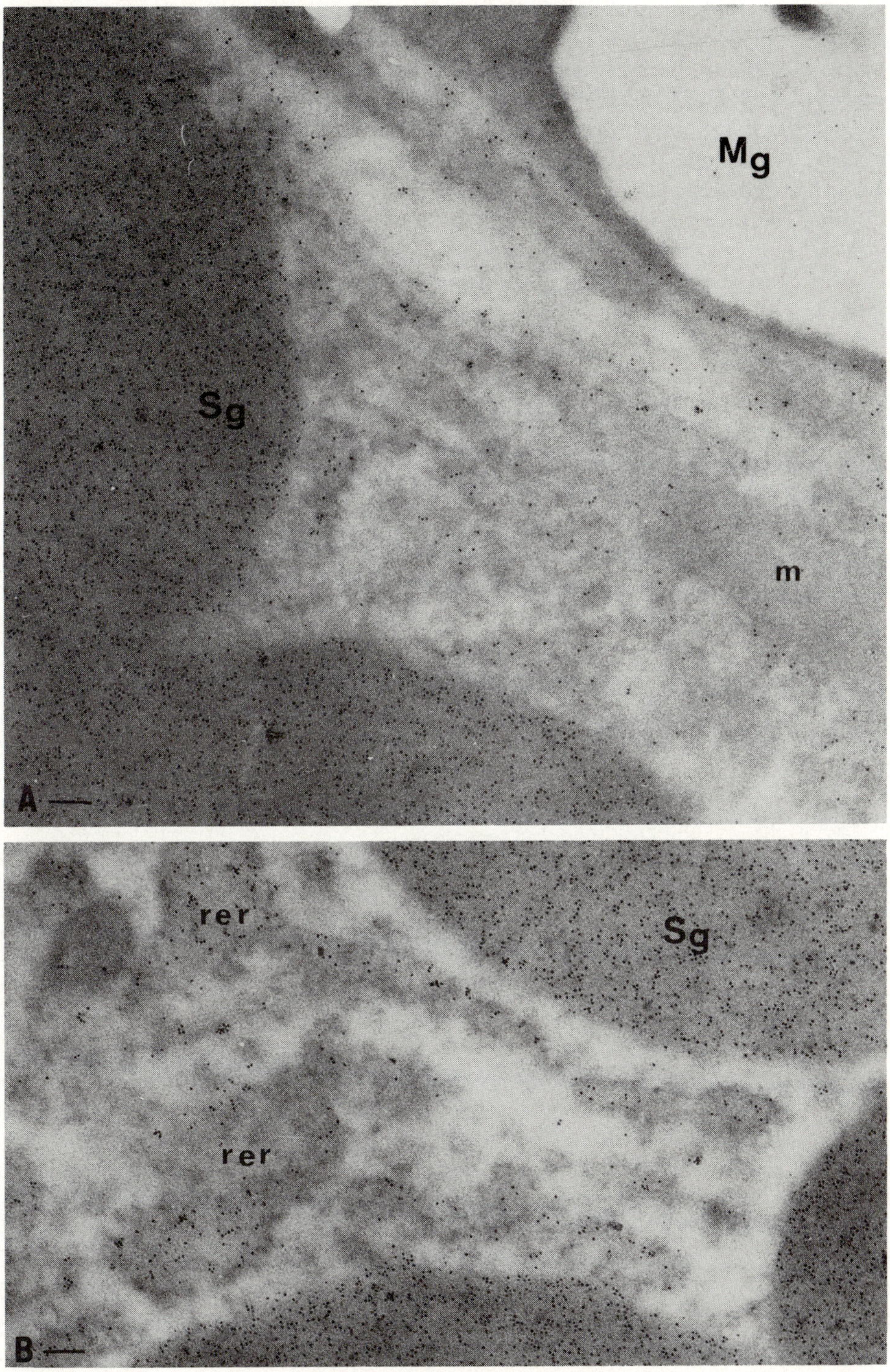

FIGURE 8. Labeling of thin sectins of laying quail oviduct embedded in glycolmethacrylate using Con A (500 μg/ml) and mannosyl ferritin (500 μg/ml). (A) secretion granules (Sg) of tubular gland cells mainly composed of ovalbumin are strongly labeled, whereas neighbor mucous granules (Mg) containing ovomucins (devoid of manosyl residues) are not labeled. (B) in tubular gland cells, rough ER cisternae (rer) in which glycosylation reactions begin are labeled, in contrast to mitochondria (m) seen in (A). Sections are contrasted with OsO_4 vapor after labeling. (Courtesy of Dr. D. Sandoz.) Bars = 100 nm.

density region. By ultrastructural cytochemistry, the Gal-BSA-HRP-containing structures were identified as vesicles or tubules heterogeneous in size and shape, and surrounded by a smooth membrane. These authors suggested that ligand-containing structures are different from Golgi complex, plasma membrane, and lysosomes. These structures seem to form distinct intracellular compartments which could be specialized in ligand receptor sorting. Furthermore, the comparative pathway of rat polymeric IgA-HRP conjugate (pIgA-HRP) and Gal-BSA-HRP, observed on the ultrastructural level in hepatocytes after injection of the ligands to rats,[122] showed that the two ligands are taken up by hepatocytes via two distinct receptor systems, and addressed by these cells to distinct destinations. Gal-BSA dissociates from the receptor in non-lysosomal or pre-lysosomal organelles, and is subsequently degraded in lysosomes, while pIgA is transferred into the bile by transcytosis.

A detailed ultrastructural qualitative and quantitative study of galactose-specific endocytosis has been conducted in cultured rat hepatocytes using Gal-BSA adsorbed on colloidal gold particles.[77] The data from this study do not support the existence of an extensive lysosomal compartment described by Haimes et al.,[120] while they confirm the existence of an important intermediate compartment in agreement with the results of Wall and Hubbard[119] and Quintart et al.[121] This intermediate compartment is composed of multivesicular bodies where Gal-BSA-gold transiently accumulates. Then, the multivesicular bodies fused with secondary lysosomes. In addition, Deschuyteneer et al.[77] reported that 10% of the internalized ligand did not pass through lysosomes,and was recycled back to the cell surface via secretory vesicles. These data strongly suggest the existence of a prelysosomal sorting mechanism of the endocytosed material.

Rat liver macrophages also possess galactose specific membrane receptors.[123] The modalities of the endocytosis mediated by this receptor, either in rat hepatocytes or in macrophages, is well documented, especially on the basis of EM studies.[79,124-126] By using colloidal gold particles of different sizes coated with galactosylated BSA, these authors have accumulated results which evidence differences in the uptake of ligands by liver parenchymal cells, macrophages, and endothelial cells.

In vivo and *in vitro,* the asiologlycoprotein receptors are initially distributed randomly in distinct plasma membrane areas in hepatocytes; they are distributed in clusters over the entire cell surface of liver macrophages (Figures 9a, b, c). On endothelial cells, ligands bind only to coated pits (Figure 9d). Fifty nm gold particles coated with Lac-BSA are able to bind to liver macrophages at 4°C, and then be taken up at 37.5°C. The uptake of particles up to 37 nm proceeds via coated pits and coated vesicles, whereas larger ligands are transported via uncoated membrane structures. At 37°C, liver parenchymal cells do not bind and do not take up ligands larger than 17 nm. However, when a preclustered arrangement of parenchymal cell receptors is experimentally induced at low temperature (4°C), 17 nm particles coated with Lac-BSA are endocytosed at 37°C. In the light of all these data, Kolb-Bachofen et al.[79] concluded that the arrangement of receptors on the membrane determines which ligands are bound and endocytosed. Even though this concept seems to be oversimplified, these results clearly indicate that the uptake of different ligands by the same receptors depends on their spatial organization. *In vitro* differentiating bone marrow macrophages of young chickens possess membrane mannose receptors able to bind mannosylated ferritin, which is endocytosed in coated pits in less than 1 min.[127] Finally, by using different glycosylated ferritins, it was shown that mannose binding sites are present in the membrane of rat liver endothelial cells[128] which also contain Gal*N*ac and Gal specific receptors.[79]

VI. NUCLEAR LECTINS AND GLYCOCONJUGATES

A. GLYCOCONJUGATES

The existence of nuclear glycoconjugates was biochemically established many years ago. It is only recently that the ultrastructural localization of these glycoconjugates has been

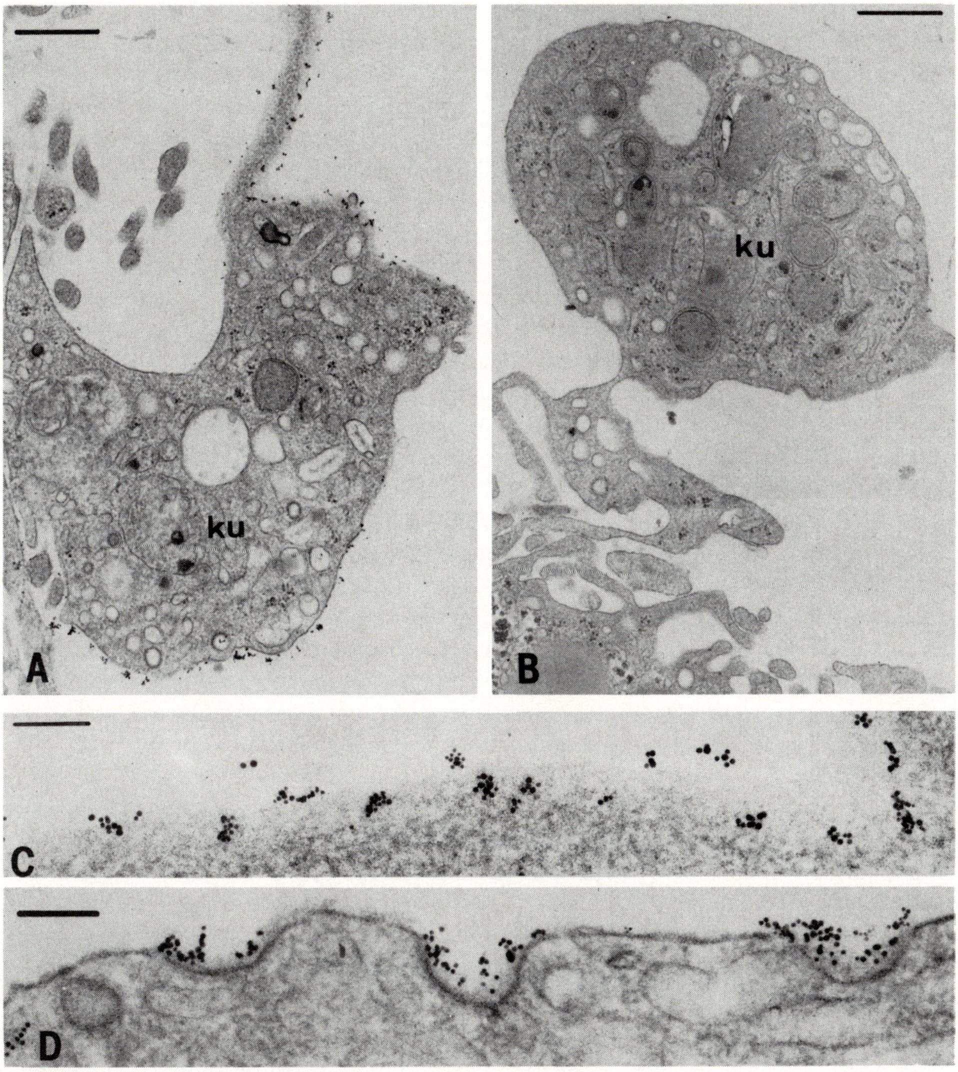

FIGURE 9. *In situ* binding of Lac-BSA-Au 5 in livers prefixed prior to ligand addition. Livers were prefixed with 0.1% glutaraldehyde before adding ligands; the distribution of bound label reflect the initial distribution of receptor activity on the liver cells. Electron-dense ligand, lactose-BSA (Lac-BSA) adsorbed onto colloidal gold particles, 5 nm in diameter (Au5), was injected into the perfusion tube. (A) Ligand is found binding to liver macrophages (ku) in preaggregated clusters all over their sinusoidal surface. (B) Ligand binding is completely abolished in the presence of Gal-*N*Ac (80 mM). (C) Part of the cell shown in (A) at higher magnification. On the tangentially cut plasma membrane the clustered binding is easily recognized. (D) Ligand binding to endothelial cells is restricted to coated pits. Bars = 0.5 μm (A,B); 0.1 μm (C).[125] (From Schlepper-Schäfer J. et al., *Exp. Cell Res.*, 165, 494, 1986. Copyright Academic Press. With permission.)

obtained, using either gold labeled Con A[129] or Con A and WGA associated with glycosylated ferritin as cytomarkers.[103] The former study provided only preliminary data concerning rat hepatocyte nuclei. The latter, performed with reptilian cells, showed that the nucleoli contain a high concentration of Con A and WGA binding sites. In addition, WGA binding sites were found inside the nuclear pores. This last result could be correlated with the fact that WGA was reported to inhibit the nuclear transport of karyophilic proteins.[130–131]

B. LECTINS

A labeling of nuclei with fluoresceinylated antibodies raised against soluble secreted lectins isolated from whole cells, suggested the existence of lectins in the nucleus.[132,133] However, the clear demonstration of the presence of lectins in the cell nucleus arose from an analysis of isolated mammalian cell nuclei incubated in the presence of different fluoresceinylated neoglycoproteins.[48,49] Observations in fluorescence microscopy showed that the lectins were preferentially located in ribonucleoprotein-enriched nuclear territories, and especially in the nucleoli. Analysis by quantitative flow cytofluorometry of nuclei (isolated according to a method in which cytoplasmic contamination is negligible) evidenced that lectins detected are actual nuclear components. In addition, the labeling intensity of nuclei seemed to depend on the physiological state of cells.[48] Recently, a colocalization of lectins and the sites of DNA synthesis (replication bands) was reported in a ciliated protozoon.[134] Ultrastructural studies of sugar-binding sites by means of gold-labeled NGPs allowed a more precise localization of nuclear lectins than fluorescence microscopy.

For instance, it is well known that nucleoli of most vertebrate cells observed in EM are composed of granules surrounding fibrillar components. So it was difficult to determine whether the nucleolar fluorescence originated from the granular shell only, or also from the internal fibrillar elements. Ultrastructural studies in reptilian cells (Figure 10) showed that mannosylated ferritin bound both the dense fibrillar component of nucleoli (known to be the site of synthesis of ribosomal precursors) and the granular one (composed of ribosomal precursors ready to be exported to the cytoplasm).[105] In segregated nucleoli obtained by treatment of HeLa cells with actinomycin D, gold-labeled NGPs containing either rhamnose or glucose bind not only to the dense fibrillar and granular components, but also to the fibrillar centers,[76] which are known to be related to the nucleolus organizing regions of the chromosomes.[135] However, in normal mammalian cells, rhamnose and glucose binding sites were restricted to the dense fibrillar and the granular components, while no binding was observed over the fibrillar centers (C. A. Bourgeois, unpublished data[141]). A possible explanation of the origin of the granule-associated lectins may be that lectins of the dense fibrillar component migrate together with other preribosomal elements from this component to the granules. This is in agreement with the fact that the dense fibrillar component is the precursor of the granular one (for a review, see Reference 136). These data strongly suggest that some lectins are associated with the ribosomal precursors as soon as the early stage of their assembly occurs. Upon incubation in the presence of mannosylated ferritin, in some reptilian cells the labeling was found to be stronger over the condensed chromatin than over territories enriched in ribonucleoproteins.[105] The presence of lectins in the condensed chromatin was also detected recently in resting human lymphocytes (J. Hubert, unpublished data[142]) and in metaphase chromosomes of some human cultured cells (C. A. Bourgeois, personal communication). In rat hepatocytes, the nuclear lectins detected with gold bead-Rha-BSA and gold bead-Glc-BSA were preferentially located both in the transcriptionally active area and in territories related to post-transcriptional events.[137]

Taken together, these data suggest (1) that nuclear lectins are ubiquitous nuclear proteins and (2), that they are localized within special nuclear structures. They are associated with areas known to be transcriptionally active and territories known to be related to post-transcription events, as well as with condensed chromatin, mitotic chromosomes, and finally, colocalized with DNA synthesis.

VII. SUMMARY AND CONCLUSION

EM studies have first provided a great deal of information allowing us to recognize the different organelles specifically involved in the uptake of molecules by the cells. Ligands labeled with markers visible in EM now allow us to trace the pathway of ligands. As shown in the present review, labeled NGPs proved to be powerful tools to analyze receptor-mediated endocytosis at the ultrastructural level. It can therefore be expected that this approach will

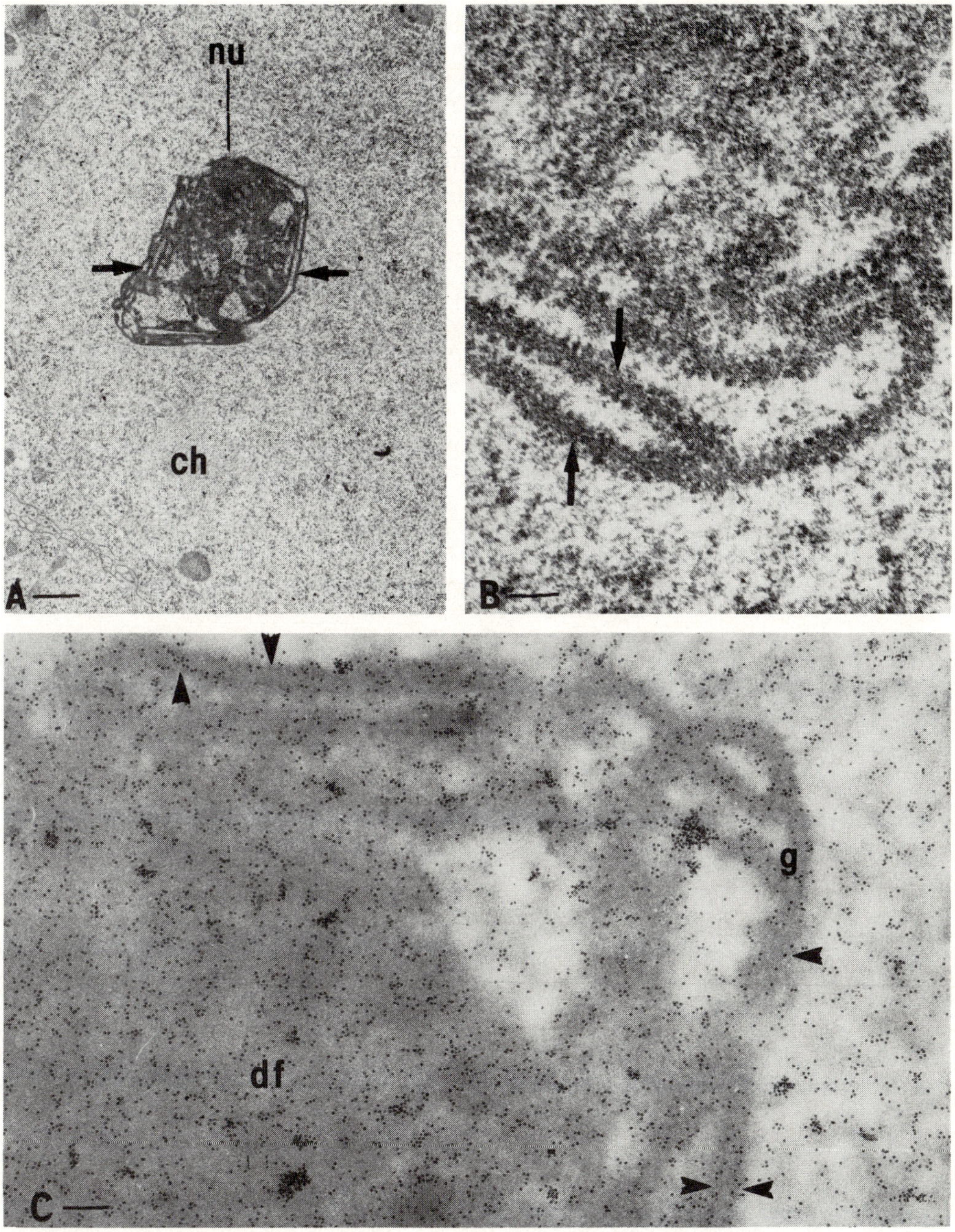

FIGURE 10. Mannose binding sites in the nucleolus of lizard ovarian follicle cells. (A) the granular component of the nucleolus (nu), related to ribosomal precursors, is organized in a cord-like fashion (arrows). ch, chromatin. (B) at higher magnification the ribosomal precursors are seen as granules (arrows) localized at both sides of the cord axis. (C) after incubation of glycol methacrylate (GMA) sections in the presence of mannosyl-ferritin, a labeling is observed at both sides of the cord axis (arrow heads) where ribosomal granules are seen in (B). df, dense fibrillar component. (A and B sections contrasted with lead citrate and uranyl acetate. (C) uncontrasted section.) Bars = 1 μm (A), 0.1 μm (B,C). (From Hubert, J. et al., *Biol. Cell,* 55(1-2), 17, 1985. With permission.)

contribute to precise further mechanisms of endocytosis, such as those related to the unbinding of the ligand-receptor complex. This is the key step leading to the recycling of receptors back to the cell surface on the one hand, and to the degradation of ligands in lysosomes on the other hand. Labeled NGPs used in ultrastructural studies could also provide useful information in the quite exciting field of investigations related to the targeting of drugs and immunostimulating factors, using NGPs as carriers.[138] Certainly, a similar approach could help us to a better understanding of glycoprotein-lectin interactions involved in cellular adhesion.[139] However, the analysis of the spatial organization of cellular organelles deduced from ultrathin sections requires serial sectioning and three-dimensional reconstructions. That must be kept in mind to avoid misleading interpretations when endocytosis is investigated in the EM.

On the basis of recent data, labeled NGPs allowed demonstration of the existence of lectin-like proteins inside the cell nucleus. Because the existence of nuclear glycoconjugates is also clearly established, it is tempting to speculate that some glycoprotein-lectin interactions could constitute an unsuspected modulation system of certain nuclear activities. The EM localization of lectins in relation to the different nuclear structures, the functions of which are known, is just beginning. It can be expected that this approach will help in the future to get new insights into the role that these nonhistone proteins might play in nuclear dynamics.

ACKNOWLEDGMENTS

We are grateful to Drs. J.Schlepper-Schäfer and D. Sandoz who supplied and allowed us to use illustrations from their work. This work has been partially supported by the Institut National de la Santé et de la Recherche Médicale (INSERM), Centre National de la Recherche Scientifique (CNRS) and Association pour la Recherche contre le Cancer (ARC) grants. A. Obrénovitch is chargée de Recherche INSERM.

REFERENCES

1. **Goldstein, I. J., Hughes, R. C., Monsigny, M., Osawa, T., and Sharon, N.,** What should be called a lectin? *Nature,* 285, 66, 1980.
2. **Lee, Y. C. and Lee, R. T.,** Neoglycoproteins as probes for binding and cellular uptake of glycoconjugates, in *The Glycoconjugates,* Horowitz M. I., Ed., Academic Press, New York, 4, 1982, 57.
3. **Hakomori, S. I.,** Glycosphingolipids in cellular interaction, differentiation and oncogenes, *Ann. Rev. Biochem.,* 50, 733, 1981.
4. **Montreuil, J.,** Glycoproteins, in *Comprehensive Biochemistry* Vol. 19B, Part 2, Neuberger, A. and Van Deenen, L. L. M., Eds., Elsevier, Amsterdam, 1982, 1.
5. **Montreuil, J.,** Spatial conformation of glycans and glycoproteins, *Biol. Cell.,* 51, 115, 1984.
6. **Evered, D. and Whelan, J., Eds.,** *Functions of the Proteoglycans,* Ciba Foundation Symp. 124, John Wiley & Sons, Chichester, U. K., 1986, 299.
7. **Fransson, L. A.,** Structure and function of cell associated proteoglycans, *Trends Biochem. Sci.,* 12, 406, 1987.
8. **Poole, A. R.,** Proteoglycans in health and disease structures and functions, *Biochem. J.,* 236, 1, 1986.
9. **Roden, L. and Horowitz, M. I.,** Structure and biosynthesis of connective tissue proteoglycans, in *The Glycoconjugates,* Vol. 2, Horowitz, M. I. and Pigman, W., Eds., Academic Press, Orlando, 1978, 3.
10. **Beyer, T. A. and Hill, R. L.,** Glycosylation pathways in the biosynthesis of non reducing terminal sequences in oligosaccharides of glycoproteins, in *The Glycoconjugates,* Vol 3, Horowitz, M. I., Ed., Academic Press, New York, 1982, 25.
11. **Datema, R., Olofsson, S., and Romero, P. A.,** Inhibitors of protein glycosylation and glycoprotein processing in viral systems, *Pharmacol. Ther.,* 33, 221, 1987.
12. **Hubbard, S. C. and Robbins, P. W.,** Synthesis of N-linked oligosaccharides of glycoproteins, *J. Biol. Chem.,* 255, 11782, 1980.

13. **Kornfeld, R. and Kornfeld, S.,** Assembly of asparagin-linked oligosaccharides, *Ann. Rev. Biochem.*, 54, 631, 1985.
14. **Schachter, H.,** Coordination between enzyme specificity and intracellular compartmentation with control of protein bound oligosaccharide biosynthesis, *Biol. Cell,* 51, 133, 1984.
15. **Farquhar, M. G.,** Progress in unraveling pathways of Golgi traffic, *Ann. Rev. Cell. Biol.*, 1, 447, 1985.
16. **Pfeffer, S. R. and Rothman, J. E.,** Biosynthetic protein transport and sorting by the endoplasmic reticulum and Golgi, *Ann. Rev. Biochem.*, 56, 829, 1987.
17. **Wieland, F. T., Gleason, M. L., Serafini, T. A., and Rothman, J. E.,** The rate of bulk flow from the endoplasmic reticulum to the cell surface, *Cell,* 50, 289, 1987.
18. **Cross, G. A. M.,** Eukaryotic protein modification and membrane attachment via phosphatidylinositol, *Cell,* 48, 179, 1987.
19. **Low, M. G. and Saltiel, A. R.,** Structural and functional roles of glycosyl phosphatidylinositol in membranes, *Science,* 239, 268, 1988.
20. **Roseman, S.,** Complex carbohydrates and intercellular adhesion, in *Biology and Chemistry of Eucaryotic Cell Surfaces,* Lee, E. Y. C. and Smith, E. E., Eds., Academic Press, New York, 1974, 317.
21. **Hemming, F. W.,** Glycosyl phosphopolyphenols, in *Glycolipids,* Wiegandl H., Ed., Elsevier, Amsterdam, 261, 1985.
22. **Marshall, R. D.,** Glycoproteins, *Ann. Rev. Biochem.*, 41, 673, 1972.
23. **Beyer, T. A., Sadler, J. E., Rearick, J. F., Paulson, J. C., and Hill, R. L.,** Glycosyltransferases and their use in assessing oligosaccharide structure and structure-function relationships, *Adv Enzymol.*, 52, 23, 1981.
24. **Ashwell, G. and Harford, J.,** Carbohydrate specific receptors of the liver, *Ann. Rev. Biochem.*, 51, 531, 1982.
25. **Brandley, B. K. and Schnaar, R. L.,** Cell surface carbohydrates in cell recognition and response, *J. Leukocyte Biol.*, 40, 97, 1986.
26. **Dennis, J. W. and Laferte, S.,** Tumor cell surface carbohydrate and the metastatic phenotype, *Cancer Metast. Rev.*, 5, 185, 1987.
27. **Liener, I. E., Sharon, N., and Goldstein, I. J.,** Eds., *The Lectins Properties, Functions and Applications in Biology and Medicine,* Academic Press, Orlando, 1986.
28. **Lis, H. and Sharon, N.,** Lectins as molecules and as tools, *Ann. Rev. Biochem.*, 55, 35, 1986.
29. **Monsigny, M., Kieda, C., and Roche, A. C.,** Membrane lectins, *Biol. Cell,* 36, 289, 1979.
30. **Monsigny, M., Kieda, C., and Roche, A. C.,** Membrane glycoproteins, glycolipids and membrane lectins as recognition signals in normal and malignant cells, *Biol. Cell,* 47, 95, 1983.
31. **Monsigny, M., Roche, A. C., and Midoux, P.,** Uptake of neoglycoproteins via membrane lectin(s) of L1210 cells evidenced by quantitative flow cytofluorometry and drug targeting, *Biol. Cell,* 51, 187, 1984.
32. **Olden, K. and Parent, J. B., Eds.,** *Vertebrate Lectins,* Van Nostrand, New York, 1987.
33. **Sharon, N.,** Carbohydrates as recognition determinants in phagocytosis and in lectin mediated killing of target cells, *Biol. Cell,* 51, 239, 1984.
34. **Morell, A. G., Irvine, R. A., Sternlieb, I., Scheinberg, I. H., and Ashwell, G.,** Physical and chemical studies on ceruloplasmin. Metabolic studies on sialic acid free ceruloplasmin *in vivo, J. Biol. Chem.*, 243, 155, 1968.
35. **Rogers, J. C. and Kornfeld, S.,** Hepatic uptake of proteins coupled to fetuin glycopeptides, *Biochem. Biophys. Res. Commun.*, 45, 622, 1971.
36. **Gabius, H. J., Engelhardt, R., Hellmann, T., Midoux, P., Monsigny, M., Nagel, G. A., and Vehmeyer, K.,** Characterization of membrane lectins in human colon carcinoma cells by flow cytofluorometry, drug targeting and affinity chromatography, *Anticancer Res.*, 7, 109, 1987.
37. **Monsigny, M., Roche, A. C., Kieda, C., Midoux, P., and Obrénovitch, A.,** Characterization and biological implications of membrane lectins in tumor, lymphoid and myeloid cells, *Biochimie,* 70, 1633, 1988.
38. **Raz, A. and Lotan, R.,** Endogenous galactoside-binding lectins: a new class of functional tumor cell surface molecules related to metastasis, *Cancer Metast. Rev.*, 6, 433, 1987.
39. **Roche, A. C., Barzilay, M., Midoux, P., Junqua, S., Sharon, N., and Monsigny, M.,** Endocytosis of glycoconjugates by Lewis lung carcinoma cells, *J. Cell Biochem.*, 22, 131, 1983.
40. **Roche, A. C., Midoux, P., Petit, C., Derrien, D., Mayer, R., and Monsigny, M.,** in *Immunomodulators and Nonspecific Host Mechanisms Against Microbial Infections,* Masihi, K. N. and Lange, W., Eds., Pergamon Press, New York, 1988, 217.
41. **Wileman, T., Charding, C. and Stahl, P.,** Receptor mediated endocytosis, *Biochem. J.*, 232, 1, 1985.
42. **Barondes, S. H.,** Developmentally regulated slime mould lectins and specific cell adhesion, in *The Molecular Basis of Cell-Cell Interaction,* Vol. 24, Lerner R. A. and Bergsma D., Eds., Alan R. Liss, New York, 1978, 491.
43. **Goldstein, I. J. and Hayes, C. E.,** The Lectins: Carbohydrate-Binding Proteins of Plants and Animals, *Adv. Carbohydr. Chem. Biochem.*, 35, 128, 1978.

44. **Barondes, S. H.,** Lectins: Their multiple endogenous cellular functions, *Ann. Rev. Biochem.,* 50, 207, 1981.
45. **Keen, J. H.,** The structure of clathrin-coated membranes: assembly and disassembly, in *Endocytosis,* Pastan I. and Willingham M. C., Eds., Plenum Press, New York, 1985, 85.
46. **Maxfield, F. R.,** Acidification of endocytic vesicles and lysosomes, in *Endocytosis,* Pastan, I. and Willingham, M. C., Eds., Plenum Press., New York, 1985, 235.
47. **Pastan, I. and Willingham, M. C.,** The pathway of endocytosis, in *Endocytosis,* Pastan, I. and Willingham, M. C., Eds., Plenum Press, New York, 1985, 1.
48. **Sève, A. P., Hubert, J., Bouvier, D., Bourgeois, C. A., Midoux, P., Roche, A. C., and Monsigny, M.,** Analysis of sugar-binding sites in mammalian cell nuclei by quantitative flow microfluorometry, *Proc. Natl. Acad. Sci. U.S.A.,* 83, 5997, 1986.
49. **Sève, A. P., Hubert, J., Bouvier, D., Bouteille, M., Mantier, C., and Monsigny, M.,** Detection of sugar binding proteins in membrane depleted nuclei, *Exp. Cell Res.,* 157, 533, 1985.
50. **Aplin, J. D. and Wriston, J. C.,** Preparation, properties and applications of carbohydrate conjugates of proteins and lipids, *Crit. Rev. Biochem.,* 10, 259, 1981.
51. **Stowell, C. P. and Lee, Y. C.,** Neoglycoprotein: the preparation and application of synthetic glycoproteins, *Adv. Carbohydr. Chem. Biochem.,* 37, 225, 1980.
52. **Goebel, W. F. and Avery, O. T.,** Chemo-immunological studies on conjugated carbohydrate proteins. The synthesis of p-aminophenol β-glucoside, p-aminophenol β-galactoside and their coupling with serum albumin, *J. Exp. Med.,* 50, 521, 1929.
53. **Avery, O. T. and Goebel, W. F.,** Chemo-immunological studies on conjugated carbohydrate proteins. II, Immunological specificity of synthetic sugar protein antigens, *J. Exp. Med.,* 50, 533, 1929.
54. **Goebel, W. F. and Avery, O. T.,** Chemo-immunological studies on conjugated carbohydrate proteins IV. The synthesis of p-amino benzyl ether of the soluble specific substance of type III *Pneumococcus* and its coupling with protein, *J. Exp. Med.,* 54, 431, 1931.
55. **Iyer, R. N. and Goldstein, I. J.,** Quantitative studies on the interaction of concanavalin A the carbohydrate binding protein of the Jack bean with model carbohydrate-protein conjugates, *Immunochemistry,* 10, 313, 1973.
56. **Privat, J. P., Delmotte, F., and Monsigny, M.,** Protein sugar interactions. Association of β-(1-4)linked *N*-acetyl-D-glucosamine oligomer derivatives with wheat germ agglutinin (lectin), *FEBS Lett.,* 46, 224, 1974.
57. **Kieda, C., Delmotte, F., and Monsigny, M.,** Preparation and properties of glycosylated cytochemical markers, *FEBS Lett.,* 76, 257, 1977.
58. **Monsigny, M., Kieda, C., Gros, D., and Schrével, J.,** New markers to visualize cell surface glycoconjugates: glycosylated horseradish peroxidase and glycosylated ferritin, in *6th Eur. Congr. Electron. Microsc.,* (Jerusalem), Vol. 2, Ben-Schaul, Y., Ed., International Publishing, Tel Aviv, 1976, 39.
59. **Monsigny, M., Kieda, C., Obrénovitch, A., and Delmotte, F.,** Glycosylated horseradish peroxidase interactions with lectins, in *Protides of the Biological Fluids,* Peeters, H., Ed., Pergamon Press, 1976, 815.
60. **Schrével, J., Kieda, C., Caigneaux, E., Gros, D., Delmotte, F., and Monsigny, M.,** Visualization of cell surface carbohydrate by a general two step lectin technique: lectin and glycosylated cytochemical markers, *Biol. Cell,* 36, 259, 1979.
61. **Buss, D. H. and Goldstein, I. J.,** Protein carbohydrate interaction. Carbohydrates containing groups for the alkylation of proteins, *J. Chem. Soc. C.,* 14, 1457, 1968.
62. **McBroom, C. R., Samanen, C. H., and Goldstein, I. J.,** Carbohydrate antigens: coupling of carbohydrates to proteins by diazonium and phenylisothiocyanate reactions, *Methods Enzymol.,* 28, 212, 1972.
63. **Jeffrey, A. M., Zopf, D. A., and Ginsburg, V.,** Affinity chromatography of carbohydrate specific immunoglobulins coupling of oligosaccharides to sepharose, *Biochem. Biophys. Res. Commun.,* 62, 608, 1975.
64. **Borch, R. F., Bernstein, M. D., and Durst, H. D.,** The cyanohydridoborate anion as a selective reducing agent, *J. Am. Chem. Soc.,* 93, 2897, 1971.
65. **Smith, D. F., Zopf, D. A., and Ginsburg, V.,** Carbohydrate antigens: coupling of oligosaccharide phenethylamine isothiocyanate derivatives to bovine serum albumin, *Methods Enzymol.,* 83, 169, 1978.
66. **Gray, G.,** The direct coupling of oligosaccharides to proteins and derivatized gels, *Arch. Biochem. Biophys.,* 163, 426, 1974.
67. **Arakatsu, Y., Ashwell, G., and Kabat, E. A.,** Immunochemical studies on dextrans. V. Specificity and cross reactivity with dextrans of the antibodies formed in rabbits to isomaltonic and isomaltotrionic acids coupled to serum albumin, *J. Immunol.,* 97, 858, 1966.
68. **Lee, Y. C., Stowell, C. P., and Krantz, M. J.,** 2-imino-2-methoxyethyl-1-thioglycoside: new reagents for attaching sugars to proteins, *Biochemistry,* 15, 3956, 1976.
69. **Krantz, M. J., Holtzman, N. A., Stowell, C. P., and Lee, Y. C.,** Attachment of thio-glycosides to proteins: Enhancement of liver membrane binding, *Biochemistry,* 15, 3963, 1976.

70. **Delmotte, F.,** Interactions: Glycanne-Protéine. Etude des Protéines Affines d'Osides Contenant de la N-Acetyl Glucosamine, Thèse d'Etat, Université d'Orléans, 1976.
71. **Rauvala, H. and Hakomori, S. I.,** Studies on cell adhesion. III the occurence of α-mannosidase at the fibroblast cell surface and its possible role in cell recognition, *J. Cell Biol.,* 88, 149, 1981.
72. **Shur, B. D.,** Cell surface glycosyltransferase activities during normal and mutant (T/T) mesenchyme migration, *Dev. Biol.,* 91, 149, 1982.
73. **Miquelis, R., Alquier, D., and Monsigny, M.,** GlcNAc specific receptor of the thyroid. Binding characteristics, identification, partial characterization and potential role, *J. Biol. Chem.,* 262, 15291, 1987.
74. **Graham, R. C. and Karnovsky, M. J.,** The early stages of absorption of injected horseradish peroxidase in proximal tubules of mouse kidney: ultrastructural cytochemistry by a new technique, *J. Histochem. Cytochem.,* 14, 291, 1966.
75. **Seligman, A. M., Karnovsky, M. J., Wasserkrug, M. D., and Hanker, J. S.,** Nondroplet ultrastructural demonstration of cytochrome oxidase activity with a polymerizing osmiophilic reagent, diaminobenzidine (DAB), *J. Cell Biol.,* 38, 1, 1968.
76. **Bourgeois, C. A., Sève, A. P., Monsigny, M., and Hubert, J.,** Detection of sugar binding sites in the fibrillar and the granular components of the nucleolus. An experimental study in cultured mammalian cells, *Exp. Cell Res.,* 172, 365, 1987.
77. **Deschuyteneer, M., Prieels, J. P., and Mosselmans, R.,** Galactose-specific adsorptive endocytosis: An ultrastructural qualitative and quantitative study in cultured rat hepatocytes, *Biol. Cell,* 50, 17, 1984.
78. **Kataoka, M. and Tavassoli, M.,** Synthetic neoglycoproteins; a class of reagents for detection of sugar-recognizing substances, *J. Histochem. Cytochem.,* 32, 1091, 1984.
79. **Kolb-Bachofen, V., Schlepper-Schäfer, S., Roos, P., Hülsman, D., and Kolb, H.,** GalNac/Gal-specific rat liver lectins: their role in recognition, *Biol. Cell,* 51, 219, 1984.
80. **Faulk, W. P. and Taylor, G. M.,** An immunocolloid method for the electron microscope, *Immunochemistry,* 8, 1081, 1971.
81. **Horisberger, M.,** Colloidal gold: a cytochemical marker for light and fluorescent microscopy and for transmission and scanning electron microscopy, in *Scanning Electron Microscopy,* Johari, O., Ed., SEM Inc., AMF O'Hare, Chicago, 1981, 9.
82. **Horisberger, M.,** Electron-opaque markers: a review, in *Immunolabeling for Electron Microscopy,* Polak, J. M. and Varndell, I., Eds., Elsevier, Amsterdam, 1984, 17.
83. **Varndell, I. M. and Polak, J. M.,** Double immunostaining procedures: techniques and applications, in *Immunolabeling for Electron Microscopy,* Polak, J. M. and Varndell, I., Eds., Elsevier, Amsterdam, 1984, 155.
84. **Crichton, R. C.,** Structure and function of ferritin, *Angew. Chem.,* 12, 57, 1972.
85. **Shinyo, S., Abe, H., and Masuda, M.,** Carbohydrate composition of horse spleen ferritin, *Biochim. Biophys. Acta,* 411, 165, 1975.
86. **Fischbach, F. A. and Anderegg, J. W.,** An X-ray scattering study of ferritin and apoferritin, *J. Mol. Biol.,* 14, 458, 1965.
87. **Haggis, G. M.,** The iron oxide core of the ferritin molecule, *J. Mol. Biol.,* 14, 598, 1965.
88. **Parmley, R. T., Denys, F. R., and Alvarez, C. J.,** Ferrocyanide enhancement of concanavalin A ferritin and cationized ferritin staining blood cell surface glycoconjugates, *Histochem. J.,* 11, 379, 1979.
89. **Ainsworth, S. K., Ito, S., and Karnovsky, M. J.,** Alkaline bismuth reagent for high resolution ultrastructural demonstration of periodate-reactive sites, *J. Histochem. Cytochem.,* 20, 995, 1972.
90. **Nicolson, G. L. and Singer, S. J.,** Ferritin-conjugated plant agglutinins as specific saccharides stains for electron microscopy: application to saccharides bound to cell membranes, *Proc. Natl. Acad. Sci. U.S.A.,* 68, 942, 1971.
91. **Drysdale, J. W. and Munro, H. N.,** Small-scale isolation of ferritin for the assay of the incorporation of ^{14}C-labeled amino acids, *Biochem. J.,* 95, 851, 1965.
92. **Bryce, C. F. A. and Crichton, R. R.,** The catalytic activity of horse spleen apoferritin, *Biochem. J.,* 133, 301, 1973.
93. **Granik, S. and Michaelis, L.,** Ferritine. II. Apoferritine de la rate de cheval, *J. Biol. Chem.,* 147, 91, 1943.
94. **Bayer, E. A., Wilchek, M., and Skutelsky, E.,** Affinity cytochemistry: The localization of lectin and antibody receptors on erythrocytes via the avidin-biotin complex, *FEBS Lett.,* 68, 240, 1976.
95. **Hoffman, K., Titus, G., Montibeller, J. A., and Finn, F. M.,** Avidin binding of carboxyl-substituted biotin and analogues, *Biochemistry,* 21, 978, 1982.
96. **Shimkus, M., Levy, J., and Herman, T. A.,** A chemically cleavable biotinylated nucleotide: Usefulnees in the recovery of protein-DNA complexes from avidin affinity columns, *Proc. Natl. Acad. Sci. U.S.A.,* 82, 2593, 1985.
97. **Schrével, J., Gros, D., and Monsigny, M.,** Cytochemistry of cell glycoconjugates, *Prog. Histochem. Cytochem.,* Gramann, W., Lojda, Z., Pearse, A. G. E., and Schiebler, H., Eds., Gustav Fischer, New York, 14, 1981.

98. **Monsigny, M., Roche, A. C., and Kieda, C.,** Lectins as tools to study cell surface membranes, in *Structure and Functions of Biological Membranes. Molecular Aspects,* Interdisciplinary training course, Commission of the European Communities, Brussels, 1978, 161.
99. **Sandoz, D., Boisvieux-Ulrich, E., and Chailley, B.,** Relationships between intramembrane particles and glycoconjugates in the ciliary membrane of the quail oviduct, *Biol. Cell,* 36, 267, 1979.
100. **Gros, D., Obrénovitch, A., Challice, C. E., Monsigny, M., and Schrével, J.,** Ultrastructural visualization of cellular carbohydrate components by means of lectins on ultrathin glycol methacrylate sections, *J. Histochem. Cytochem.,* 25, 104, 1977.
101. **Bernhard, W. and Avrameas, S.,** Ultrastructural visualization of cellular carbohydrate components by means of concanavalin A, *Exp. Cell Res.,* 64, 232, 1971.
102. **Wall, D. A., Wilson, G., and Hubbard, A. L.,** The galactose-specific recognition system of mammalian liver. The route of ligand internalization in rat hepatocytes, *Cell,* 21, 79, 1980.
103. **Sève, A. P., Hubert, J., Bouvier, D., Masson, C., Geraud, G., and Bouteille, M.,** *In situ* distribution in different cell types of nuclear glycoconjugates detected by two lectins, *J. Submicrosc. Cytol.,* 16, 631, 1984.
104. **Dingwall, C. and Lasley, R. A.,** Protein import in the cell nucleus, *Ann. Rev. Cell Biol.,* 2, 367, 1986.
105. **Hubert, J., Sève, A. P., Bouvier, D., Masson, C., Bouteille, M., and Monsigny, M.,** *In situ* ultrastructural localization of sugar binding sites in lizard gronulosa cell nuclei, *Biol. Cell,* 55, 15, 1985.
106. **Schrével, J., Kieda, C., Gros, D., Obrénovitch, A., Caigneaux, E., Delmotte, F., and Monsigny, M.,** Nouveaux marqueurs de glycoconjugués membranaires, *J. Microsc. Biol. Cell.,* 27, 23a, 1976.
107. **Hellio, R. and Ryter, A.,** Relationships between anionic sites and lectin receptors in the plasma membrane of *Dictyostelium discoideum* and their role in phagocytosis, *J. Cell Sci.,* 41, 89, 1980.
108. **Ryter, A. and Hellio, R.,** Electron microscope study of *Dictyostelium discoideum* plasma membrane and its modifications during and after phagocytosis, *Biol. Cell.,* 36, 9a, 1979.
109. **Mauchamp, B. and Hubert, M.,** Internalization of plasma membrane glycoconjugates and plasma membrane lectin into epidermal cells during pharate adult-wing development of Pieris brassicae: correlation with resorption of molting fluid component, *Biol. Cell.,* 50, 285, 1984.
110. **Rougier, M.,** Ultrastructural detection of fucosyl residues at the surface of axenically grown maize roots/ sequential use of UeA lectin and fucosyl ferritin as the specific glycosylated marker, *Eur. J. Cell Biol.,* 34, 45, 1984.
111. **Kieda, C., Roche, A. C., Delmotte, F., and Monsigny, M.,** Lymphocyte membrane lectins. Direct visualization by the use of fluoresceinyl-glycosylated cytochemical markers, *FEBS Lett.,* 99, 329, 1979.
112. **Midoux, P., Roche, A. C., and Monsigny, M.,** Estimation of the degradation of endocytosed material by flow cytofluorometry using two neoglycoproteins containing different numbers of fluorescein molecules, *Biol. Cell.,* 58, 221, 1986.
113. **Midoux, P., Roche, A. C., and Monsigny, M.,** Quantitation of the binding, uptake and degradation of fluoresceinylated neoglycoproteins by flow cytofluorometry, *Cytometry,* 8, 327, 1987.
114. **Ashwell, G. and Morell, A. G.,** The role of surface carbohydrates in the hepatic recognition and transport of circulating glycoproteins, *Adv. Enzymol.,* Meister, A., Ed., John Wiley & Sons, New York, 1974, 99.
115. **Harford, J. and Ashwell, G.,** Chemical and physical properties of the hepatic receptor for asialoglycoproteins, in *Endocytosis,* Pastan, I., and Willingham, M. C., Eds., Plenum Press, New York, 1985, 69.
116. **Hubbard, A. L. and Stukenbrok, H.,** An electron microscope autoradiographic study of the carbohydrate recognition systems in rat liver, *J. Cell Biol.,* 83, 65, 1979.
117. **Labadie, J. H., Chapman, K. P., and Aronson, N. N.,** Glycoprotein catabolism in rat liver. Lysosomal digestion of iodinated asialo-fetuin, *Biochem. J.,* 152, 271, 1975.
118. **Courtoy, P., Quintart, J. N., Limet, J. N., De Roc, C., and Baudhuin, P.,** Polymeric IgA galactose specific pathway in rat hepatocytes: Evidence for intracellular ligand sorting, in *Endocytosis,* Pastan, I. and Willingham, M. C., Eds., Plenum Press, New York, 1985, 163.
119. **Wall, D. A. and Hubbard, A. L.,** Galactose-specific recognition system of mammalian liver: receptor distribution on the hepatocyte cell surface, *J. Cell Biol.,* 90, 687, 1981.
120. **Haimes, H. B., Stockert, R. J., Morell, A. G., and Novikoff, A. E.,** Carbohydrate-specified endocytosis: localization of ligand in the lysosomal compartment, *Proc. Natl. Acad. Sci. U.S.A.,* 78, 6936, 1981.
121. **Quintart, J., Courtoy, P. J., Limet, J. N., and Baudhuin, P.,** Galactose specific endocytosis in rat liver, *Eur. J. Biochem.,* 131, 105, 1983.
122. **Limet, J. H., Quintart, J., Schneider, Y. J., and Courtoy, P. J.,** Receptor-mediated endocytosis of polymeric IgA and galactosylated serum albumin in rat liver, *Eur. J. Biochem.,* 146, 539, 1985.
123. **Roos, P. H., Kolb-Bachofen, V., Schlepper-Schäfer, J., Monsigny, M., Stockert, J., and Kolb, H.,** Two galactose specific receptors in the liver with different function, *FEBS Lett.,* 157, 253, 1983.
124. **Kempka, G. and Kolb-Bachofen, V.,** Galactose-specific receptors on liver cells. I: Hepatocyte and liver macrophage receptor differ in their membrane anchorage, *Biochim. Biophys. Acta,* 847, 108, 1985.

125. **Kolb-Bachofen, V., Schlepper-Schäfer, J., and Kolb, H.,** Receptor-mediated particle uptake by liver macrophages. The galactose-particle receptor mediates uptake via coated and also non-coated structures, *Exp. Cell Res.,* 148, 173, 1983.
126. **Schlepper-Schäfer, J., Hülsmann, D., Djovkar, A., Meyer, H. E., Herbertz, L., Kolb, H., and Kolb-Bachofen, V.,** Endocytosis via galactose receptors *in vivo, Exp. Cell Res.,* 165, 494, 1986.
127. **Rossi, G., and Himmelhoch, S.,** Binding of mannosylated ferritin to chicken marrow macrophages, *Immunobiol.,* 165, 46, 1983.
128. **Irie, S. and Tavassoli, M.,** Mapping of the rat liver endothelial membrane with lectins and glycosylated ferritin, *Am. J. Anat.,* 177, 403, 1986.
129. **Roth, J.,** Application of lectin-gold complexes for electron microscopic localization of glycoconjugates on thin sections, *J. Histochem. Cytochem.,* 31, 987, 1983.
130. **Finlay, D. R., Newmeyer, D. D., Price, T. M., and Forbes, D. J.,** Incubation of *in vitro* nuclear transport by a lectin that binds to nuclear pores, *J. Cell Biol.,* 104, 189, 1987.
131. **Yoneda, Y., Anoka, T., Imamoto-Sonobe, N., Sugawa, H., Shimanishi, Y., and Uchida, T.,** Synthetic peptides containing a region of SV40 large T antigen involved in nuclear localization direct the transport of proteins into the nucleus, *Exp. Cell Res.,* 170, 439, 1987.
132. **Beyer, E. C. and Barondes, S. H.,** Chicken tissue binding sites for a purified chicken lectin, *J. Supramol. Struct.,* 13, 219, 1980.
133. **Childs, R. A. and Feizi, T. C.,** β-Galactoside-binding lectin of human and bovine tissues, *Cell Biol. Int. Rep.,* 4, 775, 1980.
134. **Olins, D., Olins, A., Sève, A. P., Bourgeois, C. A., Hubert, J., and Monsigny, M.,** Lectin like components in the macronuclear replication bands of *Euplotes eurystomus, Biol. Cell,* 62, 95, 1988.
135. **Fakan, S. and Hernandez-Verdun, D.,** The nucleolus and the nucleolar organizer regions, *Biol. Cell,* 56, 189, 1986.
136. **Goesseus, G.,** Nuclear structure, *Int. Rev. Cytol.,* 87, 107, 1984.
137. **Bourgeois, C. A., Sève, A. P., Monsigny, M., and Hubert, J.,** Localization of lectins in the nuclear and the nucleolar structures of reptilian and mammalian cells, in *Lectins Biology, Biochemistry, Clinical Biochemistry,* Freed, D. and Bøg-Hansen, T. C., Eds., Sigma Chemical Co., St. Louis, 6, 431, 1988.
138. **Roche, A. C., Bailly, P., and Monsigny, M.,** Macrophage activation by MDP bound to neoglycoproteins: Metastasis eradication in mice, *Invas. Metast.,* 5, 218, 1985.
139. **Kieda, C. and Monsigny, M.,** Involvement of membrane sugar receptors and membrane glycoconjugates in the adhesion of 3LL subpopulations to cultured pulmonary cells, *Invas. Metast.,* 6, 347, 1986.
140. **Ferguson, M. A. J., Hamans, S. W., Divek, R. A., and Rademacker, T. W.,** Glycosyl-phosphatidylinositol moiety that anchors *Trypanosoma brucei* variant surface glycoprotein to the membrane, *Science,* 239, 753, 1988.
141. **Bourgeois, C. A.,** unpublished.
142. **Hubert, J.,** unpublished.

Chapter 14

INTERACTION BETWEEN CYTOSKELETON AND PLASMA MEMBRANE IN VIRUS MORPHOGENESIS

Wolfgang Bohn

TABLE OF CONTENTS

I. INTRODUCTION

An increasing number of morphological and biochemical data point to a general presence of a membrane skeleton in eukaryotic cells also known as "outer lamina", or "cortical network".[1-8] These membrane skeletons seem to be of fundamental importance for a variety of cell functions, such as (1) maintenance of the cell shape, (2) control of surface membrane mobility, and (3) formation of intercellular contacts.[9-13]

The structural composition and the dynamic properties of the membrane skeletons vary considerably between different cell types, making it difficult to define a general organization scheme. Furthermore, each of the main cytoskeletal components, namely actin, intermediate filaments, and microtubuli, is suspected of interacting directly or indirectly with the plasma membrane. The type of interaction even varies within different cellular compartments. For instance, the shape of the apical plasma membrane of polarized cells is determined by the organization of the microfilament system, whereas lateral contacts between neighbor cells show the involvement of intermediate filaments.[14,15]

This heterogeneous organization has led to the selection of distinct cell types for investigations of the structure and function of membrane skeletons, namely erythrocytes,[4] intestinal cells,[14,15] platelets[16,17], and the slime mold *Dictyostelium discoideum*.[2,18] Especially those cells exhibiting a highly ordered membrane skeleton structure are now very well characterized. This refers to the membrane skeletons of erythrocytes and the brush border microvilli, both of which can be isolated in pure form and are available in high quantities.

Membrane skeletons of erythrocytes, which are considered as paradigms, show a hexagonal and pentagonal network of spectrin tetramers cross-linked by junctions containing actin oligomers and band 4.1 protein.[4,19,20] This two-dimensional network is anchored to the protoplasmic membrane surface by ankyrin and band 4.1 protein, linking spectrin to the anion transporter band 3 and to glycophorin, respectively. Meanwhile, immunoreactive analogs to erythrocyte spectrin, ankyrin, and band 4.1 have been detected in a variety of nucleated cell types, suggesting a similar organization of the membrane skeletons in these cells.[21,22]

Membrane skeletons of brushborder microvilli of intestinal epithelial cells represent the second well characterized model system.[14,15] The microvilli are regularely arranged at the apical surface of these cells, and each of them contains a bundle of about 20 actin filaments, cross-linked to each other by fimbrin and the calcium sensitive protein villin.[23,24] The bundles are laterally attached to the plasma membrane at a 33 nm periodicity via 110 kD protein-calmodulin complexes.[25] These filament bundles are anchored in the terminal web composed of actin filaments, intermediate filaments, and the spectrin isoforms TW 240/260.[15] The microvilli are rigid structures, even maintaining their organization during purification.

The organization of membrane skeletons of other nonmuscle cells is far less well characterized. The structural analysis of these skeletons is significantly impeded by their isotropic organization, that is the absence of a preferred filament orientation. Furthermore, these networks exhibit a high degree of plasticity and are continuously reorganized.

For a further analysis of membrane skeletons in cultured nonmuscle cells, the infection with enveloped viruses which bud from the plasma membrane of infected cells, may be a favorable approach. The structural composition of these viruses is well known, and an increasing number of data indicate that cytoskeletal elements participate in their morphogenesis.

This article will summarize some data on virus-cytoskeleton interaction in the course of virus morphogenesis at the plasma membrane, and will describe in more detail data on assembly and budding of measles virus (Figure 1), with the emphasis on electron microscopic methods. Evidence will be presented that infection with this virus may represent a favorable model to determine regulatory mechanisms of the dynamic cytoskeleton-plasma membrane interaction.

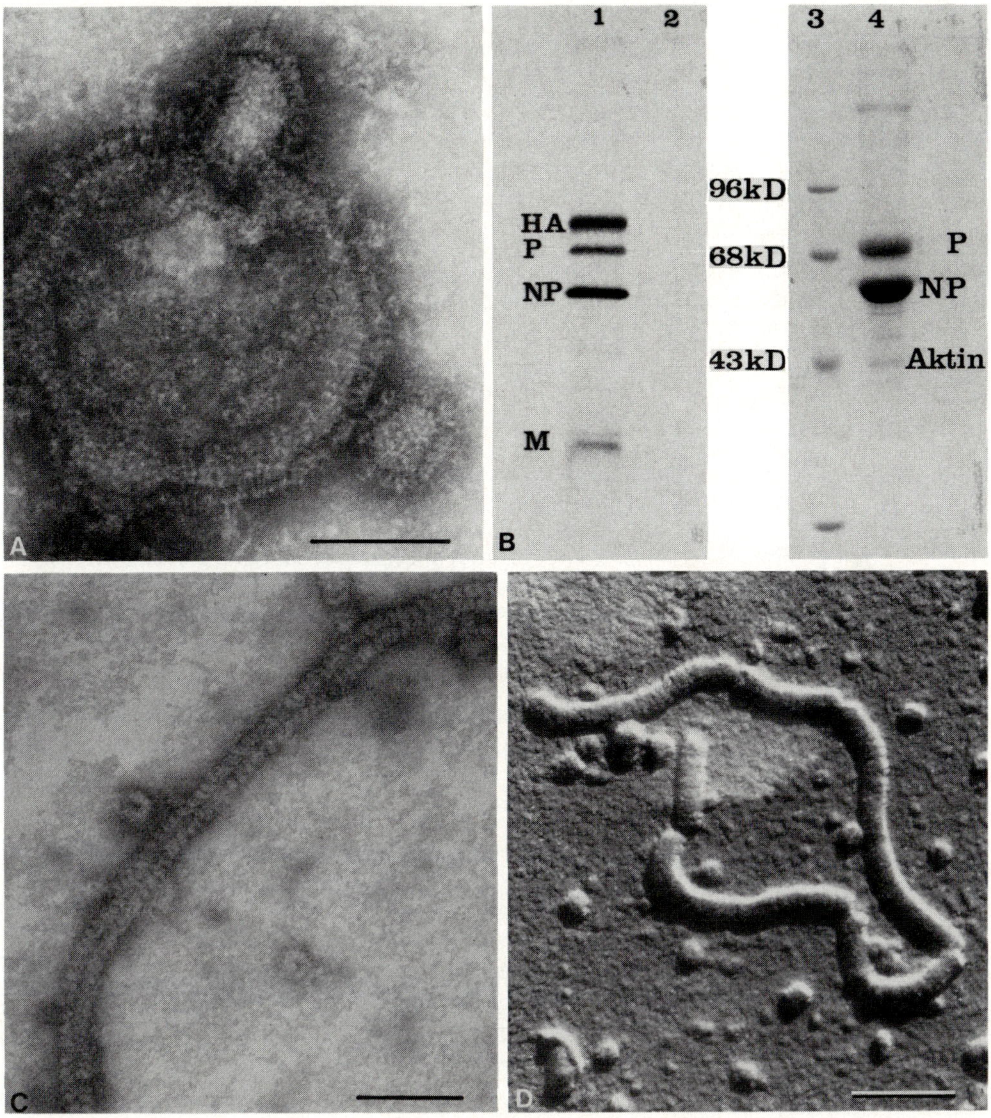

FIGURE 1. Structural and biochemical composition of measles virus. (A): Negatively stained virion; glycoprotein spikes project from the outer surface of the lipid bilayer. (B): Polypeptide composition of the virus; lane 1, polypeptides precipitated from ^{35}S-methionine labeled cells with a polyclonal anti-measles antiserum (for abbreviations, see text); lane 2, precipitation with a nonimmune serum (control); lane 3, molecular weight markers; lane 4, polypeptide composition of purified nucleocapsids (Coomassie blue staining). (C): Negatively stained nucleocapsids. (D): Nucleocapsid, freeze dried and Pt/C shadowed. Bars = 0.1 μm (A, D), 0.5 μm (C).

II. CHARACTERIZATION OF MEASLES VIRUS, USED AS A MODEL SYSTEM

A. PROTEIN COMPOSITION OF THE VIRUS PARTICLE

Enveloped RNA viruses maturating at the plasma membrane of infected cells and leaving the cell in a budding process, such as myxo-, paramyxo-, rhabdo-, retro-, and alphaviridae, exhibit two morphological subunits, the viral nucleocapsid containing the virus genome, and the enclosing lipid envelope.[26-28]

The polypeptide analysis of measles virus, a member of the paramyxoviridae, reveals at least six virus-specific proteins, a large molecular weight protein (L), the hemagglutinin (HA), the phosphoprotein (P), the nucleoprotein (NP), the fusion protein (F) and the membrane protein (M) (Figure 1B, lanes 1 and 4).[29-31] Measles virus nucleocapsids contain the nucleoprotein as the main polypeptide component and the associated L- and P- polypeptides (Figure 1B, lane 4). The latter are part of the virus-specific RNA-polymerase complex.[32] The lipid envelope contains the transmembrane glycoproteins HA and F and can be separated from the nucleocapsid by incubation with nonionic detergents.[29,30] The M-protein remains associated with the nucleocapsid in this procedure, but can be separated from it by subsequent incubation with high salt concentrations.[30] The protein composition of the lipid envelope seems to be exclusively virus-specific, whereas its lipid composition reflects that of the plasma membrane the virion is budding from.[33,34]

B. STRUCTURAL FEATURES OF THE VIRUS PARTICLE

Measles virus nucleocapsids are about 18 nm wide and 1 μm long and contain a single stranded RNA molecule of negative polarity, coding for the six proteins mentioned above. Transmission electron micrographs of negatively stained preparations reveal a helical arrangement of nucleoprotein subunits around a central channel of 5 nm in width (Figure 1C).[35,36] The tightness of the helically arranged nucleoprotein subunits can significantly be influenced by ionic conditions.[37] For instance, at low ionic conditions the nucleocapsid becomes loosely coiled. Platinum/carbon (Pt/C) shadowed samples show the three-dimensional configuration of the nucleocapsid and point to a high flexibility along its longitudinal axis (Figure 1D).

On negatively stained preparations of whole virions (Figure 1A), spikes of about 12 nm in length are visible on the surface of the lipid envelope. These spikes represent oligomers of the two transmembrane viral glycoproteins HA and F. The inner face of the lipid envelope is probably coated with the M-polypeptide which links the nucleocapsid to the envelope.[38]

C. ASSEMBLY OF VIRAL SUBUNITS ON THE PLASMA MEMBRANE

The synthesis and post-translational processing of the viral glycoproteins follow the same cellular pathways normally used by cellular glycoproteins. The viral glycoproteins are synthesized on intracellular membranes, oligomerized minutes after the synthesis, and passed through the Golgi-system. They are post-translationally modified by glycosylation, oligosaccharide trimming, and acylation.[39-42] Maturation of measles virus hemagglutinin takes about 5 h. and is a relatively slow process compared with that of vesicular stomatitis (VSV) and Sindbis virus glycoproteins.[43-45] Transport of the viral glycoproteins to the plasma membrane probably occurs without involvement of clathrin.[46]

The synthesis and assembly of viral nucleocapsids is less well characterized. Most probably, the nucleocapsids assemble in the cytoplasm and are transferred to the plasma membrane in ways independent from glycoprotein transport.[47] The nucleocapsids then attach to the protoplasmic surface of the plasma membrane and selectively associate with the transmembrane viral glycoproteins. This linkage is mediated by the M-protein, which therefore plays a key role in virus assembly.[48] Finally, the nucleocapsid-envelope complexes protrude from the cell surface and are released into the supernatant culture fluid, a process designated as budding.

Characteristically, measles virus infection does not lead to an early shutoff of the host cell protein synthesis and persistent infections showing no cytopathic effect, but continuously producing virus particles can easily be established in cell cultures.[49]

III. CELL SURFACE CHANGES CONNECTED WITH VIRUS BUDDING

The morphogenesis of enveloped viruses at the plasma membrane of infected cells leads

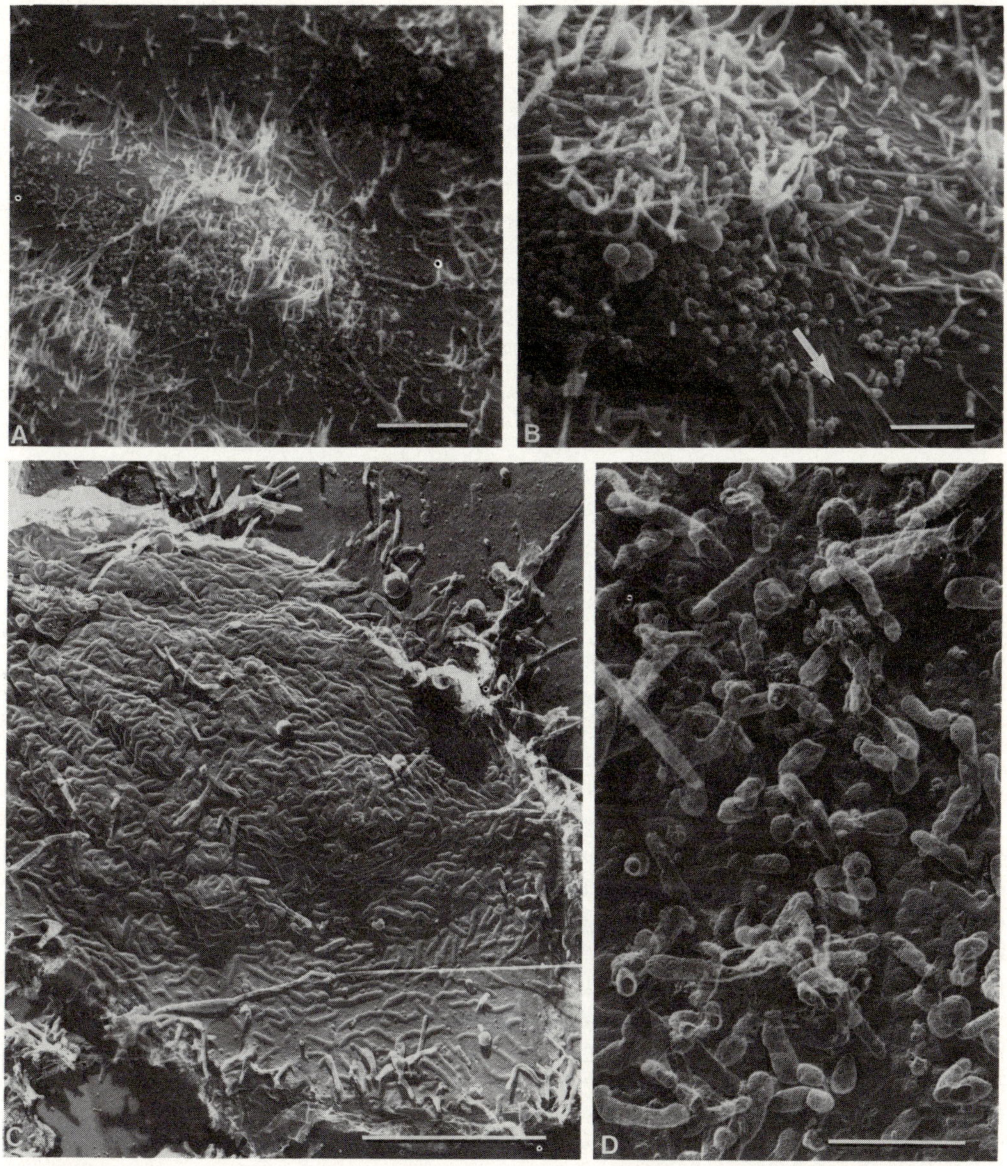

FIGURE 2. Budding of measles virus at the plasma membrane and its modification by phenothiazines. (A and B): Scanning electron micrographs of an infected HeLa cell; budding virions co-localize with cellular microspikes; strand-like configurations (arrows). (C): Infected HeLa cell incubated with trifluoperazine; microspikes and budding virions are absent; strand-like configurations have accumulated. (D): Reappearance of virus budding after transfer into drug-free culture fluid. Bars = 5 μm (A, C), 2 μm (B), 1 μm (D). (Figure 2C and 2D from Bohn, W. et al., *Virology,* 130, 44, 1983. With permission.)

to characteristic antigenic and morphological alterations at the cell surface. Scanning electron micrographs of measles virus infected cells, for instance, reveal two virus-specific structures at the cell surface, namely (1) strand-like configurations of about 1 μm in length, corresponding to the length of viral nucleocapsids underlying the plasma membrane, and (2) spherical structures with a diameter of about 200 nm (Figure 2A and B). Obviously, spherical particles predominantly colocalize with cellular microspikes, whereas the strand-like structures preferentially accumulate on areas free of microspikes. Both virus-specific structures can easily be distinguished from cellular protrusions by immunogold-labeling with antibodies to viral glycoproteins, combined with Pt/C replication of the cell surface.[50-54]

The appearance of virus-specific cell surface protrusions may indicate *per se* an involvement of cytoskeletal structures in virus morphogenesis, because changes of the cell shape are known to be controlled by the interaction between the plasma membrane and the cytoskeleton. More direct evidence for a participation of cytoskeletal elements in this process was obtained with two main experimental approaches. First, drugs interfering with the function of distinct cytoskeletal elements or influencing the cell surface morphology were tested with regard to their effect on virus budding and virus release. This approach also helped to identify individual stages in virus morphogenesis at the plasma membrane. Secondly, the structural composition of virus-specific protrusions and their association with cytoskeletal components was determined with ultrastructural and immunocytochemical investigations of isolated cytoskeletons and permeabilized cells.

IV. DRUG INDUCED MODIFICATIONS OF VIRUS BUDDING

Cytochalasins, a group of fungal metabolites, were most frequently used to determine an involvement of the cytoskeleton in virus morphogenesis. On cultured cells these drugs were shown to alter the cell surface morphology and to induce cell rounding, retraction, disruption of stress filaments, and aggregation of actin.[55] *In vitro* experiments indicated that cytochalasins bind to the barbed end of actin filaments and thereby prevent filament elongation, but also dissociation of actin monomers from this end.[56,57] These *in vitro* results suggested that the effects of cytochalasins on cultured cells result from their interference with the microfilament system, and that these drugs might be specific probes for determining the involvement of actin in complex cellular processes.

Cytochalasins were shown to interfere with the morphogenesis of various enveloped viruses, such as retroviridae,[58-60] paramyxoviridae,[61] poxviridae,[62] baculoviridae,[63] and iridoviridae.[64] Virus budding was, for instance, efficiently impaired by cytochalasin D in cells infected with the murine mammary tumor virus (MMTV), or with measles virus.[60,61] This could be correlated with a reduced amount of extracellular virus, and an increased titer of cell bound infectivity in infected cultures. The drugs most probably interfered with late steps in virus maturation at the plasma membrane, since they had little or no effect on virus polypeptide synthesis, cellular metabolism, and transport of virus components to the plasma membrane. Suggestions were made that cytochalasins impaired the interaction between viral structures and the cytoskeleton, normally required for virus release.

As a second characteristic effect, cytochalasins were shown to induce an aggregation of virus-specific antigens at the cell surface, and these patches colocalized with aggregations of actin. This effect was also noticed with cells infected with measles virus, and suggested a direct linkage between viral structures and actin filaments.[9,65] In contrast, cytochalasins had no effect on the release of influenza and vesicular stomatitis virus, arguing against a participation of actin filaments in their replication cycle.[66,67] However, this observation is not compatible with the fact that these drugs alter the distribution of these budding virions at the surface of infected cells.[68]

Drugs interfering with the microtubular system were also shown to decrease the amount of extracellular virus. On cells infected with retroviruses, viral protrusions were not visible under these conditions.[69,70] This effect most probably resulted from an interference of these drugs with the transport of virus-specific subunits to the plasma, rather than from a direct effect on virus budding itself.

As shown in our laboratory, the application of phenothiazines may represent another experimental approach to analyze the steps of virus budding at the plasma membrane, and to determine a possible involvement of cytoskeletal structures in this process.[71] Addition of the phenothiazine derivates trifluoperazine or chlorpromazine to the culture medium of HeLa cells infected with measles virus led to a complete disappearance of spherical virions from the plasma membrane, and concomitantly, cellular microspikes were no longer visible. Instead, strand-like virus-specific structures accumulated and finally covered the entire cell surface (Figure 2C). After transfer of these cells into drug-free medium, the microspikes and budding virions gradually reappeared and the normal cell surface morphology was restored within one h. (Figure 2D). Surface replicas indicated that the strand-like configurations elevated at one end from the plasma membrane and formed stub-like protrusions on the cell surface.

The inhibition of measles virus budding by phenothiazines at least did not result from an incomplete virus polypeptide synthesis or an incomplete virus assembly, as shown by immunoprecipitation and intracellular immunocytochemical staining. Furthermore, no additional protein synthesis was necessary for the genesis of viral protrusions after drug removal, since their reappearance could not be prevented by cycloheximide.

Further investigations suggested that the phenothiazines impaired the interaction of viral structures with microfilaments. This was based on the fact that cytochalasin induced redistribution of actin into patches in phenothiazine treated cells, but viral antigens at the cell surface did not comigrate (Figure 3).

The electron microscope findings indicated that phenothiazines were useful drugs to arrest and synchronize the formation of measles virus protrusions at the cell surface. The identical behavior of microspikes and viral protrusions in these experiments also supported the assumption that similar mechanisms were responsible for the formation of both structures. However, difficulties were encountered with determining the exact mechanisms of these drug effects. Initially, the effects of phenothiazines on complex cellular processes were interpreted as to result from an inhibition of calmodulin.[72,73] However, unspecific side effects of these drugs do not seem to justify this interpretation.[74,75]

V. STRUCTURAL ANALYSIS OF VIRUS-CYTOSKELETON INTERACTION

A. RETENTION OF VIRAL COMPONENTS ON CYTOSKELETONS

The application of drugs described above was shown to give first indications for a role of plasma membrane-cytoskeleton interactions in the morphogenesis of enveloped viruses. Further detailed information about the structures involved in this process were obtained by biochemical and ultrastructural investigations of isolated cytoskeletons. Cytoskeletons of cultured cells usually are prepared by extraction with nonionic detergents leaving behind insoluble structures mainly comprised of microtubuli, intermediate, and actin filaments.[76–82] As a characteristic property these cytoskeletons were shown to retain distinct plasma membrane bound components, such as cellular receptors,[83,84] histocompatibility antigen,[85] and glycoproteins,[86,87] suggesting a role of this linkage in cellular functions.

Similarly, virus-specific antigens and substructures of enveloped viruses were found on cytoskeletons of infected cells after extraction with nonionic detergents. For instance, glycoproteins of Sindbis virus, a member of the alphaviridae, were almost quantitatively retained by the "outer lamina" of detergent extracted cells.[1] The virus core precursor polypeptide pr65^{gag} of the Rauscher murine leukemia virus remained in the detergent resistent fraction of infected cells,[88] an observation which is consistent with the finding that budding retrovirions are in close association with thin filaments underlying the plasma membrane.[89] This linkage between the precursor polypeptide and the cytoskeleton is probably an essential step in retrovirus assembly and release.

Excellent studies were performed on the interaction between vaccinia virus particles and the cytoskeleton in the course of virus release from the plasma membrane. High voltage electron microscopy (see Chapter 17 by Ris and Pawley) of critical point dried cells indicated that vaccinia virions are released from microspike-like protrusions[90] which contain a core

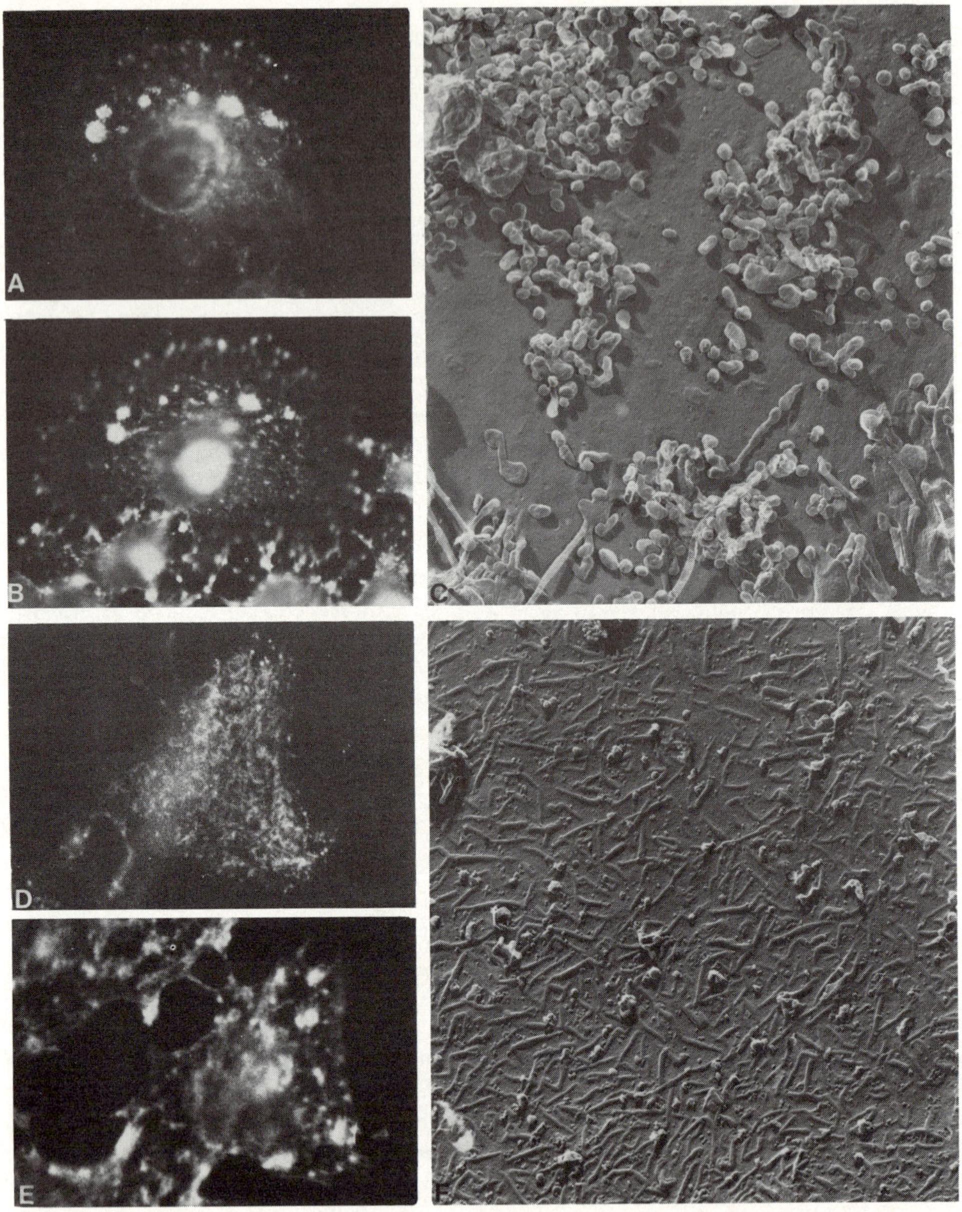

FIGURE 3. Comigration of virus antigen with actin is disrupted in phenothiazine treated cells. On infected HeLa cells solely incubated with cytochalasin B the cell surface hemagglutinin (A) largely colocalizes with actin (B) in double immunofluorescence microscopy. Surface replicas (C) reveal aggregates of budding virions (bar = 2 μm). On cells incubated with trifluoperazine and cytochalasin the cell surface hemagglutinin (D) does not comigrate with actin patches (E); surface replicas (F) reveal no aggregation of virusspecific strands. Frame size = 110 × 80 μm (A, B), 55 × 40 μm (D, E). Bars = 2 μm (C, F). (From Bohn, W. et al. *Virology*, 130, 44, 1983. With permission.)

of microfilaments and the actin filament cross-linking proteins α-actinin, filamin, and fimbrin.[91] The appearance of these protrusions was found to depend on reaching a distinct step in virus maturation.[92,93]

Numerous microspikes were also shown to appear at the surface of cells infected with frog virus,[64] an iridovirus; however, these microspikes were not further analyzed concerning their composition, thus giving only indirect evidence for an involvement of actin filaments.

As shown by our laboratory, significant amounts of measles virus polypeptides also resisted the extraction of infected cells with nonionic detergents.[94] Furthermore, the spatial distribution of plasma membrane bound viral glycoproteins was hardly affected by this procedure as observed in immunofluorescence. A radioactive binding assay indicated that about 80% of the plasma membrane associated viral glycoprotein remained in the cytoskeletal fraction. In addition, the nucleocapsid associated polypeptides were also present to a higher extent in this fraction compared with the supernatant lysis buffer. Transmission electron microscopy of Pt/C shadowed cytoskeletons revealed that measles virus structures, identified by immunogold-labeling with antibodies to hemagglutinin, were associated with cytoskeletal filaments comprising the outer part of the framework (Figure 4A, B). These filaments measured about 100 A in diameter and probably represented actin filaments. A specific binding of virus structures to actin filaments was also supported by immunofluorescence staining on cytoskeletons of cytochalasin treated cells. Cytochalasin induced the typical comigration of viral antigens with actin into patches (Figure 5), but left the organization of vimentin and cytokeratin unaffected in these cells.

The results supported the idea that measles virus structures associate with distinct cytoskeletal filaments, most probably actin filaments, during morphogenesis at the plasma membrane. The data also showed that this was not only the property of a minor subfraction of the virus, but seemed to be a normal step in measles virus morphogenesis. Furthermore, the references cited above indicate that this may generally represent a characteristic feature of enveloped viruses maturing at the plasma membrane.

However, data obtained with cytoskeletons are of limited value for the characterization of plasma membrane-cytoskeleton interactions when the lipid bilayer has been disrupted. Additional information about this interaction can be obtained by ultrastructural investigations of permeabilized cells.

B. INVESTIGATIONS OF PERMEABILIZED CELLS

Permeabilization of cells was frequently performed to get access to intracellular structures for immunocytochemical labeling in electron microscopy, without disrupting too much the organization of cellular compartments. This, for instance, was achieved by use of the nonionic detergents saponin or digitonin, both of which selectively react with cholesterol.[95,96] Since cholesterol is present at significantly higher concentrations in the plasma membrane than in intracellular membranes, this leads to a rather selective plasma membrane permeabilization.[97]

Permeabilization has also been used to achieve penetration of myosin fragments into cells for the identification of actin filaments in studies of plasma membrane-cytoskeleton interaction. The myosin fragments "heavy meromyosin" (HMM) and S_1 not only specifically bind to this filament type, but can also give information about the relative direction of actin filament growth.[98,99] Excellent micrographs of S_1 decorated actin filaments were demonstrated on shadowed cytoskeletons of brush border microvilli, showing that actin filaments terminate with their barbed, fast growing end at the tip of these structures.[15] In *Dictyostelium discoideum* both lateral and end-on associations were found between actin filaments and the plasma membrane with this procedure.[100]

Permeabilization, combined with HMM incubation, now was used in our laboratory to further characterize the role of actin filaments in measles virus morphogenesis at the plasma membrane.[94] For a controlled permeabilization the infected cells were incubated with tannic acid prior to reaction with the nonionic detergent, which prevented a complete solubilization of the plasma membrane. After incubation with HMM the cells were fixed, dehydrated with ethanol, and embedded. Ultrathin sections revealed that filaments decorated with HMM projected from a dense layer of submembraneous actin filaments into budding virus structures at the cell surface, and terminated with their barbed ends on viral nucleocapsids (Figure 4C). This for the first time not only gave direct evidence for a polar orientation of actin

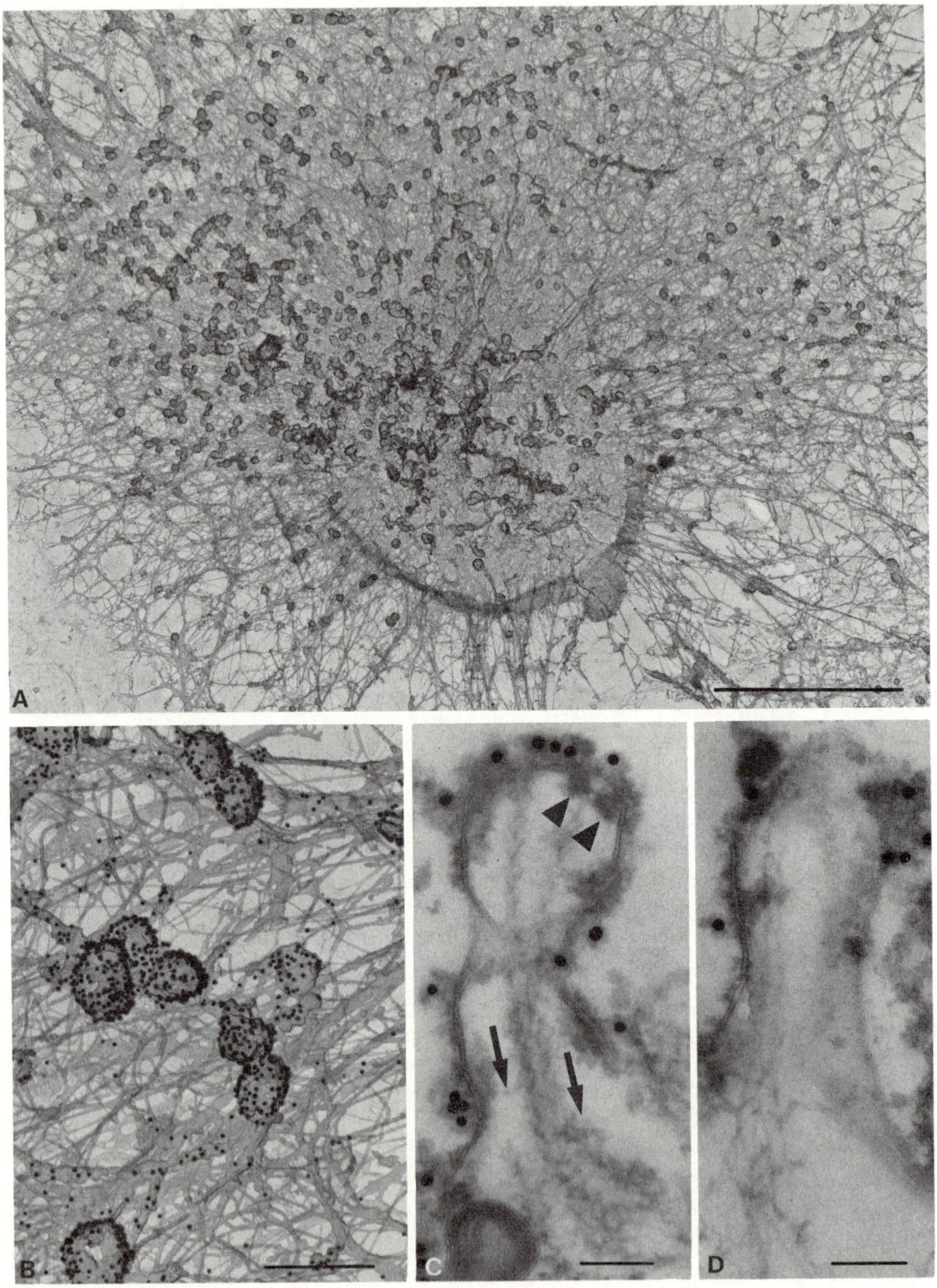

FIGURE 4. Association of measles virus with the cytoskeleton. (A, B): Carbon shadowed cytoskeletons of infected HeLa cells; virus structures were identified by immunogold labeling with antibodies to hemagglutinin. (C, D): Ultrathin sections of permeabilized cells. (C): Filaments decorated with heavy meromyosin (HMM) protrude with the barbed end into a budding virion; actin filament orientation (arrows); section profiles of viral nucleocapsids (arrowheads). (D): Control for specificity of HMM decoration; decoration is prevented by sodium pyrophosphate. Bars = 5 μm (A), 0.5 μm (B), 0.1 μm (C, D). (Figures 4C and 4D from Bohn, W. et al., *Virology*, 149, 91, 1986.)

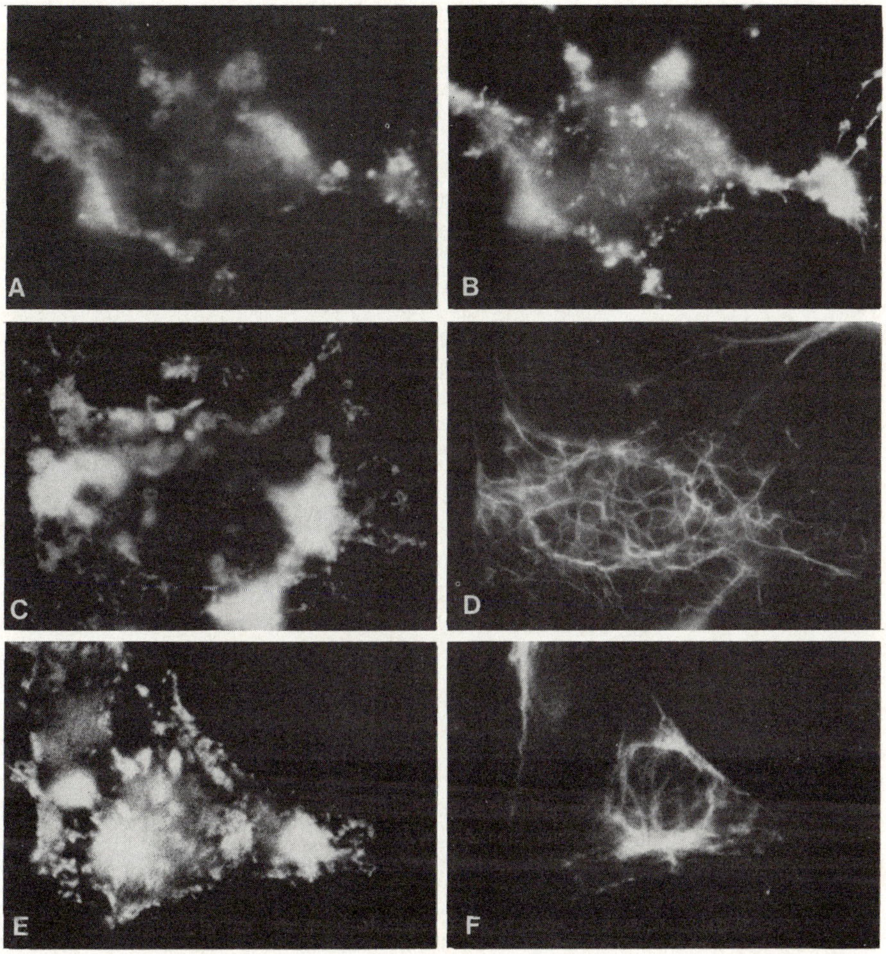

FIGURE 5. Double immunofluorescence labeling of cytoskeletons of cytochalasin treated cells. Measles virus hemagglutinin (A, C, E) has comigrated with actin (B) into patches; cytokeratin (B), and vimentin (F), exhibit no aggregation. Frame size = 47 × 33 μm. (From Bohn, W. et al., *Virology*, 149, 91, 1986.)

filaments in budding virions, but also showed that these actin filaments have the same orientation as is typical for cellular microspikes and brushborder microvilli.[14,15]

The morphological findings suggested that nucleocapsids may contain recognition signals for actin binding. This notion is supported by the fact that actin in purified paramyxovirions partitions with the nucleocapsid rather than with the envelope, upon treatment with nonionic detergents.[29] Biochemical data, showing cosedimentation of actin with the purified M-protein of a paramyxovirus,[101] and the finding that this protein is almost quantitatively retained in the detergent resistant fraction of measles virus infected cells,[94] let this polypeptide appear a putative candidate for interaction with actin.

The interpretation of the results obtained with cytoskeletons and permeabilized cells is, however, complicated by the fact that treatment with detergents leads to a significant loss of lipids and polypetides. About 45% of the total cellular protein is lost during extraction of cells with nonionic detergents.[1] Thus plasma membrane-cytoskeleton contacts demonstrated with these procedures may only represent a distinct type of interaction. These limitations may partially be overcome by analyzing specifically the structural composition of the protoplasmic surface of isolated plasma membranes.

C. THE STRUCTURAL ORGANIZATION OF THE PROTOPLASMIC SURFACE OF THE PLASMA MEMBRANE

Different technical approaches have been used to get access to the protoplasmic surface of the plasma membrane in electron microscopy.[102] Initially, cells were bound to a solid support by cationic compounds and then opened by mechanical stress, i.e., by squirting buffer on them.[103] After fixation, dehydration, critical point drying, and Pt/C shadowing, the protoplasmic surface of the plasma membrane could be visualized in the transmission electron microscope. This method has been used to demonstrate the structural and antigenic composition of the protoplasmic membrane surface of erythrocytes,[103] macrophages,[104] and cultured cells.[105,106]

Another technique for membrane isolation (''sandwich-technique'') consists in covering a cell monolayer grown on a glass coverslip with a second cationized coverslip, and separating them subsequently in buffer. The plasma membrane adherent to the cationized coverslip is then fixed, dehydrated, critical point dried, and Pt/C shadowed.[107,108] This procedure can be combined with immunogold labeling of cell surface antigens prior to membrane isolation.[107] In addition, a procedure was introduced for labeling antigens at the exoplasmic surface prior to membrane isolation, and subsequently, labeling antigens at the protoplasmic surface after membrane isolation.[108] This permits correlation of the distribution of antigens at both sides of the plasma membrane in one electron microscopic image. As an alternative method, dry-cleaving can be performed, i.e., isolating the plasma membrane of critical point dried cells with an adhesive tape.[86] The sandwich-, as well as the dry-cleaving technique, permit us to isolate selectively the apical part of the plasma membrane.

The sandwich-technique has been used for investigations on the protoplasmic surface of measles virus infected cells.[107,108] Measles virus nucleocapsids are visible as strand-like configurations on this surface of Pt/C shadowed membranes in the transmission electron microscope (Figure 6). Most remarkably the distribution of viral glycoproteins at the cell surface labeled prior to membrane isolation strongly coincides with that of nucleocapsids. However, no clear evidence could be presented so far for a characteristic organization of cytoskeletal elements on these membranes. In general, the amount of cytoskeletal elements remaining on membranes isolated with these methods proved to be considerably influenced by the intensity of squirting and washing.[102] On the other hand, any additional cytoskeletal elements remaining with the isolated membranes impede the structural resolution of the plasma membrane. Thus, this isolation procedure appears suitable for studies of the composition of the plasma membrane itself, but can give only limited information about membrane skeletons because it largely disrupts its complex three-dimensional organization.

VI. SUMMARY AND CONCLUSION

This article summarizes data concerning the interaction between the cytoskeleton and enveloped viruses in virus morphogenesis at the plasma membrane, the structural analysis of which is significantly hampered by its dynamic nature. Using measles virus as a model system, two strategies have been followed to get access to its regulating mechanisms.

1. We searched for drugs that influence the dynamic cytoskeleton-virus interaction by arresting this process at a distinct step. Each of the drug induced modifications were monitored by immunofluorescence and electron microscopy.
2. We analyzed the structural composition of isolated cellular compartments, cytoskeleton and plasma membrane, in order to obtain direct morphological evidence for an involvement of cytoskeletal elements in virus information at the plasma membrane.

For future investigations the use of virus mutants (with various defects in morphogenesis)

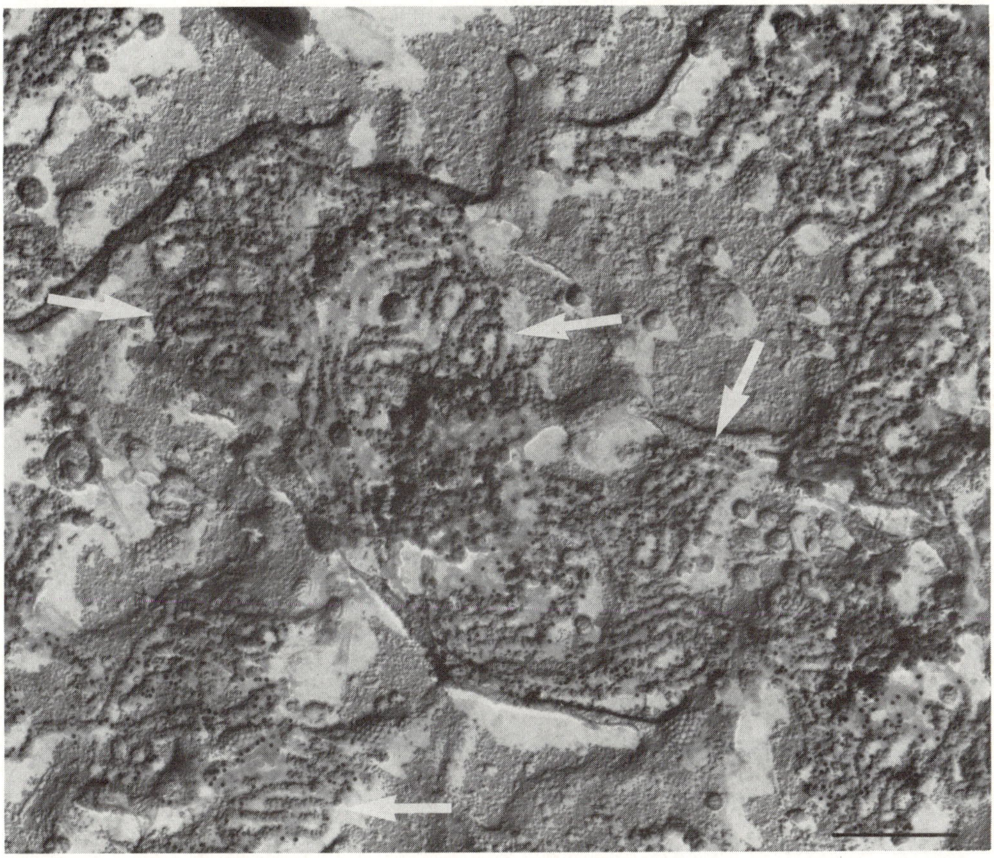

FIGURE 6. View onto the protoplasmic membrane surface of an isolated plasma membrane; measles virus nucleocapsids (arrows) on the protoplasmic membrane surface colocalize with immunogold labeled viral glycoproteins on the exoplasmic membrane surface; cytoskeletal filaments are not detectable, Bar = 0.5 μm. (Courtesy Dr. Rutter.)

and of cloned cell lines (differing in their cytoskeletal composition) seems to be a favorable experimental approach. Preliminary data obtained with subclones of the C_6 rat glioma cell line, showing either vimentin or no intermediate filaments at all, indicate that this filament-type plays no role, at least in measles virus budding, and support the assumption of a specific involvement of actin filaments in this process.[109]

The data obtained so far suggest that distinct proteins of enveloped viruses compete with cellular components in actin binding at the plasma membrane, and allow these viruses to use cellular mechanisms for a controlled morphogenesis. Thus, enveloped viruses may represent favorable models for further investigations on the complex structure and functions of membrane skeletons in nonmuscle cells.

ACKNOWLEDGMENTS

This work was supported by Gemeinnützige Hertie-Stiftung, Frankfurt/Main. The Heinrich-Pette-Institut is financially supported by Freie und Hansestadt Hamburg and Bundesministerium für Jugend, Familie, Frauen und Gesundheit. The author is grateful to G. Rutter, H. Hohenberg, and K. Mannweiler for extensive discussions on this subject.

REFERENCES

1. **Ben-Ze'ev, A., Duerr, A., Solomon, F., and Penman, S.,** The outer boundary of the cytoskeleton: a lamina derived from the plasma membrane, *Cell,* 17, 859, 1979.
2. **Luna, E. J., Fowler, V. M., Swanson, J., Branton, D., and Taylor, D. L.,** A membrane cytoskeleton from *Dictyostelium discoideum.* I. Identification and partial characterization of an actin-binding activity, *J. Cell Biol.,* 88, 396, 1981.
3. **Mescher, M., Jose, M., and Balk, S.,** Actin-containing matrix associated with the plasma membrane of the murine tumor and lymphoid cells, *Nature,* 289, 139, 1981.
4. **Branton, D., Cohen, C. M., and Taylor, J.,** Interaction of cytoskeletal proteins on the human erythrocyte membrane, *Cell,* 24, 24, 1981.
5. **Geiger, B.,** Membrane-cytoskeleton interaction, *Biochim. Biophys. Acta,* 737, 305, 1983.
6. **Schliwa, M.,** *The Cytoskeleton. An Introductory Survey,* Cell Biology Monographs, Vol. 13, Springer-Verlag, New York, 1986, 179.
7. **Moon, R. T. and McMahon, A. P.,** Composition and expression of spectrin-based membrane skeletons in nonerythroid cells, *BioEssay,* 7, 159, 1987.
8. **Apgar, J. R., Herrmann, S. H., Robinson, J. M., and Mescher, M. F.,** Triton X-100 extraction of P815 tumor cells: evidence for a plasma membrane skeleton structure, *J. Cell Biol.,* 100, 1369, 1985.
9. **Sundqvist, K. G. and Ehrnst, A.,** Cytoskeletal control of surface membrane mobility, *Nature,* 264, 226, 1976.
10. **Nicolson, G. L.,** Transmembrane control of the receptors on normal and transformed cells. I. Cytoplasmic influence over cell surface components. *Biochim. Biophys. Acta,* 457, 57, 1976.
11. **Bourguignon, L. Y. W. and Singer, S. J.,** Transmembrane interactions and the mechanism of capping of surface receptors by their specific ligand, *Proc. Natl. Acad. Sci. U.S.A.,* 74, 5031, 1978.
12. **Carraway, C. A., Jung, G., and Carraway, K. L.,** Isolation of actin-containing transmembrane complexes from ascites adenocarcinoma sublines having mobile and immobile receptors, *Proc. Natl. Acad. Sci. U.S.A.,* 80, 430, 1983.
13. **Bologna, M., Allen, R., and Dulbecco, R.,** Organization of cytokeratin bundles by desmosomes in rat mammary cells, *J. Cell Biol.,* 102, 560, 1986.
14. **Mooseker, M. S. and Pollar, T. T.,** Organization of an actin filament-membrane complex: filament polarity and membrane attachment in the microvilli of intestinal epithelial cells, *J. Cell Biol.,* 67, 725, 1975.
15. **Hirokawa, N., Tilney, L. G., Fujiwara, K., and Heuser, J. E.,** Organization of actin, myosin, and intermediate filaments in the brush border of intestinal epithelial cells, *J. Cell Biol.,* 94, 425, 1982.
16. **Lewis, J. C.,** Cytoskeleton in platelet function, in *Cell and Muscle Motility,* Vol. 5, *The Cytoskeleton,* Shay, J. W., Ed., Plenum Press, New York, 1984, 341.
17. **Fox, J. E. B. and Boyles, J. K.,** The membrane skeleton — a distinct structure that regulates the function of cells. *BioEssays,* 8, 14, 1987.
18. **Brown, S. S. and Petzold, A. S.,** Using antibodies against *Dictyostelium* membranes to identify an actin-binding membrane protein, *J. Cell Biol.,* 104, 513, 1987.
19. **Marchesi, V. T.,** Stabilizing infrastructure of cell membranes. *Ann. Rev. Cell Biol.,* 1, 513, 1985.
20. **Liu, S. C., Derick, L. H., and Palek, J.,** Visualization of the hexagonal lattice in the erythrocyte membrane skeleton, *J. Cell Biol.,* 104, 527, 1987.
21. **Burridge, K., Kelley, T., and Mangeat, P.,** Nonerythrocyte spectrin: actin-membrane attachment proteins occurring in many cell types, *J. Cell Biol.,* 95, 478, 1982.
22. **Glenney, J. R. and Glenney, P.,** Fodrin is the general spectrin-like protein in most cells whereas spectrin and the TW protein have a restricted distribution, *Cell,* 34, 503, 1983.
23. **Bretscher, A. and Weber, K.,** Villin is a major protein of the microvillus cytoskeleton which binds both G- and F-actin in a calcium-dependent manner, *Cell,* 20, 839, 1980.
24. **Glenney, J. R., Kaulfus, P., Matsudaira, P., and Weber, K.,** F-actin binding and bundling properties of fimbrin, a major cytoskeletal protein of microvillus core filaments, *J. Biol. Chem.,* 256, 9283, 1981.
25. **Caluccio, L. M. and Bretscher, A.,** Calcium-regulated cooperative binding of the microvillar 110K-calmodulin complex to F-actin: formation of decorated filaments, *J. Cell Biol.,* 105, 325, 1987.
26. **Dubois-Dalcq, M., Holmes, K. V., and Rentier, B.,** *Assembly of Enveloped RNA Viruses,* Springer-Verlag, New York, 1984.
27. **Matsumoto, T.,** Assembly of paramyxoviruses, *Microbiol. Immunol.,* 26, 285, 1982.
28. **Simons, K. and Garoff, H.,** The budding mechanisms of enveloped animal viruses, *J. Gen. Virol.,* 50, 1, 1980.
29. **Tyrell, D. L. J. and Norrby, E.,** Structural polypeptides of measles virus, *J. Gen. Virol.,* 37, 233, 1978.
30. **Stallcup, K.C., Wechsler, S. L., and Fields, B. N.,** Purification of measles virus and characterization of subviral components, *J. Virol.,* 30, 166, 1979.

31. **Vainionpää, R., Ziola, B., and Salmi, A.,** Measles virus polypeptides in purified virions and in infected cells, *Acta Pathol. Microbiol. Scand. Sect. B,* 86, 379, 1978.
32. **Hamaguchi, M., Yoshida, T., Nishikawa, K., Noruse, H., and Nagai, Y.,** Transcriptive complex of Newcastle disease virus. I. Both L and P proteins are required to constitute an active complex, *Virology,* 128, 105, 1983.
33. **Rott, R. and Klenk, H. -D.,** Structure and assembly of viral envelope, in *Virus Infection and The Cell Surface,* Poste, G. and Nicolson, G. L., Eds., North-Holland, New York, 1977, 47.
34. **Barnhart, E. R. and Ash, R. J.,** Similarities and differences between viral and cellular membranes, *Prog. Med. Virol.,* 25, 89, 1979.
35. **Waters, D. J., Hersh, R. T., and Bussell, R. H.,** Isolation and characterization of measles nucleocapsids from infected cells, *Virology,* 48, 278, 1972.
36. **Lund, G. A., Tyrell, D. L. J., Bradley, R. D., and Scraba, D. G.,** The molecular length of measles virus RNA and the structural organization of measles nucleocapsids, *J. Gen. Virol.,* 65, 1535, 1984.
37. **Heggeness, M. H., Scheid, A., and Choppin, P. W.,** Conformation of the helical nucleocapsid of paramyxoviruses and vesicular stomatitis virus: reversible coiling and uncoiling induced by changes in salt concentration, *Proc. Natl. Acad. Sci. U.S.A.,* 77, 2631, 1980.
38. **Yoshida, T., Nagai, Y., Yoshii, S., Maeno, K., Matsumoto, T., and Hoshino, M.,** Membrane (M) protein of HVJ (Sendai virus): its role in virus assembly, *Virology,* 71, 143, 1976.
39. **Katz, F. N. and Lodish, H. F.,** Transmembranal biogenesis of the vesicular stomatitis virus glycoprotein, *J. Cell Biol.,* 80, 416, 1979.
40. **Bergmann, J. E., Tokuyasu, K. T., and Singer, S. J.,** Passage of an integral membrane protein, the vesicular stomatitis virus glycoprotein, through the Golgi apparatus en route to the plasma membrane, *Proc. Natl. Acad. Sci. U.S.A.,* 76, 4350, 1981.
41. **Wickner, W. T. and Lodish, H. F.,** Multiple mechanisms of protein insertion into and across membranes, *Science,* 230, 400, 1985.
42. **Schlesinger, M. J. and Schlesinger, S.,** Domains of virus glycoproteins, *Adv. Virus Res.,* 33, 1, 1987.
43. **Kohama, T., Sato, T. A., Kobune, F., and Sugiura, A.,** Maturation of measles virus hemagglutinin glycoprotein, *Arch. Virol.,* 85, 257, 1985.
44. **Morrison, T. G. and Ward, L. J.,** Intracellular processing of the vesicular stomatitis virus glycoprotein and the Newcastle disease virus hemagglutinin-neuraminidase glycoprotein, *Virus Res.,* 1, 225, 1984.
45. **Schlesinger, M. J. and Schlesinger, S.,** Large-molecular-weight precursors of Sindbis virus proteins, *J. Virol.,* 11, 1013, 1973.
46. **Griffiths, G., Pfeiffer, S., Simons, K., and Matlin, K.,** Exit of newly synthesized membrane proteins from the trans cisterna of the Golgi complex to the plasma membrane, *J. Cell Biol.,* 101, 949, 1985.
47. **Hamaguchi, M., Nishikawa, K., Toyoda, T., Yoshida, T., Hanaichi, T., and Nagai, Y.,** Transcriptive complex of Newcastle disease virus. II. Structural and functional assembly associated with the cytoskeletal framework, *Virology,* 147, 295, 1985.
48. **Shimizu, K. and Ishida, N.,** The smallest protein of Sendai virus: its candidate function of binding nucleocapsid to envelope, *Virology,* 427, 1975.
49. **Fraser, K. B. and Martin, S. J.,** Measles virus and its biology, in *Experimental Virology,* Vol. 1, Tinsley, T. W. and Brown, F., Eds., Academic Press, New York, 1978, Chapter 7.
50. **Hohenberg, H., Bohn, W., Rutter, G., and Mannweiler, K.,** Plasma membranes detected by replica techniques, in *Science of Biological Specimen Preparation for Microscopy and Microanalysis,* Müller, M., Becker, R. P., Boyde, A., and Wolosewick, J. J., Eds., SEM Inc., AMF O'Hare, Chicago, IL, 1986, 235.
51. **Hohenberg, H.,** Replica preparation techniques in immuno-gold cytochemistry, in *Immuno-Gold Probes in Cell Biology,* Verkleij, A. and Leunissen, J. L., Eds., CRC Press, Boca Raton, in press.
52. **Mannweiler, K., Hohenberg, H., Bohn, W., and Rutter, G.,** Protein-A gold particles as markers in replica immunocytochemistry: high resolution electron microscope investigations of plasma membrane surfaces, *J. Microsc.,* 126, 145, 1982.
53. **Mannweiler, K., Bohn, W., and Rutter, G.,** Labeling of viral antigens in replica technique, in *Immuno-Gold Probes in Cell Biology,* Verkleij, A. and Leunissen, J. L., Eds., CRC Press, Boca Raton, in press.
54. **Bohn, W., Rutter, G., and Mannweiler, K.,** Production of monoclonal antibodies to measles virus proteins by immunization of mice with heated and detergent treated antigens, *Virology,* 116, 368, 1982.
55. **Cooper, J. A.,** Effects of cytochalasin and phalloidin on actin, *J. Cell Biol.,* 105, 1473, 1987.
56. **Brown, S. S. and Spudich, J. A.,** Cytochalasin inhibits the rate of elongation of actin filament fragments, *J. Cell Biol.,* 83, 657, 1979.
57. **Brown, S. S. and Spudich, J. A.,** Mechanism of action by cytochalasin: evidence that it binds to actin filament ends, *J. Cell Biol.,* 88, 487, 1981.
58. **Panem, S.,** Cell cycle-dependent inhibition of Kirsten sarcoma-leukemia virus release by cytochalasin B, *Virology,* 76, 146, 1977.

59. **Mousa, G. Y., Trevithick, J. R., Bechberger, J., and Blair, D. G.,** Cytochalasin D induces the capping of both leukaemia viral proteins and actin in infected cells, *Nature,* 274, 808, 1978.
60. **Maldarelli, F., King, N. W., and Yagi, M. J.,** Effects of cytoskeletal disrupting agents on mouse mammary tumor virus replication, *Virus Res.,* 7, 281, 1987.
61. **Stallcup, K. C., Raine, C. S., and Fields, B. N.,** Cytochalasin B inhibits the maturation of measles virus, *Virology,* 124, 59, 1983.
62. **Payne, L. G. and Kristensson, K.,** The effect of cytochalasin D and monensin on enveloped vaccinia virus release, *Arch. Virol.,* 74, 11, 1982.
63. **Volkman, L. E., Goldsmith, P. A., and Hess, R. T.,** Evidence for microfilament involvement in budded Autographa californica nuclear polyhedrosis virus production, *Virology,* 156, 32, 1987.
64. **Murti, K. G., Chen, M., and Goorha, R.,** Interaction of frog virus 3 with the cytomatrix. III. Role of microfilaments in virus release, *Virology,* 142, 317, 1985.
65. **Tyrell, D. L. J. and Ehrnst, A.,** Transmembrane communication in cells chronically infected with measles virus, *J. Cell Biol.,* 81, 396, 1979.
66. **Griffin, J. A. and Compans, R. W.,** Effect of cytochalasin B on the maturation of enveloped viruses, *J. Exp. Med.,* 150, 379, 1979.
67. **Gentry, N. and Busserau, F.,** Is cytoskeleton involved in vesicular stomatitis virus reproduction?, *J. Virol.,* 34, 777, 1980.
68. **Brown, J. C. and Salomonsky, N. L.,** Site-specific maturation of enveloped viruses in L-cells treated with cytochalasin B, *J. Cell Biol.,* 100, 357, 1985.
69. **Heine, U. I., Demsey, A. E., Tucker, R. W., and Bykowsky, A. F.,** Intracellular type A retrovirus movement associated with an intact microtubule system, *J. Gen. Virol.,* 66, 275, 1985.
70. **Satake, M. and Luftig, R. B.,** Microtubule-depolymerizing agents inhibit Moloney murine leukaemia virus production, *J. Gen. Virol.,* 58, 339, 1982.
71. **Bohn, W., Rutter, G., Hohenberg, H., and Mannweiler, K.,** Inhibition of measles virus budding by phenothiazines, *Virology,* 130, 44, 1983.
72. **Merrit, J. E., Tomlinson, S., and Brown, B. L.,** Phenothiazines inhibit prolactin secretion *in vitro*. A possible role for calmodulin in stimulus-secretion coupling in the pituitary, *FEBS Lett.,* 135, 107, 1981.
73. **Nelson, G. A., Andrews, M. L., and Karnovsky, M. J.,** Participation of calmodulin in immunoglobulin capping, *J. Cell Biol.,* 95, 771, 1982.
74. **Corps, A. N., Hesketh, T. R., and Metcalf, J. C.,** Limitations on the use of phenothiazines and local anesthetics as indicators of calmodulin function in intact cells, *FEBS Lett.,* 138, 280, 1982.
75. **McLaughlin, S. and Whitaker, M.,** Cations that alter surface potentials of lipid bilayers increase the calcium requirement for exocytosis in sea urchin eggs, *J. Physiol.,* 396, 189, 1988.
76. **Brown, S., Levinson, W., and Spudich, J. K.,** Cytoskeletal elements of duck embryo fibroblasts revealed by detergent extraction, *J. Supramol. Struct.,* 5, 119, 1976.
77. **Trotter, J. A., Foerder, B. A., and Keller, J. M.,** Intracellular fibres in cultured cells: analysis by scanning and transmission electron microscopy and by SDS-polyacrylamide gel electrophoresis, *J. Cell. Sci.,* 31, 369, 1978.
78. **Small, J. V. and Celis, J. E.,** Direct visualization of the 10nm (100A)-filament network in whole and enucleated cultured cells, *J. Cell Sci.,* 31, 393, 1978.
79. **Webster, R. E., Osborn, M., and Weber, K.,** Visualization of the same PtK2 cytoskeletons by both immunofluorescence and low power electron microscopy, *Exp. Cell Res.,* 117, 47, 1978.
80. **Schliwa, M. and Blerkom, J. V.,** Structural interaction of cytoskeletal components, *J. Cell Biol.,* 90, 222, 1981.
81. **Wolosewick, J. J. and Porter, K. R.,** Microtrabecular lattice of the cytoplasmic ground substance. Artefact or reality, *J. Cell Biol.,* 82, 114, 1979.
82. **Heuser, J. E. and Kirschner, M. W.,** Filament organization revealed in platinum replicas of freeze-dried cytoskeletons, *J. Cell Biol.,* 86, 212, 1980.
83. **Lacy, B. E. and Underhill, C. B.,** The hyaluronate receptor is associated with actin filaments, *J. Cell Biol.,* 105, 1395, 1987.
84. **Wiegant, F. A. C., Blok, F. J., Defize, L. H. K., Linnemans, W. A. M., Verkleij, A. J., and Boonstra, J.,** Epidermal growth factor receptors associated to cytoskeletal elements of epidermoid carcinoma (A431) cells, *J. Cell Biol.,* 103, 87, 1986.
85. **Koch, G. L. E. and Smith, M. J.,** An association between actin and the major histocompatibility antigen H-2, *Nature,* 273, 274, 1978.
86. **Roos, E., Spiele, H., Feltkamp, C. A., Huisman, H., Wiegant, F. A. C., Traas, J., and Mesland, D. A. M.,** Localization of cell surface glycoproteins in membrane domains associated with the underlying filament network, *J. Cell Biol.,* 101, 1817, 1985.
87. **Tarone, G., Ferracini, R., Galetto, G., and Comoglio, P.,** A cell surface integral membrane glycoprotein of 85,000 mol wt (gp85) associated with Triton X-100-insoluble cell skeleton, *J. Cell Biol.,* 99, 512, 1984.

88. **Edbauer, C. A. and Naso, R. B.,** Cytoskeleton-associated pr65^{gag} and retrovirus assembly, *Virology,* 130, 415, 1983.
89. **Damsky, C. H., Sheffield, J. B., Tuszynski, G. P., and Warren, L.,** Is there a role for actin in virus budding?, *J. Cell Biol.,* 75, 593, 1977.
90. **Stokes, G. V.,** High-voltage electon microscope study of the release of vaccinia virus from whole cells, *J. Virol.,* 18, 636, 1976.
91. **Krempien, U., Schneider, L., Hiller, G., Weber, K., Katz, E., and Jungwirth, C.,** Conditions for pox virus-specific microvilli formation studied during synchronized virus assembly, *Virology,* 113, 556, 1981.
92. **Hiller, G., Weber, K., Schneider, L., Parajsz, C., and Jungwirth, Ch.,** Interaction of assembled progeny pox viruses with the cellular cytoskeleton, *Virology,* 98, 142, 1979.
93. **Hiller, G., Jungwirth, Ch., and Weber, K.,** Fluorescence microscopical analysis of the life cycle of vaccinia virus in chick embryo fibroblasts. Virus-cytoskeleton interaction, *Exp. Cell Res.,* 132, 81, 1981.
94. **Bohn, W., Rutter, G., Hohenberg, H., and Mannweiler, K.,** Involvement of actin filaments in budding of measles virus: studies on cytoskeletons of infected cells, *Virology,* 149, 91, 1986.
95. **Ohtsuki, J., Manzi, R. M., Palade, G. E., Jamieson, J. D.,** Entry of macromolecular tracers into cells fixed with low concentrations of aldehydes, *Biol. Cell,* 31, 119, 1978.
96. **Bohn, W.,** A fixation method for improved antibody penetration in electron microscopical immunoperoxidase studies, *J. Histochem. Cytochem.,* 26, 293, 1978.
97. **Elias, P. M., Friend, D. S., and Goerke, J.,** Membrane sterol heterogeneity. Freeze- fracture detection with saponins and filipin, *J. Histochem. Cytochem.,* 27, 1247, 1979.
98. **Huxley, H. E.,** Electron microscope studies on the structure and of natural and synthetic protein filaments from striated muscle, *J. Mol. Biol.,* 281, 1963.
99. **Woodrum, D. T., Rich, S. A., and Pollard, T. D.,** Evidence for biased bidirectional polymerization of actin filaments using heavy meromyosin prepared by an improved method, *J. Cell Biol.,* 67, 231, 1975.
100. **Bennet, H. and Condeelis, J.,** Decoration with myosin subfragment-1 disrupts contacts between microfilaments and the cell membrane in isolated Dictyostelium cortices, *J. Cell Biol.,* 99, 1434, 1984.
101. **Giuffre, R. M., Tovell, D. R., Kay, C. M., and Tyrell, D. L. J.,** Evidence for an interaction between the membrane protein of a paramyxovirus and actin, *J. Virol.,* 42, 963, 1982.
102. **Nermut, M.,** The "cell monolayer technique" in membrane research, *Eur. J. Cell Biol.,* 28, 160, 1982.
103. **Lang, R. D. A., Nermut, M. V., and Williams, L. D.,** Ultrastructure of sheep erythrocyte plasma membrane and cytoskeletons bound to solid supports, *J. Cell Sci.,* 49, 383, 1981
104. **Aggeler, J., Takemura, R., and Werb, Z.,** High-resolution three-dimensional views of membrane-associated clathrin and cytoskeleton in critical-point-dried macrophages, *J. Cell Biol.,* 97, 1452, 1983.
105. **Nicol, A., Nermut, M. V., Doeinck, A., Robenek, H., Wiegand, C., and Jockusch, B.,** Labeling of structural elements at the ventral plasma membrane of fibroblasts with the immunogold technique, *J. Histochem. Cytochem.,* 35, 499, 1987.
106. **Odenwald, F. W., Arnheiter, H., Dubois-Dalcq, M., and Lazzarini, R. A.,** Stereo images of vesicular stomatitis virus assembly, *J. Virol.,* 57, 922, 1986.
107. **Rutter, G., Bohn, W., Hohenberg, H., and Mannweiler, K.,** Preparation of apical plasma membranes from cells grown on coverslips. Electron microscopic investigations of the protoplasmic surface, *Eur. J. Cell Biol.,* 39, 443, 1985.
108. **Rutter, G., Bohn, W., Hohenberg, H., and Mannweiler, K.,** Demonstration of antigens at both sides of plasma membranes in one coincident electron microscopic image. A double immunogold replica study, *J. Histochem. Cytochem.,* 1988, in press.
109. **Röser, K., Bohn, W., and Mannweiler, K.,** Characterization of a rat glioma cell subclone which lacks intermediate filaments, *Eur. J. Cell Biol.,* 46(Suppl. 22), 57, 1988.

Chapter 15

SCANNING ELECTRON MICROSCOPY OF CELL SURFACE ANTIGENS LABELED WITH COLLOIDAL GOLD MARKERS

Etienne de Harven

TABLE OF CONTENTS

I. INTRODUCTION

Dynamic events in the plasmalemma of most animal cells, which often correspond to redistribution of membrane proteins made possible because of the fluid nature of the membrane itself, are triggered by cytoskeletal elements of the subjacent cytoplasm. To study such redistribution phenomena, methods are needed to efficiently label single molecules, preferably recognized in *en face,* high resolution views of cell surfaces. To be useful in analysis, these methods should not induce molecular redistribution by themselves. They should also be compatible with adequate preservation of cell surface structural features.

We shall demonstrate in this chapter that scanning electron microscopy (SEM) in the backscattered electron (BE) imaging mode offers such a possibility when applied to the study of glutaraldehyde prefixed, colloidal gold labeled cells.

II. SCANNING ELECTRON MICROSCOPY OF COLLOIDAL GOLD PARTICLES

Colloidal gold particles ranging in diameter from 5 to 50 nm can be recognized in the conventional, secondary electron (SE) imaging mode of the SEM, as initially demonstrated by Horisberger.[1] However, the topographical contrast generated in the SE mode, and the masking effect of the necessary conductive coating, will seriously hamper observation of small particles. Moreover, the imaging contribution of secondary electrons which actually originate from electron backscattering events is difficult to evaluate.

In the BE imaging mode of SEM, contrast is not generated as a function of topography. It originates instead from collisions between electrons from the incident electron beam and elements of high atomic number[2] present on or near the sample's surface. When SEMs equipped with backscattered electron detectors became commercially available more than 10 years ago, biologists soon reported novel applications of this type of microscope, taking advantage of the presence of elements of high atomic number, like lead or osmium, in the reaction products of several classical cytochemical methods.[3-4]

In 1981, two groups of investigators used the BE imaging mode of the SEM to demonstrate colloidal gold marker particles.[5-6] The principle of the method was to avoid the use of any other element of high atomic number beside the gold maker itself, therefore ensuring significant elemental contrast of gold markers in BE images. Chemical fixation was limited to aldehyde fixatives (no osmium tetroxide), and conductive coating restricted to evaporated carbon.

In 1984, we reported on the BE imaging of 40 nm gold particles used as specific marker of an antigen expressed on the surface of human circulating granulocytes, and identified by the D2 murine IgM monoclonal antibody.[7] Several points emerged from that study.

1. The extremely low nonspecific background labeling ensured immunological specificity to the method.
2. The number of gold particles observed in BE images of a given cell was considerably greater than that seen in SE images of the same cell, indicating that attempts to demonstrate the possible quantitative value of immunogold cell surface labeling procedures should be based on BE imaging only.
3. Mixed BE/SE images permit precise correlations between surface features of the labeled cells and distribution of gold labeled sites.
4. Cell surface features are relatively well preserved, provided cells are prefixed with 0.1% glutaraldehyde for 10 min.
5. Such a glutaraldehyde prefixation completely blocks redistribution and/or internalization of gold labeled antigen/antibody complexes.

The major limitation of the method, however, resulted from the use of gold particles of 40 nm in diameter, known to have poor labeling efficiency primarily due to steric hindrance.[8-9]

Obviously, further progress was dependent on the possibility of resolving smaller gold marker particles in the BE imaging mode. Using a JEOL JSM #840 SEM equipped with a lanthanum hexaboride (LaB_6) electron gun, we resolved 13 nm gold particles on the surface of labeled cells, operating at a working distance of 8 mm and under 15 or 20 kV accelerating voltage.[10] Under such conditions, this represents perhaps the limit of the resolving power of such instruments in the BE imaging mode. With a field emission SEM, however, Walther and Müller[11] resolved single 5 nm particles, definitely bringing immunogold methodology closer to the labeling of individual molecules exposed on cell surfaces.

More recently, Hainfeld[12] resolved the undecagold complex marker, the core of which measures approximately 0.8 nm in diameter. This was achieved with a type of scanning transmission field emission microscope (STEM), not presently commercially available.

III. A BASIC GOLD LABELING METHOD FOR CELL SUSPENSIONS

A method for labeling human peripheral blood mononuclear leukocytes with colloidal gold markers will now be briefly outlined and illustrated. The method is identical to that described in previous publications.[13-14]

Leukocytes, separated from freshly drawn venous blood,[15] were resuspended in phosphate buffered saline (PBS) and allowed to attach on fragments of glass coverslip pretreated with 85,000 molecular weight poly-L-lysine hydrobromide, according to routine procedures,[16] while carefully avoiding any risk of air drying. After 30 min, the non-adherent cells were rinsed off with buffer, and the glass attached cells immediately covered for no longer than 10 min with a large drop of a 0.1% (v/v) solution of buffered glutaraldehyde (EM grade) in 0.1 *M* PBS, pH 7.2, containing 4% sucrose (total osmolality 420 mOsm), at room temperature. After one rinse with PBS containing 0.1% glycine, and two rinses with PBS supplemented with 1% ovalbumin and 1% human decomplemented AB serum as carrier proteins, the cells were incubated for 30 min with a 1/10 dilution of a murine monoclonal antibody (LEU-1 in the example illustrated here, Becton Dickinson, CA) in the presence of the same carrier proteins. Rinses in PBS/ovalbumin were then followed by a 60 min incubation with commercially available colloidal gold conjugated with affinity purified goat antimurine IgG immunoglobulin (GAM-G40, 15, or 5, from Janssen Pharmaceutica, Beerse, Belgium).[17] The cells were then rinsed with PBS, postfixed with 2.5% glutaraldehyde for 30 min, dehydrated in graded ethanol, dried at the critical point of CO_2, mounted by affixing the glass coverslip on an aluminum stub with a conductive glue, and coated by evaporation of approximately 10 to 20 nm of carbon, using a rotating/tilting stage. The samples were kept under vacuum (glass desiccator) until observation in the SEM.

A JEOL JSM #840 SEM, equipped with a lanthanum hexaboride electron source, was used throughout our studies. Optimum imaging was routinely achieved under 15 or 20 kV accelerating voltage, with 6×10^{-10} A beam current, and at 8 mm working distance. In routine studies, three pictures were taken of each cell of interest, at direct magnifications ranging from 12,000 to 16,000 ×. The first was a conventional SE image (Figure 1), the second a BE image in the reverse polarity of the BE signal (Figure 2), and the third a mixed signal image (SE + BE, normal polarity) (Figure 3).

The human peripheral blood leukocyte illustrated here has been identified as a T-derived lymphocyte, since it was positively labeled after priming the cells with the pan-T cell identifier LEU-1 (CD5) monoclonal antibody. The surface morphology was optimally demonstrated in the SE image (Figure 1); the density of the gold labeling was clearly recognized in the BE image (Figures 2 and 3), while the precise correlation between the distribution of

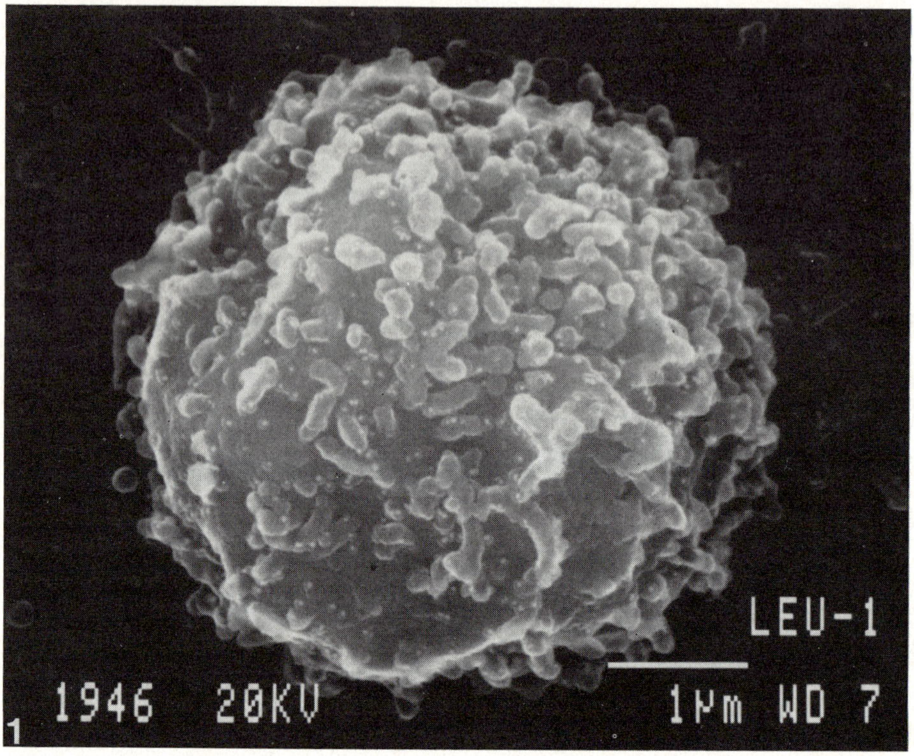

FIGURE 1. Human peripheral blood lymphocyte incubated with the LEU-1 monoclonal antibody and immunolabeled with 40 nm colloidal gold particles. Secondary electron (SE) image.

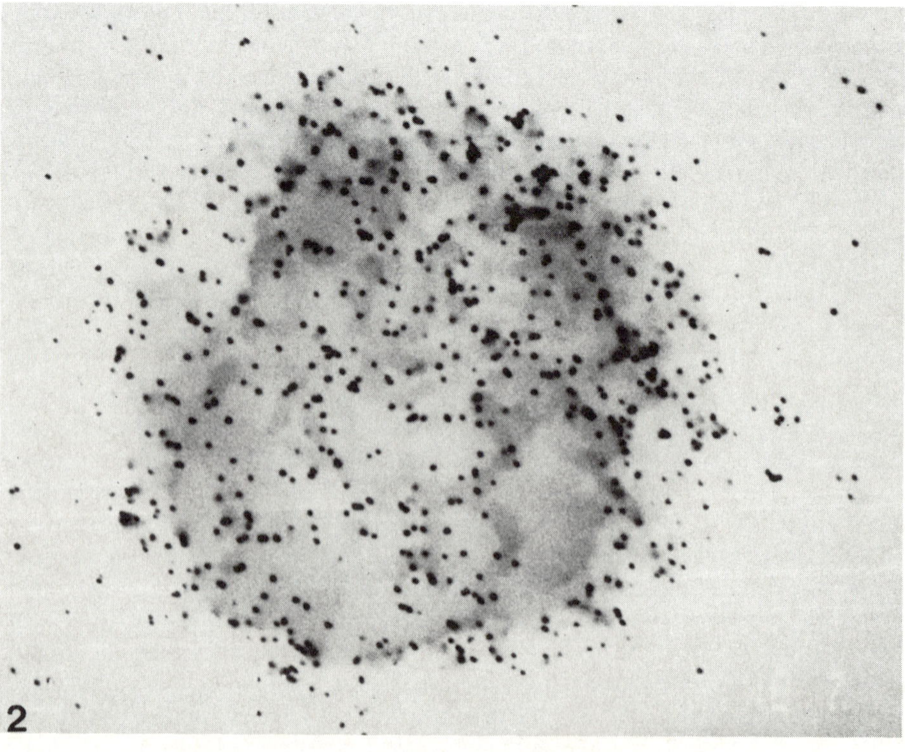

FIGURE 2. Backscattered electron (BE) image of the same cell.

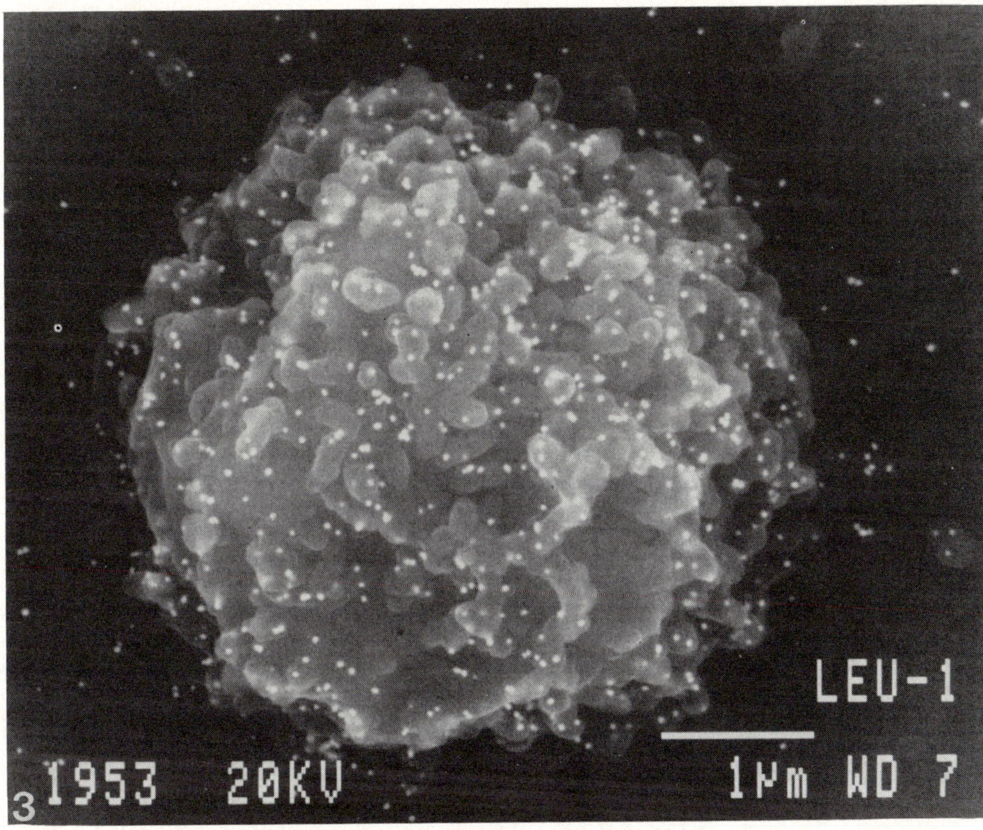

FIGURE 3. Mixed (SE + BE) image of the cell shown in Figures 1 and 2.

gold-labeled sites and surface microvilli was established only in SE + BE images (Figure 3).

The method illustrated here can be adapted to the study of tissue fragments, like mucosal biopsies, as demonstrated in our recent study of biopsies of the human bladder mucosa in cases of noninvasive bladder cancer.[18] It permitted the recognition of the phenotypic heterogeneity of the luminal cells of the urothelium, after immunogold labeling of the partially deleted blood group ABH antigens.

IV. CHOICE OF AN APPROPRIATE LIGAND

Four different ligands are available for such studies. *Staphylococcus aureus* protein A,[19] protein G,[20] immunoglobulins,[1,21] and biotin/streptavidin complexes.[22] The choice of a given ligand will affect both the specificity and the stability of the labeling.

Protein A has high affinity binding for the F_c fragments of many, but not all, immunoglobulin molecules. It has very low affinity for murine IgM immunoglobulins, considerably limiting its usefulness; therefore, in studies which are primarily dependent on the use of murine monoclonal antibodies as identifiers of target molecules.

Protein G circumvents this difficulty to a large extent: it binds to all immunoglobulins so far tested.[20] Unfortunately, both protein A and G are unlikely to provide satisfactory labeling specificity in studies of cells expressing F_c receptor on their surfaces, like lymphocytes and macrophages. Such cells almost inevitably decorate their surfaces with various immunoglobulin molecules, resulting in nonspecific protein A or G/gold labeling.

Anti-immunoglobulin molecules used as ligand for the colloidal gold marker[17] will give excellent results as far as specificity is concerned. A goat antimurine IgG or IgM can be used according to the subclass of the murine monoclonal antibody used as identifier of a given target molecule. Unfortunately, the linkage between immunoglobulins and anti-immunoglobulins is not high affinity in nature, and may be ruptured during subsequent preparatory procedural steps for example by turbulence during critical point drying. Our recent results (unpublished observation) indicate that some 40 mm gold marker particles are apparently lost during critical point drying. Comparing the density of gold labeling on air dried vs. critical point dried gold labeled cells is quite informative in this respect. Consequently, a low affinity ligand like anti-immunoglobulin is probably not optimum for quantitative studies.

Biotin/streptavidin, on the other hand, is a very high affinity complex. Biotin, a relatively ubiquitous protein in the cytoplasm of many cells, has never to our knowledge been recognized on cell surfaces. Therefore, in studies aimed at the identification of cell surface ectoprotein molecules,[23] it probably represents the ligand of choice. Indeed, and contrasting with our results with anti-immunoglobulins, we were unable to demonstrate any loss of gold label after critical point drying when the biotin/streptavidin ligand was used.

V. DOUBLE LABELING PROCEDURES

Having made an appropriate choice of ligands, and taking advantage of the availability of colloidal gold particles of different sizes, double labeling protocols can be used, as initially shown by Horisberger.[1] However, possible pitfalls must be carefully avoided: size overlap of the gold particles and labeling crossreactivity, among others.

For example, in experiments based on the labeling of antigen #1 with 5 nm gold particles, and of antigen #2 with 15 nm particles, one has to be reasonably assured that relatively monodisperse gold particles are used, and that the risk of size overlap is negligible. This can be verified with the TEM by analyzing particle size distribution histograms. In our experience, some commercially available gold conjugate preparations are excessively heterodisperse in diameter and are probably not, therefore, appropriate for double labeling experiments. Slot and Geuze[24] have described a method based on the reduction of chloroauric acid with low molecular weight tannic acid. Gold particles homogenous in size (monodisperse) are reproducibly prepared by this method and are, therefore, suitable for double labeling studies.

Cross-reactivity between the labeling of two distinct antigens can be kept at acceptably low levels by using different subclasses of immunoglobulins as identifiers of the two target antigens. Antigen #1 can be identified with an IgG molecule, and antigen #2 with an IgM molecule. Sequentially using a goat antimurine IgG adsorbed on 15 nm gold for antigen #1, and then an antimurine IgM adsorbed on 40 nm gold for antigen #2, does result in significant double labeling. Moreover, silver enhancement[25] can be used to enhance both labelings and facilitate the SEM observation at relatively low magnification. Some cross-reactivity is, however, usually observed when this method is applied.

In our experience, cross-reactivity becomes nonmeasurable when ligands of unrelated chemistry are used, for example, if antigen #1 is identified with a biotinylated monoclonal antibody followed by streptavidin/gold 15, and antigen #2 with a non-biotinylated monoclonal antibody followed by goat antimurine Ig/gold 40. Both labelings can then be silver enhanced to facilitate observation at relatively low magnification of the SEM in the BE imaging mode. This method gave us satisfactory results with no detectable level of cross-reactivity, when applied to double labeling experiments of B- and T-derived human peripheral blood lymphocytes.[26]

VI. IMAGING OF 5 nm COLLOIDAL PARTICLES

In order to increase labeling efficiency by maximizing the number of antigenic sites

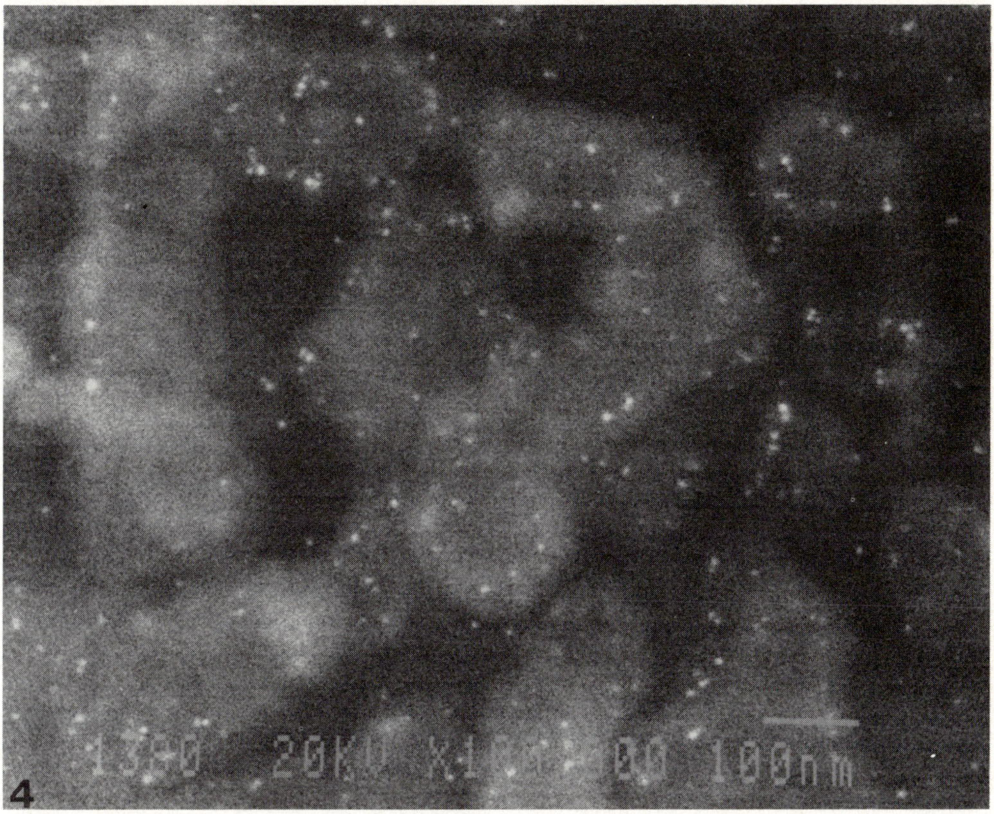

FIGURE 4. High resolution BE image of a human peripheral blood lymphocyte incubated with the LEU-4 monoclonal antibody and labeled with approximately 6 nm colloidal gold particles. The micrograph is taken with the field emission SEM (JEOL #890). Note the frequency of doublets, and also the predominant labeling of microvilli.

recognized by a given monoclonal antibody and by minimizing steric hindrance, gold particles of small size have to be used.

Gold particles 5 nm in diameter have been demonstrated using field emission SEM in the BE imaging mode.[11,27] From the point of view of methodology, Walther and Müller used a yttrium-aluminum-garnet (YAG) scintillation detector, while we used the conventional solid-state BE detector of a JEOL #890 field emission SEM.[27] Results are illustrated in Figure 4 in which part of the surface of a LEU-4 identified human T-lymphocyte is recognized. By comparison with Figure 2, it is apparent that labeling density is considerably higher. It seems also that most gold labeled sites are located on surface microvilli, suggesting that the CD3 molecule might be preferentially expressed on the surface microvilli of human T-lymphocytes. It would have been difficult to make this observation with gold marker of larger diameter, like 40 nm particles. Moreover, the frequency of gold marker doublets is intriguing. It may suggest that the resolution offered by the 5 nm gold markers is adequate to demonstrate the homodimeric nature of the receptor under investigation,[27] although this hypothesis remains to be tested in several control experiments.

Access to field emission SEMs is limited, unfortunately, by the cost of these instruments. We have demonstrated, however, that the technique of silver enhancement obviates this difficulty, since silver enhanced 5 nm gold particles can easily be resolved in the BE imaging mode of conventional SEMs equipped with a lanthanum hexaboride (LaB_6) electron source (Figure 5).

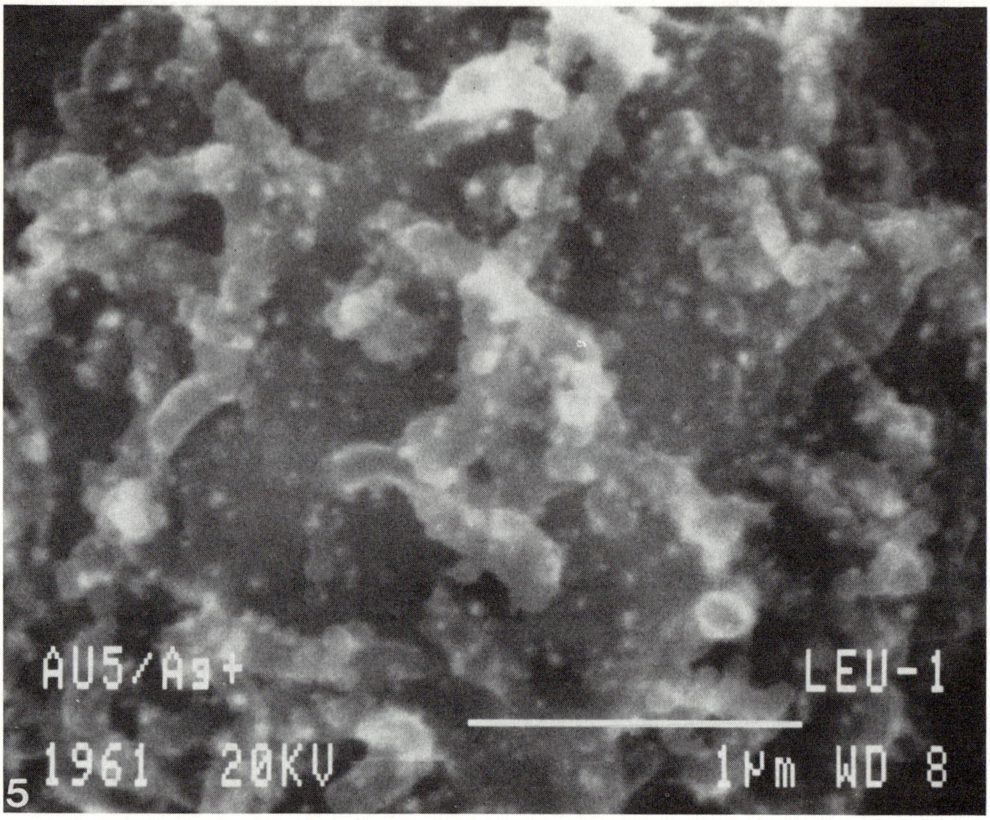

FIGURE 5. Mixed (SE + BE) image of a lymphocyte labeled with silver enhanced 5 nm colloidal gold particles. Silver enhancement has made possible the imaging of 5 nm gold particles in this picture taken with the conventional LaB_6 scanning electron microscope.

VII. QUANTIFICATION OF GOLD MARKER PARTICLES ON THE SURFACE OF LABELED CELLS

Quantification of gold marker particles of large size (i.e., 40 nm) is relatively easy in BE images of cell surfaces, but is probably biologically irrelevant.[14] For example, when receptor molecules, like the 55 kDa IL 2 low affinity receptor (Tac) expressed on the surface of activated lymphocytes are labeled with 40 nm gold markers, the number of gold labeled sites is far below that of Tac molecules as calculated by other methods.[28] Such experiments will therefore have to be repeated with 5 nm gold markers, if numbers of possible biological significance are to be obtained.

VIII. CONCLUSION

The method of colloidal gold labeling of cell surface ectoproteins has been significantly improved in recent years. This results primarily from the use of the backscattered electron imaging mode of the SEM, and from the possibility of mixing the intense BE signal generated by the gold marker with the topographical image of well preserved cell surfaces. In fact, satisfactory cell surface preservation necessitates, in most cases, a brief prefixation with glutaraldehyde. This requirement limits the usefulness of the method to those antigens which are not significantly denatured by a brief glutaraldehyde prefixation, and can, therefore, still be identified by corresponding monoclonal antibodies under these conditions.

The biological relevance of direct quantification remains to be established, although the possibility of resolving 5 nm gold particles in the BE imaging mode of the SEM brings us closer to significant quantitative observations.

Dynamic events involving either translation, emergence, or deletion of plasmalemmal ectoproteins can, in many cases, be analyzed by SEM/BE methods of immunogold labeling at the molecular level.

ACKNOWLEDGMENTS

The skillful assistance of Mrs. Hilary Christensen and of Mr. George Bootsma is gratefully acknowledged through these studies. The expertise of Mr. H. Otsuji and of Mr. T. Laudate, JEOL USA Inc., Peabody, Massachusetts, has made possible the observations with the field emission SEM. The study was supported by a research grant from the Medical Research Council of Canada.

REFERENCES

1. **Horisberger, M., Rosset, J., and Bauer, H.,** Colloidal gold granules as markers for cell surface receptors in the scanning electron microscope, *Experientia,* 31, 1147, 1975.
2. **Palluel, P.,** Composante rediffusee du rayonnement elecronique secondaire des metaux, *C. R. Acad. Sci. Paris,* 224, 1492, 1948.
3. **Becker, R. P. and Debruyn, P. P. H.,** Backscattered electron imaging of endogenous and exogenous peroxidase activity in rat bone marrow, *Scanning Electron Microsc.,* 2, 171, 1976.
4. **Soligo, D. and de Harven, E.,** Ultrastructural cytochemical localizations by backscattered electron imgaing of white blood cells, *J. Histochem. Cytochem.,* 29, 1071, 1981.
5. **Trejdosiewicz, L. K., Smolira, M. A., Hodges, G. M., Goodman, S. L., and Livingston, D. C.,** Cell surface distribution of fibronectin in cultures of fibroblasts and bladder derived epithelium: SEM-immunogold localization compared to immunoperoxidase and immunofluorescence, *J. Microsc.,* 123, 227, 1981.
6. **Sieber-Blum, M., Sieber, F., and Yamada, K. M.,** Cellular fibronectin promotes adrenergic differentiation of quail neural crest cells *in vitro, Exp. Cell Res.,* 133, 285, 1981.
7. **de Harven, E., Leung, R., and Christensen, H.,** A novel approach for scanning electron miscrocopy of colloidal gold-labeled cell surfaces, *J. Cell Biol.,* 99, 53, 1984.
8. **Horisberger, M. and Tacchini-Vonlanthen, M.,** Stability and steric hindrance of lectin-labelled gold markers in transmission and scanning electron microscopy, in *Lectins,* Vol. III, Bg-Hansen, T. C., and Spengler, G. A., Eds., Walter de Gruyter & Co., New York, 1983, 189.
9. **Voorhout, W. F., Leunissen-Bijvelt, J. J. M., Leunissen, J. L. M., and Verkleij, A. J.,** Steric hindrance in immunolabelling, *J. Microsc.,* 141, 303, 1986.
10. **Soligo, D. and de Harven, E.,** Scanning electron microscopy of blood cells, in *Science of Biological Specimen Preparation,* Muller, M., Becker, R. P., Boyde, A. and Wolosewick, J. J., Eds., SEM Inc., Chicago, 1985, 289.
11. **Walther, P. and Müller, M.,** Detection of small (5 to 15 nm) gold labelled surface antigens by backscattered electrons, in *Proc. 43rd Ann. Meet. Electron Microsc Soc. Am.,* Bailey, G. W., Ed., San Francisco Press, San Francisco, 1985, 538.
12. **Hainfeld, J. F.,** A small gold-conjugated antibody label: improved resolution for electron microscopy, *Science,* 236, 450, 1987.
13. **de Harven, E. and Soligo, D.,** Scanning electron microscopy of cell surface antigens labeled with collodial gold, *Am. J. Anat.,* 175, 277, 1986.
14. **de Harven, E., Soligo, D., and Christensen, H.,** Should we be counting immunogold marker particles on cell surfaces with the SEM? *J. Microsc.,* 146, 183, 1987.
15. **Boyum, A.,** Separation of blood leukocytes, granulocytes and lymphocytes, *Tissue Antigens,* 4, 269, 274, 1974.
16. **Murphy, J. A.,** Considerations, materials, and procedures for specimen mounting prior to scanning electron microscopic examination, *Scanning Electron Microsc.,* 2, 657, 1982.

17. **De Mey, J. R.,** The preparation of immunoglobulin gold conjugates (IGS) reagents and their use as markers for light and electron microscopy immunocytochemistry, in *Immunocytochemistry,* Cuello, A. C., Ed., John Wiley & Sons, New York, 1983, 347.
18. **de Harven, E., He, S., Hanna, W., Bootsma, G., and Connolly, J. G.,** Phenotypically heterogeneous deletion of the ABH antigen from the transformed bladder urothelium. A scanning electron microscope study, *J. Submicrosc. Cytol.,* 19, 639, 1987.
19. **Romano, E. L. and Romano, M.,** Staphylococal protein A bound to colloidal gold: a useful reagent to label antigen-antibody sites in electron microscopy, *Immunochemistry,* 14, 711, 1977.
20. **Bendayan, M.,** Introduction of the protein G-Gold complex for high-resolution immunocytochemistry, *J. Electron Microsc. Techn.,* 6, 7, 1987.
21. **De Mey, J., Moeremans, M., Geuens, G., Nuydens, R., and De Brabander, M.,** High resolution light and electron microscopic localization of tubulin with the IGS (immunogold staining) method, *Cell Biol. Int. Rep.,* 5, 889, 1981.
22. **Hopkins, C. R., Boothroyd, B., and Gregory, H.,** Identification of receptors for epidermal growth factor in the electron microscope by using an EGF/biotin/avidin-gold procedure, *Biochem. Soc. Trans.,* 7, 956, 1979.
23. **Rothman, J. E. and Lenard, J.,** Membrane asymmetry, *Science,* 195, 743, 1977.
24. **Slot, J. W. and Geuze, H. J.,** A new method of preparing gold probes for multiple-labeling cytochemistry, *Eur. J. Cell Biol.,* 38, 87, 1985.
25. **Danscher, G. and Noorgaard, J. O.,** Light microscopic visualization of colloidal gold on resin-embedded tissue, *J. Histochem. Cytochem.,* 31, 1394, 1983.
26. **de Harven, E., Soligo, D., Christensen, H., and Bootsma G.,** Double labeling of cell surface antigens with colloidal gold markers, *Histochem. J.,* in press.
27. **de Harven, E. and Christensen, H.,** High resulution immunogold labelling of cell surfaces: the hypothetical homodimeric nature of the CD5 receptor, in *Proc. 46th Ann. Meet. Electron Microsc. Soc. Am.,* 1988, 380.
28. **McGroarty, R. J., Mills, G. B., and de Harven, E.,** Ultrastructural localization of the low affinity (55kd) IL2 receptor using anti-tac monoclonal antibodies and colloidal gold after mitogen induced activation, in preparation,

Chapter 16

IMMUNOCYTOCHEMISTRY OF CONTRACTILE AND CYTOSKELETAL PROTEINS

J. V. Small

TABLE OF CONTENTS

I. INTRODUCTION

In the last 10 years or so we have witnessed, through the application of immunofluorescence microscopy, dramatic advances in the elucidation of the organization of cytoskeletal elements, mainly in cultured cells and unicellular organisms. A wide spectrum of cytoskeleton associated proteins have been found in parallel biochemical studies, and current attention is focused on defining their individual functions. Here, electron microscopy should be playing a key role, providing utrastructural information about molecular interactions *in situ*, but progress in this area, except for extremely ordered systems, has been slow. In very special cases (e.g., myosin cross-bridges in muscle and molecular links in flagella) it has been possible to identify molecules or their projections purely on morphological grounds. But for other molecules, and in particular, in less ordered systems, such direct molecular mapping has proved difficult. For this reason antibodies as indirect markers have been in common use for some time. However, after an initial phase of renewed enthusiasm, heralded by the introduction of gold-labeled probes, the field of ultrastructural immunocytochemistry has now acknowledged the possibilities and limitations of current methodology.[1,2] In short, we still have a long way to go to be able to pinpoint the precise localization of molecules in their subcellular, macromolecular environments. This is not to say that progress has been slight. On the contrary, some interesting results have been forthcoming and some of these will be the subjects of the following sections. But we should not let these obscure the obvious need for new approaches in this active field.

In this brief review I shall attempt to evaluate the recent contributions of electron microscopy to our understanding of cell structure and motility against the background of data obtained by complementary methods. Major emphasis will be placed on the application of immunocytochemical techniques for the subcellular location of contractile and cytoskeletal components.

II. MAPPING MUSCLE STRUCTURE

A. SKELETAL MUSCLE

Our present understanding of the process of muscle contraction owes much to the natural and extensive order of striated muscle that has and continues to make this tissue so amenable to structural analysis. In a now classical series of studies[3] the basic molecular architecture of the myofibril was yielded by a multifaceted approach involving X-ray diffraction, electron microscopy (EM), and biochemistry. Yet many gaps still remain in our knowledge of sarcomere organization and myofibrillogenesis. For example, attention has now been drawn to the cytoskeletal lattice that embraces the contractile elements and whose existence was generally disregarded until only recently.[4] Antibody methods have been, and are playing an important role in revealing the location of component molecules.

It is now more than 20 years ago that Ohtsuki and co-workers, in an early EM study, demonstrated the periodic spacing of troponin on the thin filament of striated muscle.[5] Here, the added density of the antibody in the ordered filament lattice was alone sufficient to mark the location of the antigen. The same approach, using a pre-embedding labeling step with permeabilized muscle fibers, has been used successfully for the localization of a number of other myofibrillar proteins: C-protein,[6,7] a related 86 kDa protein,[8] X-protein, and H-protein.[9] Antibodies directed against the myosin head subfragment S-1[10] and the myosin tail[11] have likewise been used to reveal details of myosin packing in the thick filament. The latter study is remarkable in showing the high resolution, in this case around 13 nm, achieved with the antibody marker. Even more immunolocalization studies, too numerous to mention here, have been carried out on isolated myofibrils or myofibril fragments, using direct antibody labeling and negative staining. Particularly noteworthy has been the detailed localization of the three troponin components on the surface of the thin filament.[12]

The discovery in skeletal muscle of the giant molecules, titin and nebulin, with molecular weights around $1.2 \cdot 10^6$ and M_r $0.5 \cdot 10,^6$ respectively, has raised interesting new questions about sarcomere organization[4,13] which are finding answers in immunocytochemical studies.[14] Both titin and nebulin are very insoluble proteins, but endogenous proteolytic degradation of intact titian (T) gives rise to a soluble fragment T2 (molecular weight $1 \cdot 10^6$) that has been purified and characterized (see Reference 4). Titin T2 is an enormously long molecule, around 1 μm in length. Recent detailed localization studies using monoclonal antibodies raised against eight different epitopes of T2 and against two epitopes specific for the T1 to T2 difference piece indicate that there is a bipolar arrangement of titin molecules in the sarcomere with single molecules extending from the M-region to the Z-line.[14] This work confirms and extends earlier ideas about titin (and nebulin) constituting an endosarcomeric structural lattice.[4] The clear antibody localization in distinct transverse stripes across the myofibrils (Figure 1a) was achieved by a pre-embedding labeling of unfixed muscle with a first and then a second antibody label (without an electron dense tag).[14] It is instructive to compare this data with the results obtained using gold-labeled probes as secondary markers[4,15] also in a pre-embedding routine. The scatter and lack of penetration of the label in the latter cases gives no hint of the real extent of the titin molecule.

The resort to pre-embedding labeling with unconjugated antibodies also reflects a general lack of success with postembedding techniques for the precise localization of antigens in longitudinal sections of muscle. Tokuyasu[16,17] has championed the cause of frozen sections obtained from fixed and sucrose-embedded tissue, since the embedding medium may be easily dissolved after sectioning. But even here, ferritin- or gold-conjugated antibodies appear unable to penetrate the actomyosin lattice. Figure 1b shows an example with antiactin and a secondary gold conjugated antibody; the low level of label and absence of any linear arrangement of gold particles suggests that there is access only to short filament segments that graze the surface of the section. This result is consistent with the demonstration by Bendayan et al.[1] of labeling of the surface only for cryosections of other tissues. In more loosely packed regions of the sarcomere, better penetration is evident, however, as indicated by the continuous labeling of intermediate filaments and microtubules with ferritin-[18] and gold-conjugated probes.[19]

Alpha-actinin has proved to be a popular muscle antigen for testing the suitability of nonextractable embedding media for ultrastructural immunocytochemistry.[20,31] Success with the labeling of this protein in cryosections,[16] as well as in sections of acrylic media, may be attributed to its homogeneous distribution in the Z-disc, a structure that in sections along the myofibril is always cut to yield an exposed face on both surfaces of the section. The single interesting result obtained with alpha-actinin antibodies is the demonstrated absence of alpha-actinin from the terminal myofibrillar segment abutting the myotendinous junction.[22] In his own study Tokuyasu[16] took the opportunity to use alpha-actinin labeling at the Z-disc to compare the efficiency of secondary probes, namely protein A-gold and immunoferritin. He observed that the protein A-gold reagent gave comparable labeling intensity to the ferritin probe only when a bridging antibody between the first antibody and the protein A was employed.

B. SMOOTH MUSCLE

Smooth muscle stands apart from cross-striated and obliquely-striated muscle in concealing obvious clues of its organization in the light microscope. The problem is compounded by the lability of the contractile apparatus, particularly of the myosin filaments, which readily disassemble under adverse conditions of chemical fixation, leading to obvious misgivings about the validity of images obtained by electron microscopy. Indeed, it was only through the final application of freeze-substitution methods[23,24] that final proof for the existence of thick myosin filaments in relaxed muscle *in vivo* could be provided. Fortunately, we now

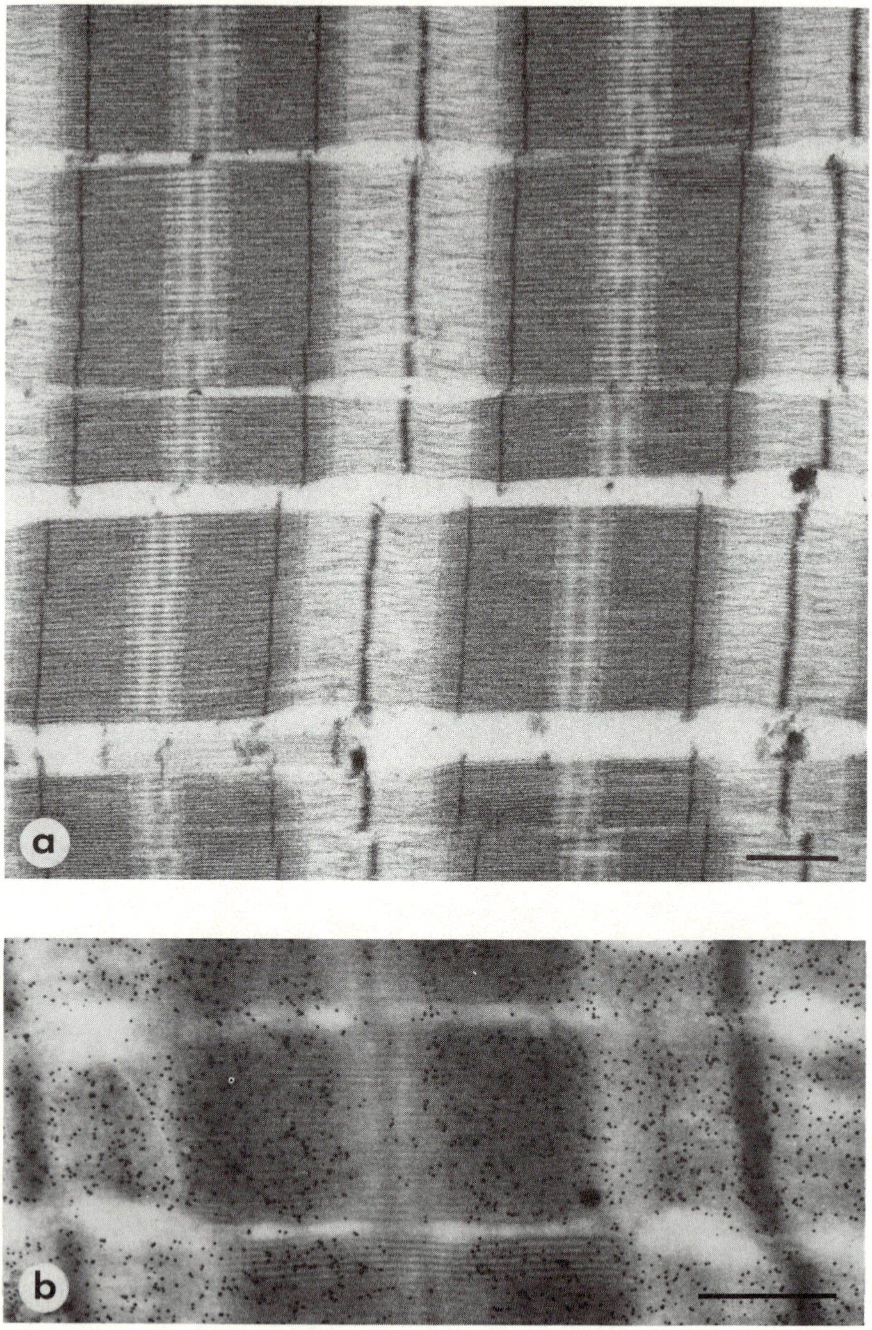

FIGURE 1 (a). Immunolocalization of one of the titin epitopes with a monoclonal antibody (T22: Fürst et al., 1988[14]) in chicken pectoralis muscle. Pre-embedding labeling with the first, followed by a second, unconjugated antibody. The sharp additional line towards the periphery of the A-bands corresponds to the antibody site. (Micrograph courtesy of Dr. D. O. Fürst.) (b) Immunolocalization of actin on an ultrathin frozen section (sucrose embedding) of striated muscle. Immunogold labeling with secondary 5 nm gold-tagged antibody. Note suboptimal labeling of actin filaments diagnostic of poor antibody access. Bars = 0,5 μm. (Micrograph courtesy of Dr. S. Watkins.)

know in some detail the factors that promote thick filament disassembly (ATP, myosin dephosphorylation, high pH, low Mg^{2+} ion concentration), and can choose conditions for chemical fixation that appear to yield a close to *in vivo* organization of the myofilaments. Even so, controversy still prevails over the three dimensional organization of the contractile apparatus,[25] differences centering on whether or not there is a minisarcomere type of arrangement of actin and myosin filaments. Here, immunoelectronmicroscopy is playing an important role in providing missing pieces of the puzzle.

Not surprisingly, alpha-actinin was the first protein to be localized immunocytochemically in smooth muscle and shown to reside in the electron dense structures, or dense bodies scattered in the cytoplasm.[26,27] On morphological grounds alone, the same structures had earlier been presumed equivalent in function to the Z-discs of striated muscle, and were thus thought to act as sites of attachment and polarity reversal of actin filaments. The localization of alpha-actinin confirmed the relatedness of dense bodies with the myofibril Z-disk. Alpha actinin was also localized along the dense, membrane associated sites, or dense plaques, thought to constitute foci of attachment of the contractile elements (presumably actin), to the cell surface. In a further study, however, these sites remained unlabeled with alpha-actinin antibodies despite positive labeling of the cytoplasmic dense bodies.[28,29] Whether or not this represents a preferential masking of sites by the embedding medium, or fixation,[29] or a lack of monospecificity of the antibodies,[26,27] is currently unclear.

In their study using frozen sections, Geiger and colleagues[26] localized the protein vinculin exclusively within the dense plaques, and in a pioneering double label approach with ferritin and imposil conjugates, showed this protein to be closer to the membrane than alpha-actinin. In a series of other studies[26,30] vinculin has been shown to be a characteristic component of sites where actin filaments associate with the cell membrane. Immunofluorescence microscopy of isolated smooth muscle cells with vinculin antibodies has demonstrated that the dense plaque structures form a set of continuous and parallel longitudinal ribs at the cell surface;[29] the distribution of these ribs presumably reflects the global geometry of membrane attachments of the myofilaments. Further proteins immunolocalised in the dense plaques are talin,[31] a protein that binds to vinculin and to the fibronectin receptor,[30] and plectin, a high molecular weight protein that appears to associate with intermediate filaments.[32] In the latter connection, we might expect the smooth muscle intermediate filament protein desmin to also reside in the dense plaques.

On this question, a point of technical interest may be cited from our own studies. We have recently made extensive use of the water soluble resin, polyvinylalcohol (PVA) in postembedding immunocytochemical analysis of smooth muscle.[28,29,33,34] In aldehyde-fixed tissue, an affinity-purified antibody directed against desmin stained the regions containing visible intermediate filaments.[34] (Figure 2a,b). However, in unfixed muscle permeabilized with a PVA-Triton mixture and embedded in PVA, we also observed extensive labeling at the cell periphery (Figure 2c) despite the general absence of morphologically recognizable intermediate filaments along the surface plaques (Figure 3a). The interesting possibility then presents itself that a nonpolymer form of intermediate filament protein (presumably desmin that becomes masked by fixation) occurs in the dense plaque regions. The general difficulties in striated muscle to find enough intermediate filaments at the myofibril Z-disc to account for the heavy labeling by desmin antibodies[35] could likewise be explained by a similar partitioning of the filament protein into polymeric and monomeric fractions. Further attention to this possibility is warranted.

Related to the problem of the organization of the contractile apparatus of smooth muscle is the question of the individual functions of the actin-associated proteins that abound in this tissue. Because of their overriding predominance, two proteins of this class, caldesmon and filamin, come to the forefront; taken together, they compare in amount with the total mass of myosin in the cell.[36] By ultrastructural immunocytochemistry of smooth muscle

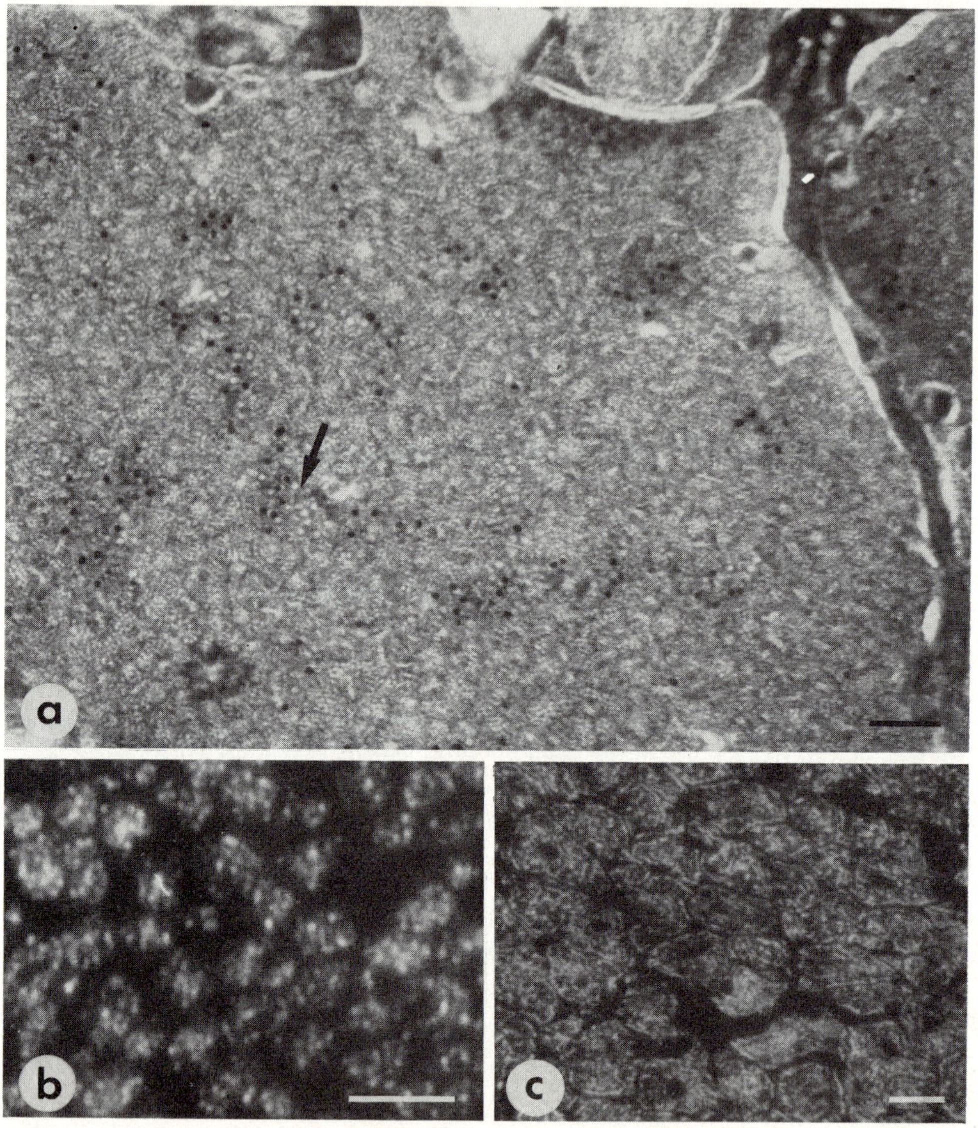

FIGURE 2. Postembedding labeling of ultrathin cross-sections of chicken gizzard smooth muscle with anti-desmin antibodies. Polyvinylalcohol embedding: negatively stained with sodium silicotungstate. (a) Immunogold labeling (10 nm gold) shows expected labeling in regions containing intermediate filaments (clear, white filaments, some indicated by arrow). (b) Ultrathin section corresponding to "a" labeled with a secondary fluorescent antibody and showing desmin label confined to intracellular patches. Fixation for a and b: paraformaldehyde-glutaraldehyde; (c) immunofluorescence image corresponding to "b", but for an ultrathin section obtained from unfixed chicken gizzard directly embedded in polyvinylalcohol. Note additional label at the cell border in the unfixed tissue (see also text). Bar: a, 0,1 μm; b, c, 5 μm.

cross sections it has been demonstrated that caldesmon and filamin are segregated in different subcellular compartments, caldesmon in the domains containing myosin,[33] and filamin in the domains containing intermediate filaments.[34] Accordingly, a domain model of smooth muscle has been proposed with intermediate-filament-actin-filament "cytoskeletal domains" segregated from "actomyosin domains".[36] Although the functional significance of this observation remains elusive, the suggestion has been put forward that filamin and caldesmon, as actin cross-linking proteins, play some role in maintaining tonic tension in smooth muscle.

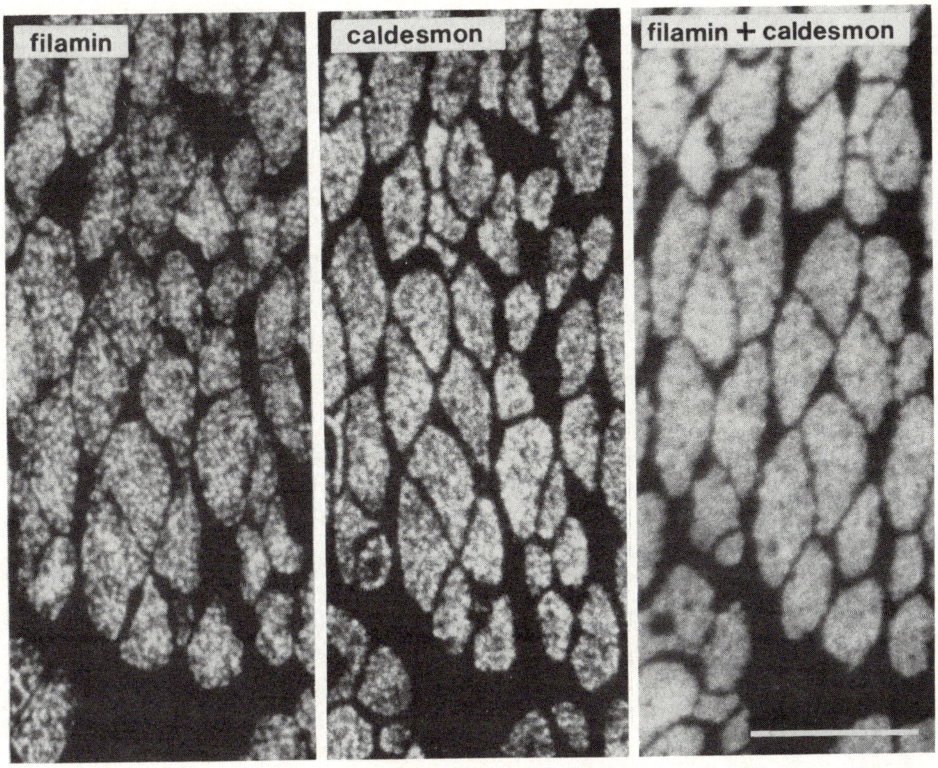

FIGURE 3. Immunofluorescence labeling of ultrathin sections of polyvinylalcohol-embedded guinea pig taenia coli smooth muscle with antibodies against the antigens indicated. Labeling indicates complementary distribution of filamin and caldesmon which was confirmed by immunoelectronmicroscopy. Bar = 10 μm.

In the context of the present discussion it is noteworthy that these latter ultrastructural observations could be largely predicted by high resolution immunofluorescence microscopy of ultrathin sections (Figure 3). Taking into account the greater sensitivity of the immunofluorescence method as compared to current immunogold procedures, at least in our hands, strong labeling of ultrathin sections in the light microscope is a logical prerequisite for successful immunoelectron microscopy with the same material.

We may conclude this section with a few remarks about the different embedding media tried for postembedding immunocytochemistry of smooth muscle. In a recent study, sections of smooth muscle embedded in three media, Lowicryl K4M, LR White, and PVA, were compared with respect to their reactivity towards a set of chosen antibodies.[37] For Lowicryl, the procedure of Lemanski et al.,[20] which had formerly been shown suitable for immunolabeling with alpha-actinin antibodies, was followed. Surprisingly, it was found that the immunoreactions with five antibodies were essentially insensitive to the embedding routine despite differences in embedding temperatures ranging from −35°C to +60°C. However, one antibody against light meromyosin showed a stronger reaction with tissue embedded in the water soluble PVA than with those embedded in either of the acrylic plastics. Whereas no gain was achieved in antigenicity by low temperature embedding in Lowicryl, the ultrastructural preservation in this material was superior to that in LR White.[37] The immunolabeling of myosin filaments with an antibody against one of the myosin light chains is shown in Figure 4, which compares tissue embedded in Lowicryl (glutaraldehyde-formaldehyde fixation) and PVA (formaldehyde fixation). With the three media listed above the visibility of myosin filaments in cross sections (glutaraldehyde-formaldehyde fixation) was better than that so far presented in frozen, sucrose infused tissue sections,[26] and actin filaments were most clearly delineated, although with poor reproducibility, in PVA.[37]

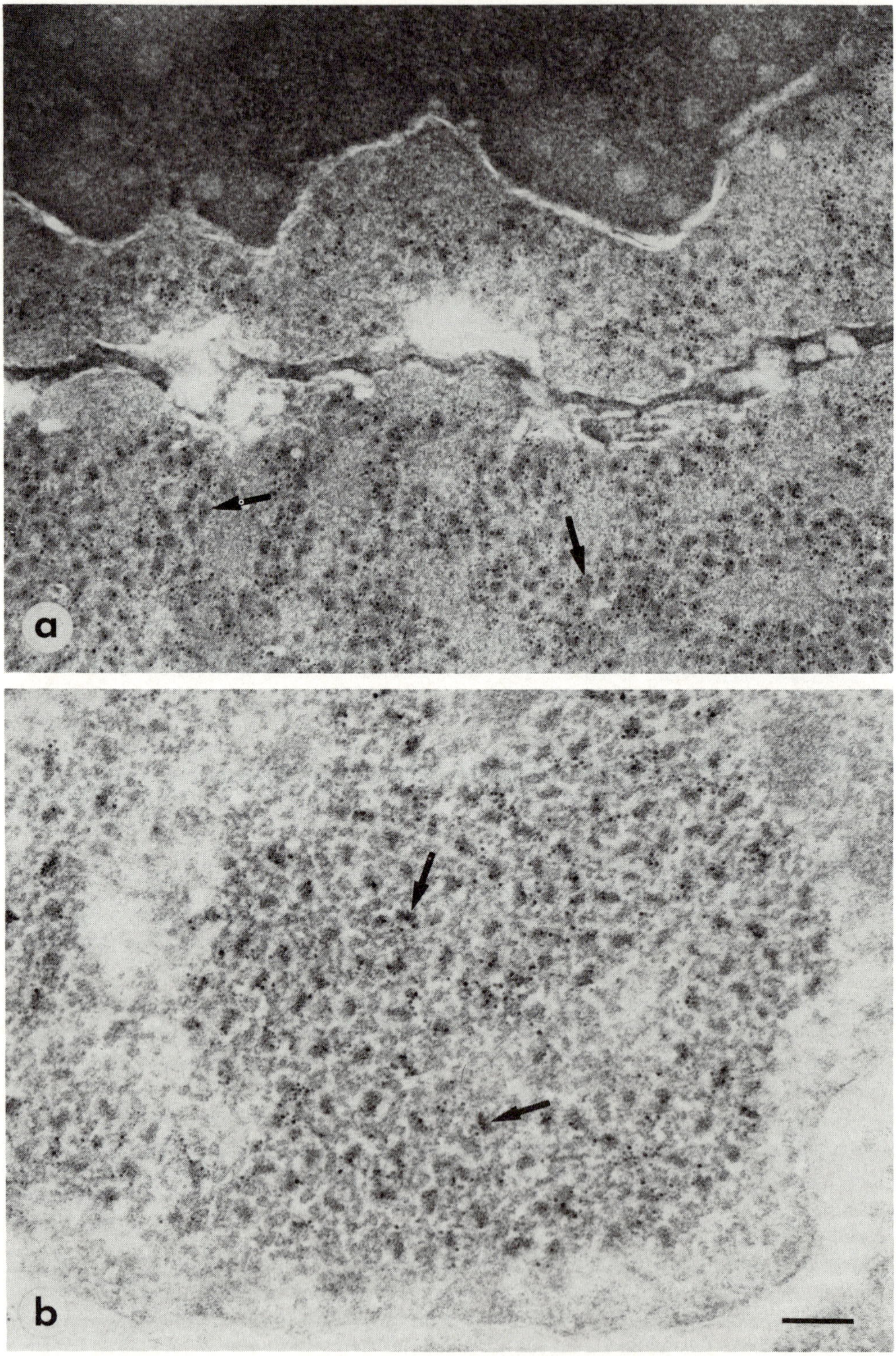

FIGURE 4. Postembedding immunogold labeling of myosin filaments (arrows) in smooth muscle cells with an anti-myosin-light chain antibody and secondary antibodies carrying a 5 nm gold tag. (a) Embedding in polyvinylalcohol (formaldehyde fixation). (b) Embedding in Lowicryl K4M (formaldehyde-glutaraldehyde fixation). Bar = 0, 1 μm. (b, courtesy of Dr. M. Maleeki.)

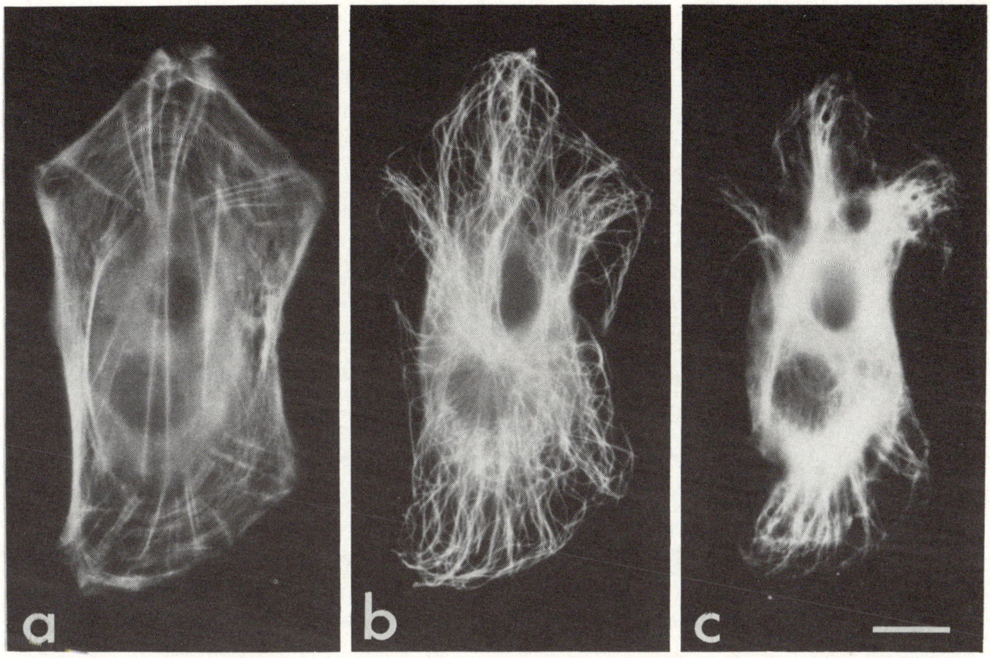

FIGURE 5. The three primary filament systems of the cytoskeleton in a human skin fibroblast: (a) actin; (b) microtubules; (c) intermediate filaments (see Reference 38 for labeling protocol). Bar = 1 μm.

We have already mentioned the doubts aired about chemically fixed smooth muscle as regards the true form of myosin in the cell. In the light of the apparent lability of the smooth muscle system, similar doubts could be raised about the presumed distributions of the other proteins we have discussed. The time is thus ripe to test these data obtained from fixed tissue by methods such as freeze-substitution immunocytochemistry that should circumvent potential rearrangements of protein components during chemical fixation. (However, see Chapter 1, by H. Plattner.)

III. PROBING THE CYTOSKELETON

A. THE PRIMARY SYSTEMS

The factors that determine and regulate subcellular organization and that transmit information from one compartment to another to direct the movement of organelles and the cell as a whole are central themes in contemporary cell biology. Here, the cytoskeleton with its coextensive arrays of filaments penetrating every corner of the cell bears all the attributes of a potential multifunctional coordinating network. In consequence, much attention has been focused on defining the individual and mutual roles of the cytoskeletal elements.

Cells in tissue culture have been popular subjects for cytoskeleton research since they display such clear and esthetically pleasing arrays of their cytoplasmic filaments. The three primary cytoskeletal systems, as visualized in the same cell in the fluorescence microscope, are shown in Figure 5.[38]

The actin system pervades more or less the entire cytoplasm, and in particular, delineates both the stationary and mobile boundaries of the cell. Most prominent in well spread and less motile cells are the large fiber bundles, or stress fibers of actin, that span the cytoplasm between sites of anchorage of the cell to the substrate. But from a functional point of view, the more diffuse forms of actin in the ruffling membranes (or leading edge), and between the stress fiber bundles, are of equal interest. These are discussed in somewhat more detail elsewhere.[36,39] The microtubular system radiates in a rather devious manner from an organizing center in the region of the centrosome towards the cell periphery. Both actin filaments and microtubules are in a state of constant flux, an expression of a dynamic equilibrium between the polymeric and monomeric forms of their respective proteins.[40,41] Typically, almost equal amounts of these proteins are split between the monomeric and polymeric pools (see Reference 36). For the intermediate filaments, the situation is somewhat different. These filaments, which express a pattern of organization not dissimilar to that shown by the microtubules (see Figure 5), are highly insoluble and more or less totally polymeric *in vivo*.

Much current attention is focused on the proteins that associate with the three filament systems, since these are presumed to mediate the interaction of the primary filaments with themselves and with other cellular elements. What has the EM contributed towards an understanding of these interactions?

B. APPROACHING MOLECULAR INTERACTIONS

An answer to the above question cannot avoid admission of present limitations in technique. While visualization of single molecules by, for example, negative staining or metal shadowing presents no severe problems, the detection of the same molecules in the macromolecular assemblies of the cell has proven difficult. Many of the proteins that associate with the cytoskeletal filaments are long, commonly flexible molecules (e.g., filamin, alpha-actinin, caldesmon, talin, MAPS (microtubule associated proteins), spectrin, fodrin, plectin, References 33, 42 to 45), but only in a few cases have they been visualized as resolvable entities on the filaments with which they associate.

Heuser and Hirokawa and their co-workers[46-51] have demonstrated in a series of impressive studies the level of detail resolvable in quick-freeze-deep-etch preparations. In the almost crystalline arrays of microtubules in protozoan axostyles, the visibility of the molecular bridging structures is remarkable.[47] Of course, molecular bridging structures are nothing new and were already described in the pioneer days of electron microscopy, in skeletal muscle[52] and flagella,[53] examples too well known to need elaboration here. It is in the less ordered systems of filament arrays, in nonmuscle cells, that particular problems are being encountered in defining and characterizing molecular interactions.

C. ACTIN-ASSOCIATED PROTEINS

The papers by Hirokawa and co-workers[50,51] on intestinal epithelial cells illustrate well the current state of the art of immunolocalization of actin-associated proteins in deep-etch preparations. Conventional electron microscopy and immunofluorescence microscopy of intestinal brush border cells has revealed two distinct cytoskeletal domains: (1) the microvilli composed of bundles of actin filaments with their associated bridging molecules, villin, fimbrin and the 110 kDa protein,[54,55] and (2) the terminal web at the base of the microvilli where other actin-associated proteins abound, among them alpha-actinin, filamin, tropomyosin and a spectrin-related protein, TW 260/240.[54,56,57] To have any chance of seeing the cytoskeletal components by the deep-etching method, Hirokawa et al.[49] had to lyse cells to remove the soluble fraction. One major class of cross links was then observed between the microvillar rootlets in impressive pictures of the terminal web, and two lines of evidence were taken to suggest that these contained myosin; they were extracted by conditions known to solubilize myosin, and they appeared to label with antimyosin antibodies. But the authors were equally aware that other cytoskeletal proteins were also lost during the initial extraction.

Subsequently, Hirokawa et al.[50] claimed to distinguish 5 nm fibrils from the 8 nm fibrils of myosin, and argued that it was these thin fibrils that were labeled with anti-fodrin (TW 260/240) antibodies. Using a conventional pre-embedding labeling routine with secondary ferritin-conjugated antibodies, Glenney et al.[57] demonstrated the general localization of TW 260/240 between the actin rootlets of the terminal web. Interestingly, in myosin depleted cells Hirokawa et al.[40] observed thin cross-links between the microvillar rootlets and the intermediate filaments abutting the terminal web region, but the nature of these links has yet to be established. A problem inherent to these localization methods, as well as to others, is that the antibodies themselves, binding to the antigen, automatically obscure the latter's identity so that a good deal of faith and judgment is required of the observer. Likewise, another problem, aired only between the lines in many of these studies, is that immunolabeling of delicate structures is complicated by distortions that readily result from the prolonged incubation and washing steps, even with fixed material. As an example, we have shown elsewhere that aldehyde fixed actin networks in fibroblasts are readily distorted even by incubation in distilled water;[58] the organization of the same networks deteriorates on incubation with antibodies in normal saline unless these are first stabilised with phalloidin.[39] Another point worthy of note is that in these preceeding studies on the brush border, methods involving postembedding immunocytochemistry have not been preferred. This is in part due to the poor visibility of the filamentous structures with these procedures (e.g., Reference 26), although improved methods of contrasting have been developed.[59]

If we restrict further discussion to the immunolocalization of cytoskeletal accessory proteins, the number of studies that have yielded information beyond that already evident in the light microscope are few. In two interesting pieces of work, Langanger and coworkers[60,61] applied pre-embedding immunogold methods to the study of actin-associated proteins in the actin stress fibers of cultured cells. To do this the cells had first to be extracted to allow antibody access; some proteins were clearly lost from the stress fibers by this procedure, but the distribution of proteins, at least as judged by light microscopy, was unchanged as compared to cells that were first fixed and then permeabilized thereafter. Two new results emerged from these studies. First, it could be shown that myosin was arranged in bipolar arrays (presumably in very short minifilaments) between adjacent alpha-actinin-containing sites in the stress fiber bundles,[61] and second, filamin appeared to be organized in a patchy fashion in concentrated foci along the length of the stress fibers.[60] The bipolar arrays of myosin were taken as consistent with the ability of stress fibers to develop tension and maintain cells in a spread out state on a substrate.

Part of our own studies have been concerned with the structural characterization of the sites of contact of a cell with its substrate *in vitro*.[62,63] The same sites or "focal contacts" represent regions where the actin cytoskeleton makes specialized attachments to the cell membrane.[30] These sites are rich in a number of proteins, including vinculin, talin, alpha-actinin, and transmembrane receptor molecules, termed integrins (see Reference 30). Although these cytoskeleton-membrane sites may be readily labeled for their associated proteins by indirect immunofluorescence microscopy, they appear inaccessible to electron dense conjugates, such as immunogold, in whole mount preparations.[62,64] Apparently, these conjugates are too large or unsuitably charged to penetrate between the actin filaments.

D. MICROTUBULES, INTERMEDIATE FILAMENTS AND THEIR ASSOCIATED PROTEINS

The results of various studies have suggested the existence of interactions of microtubules with organelles, as well as with intermediate and actin-filaments. Microtubule-associated proteins, or MAPS,[65] have been implicated as mediators of these interactions.[66] High molecular weight MAPS form visible projections on the surface of microtubules, and these have been demonstrated extensively *in situ*, again most dramatically in quick-freeze/deep-etch preparations.[48,51] In cultured cells viewed by immunofluorescence microscopy, MAPS likewise appear evenly distributed along the length of microtubules, and this has been confirmed at the ultrastructural level by the immunoperoxidase method.[67] In a timely double immuno-EM study, Shiomura and Hirokawa[68] have shown in dendrites that at least two of the MAPS (MAP1A and MAP2) are codistributed along the surface of microtubules; notably

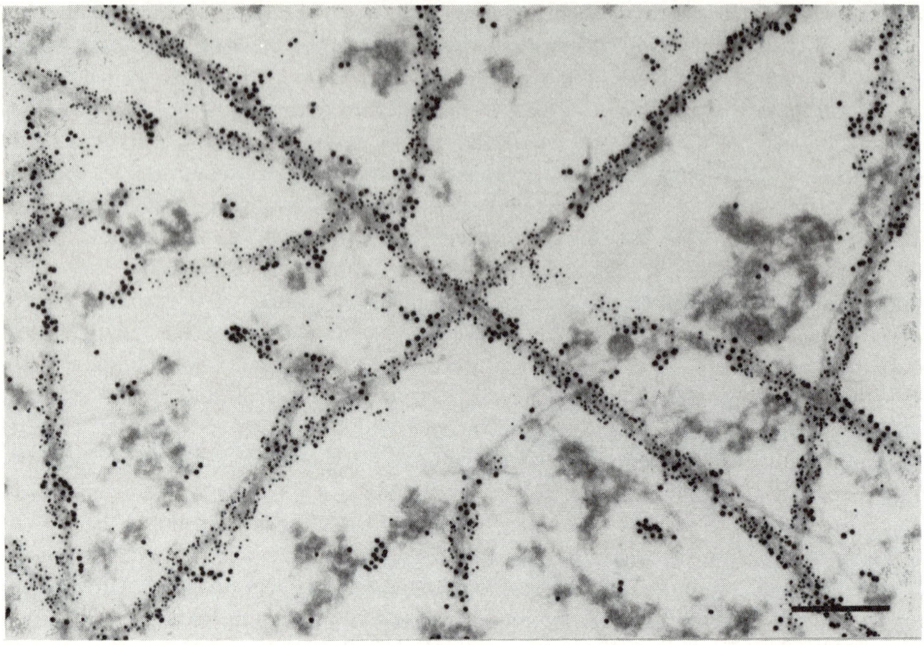

FIGURE 6. Pre-embedding, double immunogold labeling of Tyr (5 nm gold) and Glu (10 nm gold) tubulin in an interphase PtK2 cell. Bar = 0,2 μm. (From Geuens, G., *J. Cell Biol.*, 103, 1883, 1986. Courtesy of Dr. M. De Brabauder.)

this study, which utilized immunogold probes, was performed using a pre-embedding labeling procedure. These types of studies are bound to be featured in future investigations of the motile phenomena mediated by microtubule-associated proteins (e.g., MAP1c[69]) and by kinesin.[70]

Tublin undergoes a post-translational modification, a tyrosination *in vivo*, which appears to be related to the turnover of microtubules.[72] Using the immunogold staining procedure, Geuens and colleagues[71] could demonstrate the existence of both tyrosinated (Tyr) and detyrosinated (Glu) subunits in the same microtubules, which could not be detected by immunfluorescence microscopy (Figure 6). Although the relevance of these findings to microtubule dynamics *in vivo*[72] remains to be established, evidence has already been obtained which suggests that glu-tubulin confers stability to assembled microtubules.[73]

Abundant circumstantial evidence exists for the interaction of microtubules with intermediate filaments. In cultured cells, these filament systems closely overlap,[74] and the drug induced depolymerization of microtubules causes collapse of the intermediate filaments about the nucleus. Likewise, microinjection of anti-tubulin antibodies that leave the microtubules apparently intact, causes a similar collapse of the intermediate filament network.[75] Microtubule-associated proteins are also copurified with intermediate filament preparations.[66] But at the EM level, intermediate filament-microtubule interactions are less well characterized, except in such oriented systems as the axon.[48,51]

Traub[76] has reviewed extensively the proteins found associated with intermediate filaments. But the function of these proteins is just as obscure as that of the intermediate filaments themselves.[77] We may only note immunoelectron microscope data that has confirmed the direct association of these proteins with the intermediate filament class.[78-80]

We have already noted the observation of links between intermediate filaments and the actin rootlets in brush border cells,[49] and indirect evidence for interactions between actin and intermediate filaments has come from their observed colocalization in certain epithelial cells in culture.[81-83] But a demonstration of the nature of the presumed linking molecules involved in these interactions has not been performed.

Intermediate filaments also show specific associations with certain classes of membrane plaques, and a number of the proteins involved at these submembranous domains have been characterized and localized at the light and electron microscope level. This work is adequately reviewed by Cowin et al.[84] and will not be discussed here.

IV. CONCLUDING REMARKS

In the departed heyday of EM, ultrastructural morphology was of prime importance. Such criteria alone no longer satisfy the demands of many of the questions posed in contemporary cell biology. Problems of any significance call for an integrated approach in which EM features as only one of the techniques at hand. Particularly noteworthy in recent years has been the convergence of biochemical, immunological, and microscopical techniques in investigations of cellular mechanisms, and the success of their combined application.

But the techniques we have developed for immunoelectron microscopy are by no means ideal; there is ample room for improvement. In particular, the results of postembedding procedures for cytoskeletal structures have not yet come up to expectations; we have limited access of antibodies and suboptimal labeling, restricted mainly to the section surface. In pre-embedding procedures, we face protein loss and structural distortions due to the necessary extraction to permit antibody entry, and the extended incubation procedures to effect labeling. Common to all methods is the limitation set by the antibody markers themselves, many of which fail to react after the degree of chemical fixation necessary for adequate ultrastructural preservation. In this respect, not enough attention is paid to the criteria used for antibody selection. Future progress will depend on how successful we are in circumventing these technical limitations.

ACKNOWLEDGMENTS

I thank Mrs. G. Maier for typing and Mrs. M. Klappacher and Mrs. K. Koppelstätter for assistance with the illustrations. The work reviewed here was supported in part by grants from the Austrian Science Research Council and the Austrian National Bank.

REFERENCES

1. **Bendayan, M., Nanci, A., and Kan, F. W. K.,** Effect of tissue processing on colloidal gold cytochemistry, *J. Histochem. Cytochem.*, 35, 983, 1987.
2. **Kellenberger, E., Dürrenberger, M., Villiger, W., Carlemalm, E., and Wurtz, M.,** The efficiency of immunolabel on Lowicryl sections compared to theoretical predictions, *J. Histochem. Cytochem.*, 35, 959, 1987.
3. **Huxley, H. E.,** The mechanism of muscle contraction, *Science*, 64, 1356, 1969.
4. **Wang, K.,** Sarcomere-associated cytoskeletal lattice in striated muscle, review and hypothesis, in *Cell and Muscle Motility*, Vol. 6, Shay, J. W., Ed., Plenum, New York, 1985, 315.
5. **Ohtsuki, I., Masaki, I., Nonomura, Y., and Ebashi, S.,** Periodic distribution of troponin along the thin filament, *J. Biochem.*, 61, 817, 1967.
6. **Dennis, J. E., Shimizu, T., Reinach, F. C., and Fischman, D. A.,** Localization of C-protein isoforms in chicken skeletal muscle: ultrastructural detection using monoclonal antibodies, *J. Cell Biol.*, 98, 1514, 1984.
7. **Offer, G.,** C-protein and the periodicity in the thick filaments of vertebrate skeletal muscle, 37th *Cold Spring Harbor Symp. Quant. Biol.*, 87, 1973.

8. **Bähler, M., Eppenberger, H. M., and Walliman, T.,** Novel thick filament protein of chicken pectoralis muscle: the 86 kd protein, *J. Mol. Biol.*, 186, 393, 1985.
9. **Bennett, P., Craig, R., Starr, R., and Offer, G.,** The ultrastructural location of C-protein, X-protein and H-protein in rabbit muscle, *J. Muscle Res. Cell Motil.*, 7, 550, 1986.
10. **Craig, R. and Offer, G.,** Axial arrangement of crossbridges in thick filaments of vertebrate skeletal muscle, *J. Mol. Biol.*, 102, 325, 1976.
11. **Shimuzu, T., Dennis, J. E., Masaki, T., and Fischman, D. A.,** Axial arrangement of the myosin rod in vertebrate thick filaments: immunoelectron microscopy with a monoclonal antibody to light meromyosin, *J. Cell Biol.*, 101, 1115, 1985.
12. **Ohtsuki, I.,** Functional organization of the tropomin-tropomyosin system, in *Muscle Contraction: Its Regulatory Mechanisms*, Ebashi, S., Maruyama, K., and Endo, M., Eds., Springer-Verlag, Berlin, 1980, 237.
13. **Maruyama, K.,** Connectin, an elastic filamentous protein of striated muscle, *Int. Rev. Cytol.*, 104, 81, 1986.
14. **Fürst, D. O., Osborn, M., Nave, R., and Weber, K.,** The organization of titin filaments in the half-sarcomere revealed by monoclonal antibodies in immunoelectron microscopy; a map of 10 nonrepetitive epitopes starting at the Z-line extends close to the M-line, *J. Cell Biol.*, 106, 1563, 1988.
15. **Gassner, D.,** Myofibrillar interaction of blot immunoaffinity-purified antibodies against native titin as studied by direct immunofluorescence and immunogold staining, *Eur. J. Cell Biol.*, 40, 176, 1986.
16. **Tokuyasu, K. T.,** Present state of immunocryoultramicrotomy, *J. Histochem. Cytochem.*, 31, 164, 1983.
17. **Tokuyasu, K. T.,** Application of cryoultramicrotomy to immunocytochemistry, *J. Microsc.*, 143, 139, 1986.
18. **Tokuyasu, K. T.,** Visualization of longitudinally-oriented intermediate filaments in frozen sections of chicken cardiac muscle by a new staining method, *J. Cell Biol.*, 97, 562, 1983.
19. **Watkins, S. C., Samuel, J. L., Marotte, F., Bertier-Savalle, B., and Rappaport, L.,** Microtubules and desmin filaments during onset of heart hypertrophy in rats in a double immunoelectron microscope study, *Cir. Res.*, 60, 327, 1987.
20. **Lemanski, L. F., Paulson, D. J., Hill, C. S., Davis, L. A., Riles, L. C., and Lim, S.-S.,** Immunoelectron microscopic localization of alpha-actinin on Lowicryl-embedded thin-sectioned tissues, *J. Histochem. Cytochem.*, 33, 515, 1985.
21. **Slepecky, N. and Chamberlain, S. C.,** Immunoelectron microscopic and immunofluorescent localization of cytoskeletal and muscle-like contractile proteins in inner ear sensory hair cells, *Hearing Res.*, 20, 245, 1985.
22. **Tidball, J. G.,** Alpha-actinin is absent from the terminal segments of myofibrils and from subsarcolemmal densities in frog skeletal muscle, *Exp. Cell Res.*, 170, 469, 1987.
23. **Somlyo, A. V., Butler, T. M., Bond, M., and Somlyo, A. P.,** Myosin filaments have nonphosphorylated light chains in relaxed smooth muscle, *Nature*, 294, 567, 1981.
24. **Tsukita, S., Tsukita, S., Usukura, J., and Ishikawa, H.,** Myosin filaments in smooth muscle cells of the guinea pig taenia coli: a freeze-substitution study, *Eur. J. Cell Biol.*, 28, 195, 1982.
25. **Bagby, R.,** Towards a comprehensive three-dimensional model of the contractile system of vertebrate smooth muscle cells, *Int. Rev. Cytol.*, 105, 67, 1986.
26. **Geiger, B., Dutton, A. H., Tokuyasu, K. T., and Singer, S. J.,** Immunoelectron microscope studies of membrane-microfilament interactions: distribution of alpha-actinin, tropomyosin and vinculin in intestinal epithelial brush border and chicken gizzard smooth muscle cells, *J. Cell Biol.*, 91, 614, 1981.
27. **Schollmeyer, J. F., Furcht, L. T., Goll, D. E., Robson, R. M., and Stromer, M. H.,** Localization of contractile proteins in smooth muscle cells and in normal and transformed fibroblasts, in *Cell Motility*, Vol. A., Goldman, R., Pollard, T., and Rosenbaum, J., Eds., Cold Spring Harbor Laboratory, New York, 1976, 361.
28. **Small, J. V.,** Polyvinylalcohol, a water soluble resin suitable for electron microscope immunocytochemistry, in *Proc. 8th Eur. Congr. Electron Microsc.*, Vol. 3, Csanady, A., Rohlich, P., and Szabo, D., Eds., Congress Comm., Budapest, 1984, 1799.
29. **Small, J. V.,** Geometry of actin-membrane attachments in the smooth muscle cell: the localizations of vinculin and alpha-actinin, *EMBO J.*, 4, 45, 1985.
30. **Burridge, K.,** Substrate adhesions in normal and transformed fibroblasts: organization and regulation of cytoskeletal, membrane and extracellular matrix components at focal contacts, *Cancer Rev.*, 4, 18, 1986.
31. **Volberg, T., Sabenay, H., and Geiger, B.,** Spatial and temporal relationship between vinculin and talin in the developing chicken gizzard smooth muscle, *Differentiation*, 32, 34, 1986.
32. **Wiche, G., Krepler, R., Artlieb, U., Pytela, R., and Denk, H.,** Occurrence and immunolocalization of plectin in tissues, *J. Cell Biol.*, 97, 887, 1983.
33. **Fürst, D. O., Cross, R. A., De Mey, J., and Small, J. V.,** Caldesmon is an elongated, flexible molecule localized in the actomyosin domains of smooth muscle, *EMBO J.*, 5, 251, 1986.

34. **Small, J. V., Fürst, O. E., and De Mey, J.,** Localization of filamin in smooth muscle, *J. Cell Biol.*, 102, 210, 1986.
35. **Lazarides, E.,** Intermediate filaments, *Ann. Rev. Biochem.*, 51, 219, 1982.
36. **Small, J. V.,** The cytoskeleton: selected topics, in *Fibrous Protein Structure,* Squire, J. M. and Vibert, P. J., Eds., Academic Press, London, 1987, 357.
37. **Malecki, M. and Small, J. V.,** Immunocytochemistry of contractile and cytoskeletal proteins in smooth muscle: Lowicryl, LR white and polyvinylalcohol compared, *Protoplasma,* 139, 160, 1987.
38. **Small, J. V., Zobeley, S., Rinnerthaler, G., and Faulstich, H.,** Coumarin-phalloidin: a new actin probe permitting triple immunofluorescence microscopy of the cytoskeleton, *J. Cell Sci.*, 89, 21, 1988.
39. **Small, J. V.,** The actin cytoskeleton, in *Electron Microscopy Reviews,* Vol. 1, Harris, J. R. and Horne, R. W., Eds., Pergamon Journals, Oxford, 1988, 155.
40. **Pollard, J. D. and Cooper J. A.,** Actin and actin-binding proteins. A critical evaluation of mechanisms and functions, *Ann. Rev. Biochem.*, 55, 987, 1986.
41. **Schulze, E. and Kirschner, M.,** Dynamic and stable populations of microtubules in cells, *J. Cell Biol.*, 104, 277, 1987.
42. **Foisner, R. and Wiche, G.,** Structure and hydrodynamic properties of plectin molecules, *J. Mol. Biol.*, 198, 515, 1987.
43. **Glenney, J. R., Glenney, P., Osborn, M., and Weber, K.,** An F-actin and calmodulin-binding protein from isolated intestinal brush borders has a morphology related to spectrin, *Cell,* 28, 843, 1982.
44. **Molony, L., McCaslin, D., Abernethy, J., Paschal, B., and Burridge, K.,** Properties of talin from chicken gizzard smooth muscle, *J. Biol. Chem.*, 262, 7790, 1987.
45. **Tyler, J., Anderson, J. M., and Branton, D.,** Structural comparison of several actin-binding macromolecules, *J. Cell Biol.*, 85, 489, 1980.
46. **Heuser, J. E. and Kirschner, M. W.,** Filament organization revealed in platinum replicas of freeze-dried cytoskeletons, *J. Cell Biol.*, 86, 212, 1980.
47. **Heuser, J. E.,** Different structural states of a microtubule cross-linking molecule, captured by quick-freezing motile axostyles in protozoa, *J. Cell Biol.*, 103, 2209, 1986.
48. **Hirokawa, N.,** Cross-linker system between neurofilaments, microtubules, and membraneous organelles in frog axons revealed by the quick-freeze, deep-etching method, *J. Cell Biol.*, 94, 129, 1982.
49. **Hirokawa, N., Tilney, L. G., Fujiwara, K., and Heuser, J. E.,** Organization of actin, myosin and intermediate filaments in the brush border of intestinal epithelial cells, *J. Cell Biol.*, 94, 425, 1982.
50. **Hirokawa, N., Cheney, R. E., and Willard, M.,** Location of a protein of the fodrin-spectrin-TW 260/240 family in mouse intestinal brush border, *Cell,* 32, 953, 1983.
51. **Hirokawa, N., Bloom, G. S., and Vallee, R. B.,** Cytoskeletal architecture and immunocytochemical localization of microtubule-associated proteins in regions of axons associated with rapid axonal transport: the β, β'-iminodipropionitrile-intoxicated axon as a model system, *J. Cell Biol.*, 101, 227, 1985.
52. **Huxley, H. E.,** The double array of filaments in cross-striated muscle, *J. Biophys. Biochem. Cytol.*, 3, 631, 1957.
53. **Gibbons, I. R.,** Cilia and flagella of eukaryotes, *J. Cell Biol.*, 91, 107s, 1981.
54. **Bretscher, A. and Weber, K.,** Localization of actin and microfilament-associated proteins in the microvilli and terminal web of the intestinal brush border by immunofluorescence microscopy, *J. Cell Biol.*, 79, 839, 1987.
55. **Matsudaira, P. T. and Burgess, D. R.,** Identification and organization of the components in the isolated microvillus cytoskeleton, *J. Cell Biol.*, 83, 667, 1979.
56. **Drenkhahn, D. and Gröschel-Stewart, U.,** Localization of myosin, actin, and tropomyosin in rat intestinal epithelium: immunohistochemical studies at the lgiht and electron microscope levels, *J. Cell Biol.*, 86, 475, 1980.
57. **Glenney, J. R., Glenney, P., and Weber, K.,** The spectrin-related molecule, TW-260/240, cross links the actin bundles of the microvillus rootlets in the brush borders of intestinal epithelial cells, *J. Cell Biol.*, 96, 1491, 1983.
58. **Small, J. V.,** Factors affecting the integrity of actin meshworks in cultured cells, in *Cell Motility: Mechanism and Regulation,* Ishikawa, H., Hatanio, S., and Sato, H., Eds., University of Tokyo Press, Tokyo, 1985, 493.
59. **Keller, G. A., Tokuyasu, K. T., Dutton, A. H., and Singer, S. J.,** An improved procedure for immunoelectron microscopy: ultrathin plastic embedding of immunolabelled ultrathin frozen sections, *Proc. Natl. Acad. Sci. U.S.A.*, 81, 5744, 1984.
60. **Langanger, G., De May, J., Moeremans, M., Daneels, G., De Brabander, M., and Small, J. V.,** Ultrastructural localization of alpha-actinin and filamin in cultured cells with the immunogold staining (IGS) method, *J. Cell Biol.*, 99, 1324, 1984.
61. **Langanger, G., Moeremans, M., Daneels, G., Sobieszek, A., De Brabander, M., and De Mey, J.,** The molecular organization of myosin in stress fibres of cultured cells, *J. Cell Biol.*, 102, 200, 1986.

62. **Rinnerthaler, G., Geiger, B., and Small, J. V.,** Contact formation during fibroblast locomotion: involvement of membrane ruffles and microtubules, *J. Cell Biol.,* 106, 747, 1988.
63. **Small, J. V. and Rinnerthaler, G.,** Cytostructural dynamics of contact formation during fibroblast locomotion *in vitro, Exp. Biol. Med.,* 10, 54, 1985.
64. **Nicol, A., Nermut, M. V., Doeinck, A., Robenek, H., Wiegand, C., and Jockusch, B. M.,** Labeling of structural elements at the ventral plasma membrane of fibroblasts with the immunogold technique, *J. Histochem. Cytochem.,* 35, 499, 1987.
65. **Olmsted, J. B.,** Microtubule-associated proteins, *Ann. Rev. Cell Biol.,* 2, 421, 1986.
66. **Wiche, G.,** High-molecular-weight microtubules-associated proteins (MAPS): a ubiquitous family of cytoskeletal connecting links, *Trends Biochem. Sci.,* 10, 67, 1985.
67. **De Brabander, M., Bulinski, J. C., Geuens, G., De Mey, J., and Borisy, G. G.,** Immunoelectron microscopic localization of the 210,000 mol wt microtubule-associated protein in cultured cells of primates, *J. Cell Biol.,* 91, 438, 1981.
68. **Shiomura, Y. and Hirokawa, N.,** Colocalization of microtubule-associated protein IA and microtubule associated protein 2 on neuronal microtubules *in situ* revealed with double label immunoelectron microscopy, *J. Cell Biol.,* 104, 1575, 1987.
69. **Paschal, B. M. and Vallee, R. D.,** Retrograde transport by the microtubule-associated protein MAPIC, *Nature,* 330, 181, 1987.
70. **Vale, R. D.,** Intracellular transport using microtubule-based motors, *Ann. Rev. Cell Biol.,* 3, 347, 1987.
71. **Geuens, G., Gundersen, G., Nuydens, R., Cornelissen, F., Bulinski, J. C., and De Brabander, M.,** Ultrastructural colocalization of tyrosinated and detyrosinated alpha-tubulin in interphase and mitotic cells, *J. Cell Biol.,* 103, 1883, 1986.
72. **Webster, D. R., Gundersen, G. G., Bulinski, J. C. and Borisy, G. G.,** Assembly and turnover of detyrosinated tubulin, *in vivo, J. Cell Biol.,* 105, 265, 1987.
73. **Kreis, J. E.,** Microtubules containing detyrosinated tubulin are less dynamic, *EMBO J.,* 6, 2597, 1987.
74. **Geiger, B. and Singer, S. J.,** Association of microtubules and intermediate filaments in chicken gizzard cells as detected by double immunofluorescence, *Proc. Natl. Acad. Sci. U.S.A.,* 77, 4769, 1980.
75. **Blose, S. H., Meltzer, D. I., and Feramisco, J. R.,** 10 nm filaments are induced to collapse in living cells microinjected with monoclonal and polyclonal antibodies against tubulin, *J. Cell Biol.,* 98, 847, 1984.
76. **Traub, P.,** *Intermediate Filaments,* Springer-Verlag, Berlin, 1985.
77. **Geiger, B.,** Intermediate filaments looking for a function, *Nature,* 329, 392, 1987.
78. **Foisner, R., Leichtfried, F. E., Herrman, H., Small, J. V., Lawson, D., and Wiche, G.,** Cytoskeleton-associated plectin: *in situ* localization, *in vitro* reconstitution and binding to immobilized intermediate filament proteins, *J. Cell Biol.,* 106, 723, 1988.
79. **Lawson, P.,** Epinerin: a new protein associated with vimentin filaments in non-neural cells, *J. Cell Biol.,* 97, 1891, 1983.
80. **Lieska, N., Yang, H.-Y., and Goldman, R. D.,** Purification of the 300 K intermediate filament-associated protein and its *in vitro* recombination with intermediate filaments, *J. Cell Biol.,* 101, 802, 1985.
81. **Celis, J. E., Small, J. V., Mose Larson, P., Fey, S. J., De Mey, J., and Celis, A.,** Intermediate filaments in monkey kidney TC7 cells: focal centers and interrelationship with other cytoskeletal systems, *Proc. Natl. Acad. Sci. U.S.A.,* 81, 1117, 1984.
82. **Green, K. J., Geiger, B., Jones, J. C. R., Talian, J. C., and Goldman, R. D.,** The relationship between intermediate filaments and microfilaments before and during the formation of desmosomes and adherens-type junctions in mouse epidermal keratinocytes, *J. Cell Biol.,* 104, 1389, 1987.
83. **Knapp, L. W., O'Guin, W. M., and Sawyer, R. H.,** Rearrangement of the keratin cytoskeleton, after combined treatment with microtubule and microfilament inhibitors, *J. Cell Biol.,* 97, 1788, 1983.
84. **Cowin, P., Franke, W. W., Grund, C., Kapprell, H.-P., and Kartenbeck, J.,** The desmosome-intermediate filament complex, in *The Cell in Contact,* Edelman, G. M., and Thiery, J. -P., Eds., John Wiley & Sons, New York, 1985, 427.

Chapter 17

ANALYSIS OF COMPLEX THREE-DIMENSIONAL STRUCTURES INVOLVED IN DYNAMIC PROCESSES BY HIGH VOLTAGE ELECTRON MICROSCOPY AND LOW VOLTAGE HIGH RESOLUTION SCANNING ELECTRON MICROSCOPY

H. Ris and J. B. Pawley

TABLE OF CONTENTS

I. INTRODUCTION

The phenomenal development of cell biology in recent decades has amply documented that dynamic processes in cells are associated with specific structures that have evolved to carry out a particular function. Because it can image structures in the molecular range, the electron microscope has played a pivotal role in relating cellular processes to the molecular architecture of cell components. We have learned that to fully understand the biochemistry of cellular processes, it is essential to know the structures involved and the modifications they undergo in relation to these dynamic changes. The elucidation of muscle contraction could only be understood after the structure of myofibrils had been worked out. A correct structural description is usually needed in order to frame the important questions about the biochemistry of a cellular process. Indeed, in many cases certain dynamic processes were not even recognized until they were suggested by the discovery of specific structures. Specific examples will be discussed in this chapter.

While the conventional 100 kV transmission electron microscope (TEM) has thus revolutionized cell biology, it has some serious limitations. Limited penetration of the electron beam and the chromatic aberration of electromagnetic lenses restrict its application to very thin specimens.

At 100 kV the resolution is approximately 1/10 of the specimen thickness.[1] Considering that with most biological materials resolution is limited by beam damage to about 3 nm, this resolution can only be obtained in specimens less than 30 nm thick. High quality in sectioning is often almost synonymous with thinness. This concentration on thinness, however, does exact a price in terms of an understanding of the cell as a three-dimensional structure. Assuming a cell diameter of about 30 μm, a section of that thickness represents only 0.1% of the cell. Certain structures allow reconstruction from serial sections, but with many complex three-dimensional organelles (such as chromosomes) that is not possible. The effect of chromatic aberration is greatly reduced at higher accelerating voltages, so that at 1MeV the resolution limit due to chromatic aberration becomes 1/200 of the specimen thickness.[2] Depending on the density of the specimen, a 2 nm resolution can be obtained with a specimen 1.5 to 1 μm thick.[3]

The first million-volt electron microscope was built in Toulouse, France by Gaston Dupouy.[4] Several commercial 1MeV instruments became available in the late 1960's (Hitachi, Jeol, AEI). In the U.S.A. this led to the establishment of three National Resources for high voltage electron microscopy (HVEM) for biomedical research supported by the Biomedical Research Technology Program of the National Institutes of Health (NIH). In 1972, Ris established the Madison, Wisconsin HVEM Facility. Porter organized the Laboratory for HVEM in Boulder, Colorado in 1973, and Parsons the HVEM Laboratory at the New York State Department of Health (Albany) in 1978. Recently, several manufacturers have produced instruments which have less chromatic aberration and which partially compensate for the chromatic magnification error. They have operating voltages between 200 kV and 400 kV. These Intermediate Voltage Electron Microscopes (IVEM) are much less expensive than 1 MeV microscopes and fit into conventional laboratory rooms. Regional facilities with IVEM are now being organized, and this will greatly facilitate structural analysis using whole mounts and sections up to about 1 μm thick.[5]

The scanning electron microscope (SEM) was designed to give information on surface topography. Its contribution to cell architecture has so far been limited to studies of cell shape and surface topography at relatively low resolution, and to determining the surface distribution of specific membrane molecules using lectins and immunogold labels. Attempts to look at intracellular structures, while limited so far, have been encouraging, and have hinted at a powerful new way to look at cytoplasmic organization if the resolution of topographical details could be improved.[6] Theoretical considerations suggested that this was feasible by using lower voltages down to 1 kV, a field emission gun, and an improved lens to form a small probe at low voltages.[7] An instrument working between 1 kV and 30 kV and having instrumental resolution of 4 nm at 1 kV has now become available with the Hitachi S-900 SEM. The first microscope of this type to be delivered outside of Japan has recently been installed at the Madison Integrated Microscopy Resource (IMR), and is available to qualified biomedical researchers. Some examples of its application to the study of cytoplasmic structure will be described here.

II. THE MILLION VOLT ELECTRON MICROSCOPE, RATIONALE AND METHODOLOGY

A. USEFUL SPECIMEN THICKNESS. THE ROLE OF RESOLUTION, CONTRAST AND STRUCTURAL COMPLEXITY

Million volt electron microscopes were designed to decrease the chromatic aberration limit of the EM so that reasonable resolution could be obtained with thicker specimens. Fotino[3] measured actual resolution with interference fringes of gold particles evaporated onto thick sections of rat liver. He showed that 2 nm resolution could be obtained with 1 μm sections of plastic embedded tissue. However visibility of small structures in thick samples is limited by other factors, especially the contrast of the sample. In an ultrathin section few structures are smaller than the section thickness. The difference in electron scattering of the pixel column through a stained specimen and an adjacent column of unstained plastic is great, and therefore the contrast high. In a thick section, however, a small structure comprises only a small fraction of a pixel column, though the structure and the difference in electron scattering compared with an adjacent pixel column of plastic is now much smaller, and therefore the contrast of a similar structure is much reduced. As a result, HVEM micrographs often appear to be less distinct than conventional micrographs of thin sections, not because of insufficient resolution, but because of the geometry and scale of the ultrastructural details in comparison to the thinner sample.

Geometry limits usable specimen thickness in a second way which we might call image complexity. The image produced in the HVEM is a projection onto a flat surface of all structures located throughout the thickness of the sample. As section thickness goes above 500 nm it often becomes difficult to distinguish between the many overlapping structures. Although stereo images are routinely used in an attempt to overcome this problem, their usefulness is limited by the requirement that each object must be not only distinguished, but "recognized" by both eyes before the brain can assign to it a relative height. If the object is obscured or is not recognizable in one view, it cannot be isolated from the confusion by binocular stereopsis.

B. THICK SECTIONS, STEREO MICROGRAPHS, SERIAL SECTIONS

The analysis of complex cell structures from stereo micrographs of thick sections has been a major contribution of HVEM. Sections 0.25 μm to 1 μm thick are easily prepared from plastic embedded cells. We recommend the Epon-Araldite mixture of Mollenhauer because stains penetrate it easily. Up to 0.25 μm diamond knives can be used, for thicker sections we recommend glass knives. To avoid fracture lines parallel to the knife edge, the slowest cutting edge should be used. The sections can be flattened on the trough by several applications of xylene vapor. Ribbons are preferably picked up on single hole grids coated with 0.5% formvar. To facilitate stain penetration, the sections are stained before a stabilizing layer of carbon is evaporated upon them. For sections 1/4 to 1 μm thick, we use 7.5% uranyl magnesium acetate in water for 2 to 4 h at 50°C, followed by lead citrate for 20 min at room temperature.

Three dimensional information on cell structures is most easily obtained from stereoscopic micrographs. The axis-centered tilt stage on the Madison 1 MeV microscope facilitates the production of stereo pairs that are free from unwanted variations in magnification caused by changes in sample height as the specimen is tilted.

Another important application of HVEM has been in reconstructions from serial sections. The use of thicker sections greatly decreases the required number of consecutive sections, and as a result, projects are now feasible which would have been too time-consuming or technically difficult if undertaken using thin sections.[8] Furthermore, relatively inexpensive hardware and software for computer-aided reconstruction from serial sections have now become available.[9] Reconstruction of rather large volumes has thus become feasible and been especially productive in neurobiology.

C. WHOLE MOUNT PREPARATIONS

Certain cells or isolated cell components are thin enough to be imaged by HVEM without recourse to embedding and sectioning. To prepare such specimens for electron microscopy it is most important to avoid surface tension distortions during the necessary dehydration. This can be achieved by using the critical point drying method (CPD) introduced in 1951 by T. Anderson.[10] This method is very sensitive to the presence of traces of water, or intermediate fluids such as ethanol, in the final CO_2 solution. Special efforts must be made to avoid such contamination of the CO_2, or cellular components such as cytoskeletal fibers will be distorted.[11] Dehydration without surface tension artifacts can also be achieved by rapid freezing and freeze-drying (FD). With this technique, it is crucial to avoid ice crystal formation, both during the original freezing and during the sublimation of water. If the specimen should warm up before all the water is sublimated, distortions similar to those observed after faulty CPD are obtained.[12] Well preserved whole mounts are especially well suited for HVEM. The absence of embedding plastic results in excellent contrast of the most delicate structures and can yield striking stereoscopic images.

D. SECTIONS WITH EMBEDMENT REMOVED: REVERSIBLE EMBEDDING FROZEN SECTIONS

The excellent contrast obtained with CPD whole mounts led to a search for embedment plastics which could be removed from sections followed by CPD of the sections. Wolosewick[13] introduced polyethylene glycol (PEG) for this purpose. This plastic can be extracted from sections with water, the sections are then dehydrated in ethanol and CPD. It turned out that good sections are rather difficult to produce with this technique, and the published micrographs show considerable distortion of cytoplasmic structures. It is not clear at present whether this distortion occurs during plastic extraction or CPD.

Another approach was taken by Gorbsky and Borisy.[14] Tissue was embedded in polymerized polymethylmethacrylate that had been dissolved in dichloromethane. Evaporation of the solvent at room temperature produced tissues embedded in hard plastic that could easily be sectioned. The plastic was then extracted from the sections and the grids dried by CPD. This method provided excellent structural preservation with either ultrathin or thick (0.25 to 0.5 μm) sections.

Guan and Ris[15] have used thick frozen sections to obtain embedment free sections. Excellent freezing microtomes that make the preparation of frozen sections almost routine are now available. The innovation of Guan and Ris was to use CPD on the thawed sections. This assured excellent structural preservation and contrast similar to that characteristic of CPD whole mounts. The use of frozen sections has advantages especially for immunocytochemistry. Fixation can be minimal. Also organic solvents and heat for plastic polymerization can be avoided preceding the antibody labeling.

The use of whole mounts and embedment free sections will greatly expand the applicability of immunogold labeling. More antigenic surfaces are exposed in such thick preparations and stereoscopic imaging greatly facilitates the exact localization of gold markers. Until recently, lack of penetration by the gold-labeled antibody into such preparations remained an impediment to this approach, but with the smaller probes now available, this problem can be overcome.

III. LOW VOLTAGE SCANNING ELECTRON MICROSCOPE (LVSEM): RATIONALE AND METHODOLOGY

A. WHY LVSEM?

At low magnification, contrast in a secondary electron image from a scanning electron microscope (SEM) is produced by variations in surface angle and by changes in the efficiency with which the signal from different parts of the specimen is collected. The resulting image is quite analogous to a black and white photograph of a macroscopic, three-dimensional object, and as long as the analogy holds, the observer can accurately interpret a SEM image in terms of the shape of a microscopic surface being imaged. The analogy breaks down at high magnification because interaction volume of the beam striking the sample becomes appreciable compared to the surface features being imaged.

In a normal SEM operating at a beam voltage of 20 kV, the interaction volume extends for tens of micrometers in dried biological tissue. To reduce this penetration (and also to control charge accumulation on the surface of the sample), the surface is usually coated with 0 to 20 nm of a heavy metal, such as gold, which is about 100 times more dense than the dried tissue. This coating reduces penetration by a similar factor, but unfortunately, it also decorates or obscures small topographic features.

Changing the beam energy from 20 kV to 1 to 2 kV also reduces penetration by a factor of 100, and so low voltage operation (LVSEM) is another solution to the penetration problem.[7] Unfortunately, conventional SEMs cannot produce the small beam diameter needed for high resolution at 1 to 2 kV. Recently, however, instruments having both a field-emission electron source and a short focal-length final lens have overcome this problem.[16] We have used an instrument of this type, a Hitachi S-900 especially modified for low voltage operation,[17] to make the images shown here.

Compared to images from conventional SEMs, the important features of LVSEM images are that they have higher topographic contrast, less of the "penetration artifacts" that make the image of a cube look bright close to its edges or permit one to "see" a subsurface feature "through" a surface one. Not so evident, but also important, is the fact that LVSEM images can be made from samples having metal coatings that are considerably thinner than those used for observation at higher voltage. A final feature of low voltage SEM operation is that radiation damage from the beam is less, and this reduces the gross damage to the sample which might otherwise cause the sample to change shape under observation, a condition that precludes the production of interpretable stereo images.

B. SPECIMEN PREPARATION FOR LVSEM OBSERVATION OF INTRACELLULAR STRUCTURES

Several investigators have tried to use SEM to study intracellular structures, tempted by the excellent contrast and depth provided in SEM pictures. Most results were disappointing due to the limited topographic resolution of available instruments. We shall mention here those studies that contain interesting specimen preparation techniques that promise to be useful with the new LVSEM. In order to see cytoplasmic structures by SEM, the cell must be cleaved open without distorting the contents. Freeze-fracture of fixed cells is one useful approach. Tanaka[18] has published stunning stereo micrographs of intracellular membrane systems by freeze-fracture of OsO_4 fixed cells, which are then macerated for 24 to 72 h in OsO_4. This procedure extracts all materials except the cytomembranes, which are now uncovered to be viewed by SEM.

To reveal chromatin structure and cytoskeleton, Haggis et al.[19-21] introduced the freeze-fracture, thaw-fix technique. Cells are rapidly frozen, fractured, then thawed in fixative and CPD. This removes soluble proteins and salts from the cytosol and exposes chromatin fibers and cytoskeleton to be imaged by SEM.

We have successfully applied a simple method first described by Watson et al.,[22] and more recently referred to as "dry cleaving".[23-25] Cells attached to coverslips or grids are fixed and CPD. To fracture the dried cells, they are gently pressed onto pressure sensitive tape and then pulled off. This removes variable thicknesses from the top of the cell, sometimes forming small windows in the plasma membrane through which the interior of the cell can be seen, and sometimes removing almost all of the cell to expose the inside of the lower membrane with attached coated vesicles and cytoskeleton. Such views of the inside of the plasma membrane can also be obtained by treating unfixed cells by the polylysine sandwich technique. Cells grown on coverslips are rinsed in cytoskeleton-stabilizing buffer[26] and then covered with a second coverslip that has been coated with 1 mg/ml polylysine, so that it will adhere to the top surface of the cells. After a few minutes, the coverslips are pulled apart and placed into fixative. The fractured cells are then dehydrated and CPD.[27-28] Because the fracturing takes place before fixation can produce any denaturation of cytoplasmic proteins, the polylysine sandwich technique is especially useful for immunolabeling of cytoplasmic components, decoration of actin with heavy meromyosin, etc.

IV. EXAMPLES OF STRUCTURAL ANALYSIS BY HVEM AND LVSEM

To illustrate the capabilities of HVEM and LVSEM for analyzing complex three-dimensional structures relevant for cell dynamics, we describe here a few examples from research performed at the Madison IMR.

A. STRUCTURE OF THE MAMMALIAN KINETOCHORE

The movement of chromosomes during mitosis is a fundamental dynamic process in the life of all cells. This movement occurs on the mitotic spindle, a structure consisting mainly of microtubules (mt). The force separating the chromosomes and pulling them to the spindle poles is transmitted to chromosomes by the kinetochore mt which connect the spindle pole to the kinetochore on each chromosome. When the spindle forms, kinetochores bind the plus ends of mt, either short mt nucleated at the kinetochores, or mt nucleated at the spindle poles.[29-33] Recent studies have shown that both *in vivo*[34] and *in vitro*,[35] the shortening of mt which pulls chromosomes to the poles is accomplished by depolymerization at the plus end, which is attached to the kinetochore. Before we can understand how a chromosome can be pulled by depolymerization of the mt where it attaches to the chromosome, we obviously must understand the structure of the kinetochore and its relationship to mts. In mammals, birds, various groups of invertebrates, and lower plants, the mt attachment site shows a characteristic structure which was described as a trilaminar disk from studies of thin sections (for recent review).[36] The inner dense layer was recognized as part of the centromeric chromatin, while the outer disk was assumed to be nonchromatin in nature, a proteinaceous product of chromatin which served as the mt attachment site. Ris and Witt[30] reinvestigated this trilaminar structure in stereo micrographs of thick sections using HVEM. Their findings were strikingly different from the generally accepted view. The outer disk was not a solid disk, but a series of hairpin-like loops of chromatin fibers that issued from the centromeric chromatin and were aligned parallel to each other (diagram, Figure 1). Figure 2 shows an orthogonal section through the apparent disk. Instead of a solid bar, the disk appears as pairs of dots and short rods which represent cross sections through the hairpin loops. In Figure 3 we see an oblique thick section through the "disk". The parallel chromatin fibers

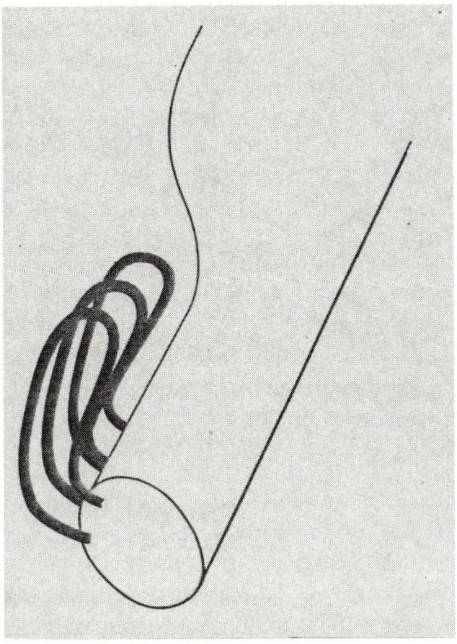

FIGURE 1. Diagram showing the arrangement of the chromatin hairpin loops in the kinetochore outer disk.

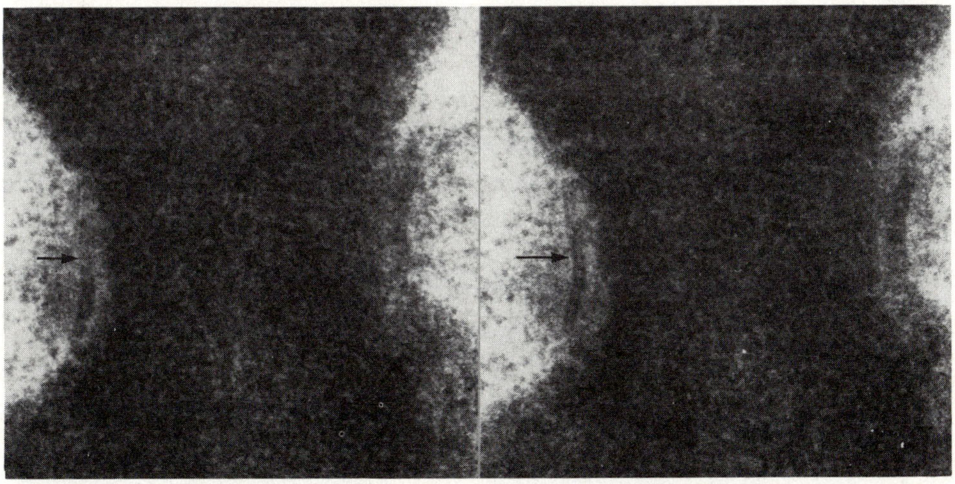

FIGURE 2. Stereopair of thick section (0.5 μm) of CHO cell metaphase chromosome treated with colcemid, fixed with 4% paraformaldehyde in 0.1 M PIPES buffer (pH 7.4), followed by 1% OsO_4. The outer disk on left shows pairs of cross sections of hairpin loops (arrow) and partially overlapping pairs of oblique sections through 10 nm fibers. Compare with Figure 1. Tilt angle 10°, magnification × 50,000; 1 MeV. (From Ris, H. and Witt, P., *Chromosoma*, 82, 153, 1981.)

connecting to the centromeric chromatin at the base are clearly visible. Figure 4 shows a cross-section of the disk, including the tops of the hairpin loops aligned side by side. Whole mounts of metaphase chromosomes with attached mt suggest that each loop is associated with one mt (Reference 30, Figure 1).

It is of interest to compare the mammalian kinetochore with the mt-chromatin relationship in the yeast *Saccharomyces cerevisiae*. Here, each chromosome is connected to a spindle pole by a single mt, which ends directly on a chromatin fiber without any structural specialization visible in electron micrographs.[37] We may consider this to represent the simplest

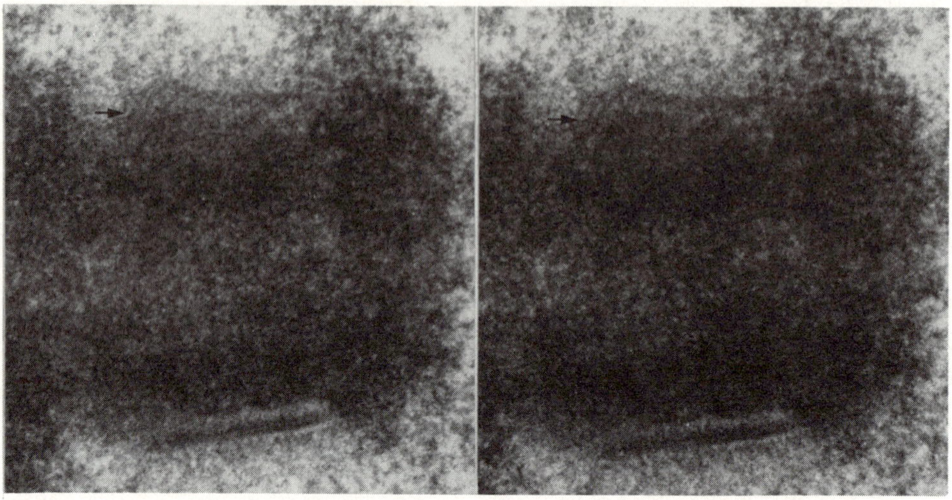

FIGURE 3. Thick section (0.25 μm) of mouse metaphase chromosome treated with colcemid, fixed with glutaraldehyde in PIPES buffer, followed by 1% OsO_4. Sections stained with 7.5% uranyl magnesium acetate and lead citrate. This section includes the level where the 10 nm fibers of the outer disk connect to the centromeric chromatin (arrow). Tilt angle 10°, magnification × 50,000; 1 MeV. (From Ris, H. and Witt, P., *Chromosoma,* 82, 153, 1981.)

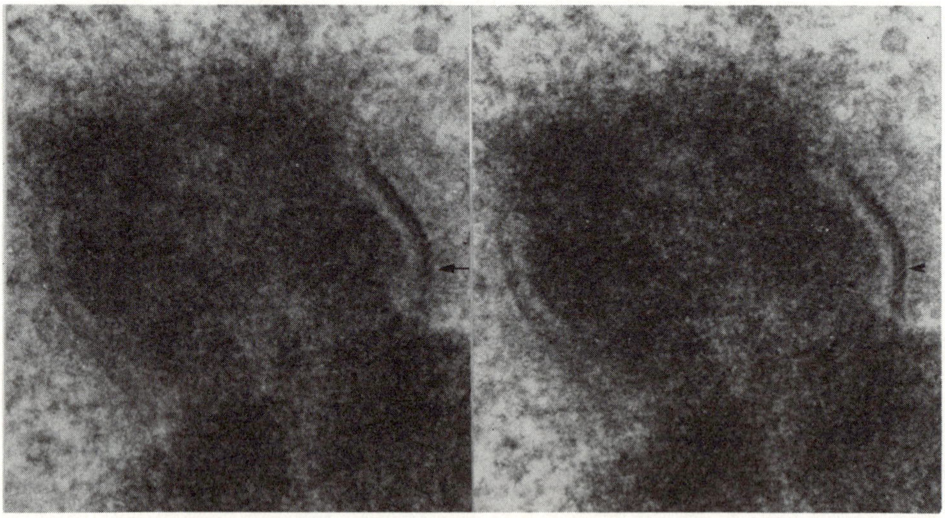

FIGURE 4. Thick section (0.25 μm) through centromere of mouse metaphase chromosome treated with colcemid. This section includes the level containing the tops of the hairpin loops which appear as parallel lines across the disk (arrow). Compare with Figure 1. Tilt angle 10°; magnification × 50,000; 1 MeV.

kinetochore type. The mammalian kinetochore may then represent a repeat of the chromatin segment that can associate with a mt. Any model that tries to explain the kinetics of mt shortening by depolymerization at the kinetochore must consider kinetochore structure as found in yeast and mouse. The diagrams in Koshland et al.[35] are not very useful, since they ignore what is known about mammalian kinetochore structure. Any model also must work in yeast where a single mt associates with a particular region of a single chromatin fiber.

B. PSEUDOPOD MOTILITY IN THE NEMATODE SPERM

Nematode sperm are unique in that they do not swim and they lack flagella. When inactive mature spermatids pass through the sprm duct, they become activated and extend

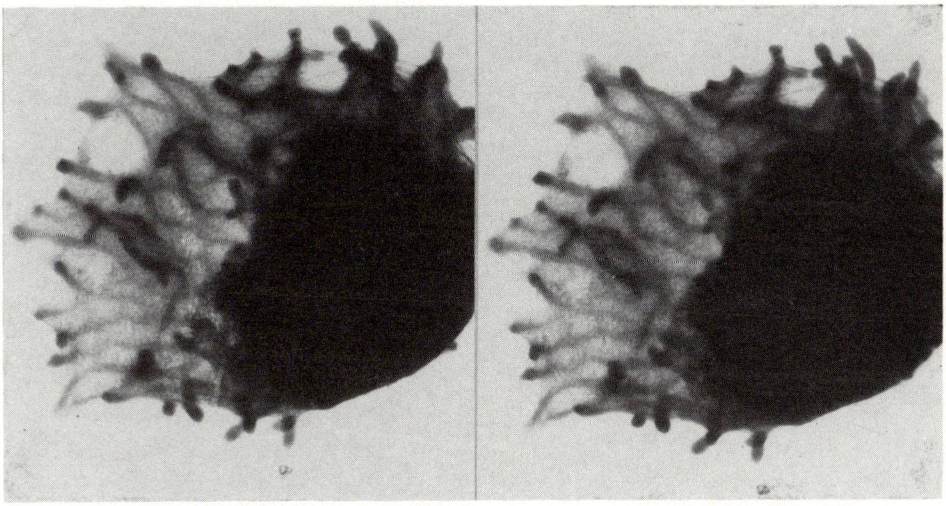

FIGURE 5. Stereomicrograph of a whole mount of an *Ascaris* sperm activated on a gold grid, fixed in 2% glutaraldehyde in 0.1 M HEPES buffer (pH 7.4) containing 0.1% Triton X-100. Post-fixed in 0.1% OsO_4, stained with 1% uranyl acetate and CPD. Note the fiber complexes connecting the tips of villipodia to the cell body. Tilt angle 12°; magnification × 3700; 1 MeV.

a pseudopod which crawls along the female reproductive tract, pulling the cell body containing nucleus, mitochondria, centrosome, etc., behind it.[38] This locomotion was first described as ameboid in analogy to the actin-myosin dependent movement of amebae and motile vertebrate cells. Further studies, however, showed that these cells contained very little actin and myosin, or lacked it totally, and cytochalasin B had no effect on locomotion.[39,40] The surface of the pseudopod is covered with finger-like projections, villipodia, which form in rows at the leading edge of the pseudopod, and move rearwards towards the cell body where they disappear. These moving villipodia attach to the substrate, and as they move backwards over the surface of the pseudopod, the cell is pulled forward.[41] Electron micrographs of thin sections indicated that the pseudopod is filled with a mass of filaments,[38,41] but until recently it was not clear whether these filaments were in any way related to the motility. Sepsenwol and Ris,[42] and Pawley et al.,[43] studied motile pseudopods in sperm of *Ascaris suum* in whole mounts and in thick sections, using HVEM. This approach revealed what was not evident from thin sections, that the filaments formed branching fiber complexes that extended from the tips of the villipodia to the cell body (Figure 5). These complexes are 150 to 250 nm thick and consist of fuzzy fibers varying in thickness from 2 to 10 nm, and perhaps representing a single thin filament that is irregularly coiled or folded into the thicker fiber. Along the fiber complex these fibers leave the complex at right angles in all directions, interdigitating with fibers of adjacent complexes and inserting on the inner surface of the pseudopod membrane (Figures 6 and 7). How do these fibers relate to villipodia motility? The complexes are visible in living cells with video-enhanced phase or differential interference contrast microscopy.[42] Analysis of video tapes indicate that the fiber complexes do not exert force on the villipodia, but move from the pseudopod edge to the cell body without changing shape or relative position, and at the same velocity as the villipodia.[43] Perhaps their function is to integrate all villipodia and the cell surface into a single system that moves from the pseudopod edge to the cell body as a unit.

The crawling of the nematode sperm is obviously a new type of cell motility. HVEM studies revealed the complex fiber system associated with the cell surface movements. This knowledge will contribute to the future elucidation of this motility system by biochemical analysis and studies on living cells.

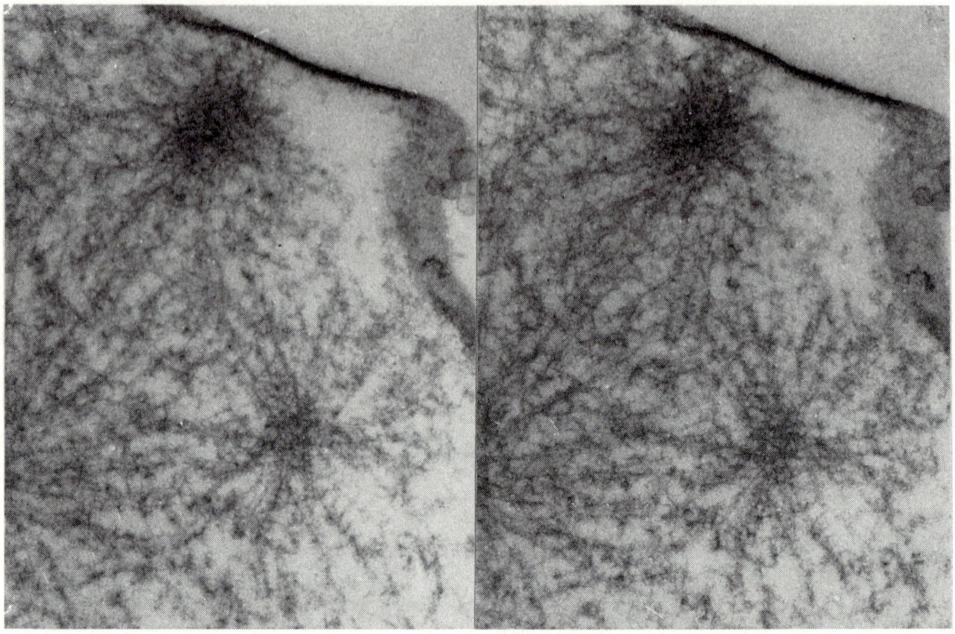

FIGURE 6. Stereomicrograph of a section 0.13 μm thick showing cross sections through fiber complexes in the pseudopod of an *Ascaris* sperm. Fixed in 2.5% glutaraldehyde in 0.1 M HEPES buffer (pH 7.4) containing 0.05% saponin and 0.2% tannic acid. Post-fixed in 0.1% OsO_4, stained with 1% uranyl acetate. The section was stained with 7.5% uranyl magnesium acetate and lead citrate. Tilt angle 30°, magnification × 40,000; 1 MeV. (From Pawley, J. P. et al., *Ann. N.Y. Acad. Sci.*, 483, 171, 1986.)

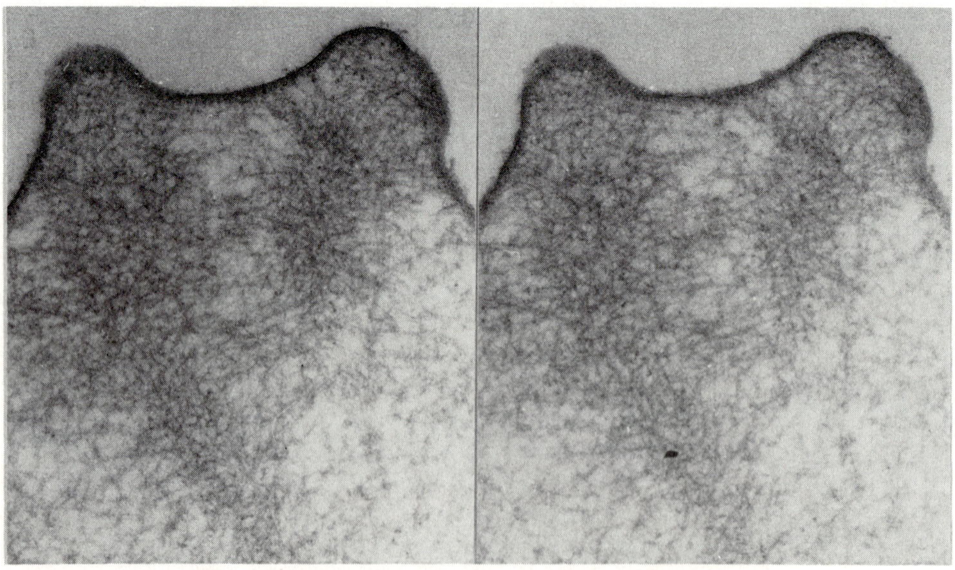

FIGURE 7. Stereo micrograph of a section 0.13 μm thick showing a longitudinal section of a fiber complex branching near the pseudopod edge and ending in two villipodia. Fixation and staining as in Figure 6. Tilt angle 30°; magnification × 40,000; 1 MeV.

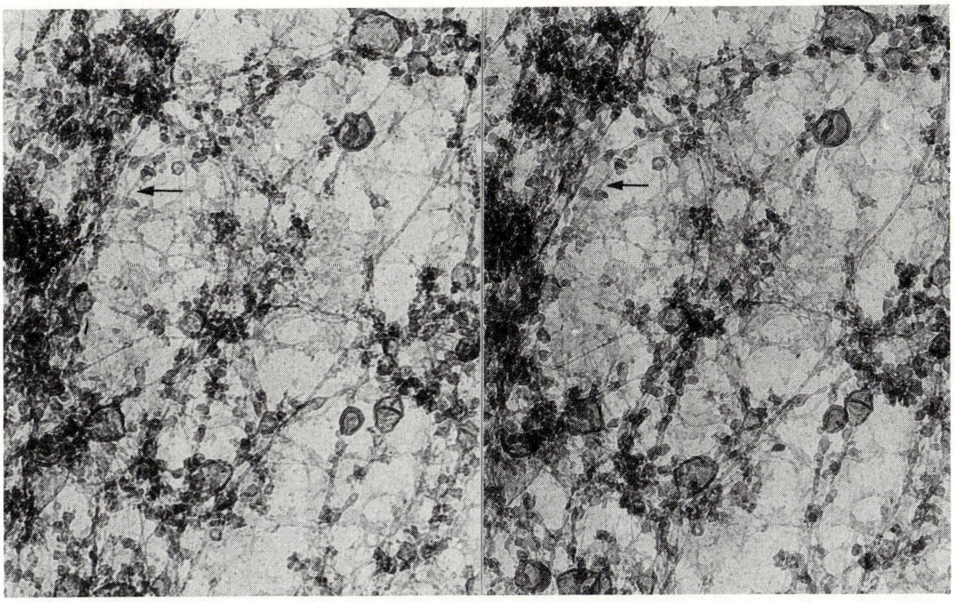

FIGURE 8. Whole mount of xanthophore with dispersed pigment. Area near the thin edge of a spread cell shows carotenoid pigment granules (40 to 50 nm in diameter) attached to 10 nm fibers (arrow). Fixed in 2% glutaraldehyde in 0.1 M HEPES buffer (pH 7.4) containing 0.05% saponin and 0.2% tannic acid, followed by 0.1% OsO_4 and 1% uranyl acetate, CPD. Tilt angle 12°; magnification × 40,000; 1 MeV.

C. PIGMENT TRANSLOCATION IN XANTHOPHORES OF THE GOLDFISH

Lower vertebrates can change the color of their skin by aggregation and dispersion of the pigment granules in their chromatophores. When the pigment is dispersed, the particular color is visible, when aggregated into a small clump inside the cell, the color vanishes. This system represents a beautiful example of directed intracellular translocation and therefore has been intensively investigated by electron microscopy and biochemical approaches.[44] It was realized early on that the direction of pigment movement was determined by the mt of the prominent cytaster, but careful EM studies and use of mt inhibitors showed that pigment translocation was not a simple mt mediated process.[45] EM studies using thin sections could not solve the question of which cytoskeletal element was responsible for pigment movement. Byers and Porter[46] first used CPD whole mounts and HVEM to look at pigment cells. They found the pigment granules enmeshed in what they called the "microtrabecular lattice" and postulated that contraction and extension of this fibrous lattice was the mechanism for pigment aggregation and dispersion. However, this "microtrabecular lattice" was later shown to be a distortion of the known cytoskeletal fibers caused by the presence of water or ethanol in the CO_2 during CPD, and not a new cytoplasmic organelle.[11] Lim and Ris[47] reinvestigated the pigment-cytoskeleton interaction in xanthophores of the goldfish. Using carefully dried whole mounts of isolated xanthophores attached to formvar carbon coated gold grids, they showed that the gold-orange carotenoid granules were attached to 10 nm filaments (Figure 8). No microtrabecular lattice was visible in such preparations, but well-preserved mt and microfilaments were present, always free of pigment. This association of the carotenoid granules with 10 nm filaments was also seen in whole-mount micrographs of homogenates of xanthophores, not only in cells with dispersed pigment, but also in cells with aggregated pigment. Immunogold labeling of xanthophore whole mounts and homogenates, using antibodies to cytokeratin, labeled the pigment associated 10 nm filaments in both preparations. It thus appears that the pigment transporting filaments belong to the class called intermediate filaments (IF). In whole mounts the pigment-filament association is only visible near the

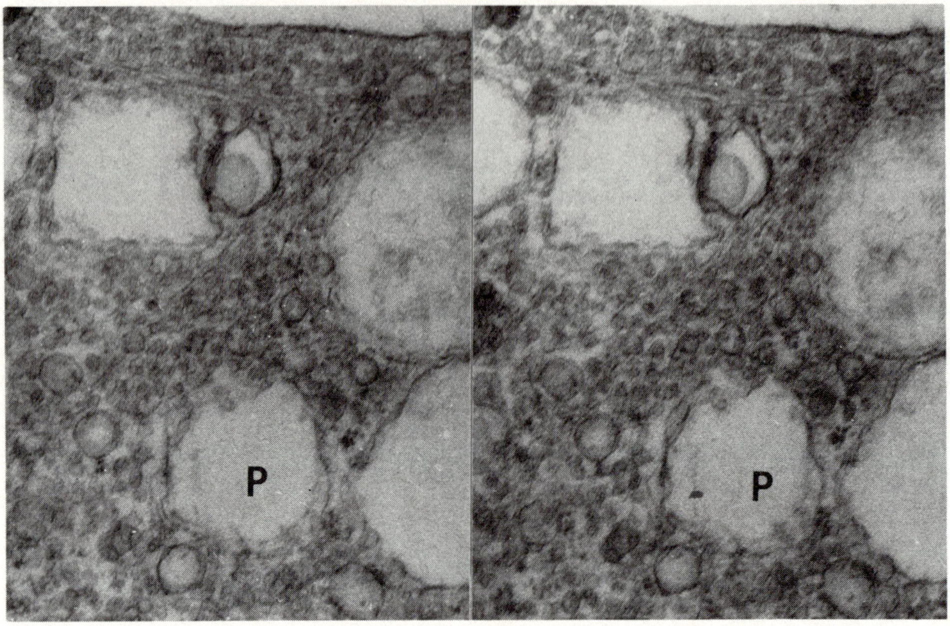

FIGURE 9. Thick section (0.25 μm) through the interior of a xanthophore with dispersed pigment, fixed as in Figure 8. Carotenoid granules and 10 nm filaments are recognized, but because of the dense packing and overlap of structures it is not possible to see the pigment-filament relationship. Tilt angle 17°; magnification × 40,000; 1 MeV.

thin edge of the cells where the pigment is more spread out. Thick sections were prepared to demonstrate that this same relationship also occurs in the internal regions of the cell. As seen in Figure 9, even stereomicrographs of 0.25 μm thick sections failed to show convincing IF with attached pigment granules. Here the LVSEM came to the rescue. Using the scotch tape technique described above on CPD Xanthophores mounted on #2 coverslips, the interior of the thicker regions of the cell became accessible for imaging. As seen in Figure 10, the cytoskeletal elements are visible with excellent topographical resolution and contrast. We recognize mt, microfilaments, and IF with attached 30 to 50 nm pigment granules (arrows). This combination of imaging well-dried whole mounts by HVEM and dry-cleaved CPD cells by LVSEM is very powerful for the analysis of complex three-dimensional structures. In the xanthophore it has demonstrated that IF are involved in the translocation of pigment granules. This is the first clear demonstration of a dynamic function for IF. The mechanism that causes the IF with attached pigment to move to the cell center and then to the periphery has yet to be elucidated, but we expect that the interaction of IF with mt on the one hand, and with the actin-myosin system on the other, will play a role. We also expect that the combination of HVEM and LVSEM will be productive in revealing the nature of this interaction.

D. SUMMARY

We have described some examples here which illustrate the advantages of using transmission electron microscopes with higher accelerating voltages to study complex three-dimensional structures in thicker specimens without sacrificing resolution. The use of thicker specimens with TEMs is mainly limited by structural overlap in densely packed cell areas. In such cases HVEM images can be complemented by images obtained with high resolution low voltage scanning electron microscopes of cells opened up with various techniques. The greater depth provided by both imaging techniques makes stereoscopic micrographs especially informative. The examples described here illustrate how the detailed knowledge of

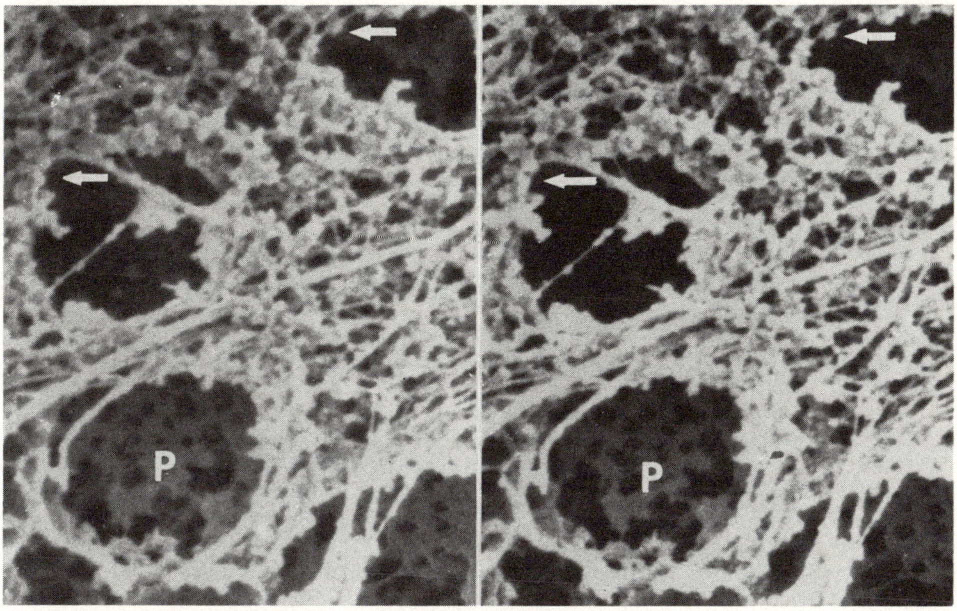

FIGURE 10. LVSEM picture of an xanthophore with dispersed pigment. The cells were fixed and CPD as in Figure 8, then dry-cleaved by touching to scotch tape. The holes (P) are spaces where pterinosomes were ripped out by the scotch tape. The densely packed cytoskeleton between the pterinosomes is clearly visible, including microtubules, microfilaments, and 10 nm filaments with attached carotenoid granules (arrows). This image demonstrates the usefulness of LVSEM for studies of cytoplasmic organization where either thin or thick sections fail to provide the information. Tilt angle 9°; magnification × 40,000, 1 kV.

the structures involved in dynamic cell processes is a necessary prerequisite for the understanding of these processes. The specialized technology described in this chapter can provide such knowledge where conventional electron microscopy of ultrathin sections has failed.

The HVEM and IMR Resources are available to qualified biomedical researchers free of charge. For information contact: Dr. Gerald Schatten, Integrated Microscopy Resource (IMR), Animal Science Building, University of Wisconsin, Madison, WI 53706, (608-262-5746); Dr. J. Richard McIntosh, Laboratory for HVEM, Campus Box 347, Boulder, Colorado 80309 (303-492-8533); Donald F. Parsons, HVEM Laboratory, Center for Laboratories and Research, New York State Department of Health, Empire State Plaza, Albany, New York, 12201 (518-474-7047).

ACKNOWLEDGMENT

The studies described here were supported by a grant to the Madison IMR from the Biomedical Research Technologies Program, DRR-NIH, (P41RR00570).

REFERENCES

1. **Cosslett, V. E.,** High voltage electron microscopy, *Q. Rev. Biophysics,* 2, 95, 1969.
2. **Cosslett, V. E.,** Perspectives in high voltage electron microscopy, *Proc. R. Soc. London., Ser. A.,* 338, 1, 1974.
3. **Fotino, M.,** Experimental studies on resolution in high-voltage transmission electron microscopy, in *Electron Microscopy in Biology,* Vol. 1, Griffith, J. D., Ed., John Wiley & Sons, New York, 1981, 89.

4. **Dupouy, G., Perrier, F., and Fabre, R.,** Microscopie électronique fonctionnant sous très haute tension, *C. R. Acad. Sci. (Paris),* 252, 627, 1961.
5. **Heath, J. P., Ho, M. H., and Peachey, L. D.,** Three-dimensional reconstruction of cell morphology using intermediate voltage electron microscopy and computer graphics, *Proc. 45th Ann. Meet. EMSA,* Bailey, G. W., Eds., San Francisco Press, San Francisco, 1987, 590.
6. **Haggis, G.H. and Bond, E. F.,** Three-dimensional view of the chromatin in freeze-fractured chicken erythrocyte nuclei, *J. Microsc.,* 115, 225, 1979.
7. **Pawley, J.,** Low voltage scanning electron microscopy, in *Electron Optical Systems,* Hren, J. J., Ed., SEM, AMF O'Hare, IL, 1984, 253.
8. **Rieder, C. L.,** Thick and thin serial sectioning for the three-dimensional reconstruction of biological ultrastructure, in *Methods in Cell Biology,* Vol. 22, Turner, J. N., Ed., Academic Press, New York, 1981, 215.
9. **Kinnamon, J. C.,** Combining HVEM and computer assisted reconstructions to elucidate the three-dimensional structure of taste buds, *Proc. 45th Ann. Meet. EMSA,* Bailey, G. W., Ed., San Francisco Press, San Francisco, 1987, 584.
10. **Anderson, T.,** Technique for the preservation of three-dimensional structure in preparing specimens for the electron microscope, *Trans. N.Y. Acad. Sci. (Ser. II),* 13, 130, 1951.
11. **Ris, H.,** The cytoplasmic filament system in critical point-dried whole mounts and plastic embedded sections, *J. Cell Biol.,* 100, 1474, 1985.
12. **Pawley, J. and Ris, H.,** Structure of the cytoplasmic filament system in freeze-dried whole mounts viewed by HVEM, *J. Microsc.,* 145, 319, 1986.
13. **Wolosewick, J. J.,** The application of polyethylene glycol (PEG) to electron microscopy, *J. Cell Biol.,* 86, 675, 1980.
14. **Gorbsky, G. and Borisy, G. G.,** Reversible embedment cytochemistry (REC): a versatile method for the ultrastructural analysis and affinity labeling of tissue sections, *J. Histochem. Cytochem.,* 34, 177, 1986.
15. **Guan, T. and Ris, H.,** Structural preservation of thick frozen sections by critical point drying, *Proc. 43rd Ann. Meet. EMSA,* Bailey, G. W., Ed., San Francisco Press San Francisco, 1985, 442.
16. **Nagatani, T. and Saito, S.,** Instrumentation for ultra-high resolution scanning electron microscopy, *Proc. 11th Int. Cong. Electron Microsc.,* Kyoto, Japan, 1987, 2101.
17. **Pawley, J.,** LVSEM: A new way of seeing biology, *Proc. 45th Ann. Meet. EMSA,* Bailey, G. W., Ed., San Francisco Press, San Francisco, 1987, 550.
18. **Tanaka, K.,** Demonstration of intracellular structures by high resolution scanning electron microscopy, *Scanning Electron Microsc.,* 2, 1, 1981.
19. **Haggis, G. H. and Bond, E. F.,** Three-dimensional view of the chromatin in freeze-fractured chicken erythrocyte nuclei, *J. Microsc.,* 115, 225, 1979.
20. **Haggis, G. H., Schweitzer, I., Hall, R., and Badon, T.,** Freeze-fracture through the cytoskeleton, nucleus and nuclear matrix of lymphocytes studied by scanning electron microscopy, *J. Microsc.,* 132, 185, 1983.
21. **Haggis, G.H.,** Freeze-fracture of cell nuclei for high resolution SEM and deep-etch TEM, *Proc. 45th Ann. Meet. EMSA,* Bailey, G. W., Ed. San Francisco Press, San Francisco, 1987, 560.
22. **Watson, J. H. L., Page, R. H., and Swedo, J. L.,** A technique for determining the interior topography of single cells, *Scanning Electron Microsc.,* I, 417, 1975.
23. **Flood, P. R.,** Dry fractioning techniques for the study of soft internal biological tissues in the scanning electron microscope, *Scanning Electron Microsc.,* I, 287, 1975.
24. **Mesland, D. A. M., Spiele, H., and Roos, E.,** Membrane associated cytoskeleton and coated vesicles in cultured hepatocytes visualized by dry-cleaving, *Exp. Cell Res.,* 132, 169, 1981.
25. **Poli, F., Pancaldi, S., Dall'Olio, G., and Vannini, G. L.,** Cytoskeletal structures in *Euglena gracilis* after Triton X-100 extraction and dry-cleaving, *Protoplasma,* 128, 28, 1985.
26. **Schliwa, M. and Blerkom, J.,** Structural interaction of cytoskeletal components, *J. Cell Biol.,* 90, 222, 1981.
27. **Isobe, Y. and Shimada, Y.,** Organization of filaments underneath the plasma membrane of developing chick skeletal muscle cells *in vitro* revealed by the freeze-dry and rotary replica method, *Cell Tissue Res.,* 244, 47, 1986.
28. **Mazia, D., Schatten, G., and Sale, W.,** Adhesion of cells to surfaces coated with polylysine, *J. Cell Biol.,* 66, 198, 1975.
29. **Witt, P., Ris, H., and Borisy, G. G.,** Origin of kinetochore microtubules in Chinese hamster ovary cells, *Chromosoma,* 81, 483, 1980.
30. **Ris, H. and Witt, P.,** Structure of the mammalian kinetochore, *Chromosoma,* 82, 153, 1981.
31. **Euteneuer, U. and McIntosh, J. R.,** Structural polarity of kinetochore microtubules in PtK1 cells, *J. Cell Biol.,* 89, 338, 1981.
32. **Rieder, C. L. and Borisy, G. G.,** The attachment of kinetochores to the prometaphase spindle in PtK1 cells. Recovery from low temperature treatment, *Chromosoma,* 82, 693, 1981.

33. **Euteneuer, U., Ris, H., and Borisy, G. G.,** Polarity of kinetochore microtubules in CHO cells recovering from colcemid block *J. Cell Biol.,* 95, 301a, 1982.
34. **Gorbsky, G. J., Sammak, P., and Borisy G. G.,** Chromosomes move poleward in anaphase along stationary microtubules that coordinately disassemble from their kinetochore ends, *J. Cell Biol.,* 104, 9, 1987.
35. **Koshland, D. E., Mitchison, T. J., and Kirschner, M. W.,** Polewards chromosome movement driven by microtubule depolymerization *in vitro, Nature,* 499, 1988.
36. **Rieder, C. L.,** The formation, structure and composition of the mammalian kinetochore and kinetochore fiber, *Int. Rev. Cytol.,* 79, 1, 1982.
37. **Peterson, J. B. and Ris, H.,** Electron microscopic study of the spindle and chromosome movement in the yeast *Saccharomyces cerevisiae, J. Cell Sci.,* 22, 219, 1976.
38. **Foor, W. E.,** Zygote formation in *Ascaris lumbricoides* (Nematoda), J. Cell Biol., 39, 119, 1968.
39. **Nelson, G. A. and Ward, S.,** Ameboid motility and actin in *Ascaris lumbricoides* sperm, *Exp. Cell Res., 131,* 149, 1981.
40. **Nelson, G. A., Roberts, T. M., and Ward, S.,** *Caenorhabditis elegans* spermatozoan locomotion: ameboid movement with almost no actin, J. Cell Biol., 92, 121, 1982.
41. **Roberts, T. M.,** Crawling *Caenorhabditis elegans* spermatozoa contact the substrate only by their pseudopods and contain 2 nm filaments, *Cell Motility,* 3, 333, 1983.
42. **Sepsenwol, S. and Ris, H.,** A new motility system in sperm of nematodes.A combined HVEM and video-enhanced light microscope approach, *Proc. 11th Int. Congr. Electron. Microsc.,* Kyoto, Japan, 2621, 1986.
43. **Pawley, J., Sepsenwol, S., and Ris, H.,** Four-dimensional microscopy of Ascaris sperm motility, *Ann. N.Y. Acad. Sci.,* 483, 171, 1986.
44. **Schliwa, M. and Euteneur, U.,** Comparative ultrastructure and physiology of chromatophores with emphasis on changes associated with intracellular transport, *Am. Zool.,* 23, 479, 1983.
45. **Schliwa, M. and Bereiter-Hahn, J.,** Pigment movements in fish melanophores: morphological and physiological studies. V. Evidence for a microtubule independent contractile system, *Cell Tissue Res.,* 158, 61, 1975.
46. **Byers, H. R. and Porter, K. R.,** Transformation in the structure of the cytoplasmic ground substance in erythrophores during pigment aggregation and dispersion, *J. Cell Biol.,* 75, 541, 1977.
47. **Lim, S. S., Ris, H., and Schnasse, B.,** Pigment granules in goldfish xanthophores are attached to intermediate filaments, *J. Cell Biol.,* 105, 37a, 1987.

Chapter 18

X-RAY MICROANALYSIS OF ION CHANGES DURING FAST PROCESSES IN CELLS, AS EXEMPLIFIED BY TRICHOCYST EXOCYTOSIS OF *PARAMECIUM CAUDATUM**

Manfred Schmitz and Karl Zierold

TABLE OF CONTENTS

I. INTRODUCTION

Quick freezing preserves the structural and elemental composition of cells at a definite point of time. Combining this method with analytical electron microscopy, for example X-ray microanalysis, provides the opportunity to measure the intracellular distribution of ions, as related to the functional state of the cell. Thus, elemental shifts and ion movements during vesicular transport and exocytosis can be studied with the time resolution of cryofixation, i.e., within less then 10 ms.

We have applied this methodical approach to study the exo-cytosis of specialized secretory organelles, called trichocysts, from *Paramecium caudatum* cells.[1-6] This system has two advantages with respect to analytical electron microscopy of dynamic processes: (1) the fast process to be studied, i.e., exocytosis of trichocysts, can be clearly identified by ultrastructural criteria; (2) this process can be synchronously triggered within a single cell.[1-6]

The preparation started by stimulating exocytosis immediately followed by cryofixation of the cells within different time intervals. Then, from the frozen cells, cryosections were prepared and studied by scanning transmission electron microscopy (STEM) and energy dispersive X-ray microanalysis (EDX). Thus, we were able to localize elemental shifts as related to the exocytosis of trichocysts from the cells. The results are discussed with respect to existing hypotheses on the mechanism of trichocyst exocytosis, and to the significance of this methodical approach for the investigation of dynamic cellular processes.

II. PREPARATION AND X-RAY MICROANALYSIS OF *PARAMECIUM CAUDATUM* CELLS

A. CELL MATERIAL

Paramecium caudatum was raised in hay infusion at room temperature. The cells were enriched by their positive photo-taxis or by centrifugation, and collected with a pipette. The preparation scheme for X-ray microanalysis is illustrated in Figure 1.

B. TIME-DEPENDENT CRYOFIXATION

Before shock freezing, one to three single cells within as little medium as possible, were mounted on modified Balzers gold planchettes with a top diameter of about 0.5 mm, as necessary for freezing and cryosectioning. These small droplets were placed into a crossbow-like device and the specimen velocity was adjusted to about 10 $m \cdot s^{-1}$, monitored for each single shot by photosensors. This arrangement allowed additional treatment of the cells by altering the extracellular medium in steps of 10, 20, or 30 ms range before freezing. We shot the planchette with the cells through a platinum wire loop with a liquid film. By placing the loop in variable distances above the coolant (Figure 1 a), different time intervals were achieved.

Exocytosis of the trichocysts was induced by shooting the cells through the liquid film containing 0.03% aminoethyldextran (AED), thus reaching a final concentration of approximately 0.015% AED in the droplet. This concentration was described to be sufficient for a high rate of synchronized exocytosis.[1,4] In some cases, 50 mM $SrCl_2$ or $BaCl_2$ were added to the medium in the liquid film containing 0.015% AED. After passing these media the cells were shock-frozen by shooting the specimens into liquid propane below 90 K[6].

C. PREPARATION AND MICROANALYSIS OF CRYOSECTIONS

Cryosections were prepared from the frozen cells by means of a dry glass knife in a Reichert FC4 cryoultramicrotome (Figure 1 b). Sectioning temperature was below 170 K, and section thickness was approximately 100 nm. The sections were mounted on Pioloform®-coated, carbon-evaporated copper grids. The grids were transferred in cold N_2 gas at a

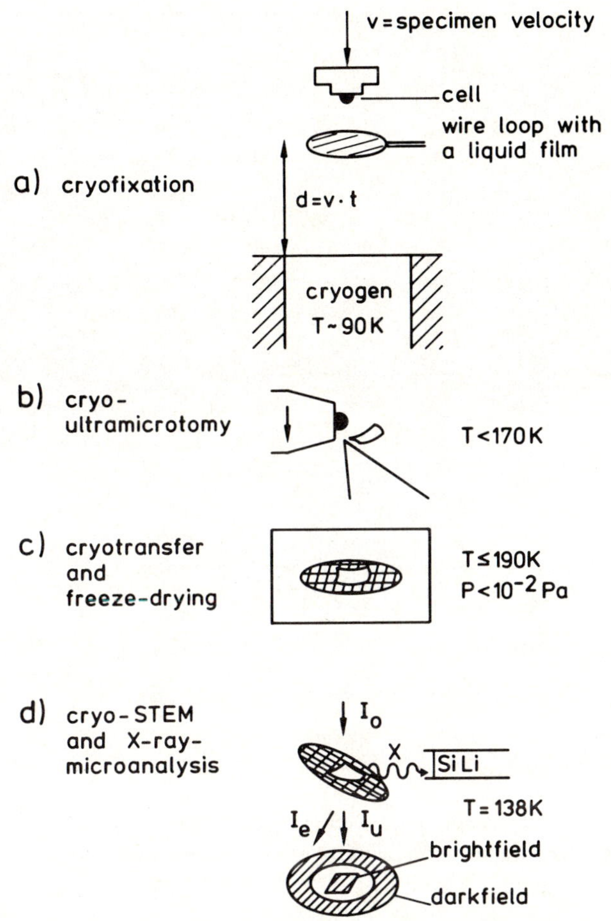

FIGURE 1. Scheme of the preparation method for EDX of trichocyst exocytosis from *Paramecium caudatum* cells. (a) Cryofixation: The specimen support with the cell is shot by a crossbow-like device into the cryogen (liquid propane). The liquid film, containing, e.g., the trigger agent for exocytosis, is placed in definite distances, d, above the cryogen. t = time between contact of the specimen with the liquid film (trigger agent) and plunging into the cryogen. (b) Cryosectioning by means of a dry glass knife in a cryoultramicrotome. (c) Cryotransfer in cold nitrogen gas atmosphere to the STEM and freeze-drying of the cryosections, placed on a Pioloform® coated, carbon-evaporated grid. (d) X-ray microanalysis in STEM: The grid bearing the cryosections is tilted by 30° towards the X-ray detector. The primary electrons with the intensity I_o excite X-rays (X) in the specimen. The intensity of the unscattered and unelastically scattered electrons (I_u) is measured by the central bright-field detector, the elastically scattered electrons (I_e) reach the annular dark-field detector. Image contrast is achieved by the ratio of the electron intensities obtained from the bright-field and dark-field detector.

temperature below 130 K to the Siemens Elmiskop ST 100 F STEM equipped with a cold stage. Freeze-drying the the cryosections was done by evacuation of the cryotransfer chamber to a pressure below 10^{-2} PA and increasing the temperature to at least 190 K (Figure 1 c). The cryosections were irradiated by scanning an electron beam of 4 nA at an accelerating voltage of 100 kV over the area of interest. During analysis in STEM, the section was kept at a temperature of 138 K. Imaging of the cryosections was done by ratio contrast; the electron signals caused by the cryosection were collected separately by the central bright-field detector and by the annular dark-field detector. Bright-field and dark-field signals were ratioed in order to enhance mass dependent contrast from the unstained cryosection. X-ray spectra were collected by means of an SiLi energy-dispersive detector (Nuclear Semiconductors) combined with a multichannel analyzer (Link Systems) (Figure 1 d). The X-ray

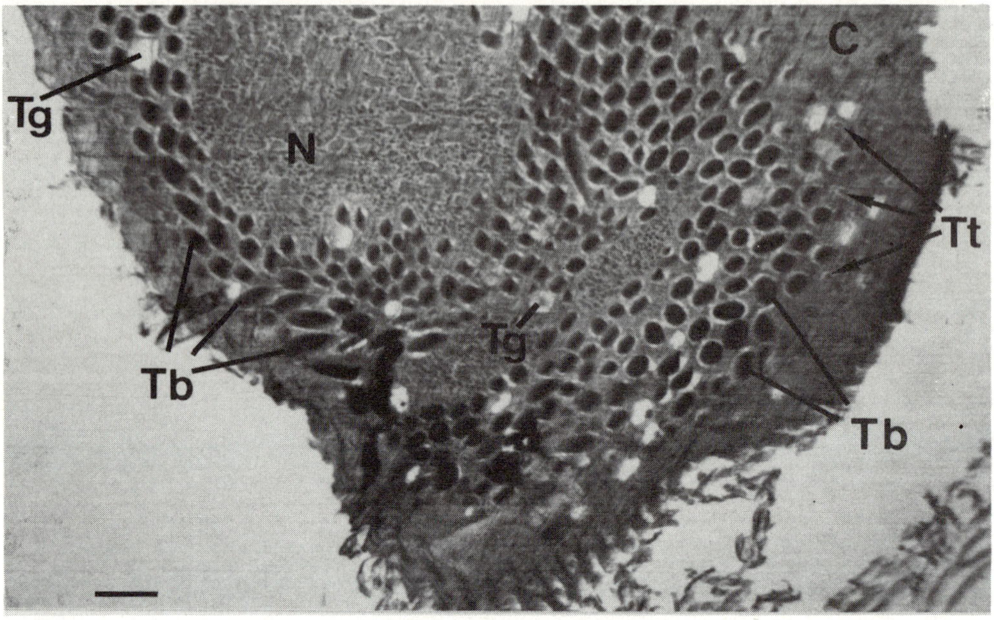

FIGURE 2. Cryosection of a peripheral region of *Paramecium caudatum*. C = cytoplasm; N = nucleus; Tb = trichocyst base; Tg = trichocyst ghost; Tt = trichocyst tip. Bar = 2.5 μm.

spectra were evaluated by the peak-to-background method, according to Hall and Gupta.[8] The detection limit is approximately 10 mMol/kg dry weight for all elements with an atomic number higher than 12. For Na and Mg the detection limit is 30 and 20 mMol/kg dry weight, respectively. The lateral analytical resolution is about 30 nm, depending on the specimen thickness. Details of preparation and analysis of cryosections are described by Zierold.[9,10]

D. INTRACELLULAR ELEMENT DISTRIBUTION

After shock-freezing without pretreatment by chemical fixation and/or cryoprotectants, cells of *Paramecium caudatum* generally exhibit the same ultrastructural arrangement as described before without triggering.[6] This corresponds to the ultrastructure found after chemical fixation, as reviewed by Vivier.[11] Additional spontaneous exocytosis, by shooting the cells through the liquid film (without AED), was not observed. In a section through the periphery of the cell at a low magnification the patterns of the pellicle with cilia, as well as the cross sectioned ''bases'' (basal regions) and ''tips'' (apical regions) of the trichocysts can be identified, and only a few empty trichocyst ghosts are detected. Cytoplasm, as well as nucleoplasm with bacteria as endosymbionts, are visible (Figure 2). A section perpendicular to the cell surface at higher magnification reveals the longitudinal shape of the radially arranged trichocysts docked at the pellicle (Figure 3). The size of the ice crystal segregation zones is negligible for EDX down to approxiamtely 30 μm below the cell surface. However, serial sections could not be obtained to reveal the complex compartmentation of membranes close to the pellicle, which are possibly involved in the movement of ions essential for exocytosis.

The cryosections of cells which were shot through a loop with mere culture medium (no AED) exhibit the same significant amounts of Na, Mg, P, S, Cl, K, and Ca in the different subcellular compartments (Figure 4 a to f) as earlier described for cells without this pretreatment.[6] The element content of the various compartments, as measured by X-ray microanalysis of freeze-dried cryosections, is compiled in Table 1. Although some data exhibit a broad standard deviation, the various intracellular compartments are characterized

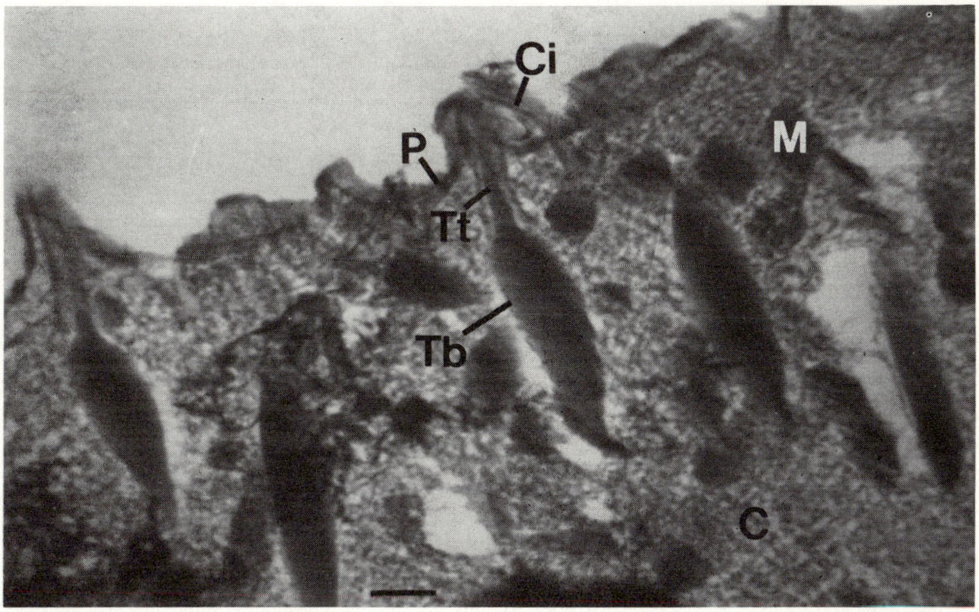

FIGURE 3. Cryosection of *Paramecium caudatum* perpendicular to the cell surface. C = cytoplasm; Ci = cilium; M = mitochondrium; P = pellicle; Tb = trichocyst base; Tt = trichocyst tip. Bar = 500 nm.

by typical patterns of their element content. In the cytoplasm, more than 500 nm apart from the pellicle, Na and Ca cannot be detected; Cl is found in low concentration, whereas Mg, P, S and K are detected at expected high levels (Figure 4a). Also, the mitochondria exhibit the same element distribution as described before, with no significant amounts of Na or Ca (Figure 4c). Peaks of Cu, Si, Au are artifacts caused by extraneous X-radiation from the electron microscope column.

In the cytoplasm adjacent to the pellicle, Cl concentrations are higher than in more central regions, and significant amounts of Na and Ca are detected (Figure 4b).

Due to the limited spatial resolution of about 30 nm, X-ray spectra obtained from intracellular regions close to the pellicle might be influenced by elements from the extracellular medium, as well as from alveolar sacs (sarcoplasmic reticulum-like flat bags with a single membrane envelope), which are also located in this region.

No Ca, low Mg, P, S, and K, but a very high concentration of Na, are present in the basis (basal regions) of trichocysts compared to cytoplasm (Figure 4d). Within this one organelle a significantly different distribution of elements is present in the apical tip of the trichocysts, which contain no detectable Mg or Cl, significantly less Na, and more S (Figure 4e) than the basal region. The element content of the remaining material in trichocyst ghosts is represented by the spectrum in Figure 4f. Significant amounts of Ca were never observed within the resting untriggered trichocysts, in their surrounding membranes, or the adjacent cytoplasm in a distance more than 500 nm apart from the pellicle.

III. EXOCYTOSIS OF TRICHOCYSTS

A. MICROANALYSIS

Exocytosis of trichocysts from *Paramecium caudatum* cells was induced as described above (section II. B). After a 10 or 20 ms exposure of the cells to AED before shock-freezing, no significant exocytosis of trichocysts was observed. After exposure to AED 30 ms before shock-freezing, synchronous exocytosis of trichocysts already docked at the

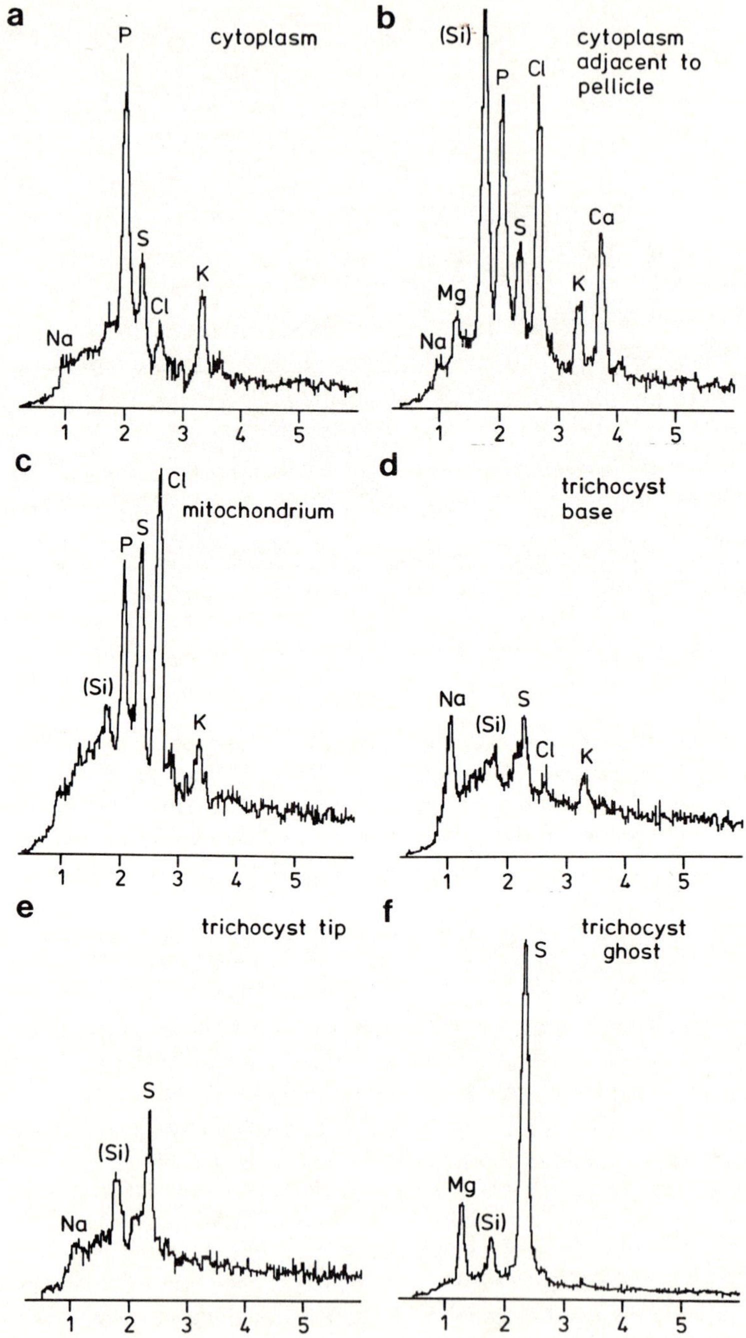

FIGURE 4. Typical X-ray spectra obtained from various intracellular compartments of *Paramecium caudatum*. The horizontal scale indicates the X-ray energy in keV. Extraneous peaks caused by the electron microscope, e.g., specimen holder, are marked by brackets like (Si). (a) Cytoplasm, (b) cytoplasm adjacent to pellicle, (c) mitochondrium, (d) trichocyst basal region, (e) trichocyst tip, (f) trichocyst ghost.

TABLE 1
Element Distribution in Various Compartments of *Paramecium caudatum* Cells

Compartment	n	Na	Mg	P	S	Cl	K	Ca
Cytoplasm	15	bdl	104±96	489±225	163±82	30±20	125±70	bdl
Cytoplasm, adjacent to pellicle	10	56±26	77±43	376±192	169±65	72±110	82±47	267±238
Mitochondrium	8	bdl	28±24	173±53	181±31	109±74	31±9	bdl
Trichocyst basal region	35	177±59	21±26	24±47	81±31	20±32	31±42	bdl
Trichocyst tip	8	41±36	bdl	319±235	200±102	bdl	63±68	bdl
Trichocyst ghost	29	bdl	821±585	bdl	1469±631	53±70	55±80	bdl

Note: Values are mMol/kg dry weight ± standard deviation; n = number of measurements; bdl indicates concentrations below the detection limit.

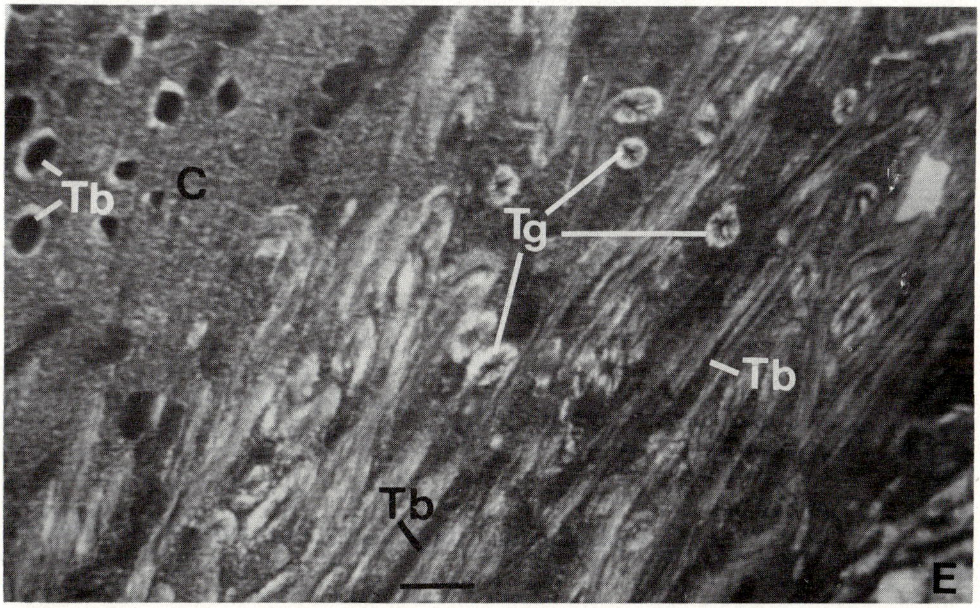

FIGURE 5. Cryosection of a *Paramecium caudatum* cell during exocytosis. The cell was frozen 30 ms after exposure to AED. C = cytoplasm; E = extracellular space; Tb = trichocyst base; Tg = trichocyst ghost. The trichocyst bases remain condensed in the inner part of the cell (left side of Figure 5), whereas they are elongated in the region close to the cell surface exposed to AED (right part of the cell). Bar = 1 μm.

pellicle was found (Figure 5), whereas the still undocked trichocysts in central regions of the cell remained in a condensed state. The element content of the condensed and of the decondensing, i.e., stretching trichocysts, remained the same as in Figure 4d, whereas the element content within the remaining material of the ghosts (after extrusion of the trichocysts) is demonstrated in Figure 4f. Na and P, present in the resting trichocysts showing the condensed state, are not detected in ghosts, but the remaining material of the ghosts includes significant amounts of Mg and S. As in resting trichocysts we could not detect any Ca, either in stretching trichocysts, or in remaining ghosts.

By sectioning deeper into the cells, more longitudinal sections of the radially arranged trichocysts were obtained, and even a section of a trichocyst during exocytosis is shown in Figure 6. In X-ray spectra obtained from this section, randomly precipitated amounts of Ca were detected within the trichocyst material during exocytosis (Figure 7a) and at the trichocyst ghost membrane (Figure 7b). The element distribution in other compartments of the cell,

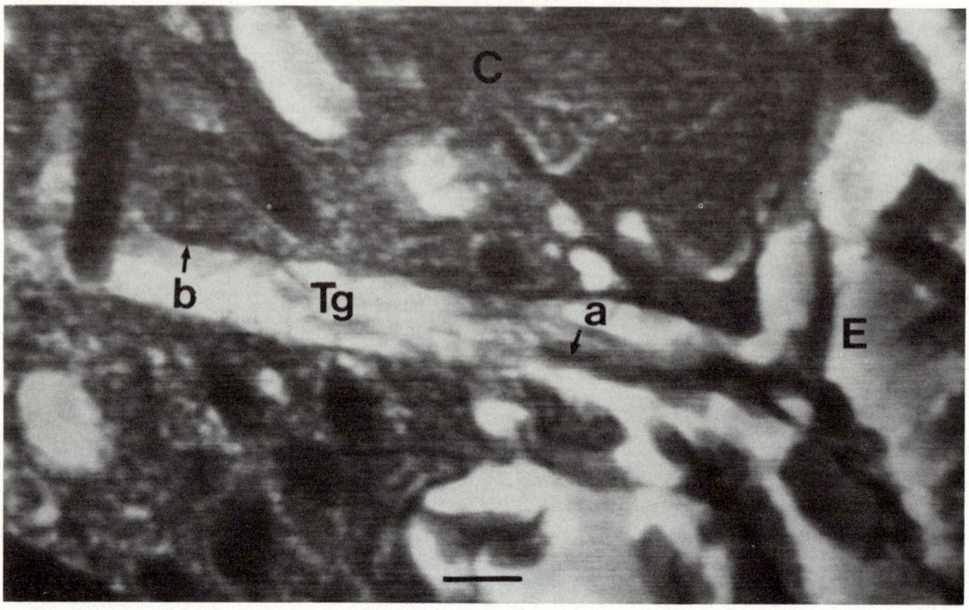

FIGURE 6. Cryosection of a *Paramecium caudatum* cell during exocytosis. C = cytoplasm; E = extracellular space; Tg = trichocyst ghost. a and b mark places of trichocyst material undergoing exocytosis and trichocyst ghost membrane, respectively, where the X-ray spectra in Figures 7a and b were obtained from, respectively. Bar = 250 nm.

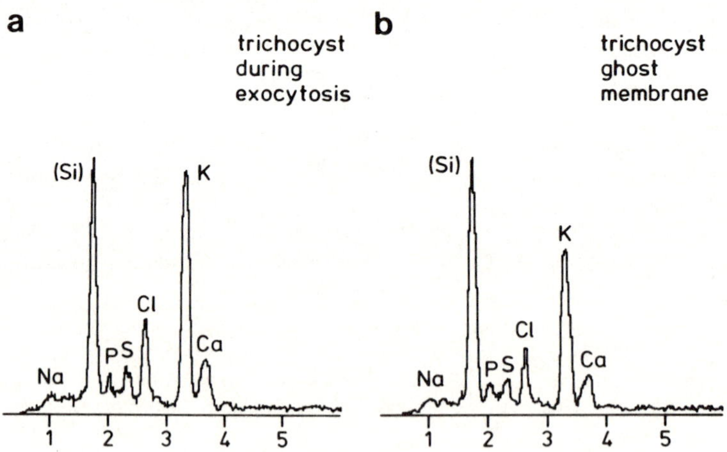

FIGURE 7. X-ray spectra of the trichocyst during exocytosis (a) and trichocyst ghost membrane (b) as marked in Figure 6. The horizontal scale indicates X-ray energy in keV. The presence of Ca additionally to the Kβ-peak of K is demonstrated by quantitative evaluation. The Ca concentrations found in areas less than 50 nm in diameter, as indicated by arrows in Figure 6, are 230 mMol/kg dry weight (Figure 7a) and 110 mMol/kg dry weight (Figure 7b). (Si) is an extraneous X-ray peak.

e.g., in the cytoplasm and in resting trichocysts, remained unchanged. We have added several divalent cations to the liquid AED film in order to test whether divalent cations such as Sr and Ba are transported into the cell during triggering by AED. Figure 8a shows the presence of Ba in the extracellular space, whereas the trichocyst ghosts have not taken up any detectable amount of Ba (Figure 8 b). Also, the other compartments of the cells never

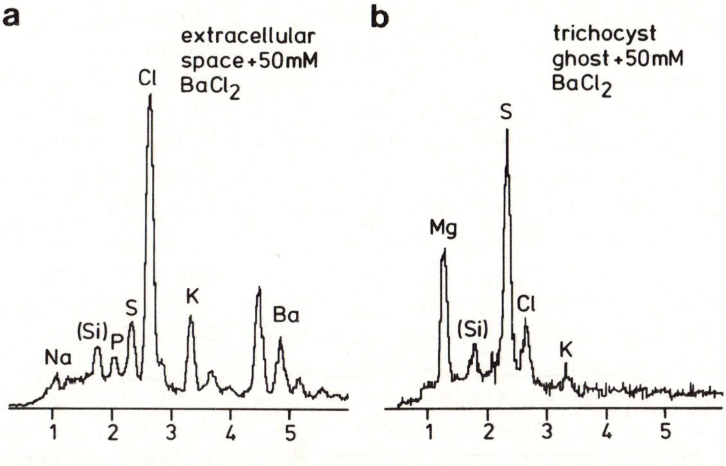

FIGURE 8. X-ray spectra from a cryosection of *Paramecium caudatum* shot through a liquid film containing AED + 50 mM $BaCl_2$, 30 ms before cryofixation. The horizontal scale denotes X-ray energy in keV. (a) extracellular space containing Ba; (b) trichocyst ghost without Ba. (Si) is an extraneous X-ray peak.

showed any uptake of the devalent cations, which were offered to the cell together with AED.

We conclude from these experiments that divalent cations involved in the process of trichocyst decondensation by stretching during exocytosis, in particular Ca, are not taken up from the extracellular space. Therefore, we assume that the Ca found in the trichocyst contents (Figure 7a) and in the trichocyst ghost membrane (Figure 7b) stems from intracellular Ca stores. Indeed, such stores are found close to the pellicle (Figure 4b).

B. HYPOTHESES

Ions are involved in various cellular processes of ciliates such as *Paramecium caudatum*. The role of ionic changes for contraction, locomotion, osmoregulation, as well as for endo- and exocytosis, has been especially investigated by a number of authors, as reviewed by Schmitz et al.[6] *Paramecium* cells generally contain more than one thousand secretory organelles, so-called trichocysts. About 90% of the trichocysts are radially arranged and docked to the cell membrane. After appropriate stimulation by AED, up to 90% of this subpopulation is extruded by exocytosis (accompanied by expansion = decondensation of the "basis" of contents) from the cells within seconds,[1] thus representing an advantageous model to study a variety of problems related to fast cellular processes such as exocytosis. First observations of the ultrastructural organization of cytoplasmic compounds in *Paramecium* were obtained by Wohlfarth-Bottermann after fixation with OsO_4 and $K_2Cr_2O_7$ (Reference 12) and the discharge of trichocysts was observed after electrical stimulation.[13] The organization of trichocysts and the rapid membrane events during exocytosis,[2,14-16] as well as the involvement of microtubules during the docking process, has been analyzed on the ultrastructural level.[17]

The stimulation of trichocyst discharge in *Paramecium aurelia* by the ionophores X-537A or A23187 was the first indication that Ca is involved in the exocytotic mechanism.[18] By adding high concentrations of Ca to the fixative or by Ca precipitation,[18-20] Ca-binding sites were localized on the inner side of the plasma membrane, of the alveolar membrane, in cytoplasmic vacuoles, along the trichocyst membrane, and within a few still condensed trichocysts. However, most of the trichocysts just undergoing exocytosis contained deposits of Ca and P on the membrane of the trichocyst tip and basis as well as on the inner lamellar sheath structures.[18] From these observations, it was argued that the membrane surrounding the trichocysts might contain an ATPase system possibly relevant for regulating ion concentrations in trichocysts, or for the triggering mechanism. An influx of Ca into the trichocysts could then occur just before exocytosis. How does this compare to previous studies on the exocytosis of trichocysts?

Several indirect methods related to the role of Ca for trichocyst function were used. For example, trichocysts isolated in the condensed state were studied.[21-23] Mutant strains with different exocytotic capability were used.[24] Exocytosis was triggered by increasing intracellular free Ca.[23] Freeze-fracture studies[25-28] have demonstrated that the docking sites of trichocysts at the cell membrane contain an array of rosette-shaped intramembranous particles. Cytochemical experiments showed the presence of a Ca^{2+}-ATPase within these docking sites.[29] It was concluded that an increase of free cytosolic intracellular Ca, estimated to be between 10^{-6} to 10^{-4} M[23], would be essential for exocytosis. Rosettes could act as Ca pumps[24] and/or Ca channels,[30] and be involved in membrane fusion. A Ca^{2+}-mobilizing fusogen, in combination with inhibition of discharge by Mg^{2+} and EGTA, produced many exocytotic openings within the cell membrane, whereas the trichocysts were still retained.[31] As monitored by EDX, tip and basal region of isolated trichocysts can accumulate Ca and Cl, when the concentration in the external medium is increased.[32] Thus, membrane fusion and release of the secretory content were considered as two distinct and independent steps, each involving Ca in sequence of exocytosis performance. Under these experimental conditions the second step (release of secretory content) is independent from the first step (membrane fusion), whereas normally the first step would automatically cause the second one. This mechanism agrees well with the requirement of exogenous Ca^{2+} necessary for exocytosis *in vivo*.[4,23]

In a series of experiments, Plattner and co-workers[1-5] have demonstrated the capability of AED to induce synchronous exocytosis in *Paramecium tetraurelia.* Since deciliated cells or ciliary mutations with "Ca^{2+}-tight" cilia responded to AED with normal performance of exocytosis as well, it was postulated that the required exogenous Ca^{2+} did not enter the cells via voltage-dependent ciliary Ca^{2+} channels.[1] By freeze-fracturing and in ultrathin sections, the initially rosette-like arranged membrane-intercalated particles (MIPs) of the docking site were studied during exocytotic membrane fusion,[2] until reinsertion of new trichocysts.[5] *In vitro* membrane fusion did not require ATP,[33] but did require an adequate assembly of molecular components to allow for the dephosphorylation by a phosphatase, acting on the 65 kDa phosphoprotein, possibly calcineurin.[34] The changes of the membrane, surrounding the still resting trichocyst, were described until formation of the ghosts.[3]

The possible role of calmodulin as stimulator of Ca pumps in *Paramecium* cell membranes related to exocytosis was first discussed by Satir et al.[35] Calmodulin was believed to be a major component of purified extruded trichocysts,[36] yet this was then dismissed[37] and calmodulin was not found to be present within the contents of intact trichocyst.[37,38] This might correlate with our finding that Ca is not enriched in trichocyst contents. Isolated and purified surface membrane complexes of *Paramecium* contained a Ca^{2+}-stimulated ATPase,[39] differing from the activity found in exocytosis-incompetent strains.[40] After stimulation, calmodulin was postulated to regulate the Ca^{2+}-dependent expansion of the trichocyst basal region[41] by the insertion of a Ca^{2+}-calmodulin complex into the trichocyst membrane, and to be regulated by pH within the trichocyst vesicles *in vivo*.[42] Differently labeled, monospecific polyclonal antibodies against calmodulin were localized on the ultrastructural level at trichocyst docking sites on the cell membrane, as well as at the cytoplasmic side on the membranes of trichocysts in the tip and basal region, but not in the contents (as proposed in References 35, 36). This also indicates the possible involvement of calmodulin in the process of exocytosis,[37] but only on the membrane level.

Opposite to Bilinski et al.,[32] our results support the hypothesis that Ca is involved in exocytosis by entering trichocyst just before discharge (Figure 7). Extracellular Ca is known to be necessary for the fusion of the trichocyst with the plasma membrane at the docking site.[31] However, as we found now, divalent cations such as Sr or Ba, offered in the extracellular medium for the same time as the AED, were not detected in trichocyst ghosts immediately after extrusion (Figure 8). Therefore, we conclude that Ca involved in the extrusion mechanism stems from intracellular Ca reservoirs, as shown in Figure 4b. Whether this Ca is located within the alveolar sacs as suggested before,[43] merely because of the structural similarity with sarcoplasmic reticulum, remains open. We found Ca during extrusion of trichocysts adjacent to their surrounding membrane (Figure 7b), and calmodulin is known to be located there as well.[37] Therefore, probably the entire trichocyst membrane is permeable for Ca. However, we cannot prove that Ca required (1) for exocytotic membrane fusion, and (2) for trichocyst contents decondensation stems only from the detected Ca pools found close to the pellicle in untriggered cells,[6] since amounts reponsible for physiological functions may be below the detection limit of EDX.

The high amounts of S and Mg detected within the remaining ghost material after discharge of trichocysts (Table 1 and Figure 4f) probably do not originate from the extracellular medium, since exogeneously added Ba or Sr were not detected. The high concentration of S and Mg could either originate from the surrounding cytoplasm or from the content of the undischarged trichocysts. The latter could be explained by the fact that EDX measures relative concentrations (mMol of the specific elements per kg dry weight). Since the condensed material of resting trichocysts contains much organic dry weight, only low levels of S and Mg can be detected there.

The basal region of resting trichocysts contains high amounts of Na not found within the tip (Figures 4d,e and Table 1), as reported previously.[6] This represents an elemental gradient within one subcellular compartment. The high concentration of Na within the basal region of trichocysts is surprising, since a Na/K ATPase was assumed to be absent from *Paramecium aurelia* cells.[44] The mechanism of accumulation of Na, as well as its possible function of exocytosis, remains obscure. Obviously, Na leaves the cell by trichocyst exocytosis. Thereby exocytosis of trichocysts could be involved in osmoregulation.

Our data prove that exocytosis of trichocysts starts within 30 ms after triggering by AED (see section IV. A for more details). This approximately agrees with electrophysiological measurements[45,46] of individual trichocyst release events.

A "fast" Ca channel with millisecond kinetics responsible for the brief Ca influx was characterized by patch clamp experiments. Although voltage-sensitive, this channel also seemed to be ion-dependent, because it showed different rates of activation and deactivation for Ca, Sr, and Ba.[47] It is probably involved in cell depolarization that precedes ciliary reversal.[48-53] Exocytosis was shown, however, to occur in the presence of AED and extracellular Ca^{2+} independent from this type of ciliary voltage-dependent Ca^{2+} channels.[1,4]

IV. X-RAY MICROANALYSIS OF DYNAMIC CELLULAR PROCESSES

Since there are only few data on EDX of fast cellular processes, we should like to present some basic comments on the now best known examples.

A. SINGLE *PARAMECIUM* CELLS

The preservation of defined functional states of *Paramecium* cells requires special attention to the preparative conditions immediately before cryofixation. This holds in particular for the fast dislocation of ions involved in exocytosis of trichocysts. On one hand, the water droplet containing the cells should be big enough to avoid drying, and thereby changing the extracellular osmolarity. Drying proceeds faster the smaller the droplet is. On the other hand, cryofixation requires a specimen as small as possible in order to reduce ice crystal damage, and thus, ion dislocation by freezing. Furthermore, cryoultramicrotomy of a single cell requires that this cell is identified within the frozen water droplet. Enriching the cells by centrifugation within several minutes before fixation would disturb their natural behavior, function, and subcellular organization. These opposite requirements were fulfilled by pipetting, under microscopical control, the cells within a water droplet of a diameter of about 0.5 mm on the specimen support of the same size. The delay between the last moment of observation of the cells and the cryofixation was approximately 5 s. Due to these problems, dried cells occasionally were frozen and processed for cryoultramicrotomy. However, such cells would be clearly eliminated, since the X-ray spectra obtained from these cells contained significantly higher concentrations of Na and Cl within the cytoplasm as compared to frozen cells surrounded by medium. This clearly indicates damage of the cell membrane.

The cooling rate achieved during cryofixation by the crossbow freezing device used was found to be 10,000 K/s (10 ms for the temperature decrease between 273 K and 173 K) as measured by thermocouples connected with a storage oscilloscope. Although the exact temperature necessary for stopping ion movements in dimensions of the microanalytical resolution of 30 nm is unknown, we assume that the time resolution for cryofixation of diffusible ions is less than 10 ms. The diffusion constant of ions (e.g., K^+, Cl^-, and Na^+) in water is about $2{\cdot}10^{-9}\ m^2{\cdot}s^{-1}$ at room temperature. According to this value, the mean free diffusion path is 2 μm within 1 ms, 6 μm within 10 ms, and 10 μm within 30 ms. Probably, the diffusion constant of the molecule AED (M_r = 40 kD) is rather less than that of the highly mobile ions like K^+, Cl^-, and Na^+. Thus, it is conceivable that one *Paramecium* cell shot through the liquid film containing AED exhibits different states of exocytosis of trichocysts at the same time. From our experiments we cannot derive the real time for triggering, although exocytosis of some trichocyst occurs within 30 ms after contact of the droplet to AED. The fact than no significant exocytosis occurred in cells within a droplet exposed to AED for less than 30 ms is probably due to the undefined thickness of the extracellular medium around the cells, which has to be penetrated by AED for triggering.

B. ISOLATED MUSCLE FIBERS

To give another example, EDX of cryosections was also used to study fast ion movements in isolated muscle fibers depending on stimulation. Wendt-Gallitelli (see Chapter 19) and Wolburg[54] described a cryofixation device which allows the freezing of isolated guinea pig heart muscle fibers in liquid propane within a few ms of stimulation. By the use of this technique, they measured over the cell membrane a decrease of Ca during contraction, as compared with the resting state. Somlyo and co-workers[55] used a similar technique to study Ca^{2+} and Mg^{2+} movements across the sarcoplasmic reticulum of frog skeletal muscle fibers *in situ*. They followed the Ca and Mg concentration in the terminal cisternae of the sarcoplasmic reticulum after tetanic shock with a time resolution of 0.6 ms (time between the last shock and contact of the fiber with the propane). These are just two other examples for the application of EDX to study fast processes related to ion shifts in cells. These and our own data show that electron probe microanalysis of cryosections from fast frozen cells may provide new and unique information on ion movements related to dynamic subcellular processes. They require, however, much skill in adapting the freezing method to the functional state of interest.

V. CONCLUSION

Freeze-dried cryosections of single *Paramecium caudatum* cells were studied by X-ray microanalysis in a scanning transmission electron microscope. Time dependent cryofixation enabled us to localize elements, in particular ions, in intracellular compartments, with a time resolution of less than 10 ms. Under our experimental conditions exocytosis of trichocysts started about 30 ms after stimulating the cells by the triggering agent AED (aminoethyldextran). Our data show that the high amount of Na present in the basal region of trichocysts is transported out of the cell by this process. The remaining trichocyst ghosts contain high levels of Mg and S. High concentrations of Ca were found in the cytoplasm adjacent to the pellicle, whereas no Ca could be detected in the cytoplasm more than 500 nm apart from the pellicle. Extracellularly offered Ba and Sr could not be detected within the cells after trichocyst exocytosis, indicating that extracellular divalent cations are not involved in the decondensation of the trichocyst secretory contents.

This study shows that X-ray microanalysis of cryosections can be a useful tool to investigate ion changes during dynamic processes in cells.

ACKNOWLEDGMENT

We thank Mrs. S. Dongard for her skillful technical assistance in preparing and analyzing cryosections from single *Paramecium* cells, as well as Mr. K. -E. Samans for constructing the freezing device.

REFERENCES

1. **Plattner, H., Matt, H., Kersken, H., Haacke, B., and Stürzl, R.,** Synchronous exocytosis in *Paramecium* cells. I. A novel approach, *Exp. Cell Res.*, 151, 6, 1984.
2. **Olbricht, K., Plattner, H., and Matt, H.,** Synchronous exocytosis in *Paramecium* cells. II. Intramembranous changes analysed by freeze-fracturing, *Exp. Cell Res.*, 151, 14, 1984.
3. **Haacke, B. and Plattner, H.,** Synchronous exocytosis in Paramecium cells. III. Rearrangement of membranes and membrane-associated structural elements after exocytosis performance, *Exp. Cell Res.*, 151, 21, 1984.
4. **Plattner, H., Stürzl., R., and Matt, H.,** Synchronous exocytosis in Paramecium cells. IV. Polyamino compounds as potent trigger agents for repeatable trigger-redocking cycles, *Eur. J. Cell Biol.*, 36, 32, 1985.
5. **Pape, R. and Plattner, H.,** Synchronous exocytosis in Paramecium cells. V. Ultrastructural adaptation phenomena during re-insertion of secretory organelles, *Eur. J. Cell Biol.*, 36, 38, 1985.
6. **Schmitz, M., Meyer, R., and Zierold, K.,** X-ray microanalysis in cryosections of natively frozen *Paramecium caudatum* with regard to ion distribution in ciliates, *Scanning Electron Microsc.*, I, 433, 1985.
7. **Wichtermann, R.,** *The Biology of Paramecium,* McGraw Hill, New York, 1953, 93.
8. **Hall, T. A. and Gupta, B. L.,** Quantification for the X-ray microanalysis of cryosections, *J. Microsc.*, 126, 333, 1982.
9. **Zierold, K.,** Preparation of cryosections for biological microanalysis, in *The Science of Biological Specimen Preparation 1985,* Müller, M., Becker, R. P., Boyde, A., and Wolosewick, J. J., Eds., Scanning Electron Microscopy, AMF O'Hare, IL, 1986, 119.
10. **Zierold, K.,** Cryoultramicrotomy, in *Cryotechniques in Biological Electron Microscopy,* Steinbrecht, R. A. and Zierold, K., Eds., Springer-Verlag, Heidelberg, 1987, 132.
11. **Vivier, E.,** Morphology, taxonomy and general biology of the genus *Paramecium,* in *Paramecium. A Current Survey,* van Wagendonk, W. J., Ed., Elsevier, North Holland, 1974, 1.
12. **Wohlfarth-Bottermann, K. E.,** Cytologische Studien II, die Feinstruktur des Cytoplasmas von *Paramecium, Protoplasma,* 49, 231, 1958.
13. **Wohlfarth-Bottermann, K. E.,** Experimentelle und elektronenoptische Untersuchungen zur Funktion der Trichozysten von *Paramecium caudatum, Arch. Protistenk.*, 98, 169, 1953.
14. **Plattner, H.,** Biologische Ultrastruktur-Forschung: Wege, Ziele und Grenzen. in *Konstanzer Universitätsreden,* Hess, G., Ed., Universitätsverlag Konstanz GmbH, Konstanz, 1981, 5.
15. **Plattner, H.,** Membrane behaviour during exocytosis, *Cell Biol. Int. Rep.*, 5, 435, 1981.
16. **Vilmart, J. and Plattner, H.,** Membrane-integrated proteins at preformed exocytosis sites, *J. Histochem. Cytochem.*, 31, 626, 1983.
17. **Plattner, H., Westphal, C., and Tiggemann, R.,** Cytoskeleton-secretory vesicle interactions during the docking of secretory vesicles at the cell membrane in *Paramecium tetraurelia cells, J. Cell Biol.*, 92, 368, 1982.

18. **Plattner, H. and Fuchs, S.**, X-ray microanalysis of calcium binding sites in *Paramecium, Histochemistry,* 45, 23, 1975.
19. **Plattner, H.**, Ciliary granule plaques: Membrane intercalated particle aggregates associated with Ca^{2+}-binding sites in *Paramecium, J. Cell Sci.*, 18, 257, 1975.
20. **Tsuchiya, T.**, Electron microscopy and electron probe analysis of the Ca-binding sites in the cilia of *Paramecium caudatum, Experientia,* 32, 1176, 1976.
21. **Anderer, R. and Hausmann, K.**, Properties and structure of isolated extrusive organelles, *J. Ultrastruct. Res.*, 60, 21, 1977.
22. **Steers, E., Beisson, J., and Marchesi, V. T.**, A structural protein extracted from the trichocyst of *Paramecium aurelia, Exp. Cell Res.*, 57, 392, 1969.
23. **Matt, H., Bilinski, M., and Plattner, H.**, Adenosinetriphosphate, calcium and temperature requirements for the final steps of exocytosis in *Paramecium* cells, *J. Cell Sci.*, 32, 67, 1978.
24. **Matt, H., Plattner, H., Reichel, K., Lefort-Tran, M., and Beisson, J.**, Genetic dissection of the final exocytosis steps in *Paramecium tetraurelia* cells: Trigger analyses, *J. Cell Sci.*, 46, 41, 1980.
25. **Bachmann, L., Schmitt, W. W., and Plattner, H.**, Improved cryofixation: Demonstrated on freeze-etched solutions, cell fractions, and micellar organisms. in *Proc. 5th Eur. Reg. Congr. Electron Microsc.*, Cosslett, V. E., Ed., The Institute of Physics, London, 1972, 244.
26. **Janisch, R.**, Pellicle of *Paramecium caudatum* as revealed by freeze-etching, *J. Protozool.*, 19, 470, 1972.
27. **Plattner, H., Miller, F., and Bachmann, L.**, Membrane specializations in the form of regular membrane-to-membrane attachment sites in *Paramecium*. A correlated freeze-etching and ultrathin-sectioning analysis, *J. Cell Sci.*, 13, 687, 1973.
28. **Beisson, J., Lefort-Tran, M., Pouphile, M., Rossignol, M., and Satir, B.**, Genetic analysis of membrane differentiation in *Paramecium*. Freeze-fracture study of the trichocyst cycle in wild-type and mutant strains, *J. Cell Biol.*, 69, 126, 1976.
29. **Plattner, H., Reichel, K., Matt, H., Beisson, J., Lefort-Tran, M., and Pouphile, M.**, Genetic dissection of the final exocytosis steps in *Paramecium tetraurelia* cells: Cytochemical determination of Ca^{2+}ATPase activity over performed exocytosis sites, *J. Cell Sci.*, 46, 17, 1980.
30. **Satir, B. H. and Oberg, S. G.**, Paramecium fusion rosettes: Possible function as Ca^{2+} gates, *Science,* 199, 536, 1978.
31. **Matt, H. and Plattner, H.**, Decoupling of exocytotic membrane fusion from protein discharge in *Paramecium* cells, *Cell Biol. Int. Rep.*, 7, 1025, 1983.
32. **Bilinski, M., Plattner, H., and Matt, H.**, Secretory protein decondensation as distinct Ca^{2+}-mediated event during the final steps of exocytosis in *Paramecium* cells, *J. Cell Biol.*, 86, 179, 1981.
33. **Vilmart-Seuwen, J., Kersken, H., Stürzl, R., and Plattner, H.**, ATP keeps exocytosis sites in a primed state but is not required for membrane fusion: An analysis with *Paramecium* cells *in vivo* and *in vitro, J. Cell Biol.*, 103, 1279, 1986.
34. **Momayezi, M., Lumpert, C. J., Kersken, H., Gras, U., Plattner, H., Krinks, M. H., and Klee, C. B.**, Exocytosis induction in *Paramecium tetraurelia* cells by exogenous phosphoprotein phosphatase *in vivo* and *in vitro*: Possible involvement of calcineurin in exocytotic membrane fusion, *J. Cell Biol.*, 105, 181, 1987.
35. **Satir, B. H., Garofalo, R. S., Gilligan, D. M., and Maihle, N. J.**, Possible functions of calmodulin in protozoa, *Ann. N.Y. Acad. Sci.*, 356, 83, 1980.
36. **Rauh, J. J. and Nelson, D. L.**, Calmodulin is a major component of extruded trichocysts from *Paramecium tetraurelia, J. Cell Biol.*, 91, 860, 1981.
37. **Momayezi, M., Kersken, H., Gras, U., Vilmart-Seuwen, J., and Plattner, H.**, Calmodulin in *Paramecium tetraurelia:* Localization from the *in vivo* to the ultrastructural level, *J. Histochem. Cytochem.*, 34, 1621, 1986.
38. **Kersken, H., Tiggemann, R., Westphal, C., and Plattner, H.**, The secretory content of *Paramecium tetraurelia* trichocysts: Ultrastructural-cytochemical characterization, *J. Histochem. Cytochem.*, 32, 179, 1984.
39. **Nogushi, M. H., Inoué, H., and Kubo, K.**, Calcium-activated adenosine triphosphatase activity in pellicles of *Paramecium caudatum, J. Biochem.*, 85, 367, 1979.
40. **Bilinski, M., Plattner, H., and Tiggemann, R.**, Isolation of surface membranes from normal and exocytotic mutant strains of *Paramecium tetraurelia*. Ultrastructural and biochemical characterization, *Eur. J. Cell Biol.*, 24, 108, 1981.
41. **Garofalo, R. S., Gilligan, D. M., and Satir, B. H.**, Calmodulin antagonists inhibit secretion in *Paramecium, J. Cell Biol.*, 96, 1072, 1983.
42. **Garofalo, R. S. and Satir, B. H.**, *Paramecium* secretory granule content: Quantitative studies on *in vitro* expansion and its regulation by calcium and pH, *J. Cell Biol.*, 99, 2193, 1984.
43. **Satir, B. H. and Wissing, S. L.**, Alveolar sacs of *Tetrahymena:* Ultrastructural characteristics and similarities to subsurface cisterns of muscle and nerve, *J. Cell Sci.*, 55, 13, 1982.

44. **Andrivon, C., Wyroba, E., and Patterson, D. J.,** Recherches préliminaires sur les ATPases de *Paramecium aurelia, J. Protozool.,* 24, 55A, 1977.

45. **Plattner, H. and Deitmer, J.,** Synchronous exo-endocytosis system: The Paramecium cell as analyzed by structural and electrophysiological methods, *Eur. J. Cell Biol.,* 36, (Suppl. 7), 51, 1985.

46. **Plattner, H.,** Synchronous exocytosis in *Paramecium* cells, in *Cell Fusion,* Sowers, A. E., Ed., Plenum Press, New York, 1987, 69.

47. **Saimi, Y. and Kung, C.,** Are ions involved in the gating of calcium channels? *Science,* 218, 153, 1982.

48. **Naitoh, Y. and Eckert, R.,** Electrical properties of *Paramecium caudatum.* All or none electrogenesis, *Z. Vgl. Physiol.,* 61, 453, 1968.

49. **Naitoh, Y. and Eckert, R.,** Ionic mechanisms controlling behaviour responses of *Paramecium* to mechanical stimulation, *Science,* 164, 963, 1969.

50. **Eckert, R.,** Bioelectric control of ciliary activity, *Science,* 176, 473, 1972.

51. **Brehm, P. and Eckert, R.,** Calcium entry leads to inactivation of calcium channel in *Paramecium, Science,* 202, 1203, 1978.

52. **Brehm, P. and Eckert, R.,** An electrophysiological study of the regulation of ciliary beating frequency in *Paramecium, J. Physiol.,* 283, 557, 1978.

53. **Satow, Y. and Kung, C.,** Voltage sensitive Ca-channels and the transient inward current in *Paramecium tetraurelia, J. Exp. Biol.,* 78, 149, 1979.

54. **Wendt-Gallitelli, M. -F. and Wolburg, H.,** Rapid freezing, cryosectioning and X-ray microanalysis on cardiac muscle preparations in defined functional states, *J. Electron Microsc. Techn.,* 1, 151, 1984.

55. **Somlyo, A. V., McClellan, G., Gonzalez-Serratos, H., and Somlyo, A. P.,** Electron probe X-ray microanalysis of post-tetanic Ca^{2+} and Mg^{2+} movements across the sarcoplasmic reticulum *in situ, J. Biol. Chem.,* 260, 6801, 1985.

Chapter 19

CORRELATED ELECTROSTIMULATION AND ELECTRON PROBE X-RAY MICROANALYSIS OF HEART MUSCLE CELLS

M. F. Wendt-Gallitelli and G. Isenberg

TABLE OF CONTENTS

I. INTRODUCTION

As the membrane of heart muscle depolarizes, Ca^{2+}-ions flow through Ca^{2+}-channels from the extra- into the intracellular space. This transfer of Ca^{2+}-ions is essential for the release of a large amount of Ca^{2+}-ions from intracellular stores, which finally activates the contraction cycle.[1]

If Ca^{2+}-influx is suppressed, contraction is abolished. Although Ca^{2+}-influx is necessary, it is rather unlikely the direct source of "activator-Ca^{2+}", for an identical Ca^{2+}-influx can activate contractions of variable extent. Frequency-dependent changes in contractility suggest that most of the activator Ca^{2+} is not derived directly from the Ca^{2+}-influx, but rather from intracellular stores. The extent to which the intracellular stores release Ca^{2+} depends on the amount of the Ca^{2+}-influx, which (1) triggers the Ca^{2+}-release, and (2) replenishes the intracellular stores.

An attempt to attribute the functional compartments to morphological structure can be made by combining electron microscopy with data of X-ray microanalysis on calcium concentrations in cell organelles. Such an attempt has been made for skeletal, heart, and smooth muscle.[2-10] Present knowledge favors identification of the compartments of Ca^{2+}-release and Ca^{2+}-uptake with the sarcoplasmic reticulum (SR). Therefore, Ca measurements in SR may evaluate its function.

Although the Ca concentration in isolated guinea pig ventricular SR vesicles has been estimated with biochemical methods to be about 90 μmol/kg wet weight, the isolation procedure may have modified structure and Ca distribution of the reconstituted microsome fractions. Information about changes in the microchemical *in situ* composition of cell organelles during activity is urgently needed for the understanding of the mechanism of regulation of contraction. The minimal requirements of a technique relevant for the contractile state of a muscle cell are the following:

1. The contraction-relaxation cycle must be interrupted at a given defined moment, i.e., the cell must be "fixed" at an exactly defined reproducible time. Thus, shock-freezing must be timed and the electrical/mechanical events must be recorded even during freezing.
2. The procedure to prepare the tissue should quantitatively preserve the distribution of calcium over distances more than the desired resolution of analysis (in the case of SR, about 25 nm).
3. It should adequately preserve the cellular fine structure and allow subsequent imaging at a suitable resolution.
4. It should be specific for the analyzed element and permit reliable quantitative measurements.

Concerning the specific problems of microanalysis, as applied to small cell compartments of freeze-fixed preparations, the following peculiarities have to be mentioned:

1. Electrical events, as ionic currents, can only be correctly characterized in single isolated cells, not in multicellular preparations. Therefore, single isolated cells are the useful model for studying the effect of controlled electrophysiological events on Ca compartmentalization in the cell. But single isolated myocytes are difficult to handle appropriately for freezing and cryocutting, particularly whem impaled with patch-clamp pipettes; the probability of being lost during one of the preparative steps is higher for single isolated myocytes than for multicellular preparations.
2. Due to the large biological variability in a population of isolated cells, the reconstruction of the physiological events of Ca^{2+}-release and -uptake in cell organelles by X-ray

and microprobe analysis (EMPA) requires a very high number of cells to be analyzed at any desired time, due to biological variability in a population of isolated cells. Therefore, EMPA is a time-demanding method.

3. If one looks to microanalysis for information about elemental composition in smallest cell compartments, such as membrane-limited SR-tubuli or cisternae, the lateral resolution of the analyzing beam in the electron microscope must be at least in the range of 20 nm.
4. EMPA cannot distinguish between free-ionized and bound elements; it detects the total amount of an element in the analyzed mass.
5. The limit of detectability of EMPA, as used in this study, is in the range of 0.5 to 1 mmol/kg dry weight, which corresponds, e.g., in heart muscle cells to 70 to 100 μmol/kg wet weight. Therefore, the small intracellular movements of free ionic Ca^{2+} in the nanomolar range remain undetected.

In this paper we describe a method that allows quick-freezing of isolated cardiac myocytes under voltage-clamp conditions at a defined moment of the Ca^{2+}-current. The method has to meet the following criteria:

1. It should allow shock-freezing of the cell at an exact time defined by electrophysiological criteria (e.g., 100 ms after activation of the Ca inward current Ic_a, which corresponds to the peak of contraction).
2. The movements required for shock-freezing should not damage the cell.
3. The preservation of the shock-frozen cell should be acceptable.
4. The cell should be delivered in a system (holder) suitable for further cryocutting.
5. The cell should be easy to locate in a reproducible manner, despite its small dimensions in relation to those of the container of the coolant.

All our trials were based on the idea that a patch electrode is suitable not only for intracellular recordings, but also for cell-transfer.

The first results demonstrate that the method has large potentiality for physiological studies, and can be easily extended to single cells isolated from other tissues, whenever information is needed about the direct influence of measured ionic currents on intracellular elemental distribution.

II. METHODS

A. PREPARATION OF MYOCYTES

Myocytes were isolated from ventricles of 250 g guinea pig by means of the pronase technique.[11] The cells were pipetted into a chamber (2 mm × 2 mm × 10 mm) placed on the stage of an inverted TV microscope. Once the cells had settled at the glass bottom of the chamber, they were continuously superfused with a prewarmed (35°C) "Tyrode solution" of the following composition in mmol/l: NaCl 150, KCl 5.4, $CaCl_2$ 3.6, MgCl 1.2, glucose 10, HEPES 10 (adjusted by NaOH to pH 7.4).

B. DATA RECORDING AND STORAGE

The voltage clamp experiments were performed with a single patch-electrode technique.[12-15] The electrodes tips had an outer diameter of 6 μm and an inner diameter of 2 to 3 μm (resistance 1 to 1.5 MOhm). The electrodes were filled with the following solution (mmol/l): 140 KCl, 3 Na_2ATP, 3 $MgCl_2$, 0.02 EGTA, 10 HEPES/KOH (pH 7.2). A PDP 11-23+ minicomputer generated stimuli or clamp pulses (0.5 Hz) and controlled the sequence of pneumatic movements that lead finally to the shock-freezing. The pulses of +5 mV started from a holding potential of −45 mV and lasted for either 0.2 s or 1 s. The resulting membrane current was digitized (12 bit resolution) and stored.

Ca-entry was calculated from I_{Ca}. The recorded current is a net current from which I_{Ca} has to be separated. I_{Ca} was approximated with the "visual estimate method", referring the net current to a time-independent line (measured 0.2 s after start of depolarization).[16] The cell volume was approximated from light microscopy and electrical measurements. Despite their different sources of error, both estimates come to similar numbers, 10 and 11 pl, respectively (n = 20 in both cases).

C. FREEZE SUBSTITUTION

For cryofixation, etc., see section III, which also deals with cryofixation. For freeze substitution[17] the cold silver holder with the frozen cell was transferred under liquid nitrogen into a 2 ml vial containing a frozen solution of 5% osmium tetroxide in acetone beneath a layer of liquid nitrogen. The specimens were held at dry ice temperature for 2 d. After warming up to room temperature, the specimens were postfixed with 1% aqueous osmium solution for 2 h, then dehydrated and embedded in araldite in the conventional mode. When the method was applied to a single isolated myocyte, a major disadvantage was the probability of the small cell getting lost during one of the several transfers necessary for fixation, dehydration, and embedding. Therefore, all transfers were undertaken with the preparation (cell within the silver holder) sitting on acetone-resistant millipore filter through which the solutions were sucked.

D. X-RAY MICROANALYSIS

A full quantitative energy-dispersive system (Link model 810-500) attached to a Siemens Elmiskop 102 transmission electron microscope was employed for the elemental analysis. The energy dispersive detector was a Si crystal of 30 mm^2 diameter and 147 eV resolution. The tilt angle was 15°, and the distance from the detector to the specimen was 15 mm. The sections were analyzed under the following conditions: 80 kV acceleration voltage, 6 nA beam current, 50 to 70 nm diameter of the analyzing spot, or alternatively, an astigmatically deformed spot with a length of 0.7 μm and a width of 7 nm[8]. The analysis time was 100 s. The "Quantem" program based on the continuum theory of Hall[18] was used to calculate the concentration of elements in mmol/kg (dry weight) with atomic number >11 (Na).

Standards—External standards were prepared in two ways. (1) albumin and phosvitin were mixed;[7,8,19] (2) salt solutions in 20% dextran (200,000 MW) were prepared. A dry mass of 2 g (salt and dextrane) was diluted to 10 ml with distilled and deionized water.[20] In the Ca-standards 150 mmol/kg (d.w.) KC1 was added, in order to have a similar counts ratio between $Ca_{k\alpha}$ and $K_{k\beta}$ peaks to that in the intracellular space. Four concentrations of each salt were used, and the ratio peak to contimuum was plotted vs. the concentrations in mmol/kg d.w. The data were analyzed using regression analysis.

The "standard factor" (FST) for calculating the concentration of each element was calculated as follows:

$$FST = [(P_a/W_{std}) \times g_{std}] / C_{std}$$

where P_a is the peak area of the element, W the background integral, $g_{std} = Z^2/A$ "g-factor", with Z = equivalence charge, A = atomic weight and C_{std} the concentration of the element in the standard (mmol/kg d.w.). Concentrations were calculated by the computer of the Link System, with a least square fitting routine included in the "Quantem" program.

III. RESULTS AND DISCUSSION

A. THE METHOD OF SHOCK-FREEZING

1. Mounting and Triggering

An isolated myocyte with clear cross striation was sucked with a patch-clamp pipette which serves for both the transfer of the cell from the chamber to a special silver holder,

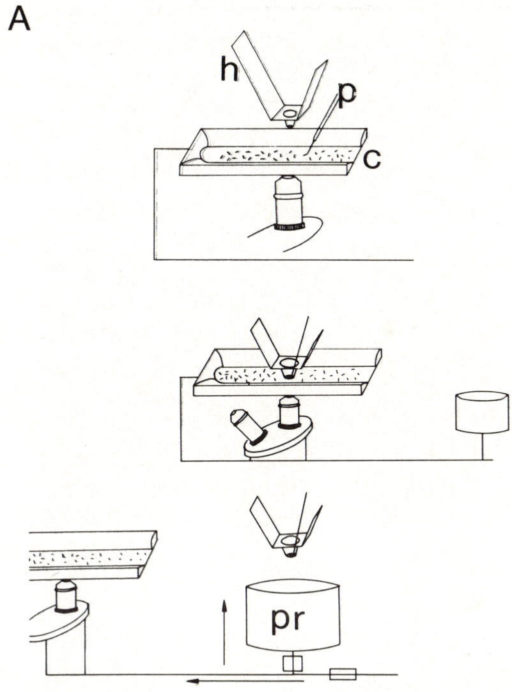

FIGURE 1. (A) Schematic drawing of the set-up for transferring and freezing the myocyte. On the top arrangement of holder (h), patch-pipette (p), chamber (c) and microscope. In the middle of the myocyte has been transferred to the holder. On the bottom, microscope and chamber have been moved away, the coolant (pr = underooled propane) is moving up to the holder with the cell. (B) A myocyte has been transferred with the patch-electrode from the chamber to the silver-holder (photograph from the TV screen). e, electrode; h, holder.

as well as for electrophysiological measurements. In the handmade silver holder, the tip of a taper, opened electrolytically, was covered with a thin pioloform film. Having selected a myocyte with clear cross striation, the Gω seal was built, but without disrupture of the patch of the membrane.

The cell was lifted from the bottom of the cryochamber to the holder and placed on the transparent pioloform film (Figure 1). Intracellular contact was then established by rupturing the patch of the membrane. After checking that the intracellular stimulation induced normal action potentials, voltage clamp was switched on. Pulse depolarizing from -45 to -5 mV evoked net membrane currents which were largely determined by the Ca inward current. The elicited shortening of the unloaded cells can be visualized on the monitor. Then keeping the holder with the cell absolutely quiet, the chamber was moved away horizontally by sliding the microscope. Before the horizontal movement stopped, the vertical movement of the coolant started and the preparation contacted the coolant (undercooled propane at 77K).

The following properties of the film helped improve our method considerably:

1. The film is transparent; one can visualize details of the cell and its sarcomere length.
2. The film prevents Tyrode-solution from streaming through the opening thereby blowing off the cell.
3. The cell can be placed on the film in such a position that optimal contact with the coolant is provided.
4. The position of the cell on the film helps the cryocutting knife to locate the cell.

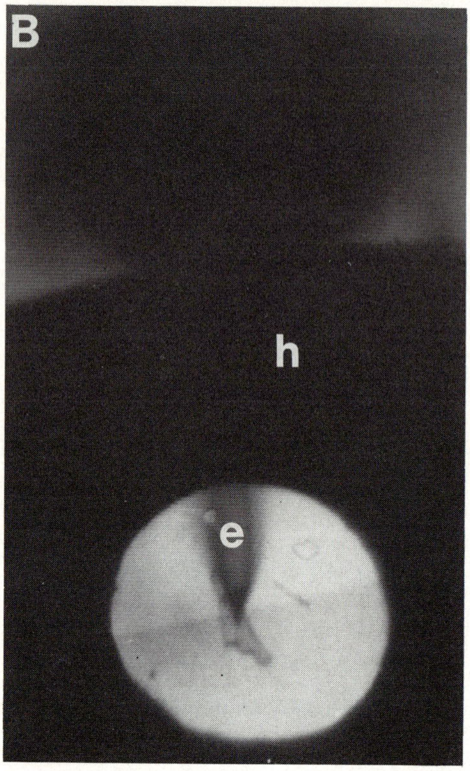

FIGURE 1B.

The holder was connected to electrical ground; it served as the reference electrode for the electrical measurements. With a clip it was fixed on a Leitz-micromanipulator. The same Leitz-micromanipulator supported the head-stage of a three-dimensional hydraulic manipulator (Narashige) that was used for the cell transfer.

2. Shock-Freezing

During the events that led finally to shock-freezing (see below), the position of the preparation was kept absolutely quiet, whereas its surrounding was moved instead. By sliding the microscope backwards, the preparation was moved out of the open end of the chamber. The movement of the sliding table was driven by 7 bar air pressure applied to pneumatic pistons.[8] The horizontal movement of 35 mm needed 105 ms (Figure 2). The movement did not disturb the electrical recordings; the record nearly superimposes on the one taken before the movement. Forty ms before the sliding table came to its final stop, the vertical movement of the coolant started. Thirteen ms later, the preparation contacted the coolant and was shut for 16 ms through the 40 mm high layer of stirred undercooled propane. After cooling to 78K, the current remained about zero with large superimposed 50 Hz disturbances. When the horizontal movement stopped, vibrations caused large current deflections.

3. Removal of the Preparation

We removed the silver holder from its clip and let if fall through the propane onto the bottom of the container. The container was removed from the piston, and the silver holder containing the cell was stored in frozen propane until further cryocutting and analysis.

B. QUALITY OF THE SHOCK-FREEZING

We like to use the ultrastructural preservation of the cell as in indication of the quality of cryo-fixation. *Freeze-substitution* (section II.C) followed by thin sectioning yielded images

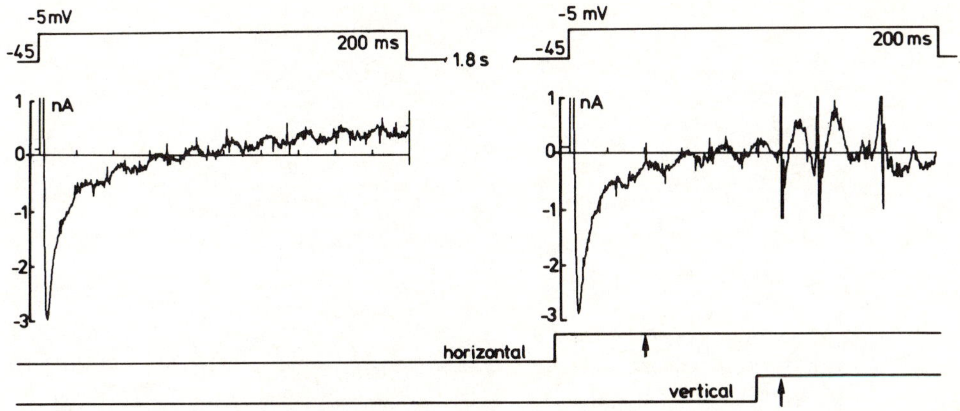

FIGURE 2. Membrane currents from an isolated myocyte as recorded before (left) and during (right) shock-freezing. 200 ms long pulses were applied at 0.2 Hz. Note that the start of the horizontal movement (left arrow) does not disturb the cell. The impact of the cell with the coolant (right arrow) evoked large current artifacts.

with sufficient contrast uniform throughout the thickness of the cell (12 μm). Unit membranes of different organelles were well resolved in regions without cryoartifacts. Figure 3 shows electron micrographs from a single cell which is representative for ten others. Cryoartifacts appeared with a gradient of increasing severity. They were absent at the surface of the cell touching the pioloform film. Cryoartifacts started about 2 μm inside the cell, i.e., cutting sections out of the middle of the cell resulted in worse ultrastructural preservations than the sections shown in Figure 3. Cryoartifacts manifested themselves by disturbing the orderly geometry of the contractile material and consisted of clefts of increasing size, usually oriented along the longitudinal cell axis. The myofilaments and the granular material in the SR tended to clump together. The mitochondrial membranes became annular and their interior became foamy in appearance.

At regions where cryoartifacts were undetectable (Figure 3a, d) the arrangement of the myofilaments seemed to be comparable to that known from glutaraldehyde-fixed specimen. The thick and thin filaments ran in a regular pattern, and crossbridges with their periodicities were resolved. The architecture of the Z-lines seemed to be undisturbed. Also, the SR was essentially similar to the corresponding images from conventionally fixed specimen. The isolated cells showed T-tubuli with adjacent junctional SR (Figure 3b). Also, the sarcolemma was faced with vesicles of peripheral junctional SR. The interior of the SR seemed to be filled with an electron dense material. Most of our shock-frozen cells showed a 2 to 5 μm thick zone of minimal cryoartifacts.

The glass electrode did not limit the quality of freezing. The absence of ice crystals suggests that the rate of freezing has been greater than 10,000°C/s (Reference 21). The 5 μm width of the zone is of the same order of the 2 to 5 μm evaluated by Reference 22 and therefore, not unexpectedly limited.

C. CRYOSECTIONING

The silver holder was mounted and fixed under liquid nitrogen in an aluminum holder constructed for the LKB cryoultramicrotome. The specimens were cut at temperatures between 113 and 173K as thin as possible (100 to 200 nm). The sections include part of the silver holder as well as the Tyrode solution surrounding the cell. The cryosections were picked up dry on nickel grids (50 or 75 meshes), freeze-dried, carbon-coated, and finally analyzed (details in Reference 8).

The procedure for cryosectioning an isolated myocyte embedded in frozen Tyrode so-

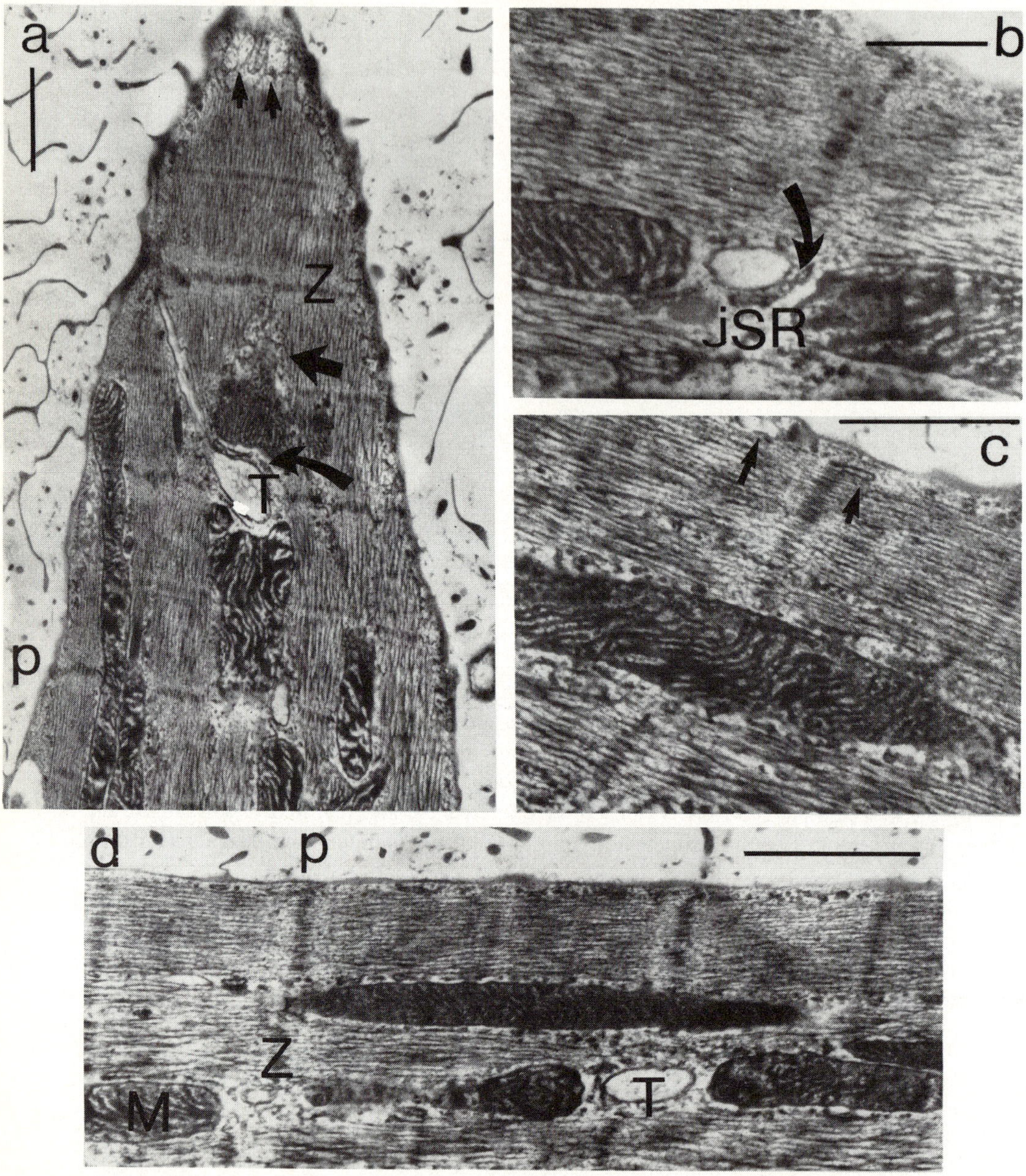

FIGURE 3. Longitudinal sections of a ventricular myocyte shock-frozen and freeze-substituted. P, plasmalemma; M, mitochondria; jSR, junctional SR; T, t-tubuli; Z, Z-line. (a) Minimal cryoartifacts on the left. Some distorsion of the myofilaments on the right, which are the first signals of ice crystals damage (cryoartifacts). Note the absence of a glycocalix on the external site of the plasmalemma. Bar, 1 μm. Curved arrows, jSR; short arrows, subsarcolemmal SR; thick arrows, free network of Sr. (b) Junctional SR, indicated by the curved arrow, is present at one side of the t-tubulus. Bar: 0.2 μm. (c) Small subsarcolemmal vesicles of the SR are clearly recognizable (arrows). Note the well preserved arrangement of myofilaments and the structure of the Z-lines. Bar; 0.2 μm. (d) Minimal cryoartifacts in the superficial zone indicated by the perfectly parallel arrangement of the filaments and the preservation of the superficial mitochondria. Note the sharp border at the sarcolemma. Bar, 1μm.

lution is difficult; the moving knife, having cut through the frozen Tyrode solution, pushes against the cell, which thereby can spring away from the silver holder. Somethimes, thick sections split off the block. This is not surprising since the various materials (silver, Tyrode solution, cell) have different ductility, which makes the cutting more similar to a process of fracturing. Due to their small dimensions, the cryosections were often blown away from the grid, when minimal turbulences occured in the nitrogen-atmosphere of the cryochamber. Therefore, after 4 to 6 sections had been collected successfully on the grid, they were covered with a thin pioloform film and mounted on a nickel ring which was stamped from a grid. The sandwich with the cryosections was transferred to the brass container (see Reference 8), freeze-dried and carbon-coated. Problems were often encountered with the transfer of the sections from the knife to the grid; electrostatical charge raised the small sections onto the nickel balk of the grid, which made them useless for further analysis (high nickel peak). Glow discharge of the grid prior to sectioning did not solve the problem satisfactorily.

D. ELEMENTAL ANALYSIS: FIRST EXAMPLES OF CALCIUM DISTRIBUTION IN RESTED AND STIMULATED MYOCYTES

1. Definition of the Analyzed Compartments

For elemental analysis the electron beam was focused to a diameter of about 50 nm. Thus, we could analyze the elements in the following compartments:

1. The cytoplasm: the electron beam was located in the space within the sarcomeres. Most of this compartment was occupied by the myofilaments (A-band). The analysis of recognizable structure, such as the Z-lines, was avoided.
2. The cell border: the electron beam was deformed astigmatically to 10 nm·1 μm, and it was placed as close as possible onto the inner side of the surface membrane (T-tubuli excluded). The measured concentrations contained contributions from elements bound to the inner side of the sarcolemma, from elements stored in peripheral SR when it was closely attached to the sarcolemma, and elements in the fringe of cytoplasm close to the cell border.
3. Junctional SR (jSR) adjacent to the transversal tubuli: the analyzing beam was a circular spot of 50 nm diameter. As in the case of the cell border, elements bound to the membranes, stored in the jSR, and located in the fringe of cytoplasm, were analyzed.
4. The mitochondria were analyzed with a circular beam-spot of 70 nm.

2. The Concentrations Measured by EMPA

These are expressed in mmoles/kg d.w., since we analyze elements in freeze-dried cryosections in which the volume of the cell compartment is reduced. But the concentrations measured by microanalysis should be converted into mmols/kg w.w., since we attempt the comparison with concentrations $D[Ca]_i$ calculated on the basis of I_{Ca} and cell volume. When one assumes for cytoplasm and mitochondria a water content of 75% and 68%, respectively, 500 mmols K/kg d.w. measured by the X-ray microprobe in the cytoplasm correspond to 125 mmoles K/kg w.w.; whereas 390 mmol K/kg d.w. in mitochondria correspond likewise to 125 mmoles/kg w.w.

3. The Elemental Distribution at Rest

This is shown in Table 1; data from thin multicellular guinea pig ventricular trabeculae were listed for comparison. Generally, the elemental distribution was similar in the single cells and the multicellular preparations. In trabeculae the [Ca] in all four compartments significantly exceeded zero. In the few isolated myocytes frozen after prolonged rest analyzed until now, Ca-values, with high SE, were not distinguishable from zero.

4. The Ca-Distribution During 0.2 s Short Pulses

This is shown in Fig 4. The series of five pulses (0.5 Hz) was followed by a sixth pulse

TABLE 1
Comparison Between the Elemental Distribution in Single Isolated Ventricular Myocytes and Thin (100 to 200 μm Diameter) Ventricular Trabeculae of the Guinea-Pig Heart.[a]

	Na	Mg	P	Cl	K	Ca
Cytoplasm						
Single cells n = 13	63	27	387	94	401	−2
	10	6	43	14	37	2
Trabeculae n = 48	71	58	232	145	459	0.6
	10	16	30	22	41	0.3
Mitochondria						
Single cells n = 10	24	22	485	72	330	−2
	8	5	39	13	42	2
Trabeculae n = 25	13	56	473	57	391	1
	8	6	7	1	10	1
Sarcoplasmic reticulum						
Single cells n = 12	97	18	389	190	376	1
	22	7	43	37	32	2
Trabeculae n = 31	69	60	335	141	469	1.8
	22	7	58	20	60	1
Cell border						
Single cells n = 10	177	12	298	384	288	−0.8
	33	3	19	47	22	2.4
Trabeculae n = 42	358	17	266	345	349	12
	101	4	45	63	41	3

[a] The isolated myocytes and the small muscles were frozen in comparable condition of rest. The data (means ± SE) of single cells were obtained from two frozen cells, that of trabeculae from three muscles. Data were collected in the following compartments: (1) Cytoplasm. In longitudinal sections the beam (50 to 70 nm diameter) was located on the sarcomers, or the A-region, distant from the Z-line, but regardless of the position within the myofilaments. (2) Mitochondria. The focused beam was located within single mitochondria. (3) Sarcoplasmic reticulum: In the isolated myocytes the elemental composition of junctional SR, adjacent to the t-tubules was analyzed with a beam of 50 to 70 nm diameter. Elements bound to the membranes, stored in jSR and such located in the fringe of cytoplasm were included in the analysis and could not be analyzed separately. (4) Cell border: The electron beam was focused astigmatically to 10 nm × 1 μm, and it was placed as close as possible onto the inner side of the surface membrane (T-tubules excluded). Therefore, in the spectra are included elements bound to the inner side of the sarcolemma, elements stored in those parts of the jSR that were closely attached to them (peripheral jSR), and elements in the fringe of cytoplasm close to the cell border.

The specimen-continuum, corrected for elements extraneous to the preparation, was of 445 counts per 100 s ± 22 in the cytoplasm, 721 counts per 100 s ± 48 in the mitochondria, 336 counts per 100 s ± 29 on the SR, and 339 ± 59 counts on the cell border. The differences in the unspecific X-ray radiation (continuum) correspond to the different density of the analyzed volume, due to the original different water content of the various compartments.

Data from Wendt-Gallitelli, unpublished.

during which the cell was shock-frozen 110 ms after start of depolarization. This moment corresponds to the time when the contraction peaked. The amount of Ca-influx associated with the 5 preceding Ca-currents corresponded to a charge of 5.1 pC, or to a fictitious increment in total intracellular $[Ca]_i$ ($D[Ca]_i$) of 473 μmol/kg w.w. During the 110 ms just before shock-freezing, I_{Ca} added an additional $D[Ca]_i$ of 73 μmol/kg w.w. Thus, the clamp-steps should have imposed $D[Ca]_i = 546$ μmol/kg w.w. One hundred ten ms after start of depolarization calcium, in the cytoplasm [Ca]c was analyzed to be 125 μmol/kg w.w. (0.5

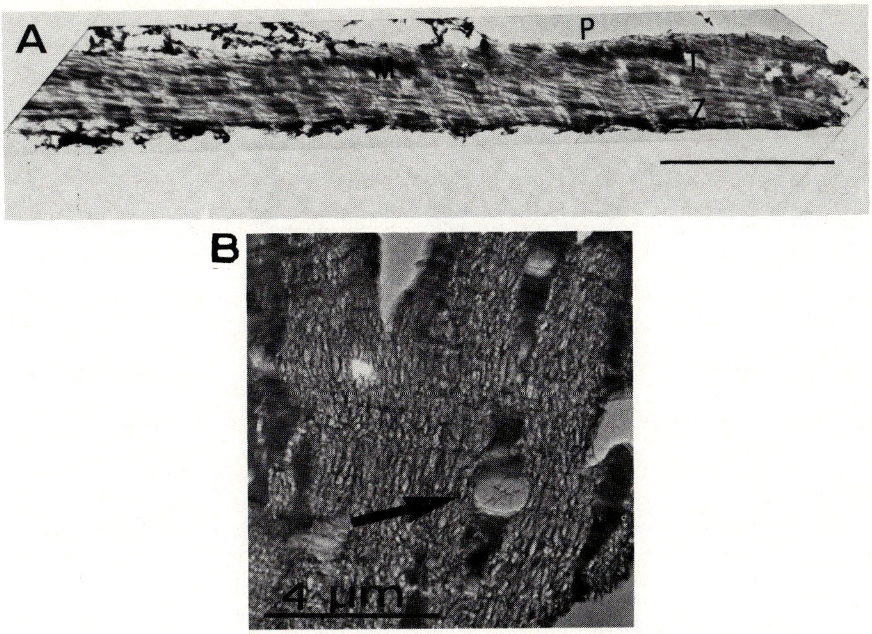

FIGURE 4. Freeze-dried cryosections of two single isolated myocytes. (A) Longitudinal section of a cell that lays on the pioloform film perfectly parallel to the knife edge of the cryomicrotome. Despite the low contrast and the thickness of the section, sarcomeres with Z-lines, (Z), mitochondria (M), and T-tubuli (T) can be identified. During cryocutting the section became deformed. Bar, 8 μm. (B) Thinner, internal (3 μm from the cell surface) cryosection of another isolated myocyte. Arrow, junctional SR.

± 0.1 mmol/kg d.w.). This concentration is almost twice as large as the fictitious increment D[Ca]c = 73 μmol/kg w.w. as it is calculated fromI_{ca} just before freezing. Thus, $[Ca]_c$ has partially to stem from other sources, i.e., from the SR. The analyzed $[Ca]_c$ is low in comparison with the $D[Ca]_c$ = 546 μmol/kg w.w., which is calculated from the postrest history of 6 depolarizations. Since calcium on the "cell border" $[Ca]_{SL}$, and calcium in jSR $[Ca]_{SR}$ are in the millimolar range, we suggest that the Ca^{2+} influx has loaded the Ca-stores.

5. The Ca-Distribution During 1 s Long Pulses

Data are shown in Figures 4 and 5. The contractility was potentialed with 4 prepulses of 1 s duration(0.5 Hz),and the cell was shock-frozen 800 ms after start on the fifth pulse. Within these 800 ms the cell had almost relaxed, i.e., only a slight tonic component of contraction was maintained (compare Reference 11). With 87 μmol/kg w.w. [Ca]c was correspondingly low. The 87 μmol/kg w.w. are smaller than D[Ca]i of 137 μmol/kg w.w. which is calculated fromthe 300 pC charge transported in the 800 ms just before freezing, obviously most of this Ca has been sequestered into compartments (or extruded). The Ca-influx during the preceding 5 long pulses loaded the comaprtments, i.e., 320 μmol/kg w.w., were measured in the space close to the inner side of the sarcolemma containing the peripheral SR. Even in the mitochondria, 412 μmol/kg w.w. $[Ca]_M$ were measured.

IV. CONCLUSIONS

We found advantages and disadvantages in comparing the present method of quick-freezing isolated single ventricular myocytes with the method on the thin multicellular trabeculae.[4-9] Disadvantageous for the single cell are the incomplete characterization of its contractile state (unloaded shortening) and the impossibility of measuring mechanical apra-

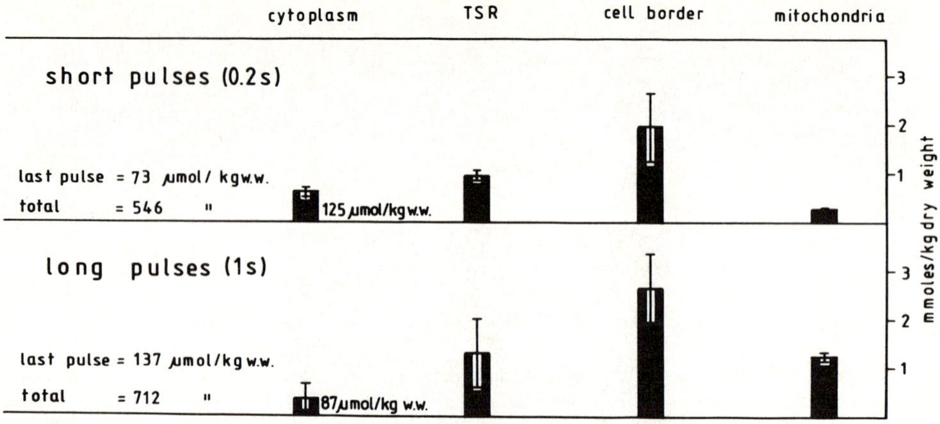

FIGURE 5. Ca-distribution (EMPA) in isolated myocytes (n = 5). Top: cell frozen 110 ms after start of depolarization to +5 mV which corresponds to the time-to-peak of contraction. The number of preceding pulses was 5 (200 ms to +5 mV, 0.5 Hz). Bottom: myocyte frozen 800 ms after start of depolarization; at this time the cell has already relaxed by 90%. The number of preceding pulses was 4 (1 s to +5 mV, 0.5 Hz). The analyzed compartments were: cytoplasm, junctional SR adjacent to t-tubuli, cell border which includes jSR adjacent to the sarcolemms, sarcolemma, a fringe of peripheral cytoplasm, mitochondria. The total Ca increment D$[Ca]_i$ due to the Ca-inward current is indicated on the left.

meters during quick-freezing itself. The patch electrodes can cause some peculiar problesm, such as "cell dialysis", due to diffusion of electrolytes filling the patch pipette into the cytoplasm after rupture of the patch of the membrane. Presently we have limited the period of "opening" of the cell membrane and the moment of freezing to less than half a minute, which is short in comparison to the time required for "cell dialysis".

Application of the EMPA method to a single isolated cell can avoid some problems of cell to cell heterogeneity which are inherent to multicellular specimen. The force, generated by a population of e.g., 100 cells, can differ between cells of the superficial cell layer (which is the subject of analysis) and those located deeper within the strand. In case of pharmacological interventions, some of the cells may have responded with a maximal iontropic effect, others with submaximal or even toxic effects.[7]

Our first elementary analysis of isolated ventricular cells quick-frozen under voltage-clamp conditions showed that the distribution of the elements was similar to that in multicellular ventricular trabeculae (Table 1). In this paper we have mainly focused on the instrumental aspects of a method which now seems to have reached the end of a lengthy development. We hope, in the near future, to publish a report on the relationship between Ca entering through the sarcolemma and Ca loading of intracellular compartments.

REFERENCES

1. **Fabiato, A.,** Calcium-induced release of calcium from the cardiac sarcoplasmic reticulum, *Am. J. Physiol.*, 245, C1, 1985.
2. **Somlyo, A. V., Shuman, H., and Somlyo, A. P.,** Elemental distribution in striated-muscle and the effects of hypertonicity. electron probe analysis of cryosections, *J. Cell. Biol.*, 74, 828, 1977.
3. **Kitazawa, T., Shuman, H., and Somlyo, A. P.,** Quantitative electron probe analysis: problems and solutions, *Ultramicrosc.*, 11, 251, 1983.
4. **Wendt-Gallitelli, M. F.,** Presystolic calcium-loading of the sarcoplasmic reticulum influences time to peak force of contractions. X-ray microanaylsis on rapidly frozen guinea-pig ventricular muscle preparations, *Basic. Res. Cardiol.*, 80, 617, 1985.

5. **Wendt-Gallitelli, M. F.,** Ca pools involved in the regulation of cardiac contraction under positive inotropy. X-ray microanalysis on rapidly frozen ventricular muscles of the guinea-pig, *Basic Res. Cardiol.,* 81, (Suppl. 1), 25, 1986.
6. **Wendt-Gallitelli, M. F. and Isenberg, G.,** Extra and intracellular lanthanum: modified calcium distribution, inward currents and contractility in guinea-pig ventricular preparations, *Pfluegers Arch.,* 405, 310, 1985.
7. **Wendt-Gallitelli, M. F. and Jacob, R.,** Effects of non-toxic doses of ouabain and sodium, potassium calcium distribution in guinea-pig papillary muscle. Electron probe microanalysis, *Basic Res. Cardiol.,* 79, (Suppl.), 79, 1984.
8. **Wendt-Gallitelli, M. F. and Wolburg, H.,** Rapid freezing, cryosectioning and X-ray microanalysis on cardiac muscle preparations in defined functional state, *J. Electron Microsc. Techn.,* 1, 151, 1984.
9. **Wendt-Gallitelli, M. F., Jacob, R., and Wolburg, H.,** Intracellular membranes as boundaries for ionic distribution in guinea-pig heart muscle in different defined electromechanical coupling states, *Z. Naturforsch.,* 37c, 712, 1982.
10. **Wheeler-Clark, E. S. and Thormey, J. McD.,** Electron probe X-ray microanalysis of sarcolemma and junctional sarcoplasmic reticulum in rabbit papillary muscles: low sodium-induced calcium alterations, *Circ. Res.,* 60, 246, 1987.
11. **Bendukidze, Z., Isenberg, G., and Kockner, U.,** Ca-tolerant guinea-pig ventricular myocytes as isolated in the presence of 250 μM free calcium, *Basic Res. Cardiol.,* 80, 13, 1985.
12. **Isenberg, G. and Klockner, U.,** Glycocalix is not required for slow inward current in isolated rat heart myocytes, *Nature,* 184, 358, 1980.
13. **Isenberg, G. and Klickner, U.,** Calcium currents of isolated bovine ventricular myocytes are fast and of large amplitude, *Pfluegers Arch.,* 395, 30, 1982.
14. **Isenberg, G., Klickner, U., Mascher, D., and Ravens, U.,** Changes in contractility and membrane currents as studied with a single patch electrode whole cell clamp technique, in *Electrophysiology of Single Cardiac Cells,* Noble, D., and Powell, T., Eds., Academic Press, Orlando, 1987, 25.
15. **Marty, A. and Neher, E.,** Tight-seal whole-cell recording, in *Single Channel Recording,* Sakman, B. and Neher, E., Eds., Plenum Press, New York, 1983, 107.
16. **McDonald, T. F. and Trautwein, W.,** Membrane currents in cat myocardium: separation of inward and outward current, *J. Physiol.,* 274, 193, 1978.
17. **Bullivant, S.,** Freeze-substitution and supporting techniques, *Lab. Invest.,* 14, 440, 1985.
18. **Hall, T. A.,** Biological X-ray microanalysis, *J. Microsc.,* 117, 145, 1979.
19. **Shumann, H., Somlyo, A. V., and Somlyo, A. P.,** Quantitative electron probe microanalysis of biological thin sections: methods and validity, *Ultramicrosc.,* 1, 317, 1976.
20. **Hagler, H. K., Lopez, L. E., Flores, J. S., Lundswick, R. J., and Buja, L. M.,** Standards for quantitive energy dispersive X-ray analysis of biological cryosections: validation and application to studies of myocardium, *J. Microsc.,* 131, 221, 1983.
21. **Heuser, J. E., Reese, T. S., Dennis, M. J., Jan, L. I., and Evans, L.,** Synaptic vesicles exocytosis captured by quick-freezing and correlated with quantal transmitter release, *J. Cell. Biol.,* 81, 275, 1979.
22. **Nassar, R., Wallace, N. R., Taylor, I., and Sommer, J. S.,** The quick-freezing of single intact muscle fibers at known time intervals following electrical stimulation, *Scanning Electron Microsc.,* 1, 309, 1986.

INDEX

A

B

C

D

E

H

M

N

Q

R

S

T

U

V

W

X

Z